STUDENT STUDY GUIDE
FOR CAMPBELL'S

BIOLOGY

MARTHA R. TAYLOR
Cornell University

FOURTH EDITION

The Benjamin/Cummings Publishing Company, Inc.

Menlo Park, California • Reading, Massachusetts
New York • Don Mills, Ontario • Wokingham, U.K • Amsterdam
Bonn • Paris • Milan • Madrid • Sydney • Singapore • Tokyo
Seoul • Taipei • Mexico City • San Juan, Puerto Rico

Sponsoring Editor: Johanna Schmid

Associate Editor: Kim Johnson

Editorial Assistant: Tabinda Khan

Senior Production Editor: Larry Olsen

Copyeditor: Marjorie Wexler

Cover Designer: Yvo Riezebos

Senior Manufacturing Coordinator: Merry Free Osborn

Illustrator: Val Felts

Page Layout: Richard Kharibian

Proofreader: Anna Reynolds Trabucco

Cover Photo: Copyright © John Sexton

ISBN 0-8053-1942-5

1 2 3 4 5 6 7 8 9 10—CRK— 99 98 97 96 95

Credits

The Benjamin/Cummings Publishing Company, Inc.
2725 Sand Hill Road
Menlo Park, California 94025

CONTENTS

PREFACE   v

CHAPTER 1
Introduction: Themes in the Study of Life   1

UNIT I   The Chemistry of Life   5

CHAPTER 2
The Chemical Context of Life   5

CHAPTER 3
Water and the Fitness of the Environment   11

CHAPTER 4
Carbon and the Molecular Diversity of Life   16

CHAPTER 5
The Structure and Function of Macromolecules   20

CHAPTER 6
An Introduction to Metabolism   27

UNIT II   The Cell   35

CHAPTER 7
A Tour of the Cell   35

CHAPTER 8
Membrane Structure and Function   45

CHAPTER 9
Cellular Respiration: Harvesting Chemical
   Energy   52

CHAPTER 10
Photosynthesis   60

CHAPTER 11
The Reproduction of Cells   68

UNIT III   The Gene   75

CHAPTER 12
Meiosis and Sexual Life Cycles   75

CHAPTER 13
Mendel and the Gene Idea   81

CHAPTER 14
The Chromosomal Basis of Inheritance   90

CHAPTER 15
The Molecular Basis of Inheritance   97

CHAPTER 16
From Gene to Protein   103

CHAPTER 17
Microbial Models: The Genetics of Viruses and
   Bacteria   111

CHAPTER 18
Genome Organization and Expression in
   Eukaryotes   120

CHAPTER 19
DNA Technology   126

UNIT IV   Mechanisms of Evolution   137

CHAPTER 20
Descent With Modification: A Darwinian View of
   Life   137

CHAPTER 21
The Evolution of Populations   143

CHAPTER 22
The Origin of Species   151

CHAPTER 23
Tracing Phylogeny: Macroevolution, the Fossil
   Record, and Systematics   157

UNIT V   The Evolutionary History of Biological Diversity   165

CHAPTER 24
Early Earth and the Origin of Life   165

CHAPTER 25
Prokaryotes and the Origins of Metabolic Diversity   170

CHAPTER 26
The Origins of Eukaryotic Diversity   177

CHAPTER 27
Plants and the Colonization of Land   184

CHAPTER 28
Fungi   193

CHAPTER 29
Invertebrates and the Origin of Animal Diversity   198

CHAPTER 30
The Vertebrate Genealogy   209

UNIT VI   Plants: Form and Function   221

CHAPTER 31
Plant Structure and Growth   221

CHAPTER 32
Transport in Plants   228

CHAPTER 33
Plant Nutrition   234

CHAPTER 34
Plant Reproduction and Development   239

CHAPTER 35
Control Systems in Plants   247

UNIT VII   Animals: Form and Function   255

CHAPTER 36
An Introduction to Animal Structure and Function   255

CHAPTER 37
Animal Nutrition   260

CHAPTER 38
Circulation and Gas Exchange   267

CHAPTER 39
The Body's Defenses   278

CHAPTER 40
Controlling the Internal Environment   288

CHAPTER 41
Chemical Signals in Animals   299

CHAPTER 42
Animal Reproduction   308

CHAPTER 43
Animal Development   317

CHAPTER 44
Nervous Systems   326

CHAPTER 45
Sensory and Motor Mechanisms   336

UNIT VIII   Ecology   347

CHAPTER 46
An Introduction to Ecology: Distribution and Adaptations of Organisms   347

CHAPTER 47
Population Ecology   355

CHAPTER 48
Community Ecology   362

CHAPTER 49
Ecosystems   370

CHAPTER 50
Behavior   378

ANSWER SECTION   387

Another name for this *Student Study Guide* for Neil Campbell's *Biology*, Fourth Edition, could be "A Student Structuring Guide." The purpose of this guide is to help you to structure and organize your developing knowledge of biology and to create your own personal understanding of the topics covered in the text. Biology is a visual science, and this fourth edition of the *Student Study Guide* includes many diagrams and pictures taken from *Biology*, Fourth Edition, to help you see and understand the relationships among parts of structures and processes. *Interactive Questions*, which are new to this edition, are designed to help you develop an active learning approach to your study of biology.

The four divisions of each study guide chapter are as follows:

- The *Framework* identifies the overall picture; it provides a conceptual framework into which the chapter information fits.

- The *Chapter Review* is a condensation of each textbook chapter.

- Interspersed in this summary are *Interactive Questions* that help you stop and synthesize the material just covered. You are asked to complete tables, label diagrams, respond to short answer or essay questions, and complete or construct concept maps.

- The *Structure Your Knowledge* section directs you to organize and relate the main concepts of the chapter.

- In the *Test Your Knowledge section*, you are provided with objective questions to test your understanding. The multiple choice questions presented in each chapter ask you to choose the best answer. Some answers may be partly correct; almost all choices have been written to test your ability to think and discriminate among alternatives. Make sure you understand why the

other choices are incorrect as well as why the correct answer is correct.

Suggested answers to all questions are provided in the *Answer Section* at the end of the book.

What are the concept maps that appear throughout this *Study Guide*? A concept map is a diagram that shows how ideas are organized and related. The *structure* of a concept map is a hierarchically organized cluster of concepts, enclosed in boxes and connected with labeled lines that explicitly show the relationships among the concepts. The *function* of a concept map is to help you structure your understanding of a topic and create meaning. The *value* of a concept map is in the thinking and evaluating required to create a map.

Developing a concept map for a group of concepts requires that you evaluate the relative importance of the concepts (which are most inclusive and important? which are less important and subordinate to other concepts?), arrange the concepts in a meaningful cluster, and draw connections between them that help you make their meanings explicit. (Additional information about concept mapping may be found in *Learning How to Learn*, Joseph D. Novak and D. Bob Gowin, Cambridge University Press, 1984.)

This book uses concept maps in several ways. A map of a chapter may be presented in the *Framework* section to show the organization of the key concepts in that chapter. An *Interactive Question* may provide a skeleton concept map, with some concepts provided and boxes for you to complete. This technique, which is new to this fourth edition, is intended to help you become more familiar with concept maps and to illustrate one possible approach to organizing the concepts of a particular section. You will also be asked to develop your own concept maps on certain subsets of ideas. In these cases, the answer section will present *suggested* concept maps. A concept map is an individual picture of your understanding at the time you make the map. As your understanding of an area develops, your concept map will evolve—sometimes becoming more complex and interrelated, sometimes becoming simplified and streamlined. Do not look to the answer section for the "right" concept map. *After* you have organized your own thoughts, look at the answer map to make sure you have included the key concepts (although you may have added more), to check that the connections you have made are reasonable, and perhaps to see another way to organize the information.

Biology is a fascinating and broad subject. Campbell's *Biology* is filled with terminology and facts organized in a manner that will help you build a conceptual framework of the major themes of modern biology. This *Student Study Guide* is intended to help you learn and recall information and, most importantly, to encourage and guide you as you develop your own understanding of and appreciation for biology.

Martha R. Taylor

CHAPTER 1

INTRODUCTION:
THEMES IN THE STUDY OF LIFE

FRAMEWORK

This chapter outlines broad themes that unify the study of biology and describe the scientific construction of biological knowledge. A course in biology is neither a vocabulary course nor a classification exercise for the diverse forms of life. Biology is a collection of facts and concepts structured within theories and organizing principles. Recognizing the common themes within biology will help you to structure your knowledge of this fascinating and challenging study of life.

CHAPTER REVIEW

Biology, the scientific study of life, is an extension of our innate interest in life in its diverse forms. The scope of biology is immense, spanning the submicroscopic level to the complex web of ecosystems from the present back through nearly 4 billion years of evolutionary history. Recent advances in research methods have aided the efforts of a large number of contemporary biologists working throughout the many subfields of biology to achieve an explosion of information. A beginning student can make sense of this expanding body of knowledge by focusing on a few enduring themes that unify the study of biology.

■ Life is organized on many structural levels

Biological organization is based on a hierarchy of structural levels, with each level—from atoms, biological molecules, organelles, cells, tissues, organs, organ systems, organisms, populations, and communities to ecosystems—building on the levels below it. The study of biology includes understanding these various levels of organization and the interactions among them.

■ Each level of biological organization has emergent properties

The study of biology balances the pragmatic reductionist strategy, which breaks down complex systems to simpler components, with holism, an approach in which the higher organizational levels of life are studied to comprehend the properties that arise from the interactions and integration of the system's parts.

Interactions among components at each level of organization lead to the emergence of novel properties at the next level: the whole is greater than the sum of its parts. The concept of emergent properties of life does not support the doctrine of vitalism, which claims that life is driven by forces that defy explanation. Rather, the principles of physics and chemistry can be used to explain the properties of life that arise from the hierarchy of structural organization.

The characteristics or properties of life include order, reproduction, growth and development directed by heritable programs, energy utilization, responsiveness to the environment, homeostasis, and evolutionary adaptation.

■ Cells are an organism's basic units of structure and function

The cell is the simplest structural level capable of performing all the activities of life. Hooke first described and named cells in 1665 when he observed a slice of cork with a simple microscope. Leeuwenhoek, a contemporary, developed lenses that permitted him to view the world of microscopic organisms. In 1839, Schleiden and Schwann concluded that all living things consist of cells. This cell theory now includes the idea that all cells come from other cells.

The recent advent of the electron microscope has revealed the complex structural organization of cells. Two major types of cells are recognized. The simpler prokaryotic cell, found only as a bacterium, lacks both a nucleus to enclose its DNA and most membrane-bound organelles. The eukaryotic cell, with its nucleus and numerous cytoplasmic compartments, is typical of all other living organisms.

■ The continuity of life is based on heritable information in the form of DNA

The biological instructions for the development and functioning of organisms are coded in the arrangement of the four nucleotides in DNA molecules and transmitted from parent to offspring in units of inheritance called genes. All forms of life use essentially the same genetic code.

■ A feeling for organisms enriches the study of life

Organisms are the units of life. The study of biology begins with a feeling for organisms—an attraction to the living world around us. Good scientists have a deep understanding of and appreciation for the organisms they work with. The organism sets the context for studies that may range in focus from molecular to ecosystem biology.

■ Structure and function are correlated at all levels of biological organization

The form of a biological structure gives information about its function, and a study of function provides insight into structural organization. The principle that form fits function is illustrated at all levels of biological organization.

■ Organisms are open systems that interact continuously with their environments

Organisms affect and are affected by the physical and biological environments they interact with. Within an ecosystem, nutrients cycle between the abiotic and biotic components, and energy flows from sunlight to photosynthetic organisms (producers) to consumers and exits in the form of heat.

■ Diversity and unity are the dual faces of life on Earth

About 1.5 million species, out of an estimated total of 5 to 30 million, have been identified and named. Taxonomy is the branch of biology that names organisms and classifies species into hierarchical groups. At the broadest unit of classification, life forms have been organized into five kingdoms: Monera, Protista, Plantae, Fungi, and Animalia.

The bacteria of the kingdom Monera (or the kingdoms Archaebacteria and Eubacteria in the six-kingdom scheme) are distinguished on the basis of their simple prokaryotic cell type. All other kingdoms have eukaryotic cells. Kingdom Protista contains mostly unicellular or simple multicellular forms. The other three kingdoms contain multicellular organisms characterized to a large extent by their mode of nutrition. Plants are photosynthetic, fungi are mostly decomposers that absorb their nutrients from dead organisms or organic wastes, and animals obtain their food by ingestion.

Within this diversity, living forms share a universal genetic code and similarities in cell structure.

■ Evolution is the core theme of biology

Evolution connects all of life by common ancestry. The history of living forms extends back over 3 billion years to the ancient prokaryotes.

In *The Origin of Species*, published in 1859, Charles Darwin presented his case for evolution, or "descent with modification," that present forms evolved from a succession of ancestral forms guided by a mechanism called natural selection. Darwin synthesized the concept of natural selection by drawing an inference from two observations: individuals vary in many heritable traits, and individuals with traits best suited for an environment leave a larger proportion of offspring than do less fit individuals. This differential reproductive success results in the gradual accumulation of adaptations within a population.

According to Darwin, new species originate when unique traits and combinations of traits that arise randomly are selected for by differing environments. Evolution makes sense of both the unity and diversity of life.

■ Science as a process of inquiry often involves hypothetico-deductive thinking

Science, which is a way of knowing, involves asking questions about nature and believing that those questions are answerable.

The scientific process is based on hypothetico-deductive thinking. A hypothesis is a tentative explanation for an observation or question. Deductive reasoning proceeds from the general to the specific, from a general hypothesis to specific predictions of results if the general premise is true. A hypothesis is usually tested by performing experiments or making observations to see whether predicted results occur.

There are five important aspects of hypotheses: hypotheses are possible causes or explanations; hypotheses reflect past experience with similar situations; having multiple hypotheses is good science; hypotheses must be testable using the hypothetico-deductive method; and hypotheses can be rejected but not proven.

The scientific process makes use of controlled experiments in which subjects are divided into an experimental group and a control group. Both groups are treated alike except for the one variable that the experiment is trying to test. Adequate sample size and control of other variables are important components of an experiment.

The experiments of Reznick and Endler illustrate the hypothetico-deductive approach. By maintaining control populations and following generations of guppies transplanted from sites with pike-cichlids to sites with killifish for 11 years, they were able to conclude that natural selection due to differential predation was the most likely explanation for differences in life history characteristics between guppy populations.

Science is characterized as progressive and self-correcting. Scientists build on the work done by others, refining or refuting their ideas, both cooperating and competing with each other.

Facts, in the form of observations and experimental results, are prerequisites of science, but it is the new ways of organizing and relating those facts that advance science. A theory is broader in scope than a hypothesis and is supported by a large body of evidence.

■ Science and technology are functions of society

Science and technology are interwoven as science uses the new instruments and techniques of technology to extend knowledge, and technology uses the information generated by science to guide the development of technological advances. The fields of molecular biology and genetic engineering illustrate this relationship. Technology contributes to our standard of living and has made it possible for the human population to expand rapidly but at a price of severe environmental consequences, including deforestation, acid rain, nuclear accidents, toxic wastes, and increased rates of extinction. Solutions to these problems will involve politics, economics, and cultural values as well as science and technology.

■ Biology is a multidisciplinary adventure

Biology is a demanding science—partly because living systems are so complex and partly because biology incorporates concepts from chemistry, physics, and math. This book presents a wealth of information. The basic themes of biology will help you understand, appreciate, and structure your growing knowledge of biology.

STRUCTURE YOUR KNOWLEDGE

1. This chapter presents unifying themes of biology. Briefly describe each of these in your own words:
   a. hierarchy of organization
   b. emergent properties
   c. cellular basis of life
   d. heritable information
   e. a feeling for organisms
   f. correlation of structure and function
   g. interaction of organisms with their environment
   h. unity and diversity
   i. evolution
   j. science as a process
   k. science and technology
   l. biology is multidisciplinary

THE CHEMISTRY OF LIFE

CHAPTER 2

THE CHEMICAL CONTEXT OF LIFE

FRAMEWORK

This chapter considers the basic principles of chemistry that explain the behavior of atoms and molecules and that form the basis for our modern understanding of biology. Emergent properties are associated with each new level of structural organization as the subatomic particles—protons, neutrons, and electrons—are organized into atoms and atoms are combined by covalent or ionic bonds into molecules.

■ **INTERACTIVE QUESTION 2.1**

On Earth, mass and weight may be considered synonymous.

a. Define mass.

b. Define weight.

CHAPTER REVIEW

■ **Matter consists of chemical elements in pure form and in combinations called compounds**

Matter is anything that takes up space and has mass. The basic forms of matter are **elements**, substances that cannot be chemically broken down to other types of matter. Chemists have identified 92 naturally occurring elements. A **compound** is made up of two or more elements combined in a fixed ratio. A compound usually has characteristics quite different from its constituent elements, an example of the emergence of novel properties in higher levels of organization.

■ **Life requires about 25 chemical elements**

Carbon, oxygen, hydrogen, and nitrogen make up 96% of living matter. The remaining 4% is composed of the seven elements included in the list on the following page. Some elements, like iron (Fe) and iodine (I), may be required in very minute quantities and are called **trace elements**.

■ **Atomic structure determines the behavior of an element**

An **atom** is the smallest unit of an element retaining the physical and chemical properties of that element. Each element has its own unique type of atom.

■ INTERACTIVE QUESTION 2.2

The following table lists the symbols for the most common elements according to percentage of human body weight. Fill in the names of these elements.

Element	Symbol
	O
	C
	H
	N
	Ca
	P
	K
	S
	Na
	Cl
	Mg

Subatomic Particles   Three stable subatomic particles are important to our understanding of atoms. Uncharged **neutrons** and positively charged **protons** are packed tightly together to form the nucleus of an atom. Negatively charged **electrons** orbit rapidly about the nucleus.

Protons and neutrons have a similar mass of about 1.7×10^{-24} g or 1 **dalton** each. A dalton is the measurement unit for atomic mass. Electrons have negligible mass.

Atomic Number and Atomic Weight   Each element has a characteristic **atomic number**, or number of protons in the nucleus of its atom. Unless otherwise indicated, the number of protons in an atom is equal to the number of electrons, and the atom has a neutral electrical charge. A subscript to the left of the symbol for an element indicates its atomic number; a superscript indicates mass number. The **mass number** is equal to the number of protons and neutrons in the

■ INTERACTIVE QUESTION 2.3

The difference between the mass number and the atomic number of an atom is equal to the number of _____. An atom of phosphorus, $^{31}_{15}P$, contains ___ protons, ___ electrons, and ___ neutrons. The atomic weight of phosphorus is _____.

nucleus and approximates the mass of an atom of that element in daltons. The term **atomic weight** is often used to refer to the mass of an atom.

Isotopes   Although the number of protons is constant, the number of neutrons can vary among the atoms of an element, creating different **isotopes**. Some isotopes are unstable, or **radioactive**; their nuclei spontaneously decay, giving off particles and energy.

Radioactive isotopes are important tools in biological research and medicine because they can be introduced into organisms and detected in minute quantities. Techniques using scintillation counters or autoradiography can determine the quantity and location of radioactively labeled molecules within a cell or tissue. Too great an exposure to radiation from decaying isotopes poses a significant hazard to life.

Energy Levels   The nucleus of an atom is extremely small compared to the large area in which the electrons orbit. The chemical behavior of an atom—the type of interactions it has with other atoms—is determined by the number and location of its electrons.

Energy is defined as the ability to do work. **Potential energy** is energy stored in matter as a consequence of the relative position of masses. Matter naturally tends to move toward a more stable lower level of potential energy and requires the input of energy to return to a higher potential energy.

The potential energy of electrons increases as their distance from the positively charged nucleus increases. Electrons can orbit in several different potential energy states, called **energy levels** or **electron shells**, surrounding the nucleus. Electrons may shift from one energy level to another; such changes of the potential energy of electrons occur in steps of discrete amounts.

■ INTERACTIVE QUESTION 2.4

To move to a shell farther from the nucleus, an electron must (absorb/release) energy; energy is (absorbed/released) when an electron moves to a closer shell. (Circle correct terms.)

Electron Orbitals   The three-dimensional space or volume within which an electron is found 90% of the time is called an **orbital**. No more than two electrons can occupy the same orbital. The first electron shell can contain two electrons in a single spherical orbital, called the 1s orbital. The second electron shell can hold a maximum of eight electrons in its four orbitals, which are a 2*s* spherical orbital and three dumbbell-shaped *p* orbitals located along the *x*, *y*, and *z* axes. The orbitals of higher electron shells are identified as 3*s* and 3*p*, and so on.

Electron Configuration and Chemical Properties
The chemical behavior of an atom is a function of its electron configuration—in particular, the number of **valence electrons** in its outermost energy shell, or **valence shell**. A valence shell of eight electrons is complete, resulting in an unreactive or inert atom. Atoms with incomplete valence shells are chemically reactive because of their unpaired electrons. The periodic table of elements is arranged in order of the sequential addition of electrons to orbitals. Atoms with the same number of electrons in their valence shell have similar chemical properties and are located in the same column of the periodic table.

■ **INTERACTIVE QUESTION 2.5**

Draw the electron configuration for the following atoms. Which of these atoms would you expect to have similar chemical properties? Why?

a. $_7$N   **b.** $_{15}$P   **c.** $_8$O   **d.** $_{17}$Cl

■ **INTERACTIVE QUESTION 2.6**

Fill in the blanks in the following concept map to help you review the atomic structure of atoms.

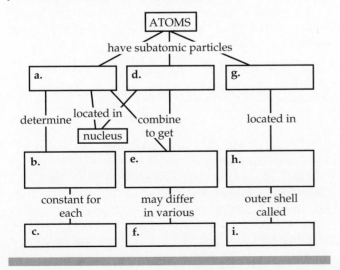

■ Atoms combine by chemical bonding to form molecules

Atoms with incomplete valence shells can either share electrons with or completely transfer electrons to or from other atoms such that each atom is able to complete its valence shell. These interactions usually result in attractions, called **chemical bonds**, that hold the atoms together.

Covalent Bonds   When two atoms share a pair of valence electrons, a **covalent bond** is formed. Their valence shells overlap, and the electrons orbit both nuclei. A **structural formula**, such as H–H, indicates both the number and type of atoms and also the bonding within a molecule. A **molecular formula**, such as O_2, indicates only the kinds and numbers of atoms in a molecule. In an oxygen molecule, two pairs of valence electrons are shared between oxygen atoms, forming a **double covalent bond**. A compound, such as H_2O, is a molecule formed from more than one kind of element.

The **valence,** or bonding capacity, of an atom is a measure of the number of covalent bonds it must form in order to complete its outer shell.

■ **INTERACTIVE QUESTION 2.7**

What are the valences of the four most common elements of living matter?

a. hydrogen ___   **b.** oxygen ___

c. nitrogen ___   **d.** carbon ___

Electronegativity is the attraction of an atom for shared electrons. If the atoms in a molecule have similar electronegativities, the electrons remain equally shared between the two nuclei, and the covalent bond is said to be **nonpolar**. If one element is more electronegative, it pulls the shared electrons closer to itself, creating a **polar covalent bond**. This unequal sharing of electrons results in a slight negative charge associated with the more electronegative atom and a slight positive charge associated with the atom from which the electrons are pulled.

Ionic Bonds   If two atoms are very different in their attraction for the shared electrons, the more electronegative atom may completely transfer an electron from another atom, forming an **ionic bond** in which atoms are held together because of the attraction of their opposite charges. This transfer of an electron results in the formation of charged atoms or molecules called **ions**. The atom that lost the electron is a positively charged **cation**. The negatively charged atom that gained the electron is called an **anion**. Ionic compounds, called salts, often exist as three-dimensional crystalline lattice arrangements held together by electrical attraction. The number of ions present in a salt crystal is not fixed, but the atoms are present in specific ratios. Salts have strong ionic bonds when dry, but the crystal dissolves in water.

Ion also refers to entire covalent molecules that are electrically charged. Ammonium (NH_4^+) is a cation; this covalently bonded molecule is missing one electron.

The sharing of electrons between atoms in the completion of their valence shells can be thought of as falling on a continuum from nonpolar covalent bonds, through polar covalent bonds, to ionic bonds in which electrons are shared so unequally as to be actually transferred from one atom to another.

■ **INTERACTIVE QUESTION   2.8**

Calcium ($_{20}$Ca) and chlorine ($_{17}$Cl) can combine to form the salt calcium chloride. Based on the number of electrons in their valence shells and their bonding capacities, what would the molecular formula for this salt be? **a.** _____ Which atom becomes the cation? **b.** _____

■ Weak chemical bonds play important roles in the chemistry of life

Weak bonds, such as ionic bonds in water, form temporary interactions between molecules and are involved in many biological signals and processes. Weak bonds within large molecules such as proteins help to create the three-dimensional shape and resulting activity of these molecules.

Hydrogen Bonds   When a hydrogen atom is covalently bonded with an electronegative atom and thus has a partial positive charge, it can be attracted to another electronegative atom and form a **hydrogen bond**.

■ A molecule's biological function is related to its shape

A molecule has a characteristic size and shape that affects how it interacts with other molecules. For example, receptor molecules on the surface of cells are keyed to the specific shape of signal molecules.

■ **INTERACTIVE QUESTION   2.9**

Sketch a water molecule, showing its shape and the electron shells with the covalently shared electrons. Indicate the areas with slight negative and positive charges that enable a water molecule to form hydrogen bonds.

■ Chemical reactions change the composition of matter

Chemical reactions involve the making or breaking of chemical bonds in the transformation of matter into different forms. Matter is conserved in chemical reactions; the same number and kinds of atoms are present in both **reactants** and **products**, although the rearrangement of electrons and atoms causes the properties of these molecules to be different.

■ **INTERACTIVE QUESTION   2.10**

Fill in the missing coefficients for respiration, the conversion of glucose and oxygen to carbon dioxide and water, so that all atoms are conserved in the chemical reaction.

$$C_6H_{12}O_6 + \_\_ O_2 \longrightarrow \_\_CO_2 + \_\_ H_2O$$

Most reactions are reversible—the products of the forward reaction can become reactants in the reverse reaction. Increasing the concentrations of reactants speeds up the rate of a reaction. **Chemical equilibrium** may be reached when the forward and reverse reactions proceed at the same rate, and the relative concentrations of reactants and products no longer change.

■ Chemical conditions on the early Earth set the stage for the origin and evolution of life

Most astronomers believe that all matter was once concentrated in a giant mass that blew apart with a "big bang" 10 to 20 billion years ago. Our sun formed about 5 billion years ago when most of the swirling matter in the cloud of dust that formed our solar system condensed in the center. Earth and the rest of the planets formed about 4.6 billion years ago as gravity attracted the remaining dust and ice to a few kernels.

Geologists believe that the initially cold Earth went through a molten period, when its components sorted into layers of different densities. The continents are borne on plates of crust that float on the flexible mantle, which surrounds a dense core.

Scientists speculate that Earth's first atmosphere of hot hydrogen gas escaped from the planet and was replaced by gases from volcanoes. This early atmosphere probably consisted mostly of water vapor, carbon monoxide, carbon dioxide, nitrogen, methane, and ammonia. The first seas were formed by torrential rains. Lightning, volcanic activity, and UV radiation were quite intense. Chemical evolution in such an environment is believed to have given rise to life.

STRUCTURE YOUR KNOWLEDGE

Take the time to write out or discuss your answers to the following questions. Then refer to the suggested answers at the end of the book.

1. Fill in the following chart for the major subatomic particles of an atom.

Particle	Charge	Mass	Location

2. Atoms can have various numbers associated with them.
   a. Define the following and show where each of them is placed relative to the symbol of an element such as C: atomic number, mass number, atomic weight.
   b. Define valence.
   c. Which of these four numbers is most related to the chemical behavior of an atom? Explain.

3. Explain what is meant by saying that the sharing of electrons between atoms falls on a continuum from covalent bonds to ionic bonds.

TEST YOUR KNOWLEDGE

MULTIPLE CHOICE: *Choose the one best answer.*

1. Each element has its own characteristic atom in which
   a. the atomic weight is constant.
   b. the atomic number is constant.
   c. the mass number is constant.
   d. two of the above are correct.
   e. all of the above are correct.

2. Isotopes can be used in studies of metabolic pathways because
   a. their half-life allows a researcher to time an experiment.
   b. they are more reactive.
   c. the cell does not recognize the extra protons in the nucleus, so isotopes are readily used in metabolism.
   d. their location or quantity can be experimentally determined because of their radioactivity.
   e. their extra neutrons produce different colors that can be traced through the body.

3. In a reaction in chemical equilibrium,
   a. the forward and reverse reactions are occurring at the same rate.
   b. the reactants and products are in equal concentration.
   c. the forward reaction has gone further than the reverse reaction.
   d. there are equal numbers of atoms on both sides of the equation.
   e. a, b, and d are correct.

4. Oxygen has eight electrons. You would expect the arrangement of these electrons to be:
   a. eight in the second energy shell, creating an inert element.
   b. two in the first energy shell and six in the second, creating a valence of six.
   c. two in the $1s$ orbital and two each in the three $2p$ orbitals, creating a valence of zero.
   d. two in the first energy shell, four in the second, and two in the third, creating a valence of two.
   e. two in the first energy shell and six in the second, creating a valence of two.

5. A covalent bond between two atoms is likely to be polar if
   a. one of the atoms is much more electronegative than the other.
   b. the two atoms are equally electronegative.
   c. the two atoms are of the same element.
   d. the bond is part of a tetrahedrally shaped molecule.
   e. one atom is an anion.

6. A triple covalent bond would
   a. be very polar.
   b. involve the bonding of three atoms.
   c. involve the bonding of six atoms.
   d. produce a triangularly shaped molecule.
   e. involve the sharing of six electrons.

7. A cation
   a. has gained an electron.
   b. can easily form hydrogen bonds.
   c. is more likely to form from an atom with a high valence number of electrons.
   d. has a positive charge.
   e. Both c and d are correct.

8. What types of bonds are identified in the following illustration of a water molecule interacting with an ammonia molecule?

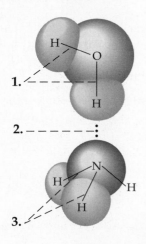

   a. Bonds 1 are polar covalent bonds, bond 2 is a hydrogen bond, and bonds 3 are nonpolar covalent bonds.
   b. Bonds 1 and 3 are polar covalent bonds and bond 2 is a hydrogen bond.
   c. Bonds 1 and 3 are polar covalent bonds and bond 2 is an ionic bond.
   d. Bonds 1 and 3 are nonpolar covalent bonds and bond 2 is a hydrogen bond.
   e. Bonds 1 and 3 are polar covalent bonds and bond 2 is a nonpolar covalent bond.

9. The early atmosphere of the earth when life first evolved was created by
   a. the "big bang."
   b. the separation of layers of different densities.
   c. the evaporation of water from the torrential rains.
   d. gases released from volcanoes.
   e. the production of oxygen by early photosynthetic bacteria.

10. The ability of morphine to mimic the effects of the body's endorphins is due to
    a. a chemical equilibrium developing between morphine and endorphins.
    b. the one-way conversion of morphine into endorphin.
    c. molecular shape similarities that allow morphine to bind to endorphin receptors.
    d. the similarities between morphine and heroin.
    e. hydrogen bonding and other weak bonds forming between morphine and endorphins.

The six elements most common in living organisms are:

$$^{12}_{6}C \quad ^{16}_{8}O \quad ^{1}_{1}H \quad ^{14}_{7}N \quad ^{32}_{16}S \quad ^{31}_{15}P$$

Use this information to answer questions 11 through 16.

11. How many electrons does phosphorus have in its valence shell?
    a. 5   c. 8   e. 16
    b. 7   d. 15

12. What is the atomic weight of phosphorus?
    a. 15   c. 31   e. 62
    b. 16   d. 46

13. A radioactive isotope of carbon has the mass number 14. How many neutrons does this isotope have?
    a. 6   c. 8   e. 14
    b. 7   d. 12

14. How many covalent bonds is a sulfur atom most likely to form?
    a. 1   c. 3   e. 5
    b. 2   d. 4

15. Based on electron configuration, which of these elements would have chemical behavior most like that of oxygen?
    a. C   c. N   e. S
    b. H   d. P

16. How many of these elements are found next to each other (side by side) on the periodic table?
    a. one group of two
    b. two groups of two
    c. one group of two and one group of three
    d. one group of three
    e. all of them

CHAPTER 3

WATER AND THE FITNESS OF THE ENVIRONMENT

FRAMEWORK

Hydrogen bonding between polar water molecules creates a cohesive liquid with a high specific heat and high heat of vaporization, both of which help to regulate environmental temperature. The polarity of water makes it a versatile solvent. An organism's pH may be regulated by buffers. Acid precipitation poses a serious environmental threat.

CHAPTER REVIEW

Water makes up 70% to 95% of the cell content of living organisms and covers 75% of the Earth's surface. Its unique properties make the external environment fit for living organisms and the internal environment of organisms fit for the chemical and physical processes of life.

■ **The polarity of water molecules results in hydrogen bonding**

A water molecule consists of two hydrogen atoms each covalently bonded to a more electronegative oxygen atom. This **polar molecule** has a right-angle shape with a slight positive charge on each hydrogen atom and a slight negative charge associated with the oxygen. Hydrogen bonds form between the hydrogen atom of one water molecule and the oxygen atom of another molecule, creating a higher level of structural organization and leading to the emergent properties of water.

■ **INTERACTIVE QUESTION   3.1**

Draw the four water molecules that can hydrogen-bond to this water molecule. Indicate the slight negative and positive charges that account for the formation of hydrogen bonds.

■ **Organisms depend on the cohesion of water molecules**

Liquid water is unusually cohesive due to the constant forming and re-forming of hydrogen bonds that hold the molecules together. This **cohesion** creates a more structurally organized liquid and enables water to move against gravity in plants. The **adhesion** of water molecules to the walls of plant vessels also contributes to water transport. Hydrogen bonding between water molecules produces a high **surface tension** at the interface between water and air.

■ Water contributes to Earth's habitability by moderating temperatures

Heat and Temperature   In a body of matter, **heat** is a measure of the total quantity of **kinetic energy**, the energy associated with the movement of atoms and molecules. **Temperature** measures the average kinetic energy of the molecules in a substance.

Temperature is measured using a **Celsius scale**. Water at sea level freezes at 0°C and boils at 100°C. Heat is measured by the **calorie** (cal). A calorie is the amount of heat energy it takes to raise 1 g of water 1°C. A **kilocalorie** (kcal) is 1000 calories, the amount of heat required to raise 1 kg of water 1°C. A **joule** (J) equals 0.239 cal; a calorie is 4.184 J.

Water's High Specific Heat   **Specific heat** is the amount of heat absorbed or lost when 1 g of a substance changes its temperature by 1°C. Water's specific heat of 1 cal/g/°C is unusually high compared with that of other common substances; water must absorb or release a relatively large quantity of heat in order for its temperature to change. Heat must be absorbed to break hydrogen bonds before water molecules can move faster and the temperature can rise, and conversely, heat is released when hydrogen bonds form as the temperature of water drops. The ability of large bodies of water to stabilize air temperature is due to the high specific heat of water. The high proportion of water in the environment and within organisms keeps temperature fluctuations within limits that permit life.

Evaporative Cooling   The transformation from a liquid to a gas is called vaporization or evaporation and happens when molecules with sufficient kinetic energy overcome their attraction to other molecules and escape into the air as gas. The addition of heat increases the rate of evaporation by increasing the kinetic energy of molecules. The heat of vaporization is the quantity of heat that must be absorbed for 1 g of a liquid to be converted to a gas. Water has a high heat of vaporization (540 cal/g) because a large amount of heat is needed to break the hydrogen bonds holding water molecules together. This property of water helps moderate the climate on Earth as solar heat is dissipated from tropical seas during evaporation and heat is released when moist tropical air condenses to form rain.

As a substance vaporizes, the liquid left behind loses the kinetic energy of the escaping molecules and cools down. **Evaporative cooling** helps to protect terrestrial organisms from overheating and contributes to the stability of temperatures in lakes and ponds.

■ Oceans and lakes don't freeze solid because ice floats

As water cools below 4°C, it expands. By 0°C, each water molecule becomes hydrogen-bonded to four other molecules, creating a crystalline lattice that spaces the molecules apart. Ice is less dense than liquid water, and therefore, it floats. The floating ice insulates the liquid water below.

■ **INTERACTIVE QUESTION   3.2**

The following concept map is one way to show how the breaking and forming of hydrogen bonds is related to temperature regulation. Fill in the blanks and compare your choice of concepts to those given in the answer section. Or, better still, create your own map to help you understand how water stabilizes temperature.

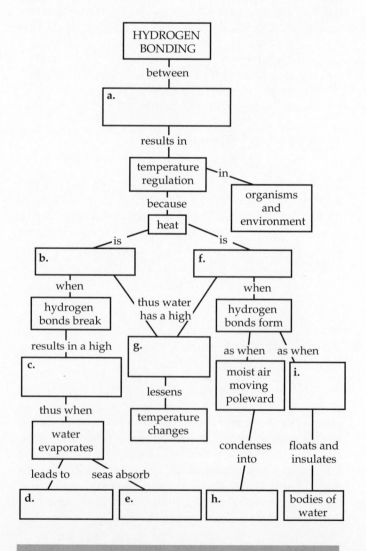

■ Water is the solvent of life

A **solution** is a homogeneous mixture of two or more substances; the dissolving agent is called the **solvent**, and the substance that is dissolved is the **solute**. An **aqueous solution** is one in which water is the solvent. The positive and negative regions of water molecules are attracted to oppositely charged ions or partially charged regions of polar molecules. Thus, solute molecules become surrounded by water molecules and dissolve into solution.

Hydrophilic and Hydrophobic Substances Ionic and polar substances are **hydrophilic**; they have an affinity for water due to electrical attractions and hydrogen bonding. Nonpolar and non-ionic compounds are **hydrophobic**; they will not mix with or dissolve in water.

Solute Concentration in Aqueous Solutions Most of the chemical reactions of life take place in water. A **mole** (mol) is the amount of a substance that has a mass in grams numerically equivalent to its **molecular weight** (sum of the weight of all atoms in the molecule) in daltons. A mole of any substance has exactly the same number of molecules—6.02×10^{23}, called Avogadro's number. The **molarity** of a solution (abbreviated *M*) refers to the number of moles of a solute dissolved in 1 liter of solution.

■ INTERACTIVE QUESTION 3.3

a. How many grams of lactic acid ($C_3H_6O_3$) are in a 0.5 *M* solution of lactic acid? (^{12}C, ^{1}H, ^{16}O)

b. How many grams of salt (NaCl) must be dissolved in water to make 2 liters of a 2 *M* salt solution? (^{23}Na, ^{34}Cl)

■ Organisms are sensitive to changes in pH

Dissociation of Water Molecules A water molecule can dissociate into a **hydrogen ion**, H^+ (which binds to another water molecule to form a hydronium ion, H_3O^+), and a **hydroxide ion**, OH^-. Although reversible and statistically rare, this dissociation into the highly reactive hydrogen and hydroxide ions has important biological consequences.

Acids and Bases In pure water, the concentrations of H^+ and OH^- ions are the same; both are equal to 10^{-7} *M*. When acids or bases dissolve in water, the H^+ and OH^- balance shifts. An **acid** adds H^+ to a solution, whereas a **base** reduces H^+ in a solution by accepting hydrogen ions or by adding hydroxide ions (which then combine with H^+ and thus remove hydrogen ions). A strong acid or strong base dissociates completely when mixed with water. A weak acid or base reversibly dissociates, releasing or binding H^+. A solution with a higher concentration of H^+ than of OH^- is considered acidic. A basic solution has a higher concentration of OH^- than of H^+.

The pH Scale In any solution, the product of the $[H^+]$ and $[OH^-]$ is constant at 10^{-14} *M*^{2}. Brackets, [ ], indicate molar concentration. In a neutral solution, $[H^+]$ and $[OH^-]$ are both equal to 10^{-7} *M*. If the $[H^+]$ is higher, then the $[OH^-]$ is lower, due to the tendency of excess hydrogen ions to combine with the hydroxide ions in solution and form water. Likewise, an increase in $[OH^-]$ causes an equivalent decrease in $[H^+]$. If $[OH^-]$ is equal to 10^{-10} *M*, then $[H^-]$ will equal 10^{-4} *M*.

The logarithmic **pH scale** compresses the range of hydrogen and hydroxide ion concentrations, which can vary in different solutions by many orders of magnitude. The pH of a solution is defined as the negative log (base 10) of the $[H^+]$: pH = $-$log $[H^+]$. For a neutral solution, $[H^+]$ is 10^{-7} *M*, and the pH equals 7. As the $[H^+]$ increases in an acidic solution, the pH value decreases. The difference between each unit of the pH scale represents a tenfold difference in the concentration of $[H^+]$ and $[OH^-]$.

Buffers Most cells have an internal pH close to 7. **Buffers** within the cell maintain a constant pH by accepting excess H^+ ions or donating H^+ ions when H^+ concentration decreases. Weak acid–base pairs that reversibly bind hydrogen ions are typical of most buffering systems.

■ INTERACTIVE QUESTION 3.4

The carbonic acid/bicarbonate system is an important biological buffer. Label the molecules and ions in this equation and indicate which is the H^+ donor and acceptor. In which direction will this reaction proceed **(a)** when the pH of a solution begins to fall and **(b)** when the pH rises above normal levels?

$$H_2CO_3 \leftrightarrows HCO_3^- + H^+$$

■ Acid precipitation threatens the fitness of the environment

Acid precipitation, with a pH lower than normal pH 5.6, is due to the reaction of water in the atmosphere with the sulfur oxides and nitrogen oxides released by the combustion of fossil fuels. Lowering the pH of the soil solution affects the solubility of minerals needed by plants. In lakes and ponds, a lowered pH and the accumulation of minerals leached from the soil by acid rain harm many species of fishes, amphibians, and aquatic invertebrates. The reduction of acid precipitation depends on the development of industrial controls and antipollution devices.

STRUCTURE YOUR KNOWLEDGE

1. Fill in the following table that summarizes the properties of water that contribute to the fitness of the environment for life.

2. To become proficient in the use of the concepts relating to pH, develop a concept map to organize your understanding of the following terms: pH, $[H^+]$, $[OH^-]$, acidic, basic, neutral, buffer, 1–14, acid–base pair. Remember to label connecting lines and add additional concepts as you need them. *A suggested concept map is given in the answer section, but remember that your concept map should represent your own understanding. The value of this exercise is in your process of organizing these concepts for yourself.*

TEST YOUR KNOWLEDGE

MULTIPLE CHOICE: *Choose the one best answer.*

1. Each water molecule is capable of forming
   a. one hydrogen bond.
   b. three hydrogen bonds.
   c. four hydrogen bonds.
   d. one covalent bond and two hydrogen bonds.
   e. one covalent bond and three hydrogen bonds.

2. The polarity of water molecules
   a. promotes the formation of hydrogen bonds.
   b. helps water to dissolve nonpolar solutes.
   c. lowers the heat of vaporization and leads to evaporative cooling.
   d. creates a crystalline structure in liquid water.
   e. does all of the above.

3. What accounts for the movement of water up xylem vessels in a plant?
   a. cohesion
   b. hydrogen bonding
   c. adhesion
   d. hydrophilic vessel walls
   e. all of the above

4. Climates tend to be moderate by large bodies of water because
   a. a large amount of solar heat is absorbed by the gradual rise in temperature of the water.
   b. the gradual cooling of the water releases heat to the environment.
   c. the high specific heat of water helps to regulate air temperatures.

Property	Explanation of Property	Example of Benefit to Life
a.	Hydrogen bonds hold molecules together and adhere them to hydrophilic surfaces.	b.
High specific heat	c.	Temperature changes in environment and organisms are moderated.
d.	Hydrogen bonds must be broken for water to evaporate.	e.
f.	Water molecules with high kinetic energy evaporate; remaining molecules are cooler.	g.
Ice floats	h.	i.
j.	k.	Most chemical reactions in life involve solutes dissolved in water.

    d. a great deal of heat is absorbed and released by the breaking and forming of hydrogen bonds.

    e. of all of the above.

5. Temperature is a measure of
    a. specific heat.
    b. average kinetic energy of molecules.
    c. total kinetic energy of molecules.
    d. Celsius degrees.
    e. joules.

6. Evaporative cooling is a result of
    a. a low heat of vaporization.
    b. a high heat of melting.
    c. a high specific heat.
    d. a reduction in the average kinetic energy of the liquid remaining after molecules enter the gaseous state.
    e. release of heat caused by the breaking of hydrogen bonds when water molecules escape.

7. Ice floats because
    a. air is trapped in the crystalline lattice.
    b. the formation of hydrogen bonds releases heat; warmer objects float.
    c. it has a smaller surface area than liquid water.
    d. it insulates bodies of water so they do not freeze from the bottom up.
    e. hydrogen bonding spaces the molecules farther apart, creating a less dense structure.

8. The molarity of a solution is equal to
    a. Avogadro's number of molecules in 1 liter of solvent.
    b. the number of moles of a solute in 1 liter of solution.
    c. the molecular weight of a solute in 1 liter of solution.
    d. the number of solute particles in 1 liter of solvent.
    e. 342 g if the solute is sucrose.

9. A solution with a pH of 2, compared to a solution with pH 4,
    a. is twice as acidic.
    b. is 100 times more acidic.
    c. is 1000 times more acidic.
    d. has two times more $[OH^-]$.
    e. is two orders of magnitude more basic.

10. A buffer
    a. changes pH by a magnitude of 10.
    b. absorbs excess OH^-.
    c. releases excess H^+.
    d. is often a weak acid–base pair.
    e. maintains pH at a value of 7.

11. Which of the following is least soluble in water?
    a. polar molecules
    b. nonpolar compounds
    c. ionic compounds
    d. hydrophilic molecules
    e. anions

12. Which would be the best method for reducing acid precipitation?
    a. Raise the height of smoke stacks so that exhaust enters the upper atmosphere.
    b. Add buffers and bases to bodies of water whose pH has dropped.
    c. Use coal burning generators rather than nuclear power to produce electricity.
    d. Increase emission control standards for factories and automobiles.
    e. Reduce the concentration of heavy metals in industrial exhaust.

13. What bonds must be broken for water to vaporize?
    a. polar covalent bonds
    b. nonpolar covalent bonds
    c. hydrogen bonds
    d. ionic bonds
    e. polar covalent and hydrogen bonds

14. How would you make a 0.1 *M* solution of glucose ($C_6H_{12}O_6$)? The mass numbers for these elements are C = 12, O = 16, H = 1.
    a. Mix 6 g C, 12 g H, and 6 g O in 1 liter of water.
    b. Mix 72 g C, 12 g H, and 96 g O in 1 liter of water.
    c. Mix 18 g of glucose with enough water to make 1 liter of solution.
    d. Mix 29 g of glucose with enough water to yield 1 liter of solution.
    e. Mix 180 g of glucose with enough water to yield 1 liter of solution.

15. How many molecules of glucose would be in the solution in question 14?
    a. 0.1   **c.** 60   **e.** 6×10^{22}
    b. 6   **d.** 6×10^{23}

FILL IN THE BLANKS: *Complete the following table on pH.*

$[H^+]$	$[OH^-]$	pH	Acidic, Basic, or Neutral?
	10^{-11}	3	acidic
10^{-8}		8	
10^{-12}			
	10^{-7}		
		1	

CHAPTER 4

CARBON AND THE MOLECULAR DIVERSITY OF LIFE

FRAMEWORK

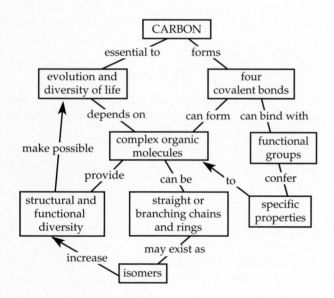

CHAPTER REVIEW

■ Organic chemistry is the study of carbon compounds

Organic chemistry is the study of carbon-containing molecules. Early organic chemists could not synthesize the complex molecules found in living organisms and therefore attributed the existence of life and the formation of these molecules to a life force independent of physical and chemical laws, a belief known as vitalism. **Mechanism,** the philosophy underlying modern organic chemistry, holds that physical and chemical laws and explanations are sufficient to account for all natural phenomena, even the evolution of life.

■ Carbon atoms are the most versatile building blocks of molecules

Carbon has six electrons. To complete its valence shell, carbon forms four covalent bonds with other atoms. This tetravalence is at the center of carbon's ability to form large and complex molecules with characteristic three-dimensional shapes and properties. When carbon forms four single covalent bonds, it is shaped like a tetrahedron. Carbon's covalent bonds can include single, double, or even triple bonds.

■ Variation in carbon skeletons contributes to the diversity of organic molecules

Carbon atoms readily bond with each other, producing chains or rings of carbon atoms. These molecular backbones can vary in length, branching, placement of double bonds, and location of atoms of other elements. The simplest organic molecules are **hydrocarbons,** consisting of only carbon and hydrogen. Fossil fuels are composed of hydrocarbons. The nonpolar C–H bonds in hydrocarbon chains account for their hydrophobic behavior.

■ **INTERACTIVE QUESTION   4.1**

From what you know about the valences of C, H, and O, sketch the structural formulas for the following molecules: C_3H_8O (propanol) and C_2H_4 (ethene).

Isomers   **Isomers** are compounds with the same molecular formula but different structural arrangements and, thus, different properties. **Structural isomers** differ in the arrangement of atoms and often in the location of double bonds. **Geometric isomers** have the same sequence of covalently bonded atoms but differ in spatial arrangement due to the inflexibility of double bonds. **Enantiomers** are molecules that are mirror images of each other. An asymmetric carbon is one that is covalently bound to four different kinds of atoms or groups of atoms. Due to the tetrahedral shape of the asymmetric carbon, the four groups can be attached in spatial arrangements that are not superimposable on each other. Enantiomers are left- and right-handed versions of each other and can differ greatly in their biological activity.

■ INTERACTIVE QUESTION   4.2

Identify the structural isomers, geometric isomers, and enantiomers from the following compounds.

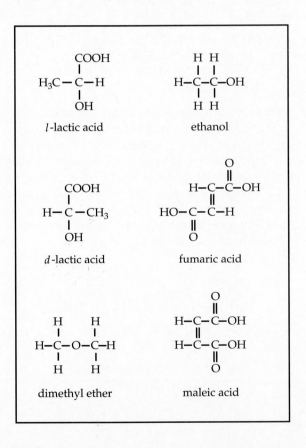

Functional groups also contribute to the molecular diversity of life

The properties of organic molecules are determined by groups of atoms, known as **functional groups**, which bond to the carbon skeleton and behave consistently from one carbon-based molecule to another. Because the functional groups considered here are hydrophilic, they increase the solubility of organic compounds in water.

The **hydroxyl group** consists of an oxygen and hydrogen (–OH) covalently bonded to the carbon skeleton. The polarity of the hydroxyl group, due to the electronegative oxygen, increases a compound's solubility in water. Organic molecules with hydroxyl groups are called **alcohols**, and their names often end in -*ol*.

Carbonyl groups consist of a carbon double-bonded to an oxygen (>CO). If the carbonyl group is at the end of the carbon skeleton, the compound is called an **aldehyde**. Otherwise, the compound is called a **ketone**.

A **carboxyl group** consists of a carbon double-bonded to an oxygen and also attached to a hydroxyl group (–COOH). Compounds with a carboxyl group are called **carboxylic acids** or organic acids because they tend to dissociate to release H^+.

An **amino group** consists of a nitrogen atom bonded to two hydrogens (–NH$_2$). Compounds with an amino group, called **amines**, can act as bases. The nitrogen, with its pair of unshared electrons, can attract a hydrogen ion, becoming –NH$_3^+$. Amino acids, the building blocks of proteins, contain both an amino and a carboxyl group.

The **sulfhydryl group** consists of a sulfur atom bonded to a hydrogen (–SH). Interactions of these groups help stabilize the structure of many proteins. **Thiols** are compounds containing sulfhydryl groups.

A **phosphate group** is bonded to the carbon skeleton by an oxygen attached to a phosphorus atom that is bonded to three other oxygen atoms (–OPO$_3^{-2}$). An organic compound with a phosphate ion can store energy and transfer it to another molecule.

The chemical elements of life: *a review*

Carbon, oxygen, hydrogen, nitrogen, and smaller quantities of sulfur and phosphorus, all capable of forming strong covalent bonds, are combined into the complex organic molecules of living matter. The versatility of carbon in forming four covalent bonds, linking readily with itself to produce chains and rings, and binding with other elements and functional groups makes possible the incredible diversity of organic molecules.

STRUCTURE YOUR KNOWLEDGE

1. Construct a concept map that illustrates your understanding of the characteristics and significance of the three types of isomers. *A suggested map is in the answer section. Comparing and discussing your map with that of a study partner would be most helpful.*

2. Fill in the following table on the functional groups.

Functional Group	Molecular Formula	Names and Characteristics of Organic Compounds Containing Functional Group
	–OH	
		Aldehyde or ketone; polar group
Carboxyl		
	–NH₂	
		Thiols; cross-links stabilize protein structure
Phosphate		

TEST YOUR KNOWLEDGE

MULTIPLE CHOICE: *Choose the one best answer.*

1. The tetravalence of carbon most directly results from
   a. its tetrahedral shape.
   b. its very slight electronegativity.
   c. its four electrons in the valence shell that can form four covalent bonds.
   d. its ability to form single, double, and triple bonds.
   e. its ability to form chains and rings of carbon atoms.

2. Hydrocarbons are not soluble in water because
   a. they are hydrophilic.
   b. the C–H bond is very nonpolar.
   c. they do not ionize.
   d. they store energy in the many C–H bonds along the carbon backbone.
   e. they are lighter than water.

3. Which of the following is *not* true of an asymmetric carbon atom?
   a. It is attached to four different atoms or groups.
   b. It results in right- and left-handed versions of a molecule.
   c. It can result in enantiomers.
   d. Its configuration is in the shape of a tetrahedron.
   e. It can result in geometric isomers.

4. A reductionist approach to considering the structure and function of organic molecules would be based on
   a. mechanism.
   b. holism.
   c. determinism.
   d. vitalism.
   e. emergent properties.

5. The functional group that can cause an organic molecule to act as a base is
   a. –COOH.   c. –SH.   e. –OPO₃⁻².
   b. –OH.   d. –NH₂.

6. The functional group that confers acidic properties on organic molecules is
   a. –COOH.   c. –SH.   e. –OPO₃⁻².
   b. –OH.   d. –NH₂.

7. Which is *not* true about structural isomers?
   a. They have different chemical properties.
   b. They have the same molecular formula.
   c. Their atoms and bonds are arranged in different sequences.
   d. They are a result of restricted movement around a carbon double bond.
   e. Their possible numbers increase as carbon skeletons increase in size.

8. How many asymmetric carbons are there in the sugar galactose?
   a. 1   c. 3   e. 5
   b. 2   d. 4

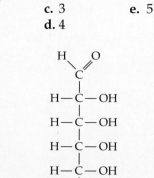

MATCHING: *Match the formulas below to the terms at the right. Choices may be used more than once; more than one right choice may be available.*

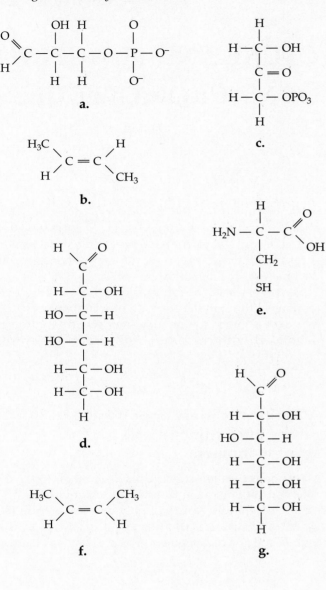

a.

b.

c.

d.

e.

f.

g.

_____ 1. structural isomers

_____ 2. geometric isomers

_____ 3. enantiomers

_____ 4. carboxylic acid

_____ 5. can make cross-link in protein

_____ 6. alcohol

_____ 7. aldehyde

_____ 8. amino acid

_____ 9. organic phosphate

_____ 10. hydrocarbon

_____ 11. amine

_____ 12. ketone

CHAPTER 5

THE STRUCTURE AND FUNCTION OF MACROMOLECULES

FRAMEWORK

The central ideas of this chapter are that molecular function relates to molecular structure and that the diversity of molecular structure is the basis for the diversity of life. Combining a small number of monomers or subunits into unique sequences and three-dimensional structures creates a huge variety of macromolecules. The table below briefly summarizes the major characteristics of the four classes of macromolecules.

CHAPTER REVIEW

Smaller organic molecules are joined together to form carbohydrates, lipids, proteins, and nucleic acids. These giant molecules, called **macromolecules**, represent another level in the hierarchy of biological organization, and their functions derive from their complex and unique architectures.

■ Most macromolecules are polymers

Polymers are large molecules formed from the linking together of many similar or identical small molecules, called **monomers**. Monomers are joined by **condensation reactions** (or dehydration reactions), in which one monomer provides a hydroxyl (-OH) and the other contributes a hydrogen (-H) to form a water molecule, and a covalent bond between the monomers is formed. Energy is required to join monomers, and the process is facilitated by enzymes.

Hydrolysis is the breaking of bonds between monomers through the addition of water molecules. A hydroxyl is joined to one monomer while a hydrogen is bonded with the other. Enzymes also control hydrolysis.

■ A limitless variety of polymers can be built from a small set of monomers

Macromolecules are constructed from about 40 to 50 common monomers and a few rarer molecules. The seemingly endless variety of polymers arises from the essentially infinite number of possibilities in the sequencing and arrangement of these basic subunits.

■ Organisms use carbohydrates for fuel and building material

Carbohydrates include sugars and their polymers.

Monosaccharides Monosaccharides have the general formula of $(CH_2O)_n$. The number of these units

Class	Monomers	Functions
Carbohydrates	Monosaccharides	Energy, raw materials, energy storage, structural compounds
Lipids	Glycerol, fatty acids, steroids	Energy storage, membranes, steroids, hormones
Proteins	Amino acids	Enzymes, transport, movement, receptors, defense, structure
Nucleic acids	Nucleotides	Heredity, code for amino acid sequence

forming a sugar varies from three to seven, with hexoses ($C_6H_{12}O_6$), trioses, and pentoses found most commonly. A sugar can be an aldose or a ketose, depending on the location of the carbonyl group. The other carbon atoms characteristically all bear hydroxyl groups. Additional diversity among sugar molecules is provided by the spatial arrangement of parts around asymmetric carbons. These small structural differences affect the recognition and interaction of molecules within cells. In aqueous solutions, most monosaccharides are in a ring structure, with carbon number 1 bonded to the oxygen attached to carbon number 5.

Glucose is broken down to yield energy in cellular respiration. Monosaccharides serve also as the raw materials for synthesis of other small organic molecules and as monomers that are synthesized into storage or structural molecules.

Disaccharides A **glycosidic linkage** is a covalent bond formed between two monosaccharides. Maltose, a **disaccharide** formed from two glucose molecules, has a 1–4 glycosidic linkage between the number one carbon of one glucose and the number four carbon of the other glucose. Sucrose, or table sugar, consists of a glucose and a fructose molecule joined by a 1–2 linkage.

Polysaccharides **Polysaccharides** are storage or structural macromolecules made from a few hundred to a few thousand monosaccharides. **Starch**, a storage molecule in plants, is a polymer made of glucose molecules joined by 1–4 linkages that give starch a helical shape. Most animals have enzymes to hydrolyze plant starch into glucose. Animals produce **glycogen**, a highly branched polymer of glucose, as their energy storage form.

Cellulose, the major component of plant cell walls, is the most abundant organic compound on Earth. It differs from starch by the configuration of the ring form of glucose and the resulting geometry of the glycosidic bonds. In a plant cell wall, hydrogen bonds between hydroxyl groups hold parallel cellulose molecules together to form strong microfibrils.

Enzymes that digest starch are unable to hydrolyze the β linkages of cellulose. Only a few organisms (some bacteria, microorganisms, and fungi) have enzymes that can digest cellulose.

Chitin is a structural polysaccharide formed from modified amino sugar monomers and found in the exoskeleton of arthropods and the cell walls of many fungi.

■ Lipids are mostly hydrophobic molecules with diverse functions

Fats, phospholipids, and steroids are a diverse assemblage of macromolecules that are classed together as **lipids** based on their hydrophobic behavior.

Fats **Fats** are composed of fatty acids attached to the three-carbon alcohol, glycerol. A **fatty acid** consists of a long hydrocarbon tail with a carboxyl group at the "head" end.

■ **INTERACTIVE QUESTION 5.1**

Fill in the following concept map that summarizes this section on carbohydrates.

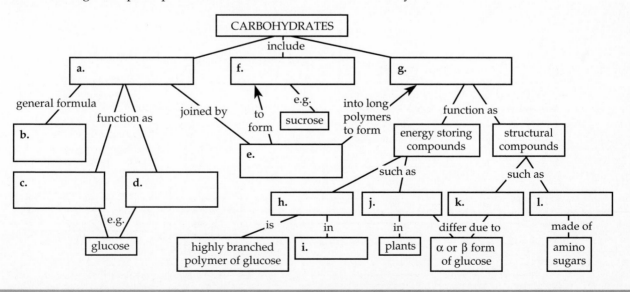

A **triacylglycerol**, or fat, consists of three fatty acids, each linked to glycerol by an ester linkage, a bond that forms between a hydroxyl and a carboxyl group. Triglyceride is another name for fats.

Fatty acids with double bonds in their carbon skeletons are called **unsaturated**. The double bonds create a kink in the shape of the molecule and prevent the fat molecules from packing closely together and becoming solidified at room temperature. **Saturated fatty acids** have no double bonds in their carbon skeletons. Most animal fats are saturated and solid at room temperature. The fats of plants and fishes are generally unsaturated and are called oils. Diets rich in saturated fats have been linked to cardiovascular disease.

Fats are excellent energy storage molecules, containing twice the energy reserves of carbohydrates such as starch. Adipose tissue, made of fat storage cells, also cushions organs and insulates the body.

Phospholipids   **Phospholipids** consist of a glycerol linked to two fatty acids and a negatively charged phosphate group, to which other small molecules may be attached. The phosphate head of this molecule is hydrophilic and water soluble, whereas the two fatty acid chains are hydrophobic. This unique structure of phospholipids makes them ideal constituents of cell membranes. Arranged in a bilayer, the hydrophilic heads face toward the aqueous solutions inside and outside the cell, and the hydrophobic tails mingle in the center of the membrane, forming a boundary between the cell and its external environment.

■ **INTERACTIVE QUESTION   5.2**

Phospholipids can self-assemble in an aqueous environment into a rounded droplet called a micelle. Sketch a micelle, and label the hydrophilic head and hydrophobic tail of one of the phospholipids.

Steroids   **Steroids** are a class of lipids distinguished by four connected carbon rings with various functional groups attached. **Cholesterol** is an important steroid that is a common component of animal cell membranes and a precursor for most other steroids, including many hormones.

■ **Proteins are the molecular tools for most cellular functions**

Proteins serve such diverse functions as structural support (fibers), storage of amino acids, transport

■ **INTERACTIVE QUESTION   5.3**

Fill in this concept map to help you organize your understanding of lipids.

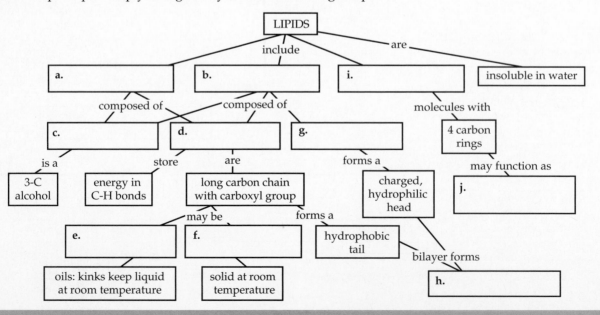

(hemoglobin), internal coordination (hormones), movement (actin and myosin in muscle), defense (antibodies), reception of chemical signals, and enzymatic control of chemical reactions. Proteins consist of one or more polypeptide chains folded into a unique three-dimensional shape or **conformation.**

A polypeptide is a polymer of amino acids connected in a specific sequence

Most **amino acids** are composed of an asymmetric carbon bonded to a hydrogen, a carboxyl group, an amino group, and a variable side chain called the R group. The R group confers the unique physical and chemical properties of each amino acid. Side chains may be either nonpolar and hydrophobic, or polar or charged and thus hydrophilic. Acidic side chains contain carboxyl groups and are usually negatively charged; basic side chains contain an amino group and are usually positively charged.

A **peptide bond** links the amino group of one amino acid with the carboxyl group of another, yielding a polymer of amino acids called a **polypeptide.** The polypeptide chain has a free carboxyl group at one end and a free amino group at the other. Polypeptides vary in length from a few to a thousand or more amino acids.

A protein's function depends on its specific conformation

Proteins have unique three-dimensional shapes created by the twisting or folding of one or more polypeptide chains. The unique conformation of a protein, which results from its sequence of amino acids, enables it to recognize and bind to other molecules.

Four Levels of Protein Structure There are three structural levels in the conformation of a protein. A fourth level may be present when a protein consists of more than one polypeptide chain.

Primary structure is the unique, genetically coded sequence of amino acids within a protein. Even a slight deviation from the sequence of amino acids can severely affect a protein's function by altering the protein's conformation. In the early 1950s, Sanger determined the primary structure of insulin through the laborious process of hydrolyzing the protein into small peptide chains, using chromatography to separate the small pieces, determining their amino acid sequences, and then overlapping the sequences of small fragments created with different agents to reconstruct the whole polypeptide. Most of these steps are now automated.

Secondary structure involves the coiling or folding of the polypeptide backbone, stabilized by hydrogen bonds between the electronegative oxygen of one peptide bond and the weakly positive hydrogen attached to a nitrogen of another bond. An **alpha (α) helix** is a delicate coil produced by hydrogen bonding between every fourth peptide bond. A **pleated sheet** is also held by repeated hydrogen bonds along the protein's backbone. This secondary structure forms when the polypeptide chain folds back and forth or when regions of the chain lie parallel to each other.

Interactions between the various side chains of the constituent amino acids produce a protein's **tertiary structure. Hydrophobic interactions** between nonpolar side groups in the center of the molecule, hydrogen bonds, and ionic bonds between negatively and positively charged side chains produce the stable and unique shape of the protein. Strong covalent bonds, called **disulfide bridges**, may occur between the sulfhydryl side groups of cysteine monomers that have been brought close together by the folding of the polypeptide.

Quaternary structure occurs in proteins that are composed of more than one polypeptide chain. The individual polypeptide subunits are held together in a precise structural arrangement.

■ **INTERACTIVE QUESTION 5.4**

In the following diagram of a portion of a polypeptide, label the types of interactions that are shown. What level of structure are these interactions producing?

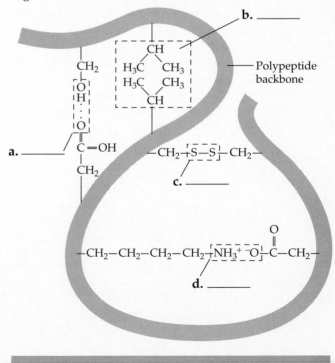

What Determines Protein Conformation? Protein conformation is dependent upon the interactions among the amino acids making up the polypeptide chain and usually arises spontaneously as soon as the protein is synthesized in the cell. These interactions can be disrupted by changes in pH, salt concentration, temperature, or other aspects of the environment, and the protein may **denature**, losing its native conformation and thus its function.

■ **INTERACTIVE QUESTION 5.5**

a. Why would a change in pH cause a protein to denature?

b. Why would transfer to a nonpolar organic solvent (such as ether) cause denaturation?

c. A denatured protein may re-form to its functional shape when returned to its normal environment. What does that indicate about a protein's conformation?

The Protein-Folding Problem The amino acid sequences of hundreds of proteins have been determined. Using the technique of X-ray crystallography, coupled with computer modeling and graphics, biochemists have established the three-dimensional shape of many of these molecules. Researchers have developed methods for following a protein through its intermediate states on the way to its final form and have discovered **chaperone proteins** that assist the folding of other proteins. As protein folding becomes better understood, scientists are learning to devise amino acid sequences for proteins that will be able to perform predetermined functions.

■ **INTERACTIVE QUESTION 5.6**

Now that you have gained more experience with concept maps, create your own map to help you organize the key concepts you have learned about proteins. Try to include the concepts of structure and function and look for cross-links on your map. One version of a protein concept map is included in the answer section, but remember that the real value is in the thinking process you must go through to create your own map.

■ **Nucleic acids store and transmit hereditary information**

Genes are the units of inheritance that determine the primary structure of proteins. **Nucleic acids** are macromolecules that carry and translate this code. **DNA, deoxyribonucleic acid**, is the genetic material that is inherited from one generation to the next and is reproduced in each cell of an organism. **RNA, ribonucleic acid**, reads the instructions coded in DNA and directs the synthesis of proteins, the ultimate enactors of the genetic program. In a eukaryotic cell, DNA resides in the nucleus and messenger RNA carries the instructions for protein synthesis to ribosomes located in the cytoplasm.

■ **INTERACTIVE QUESTION 5.7**

Show the flow of genetic information in a cell.

_____ → _____ → _____

■ **A DNA strand is a polymer with an information-rich sequence of nucleotides**

Nucleic acids are polymers of **nucleotides**, monomers that consist of a pentose (five-carbon sugar) covalently bonded to a phosphate group and a nitrogenous base. In DNA, the pentose is deoxyribose. There are two families of nitrogenous bases: **Pyrimidines**, including cytosine (C) and thymine (T), are characterized by six-membered rings of carbon and nitrogen atoms. **Purines**, adenine (A) and guanine (G), add a five-membered ring to the pyrimidine ring.

Nucleotides are linked together into a DNA polymer by phosphodiester linkages, which join the phosphate of one nucleotide with the sugar of the next. The nitrogenous bases extend from this repeating sugar–phosphate backbone. The unique sequence of bases in a gene codes for the specific amino acid sequence of a protein.

■ **Inheritance is based on precise replication of DNA**

DNA molecules consist of two chains of nucleotides spiraling around an imaginary axis in a **double helix**. In 1953, Watson and Crick first proposed this double helix arrangement, which consists of two sugar–phosphate backbones on the outside of the helix with their nitrogenous bases pairing and hydrogen-bonding together in the inside. Adenine pairs with only thymine; guanine always pairs with cytosine. Thus, the sequences of nitrogenous bases on the two strands of DNA are complementary. Because of this specific base-pairing property, DNA can replicate itself and precisely copy the genes of inheritance.

■ INTERACTIVE QUESTION 5.8

Take the time to create a concept map that summarizes what you have just reviewed about nucleic acids. Compare your map with that of a study partner or explain it to a friend. One version of a map on nucleic acids is included in the answer section. Refer to Figures 5.27 and 5.28 in your textbook to help you visualize nucleotides, nucleotide polymers, and the double helix of DNA.

■ ## We can use DNA and proteins as tape measures of evolution

Genes form the hereditary link between generations. Closely related members of the same species share many common DNA sequences and proteins. More closely related species have a larger proportion of their DNA and proteins in common. This "molecular genealogy" provides evidence of evolutionary relationships.

STRUCTURE YOUR KNOWLEDGE

1. Describe the four structural levels in the conformation of a protein.

2. Identify the type of monomer or group shown by these formulae. Then match the chemical formulae with their description. Answers may be used more than once.

    _____ 1. molecules that would combine to form a fat

    _____ 2. molecule that would be attached to other monomers by a peptide bond

    _____ 3. molecules or groups that would combine to form a DNA nucleotide

    _____ 4. molecules that are carbohydrates

    _____ 5. molecule that is a purine

    _____ 6. monomer of a protein

    _____ 7. groups that would be joined by phosphodiester bonds

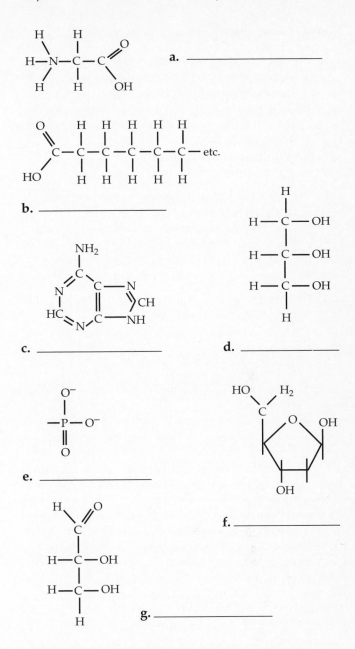

a. _____

b. _____

c. _____   d. _____

e. _____

f. _____

g. _____

TEST YOUR KNOWLEDGE

MATCHING: *Match the molecule with its class of macromolecules.*

    _____ 1. glycogen   A. carbohydrate

    _____ 2. cholesterol   B. lipid

    _____ 3. RNA   C. protein

    _____ 4. collagen   D. nucleic acid

    _____ 5. hemoglobin

    _____ 6. a gene

    _____ 7. triacylglycerol

    _____ 8. enzyme

    _____ 9. cellulose

    _____ 10. chitin

MULTIPLE CHOICE: *Choose the one best answer.*

1. Polymerization is a process that
   a. creates bonds between amino acids in the formation of a peptide chain.
   b. involves the removal of a water molecule.
   c. links the phosphate of one nucleotide with the sugar of the next.
   d. requires a condensation reaction.
   e. involves all of the above.

2. Which of the following is *not* true of pentoses?
   a. They are found in nucleic acids.
   b. They can occur in a ring structure.
   c. They have the formula $C_5H_{12}O_5$.
   d. They have hydroxyl and carbonyl groups.
   e. They may be an aldose or a ketose.

3. Disaccharides can differ from each other in all of the following ways *except*
   a. in the number of their monosaccharides.
   b. as enantiomers.
   c. in the monomers involved.
   d. in the location of their glycosidic linkage.
   e. in their structural formulae.

4. Which of the following is *not* true of cellulose?
   a. It is the most abundant organic compound on Earth.
   b. It differs from starch because of the configuration of glucose and the geometry of the glycosidic linkage.
   c. It may be hydrogen-bonded to neighboring cellulose molecules to form microfibrils.
   d. Few organisms have enzymes that hydrolyze its glycosidic linkages.
   e. Its monomers are amino sugars.

5. Plants store most of their energy as
   a. glucose.
   b. glycogen.
   c. starch.
   d. sucrose.
   e. cellulose.

6. What happens when a protein denatures?
   a. It loses its primary structure.
   b. It loses its secondary and tertiary structure.
   c. It becomes irreversibly insoluble and precipitates.
   d. It hydrolyzes into component amino acids.
   e. Its hydrogen bonds, ionic bonds, hydrophobic interactions, disulfide bridges, and peptide bonds are disrupted.

7. The alpha helix of proteins is
   a. part of the tertiary structure and is stabilized by disulfide bridges.
   b. a double helix.
   c. stabilized by hydrogen bonds and commonly found in fibrous proteins.
   d. found in some regions of globular proteins and stabilized by hydrophobic interactions.
   e. a complementary sequence to messenger RNA.

8. A fatty acid that has the formula $C_{16}H_{32}O_2$ is
   a. saturated.
   b. unsaturated.
   c. branched.
   d. hydrophilic.
   e. part of a steroid molecule.

9. Three molecules of the fatty acid in question 8 are joined to a molecule of glycerol ($C_3H_8O_3$). The resulting molecule has the formula
   a. $C_{48}H_{96}O_6$.
   b. $C_{48}H_{98}O_9$.
   c. $C_{51}H_{102}O_8$.
   d. $C_{51}H_{98}O_6$.
   e. $C_{51}H_{104}O_9$.

10. Pleated sheets are characterized by
    a. disulfide bridges between cysteine amino acids.
    b. back-and-forth folds of the polypeptide chain held together by hydrophobic interactions.
    c. folds stabilized by hydrogen bonds between segments of polypeptide chains.
    d. membrane sheets composed of phospholipids.
    e. hydrogen bonds between adjacent cellulose molecules.

FILL IN THE BLANKS

1. The man who determined the amino acid sequence of insulin was _____ .

2. Cytosine always pairs with _____ .

3. Adenine and guanine are _____ .

4. A pentose joined to a nitrogenous base and a phosphate group is called a _____ .

5. The conformation of a protein is determined by its _____ .

6. Proteins with more than one peptide chain have _____ structure.

7. The energy storage molecule of animals is _____ .

8. Membranes are composed of a bilayer of _____ .

9. The linkages between amino acids in a protein are called _____ .

10. A technique used to determine the three-dimensional structure of macromolecules is _____ .

CHAPTER 6

AN INTRODUCTION TO METABOLISM

FRAMEWORK

This chapter considers metabolism, the totality of the chemical reactions that take place in living organisms. Two key topics are emphasized: the energy transformations that underlie all chemical reactions and the role of enzymes in the "cold chemistry" of the cell. Enzymes are biological catalysts that lower the activation energy of a reaction and thus greatly speed up metabolic processes. An enzyme is a three-dimensional protein molecule with an active site specific for its substrate. Intricate control and feedback mechanisms produce the metabolic integration necessary for life.

The reactions in a cell either release or consume energy. The following illustration summarizes some of the components of the energy changes within a cell.

CHAPTER REVIEW

■ **The chemistry of life is organized into metabolic pathways**

Metabolism includes the thousands of precisely coordinated, complex, efficient, and integrated chemical reactions in an organism. These reactions are ordered into metabolic pathways—sequenced and branching routes controlled by enzymes. Through these pathways the cell transforms and creates the organic molecules that provide the energy and material for life.

Catabolic pathways release the energy stored in complex molecules through the breaking down of these molecules into simpler compounds. **Anabolic**

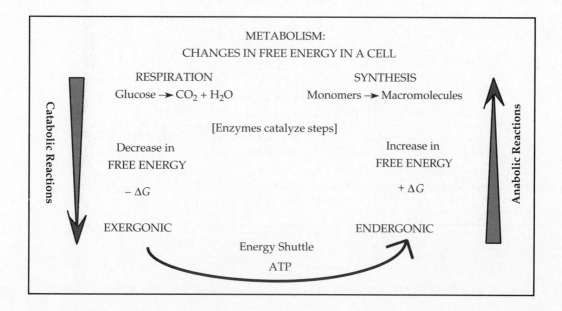

METABOLISM:
CHANGES IN FREE ENERGY IN A CELL

RESPIRATION
Glucose → CO_2 + H_2O

SYNTHESIS
Monomers → Macromolecules

[Enzymes catalyze steps]

Catabolic Reactions

Decrease in
FREE ENERGY

$-\Delta G$

EXERGONIC

Increase in
FREE ENERGY

$+\Delta G$

ENDERGONIC

Anabolic Reactions

Energy Shuttle
ATP

pathways require energy to combine simpler molecules into more complicated ones. This energy is often supplied through energy coupling, the linking of catabolic and anabolic pathways in a cell. Energy is involved in all metabolic processes. Thus, the study of energy transformations, called **bioenergetics**, is central to the understanding of metabolism.

Organisms transform energy

Energy has been defined as the capacity to do work, to move matter against an opposing force, to rearrange matter. **Kinetic energy** is the energy of motion, of matter that is moving. This matter does its work by transferring its motion to other matter. **Potential energy** is the capacity of matter to do work as a consequence of its location or arrangement. Chemical energy is a form of potential energy stored in the arrangement of atoms in molecules.

Energy can be converted from one form to another. Plants convert light energy (a form of kinetic energy) to the chemical energy in sugar, and cells release this potential energy to drive cellular processes.

The energy transformations of life are subject to two laws of thermodynamics

Thermodynamics is the study of energy transformations. The **first law of thermodynamics** states that energy can be neither created nor destroyed. Energy can be transferred between matter and transformed from one kind to another, but the total energy of the universe is constant. According to this principle of the *conservation of energy,* chemical reactions that either require or produce energy are merely transforming a set amount of energy into a different form.

The **second law of thermodynamics** states that every energy transformation or transfer results in an increasing disorder within the universe. **Entropy** is used as a measure of disorder or randomness.

A system, such as a cell, may become more ordered, but it does so with an attendant increase in the entropy of its surroundings. A cell can use highly ordered organic molecules as a source of the energy needed to create and maintain its own organized structure, but it returns heat and the simple molecules of carbon dioxide and water to the environment.

In any energy transformation or transfer, some of the energy is converted to heat, a less-ordered kinetic energy. The quantity of energy in the universe may be constant, but its quality is not.

Organisms live at the expense of free energy

A spontaneous process is a change that occurs without the input of external energy and that results in greater stability, reflected as an increase in entropy.

Free Energy: A Criterion for Spontaneous Change
Free energy, a measure of the instability of a system, can be defined as the portion of a system's energy available to perform work when the system's temperature is uniform. Free energy (G) depends on the total energy of a system (H) and its entropy (S), such that G

■ INTERACTIVE QUESTION   6.1

Fill in this concept map that summarizes the key concepts of energy.

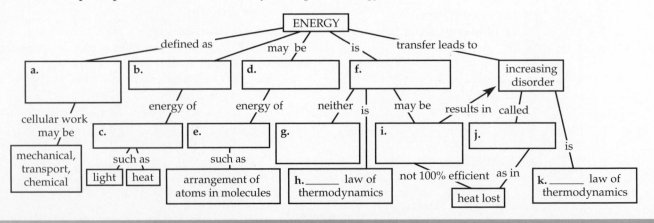

$= H - TS$. Absolute temperature (°Kelvin, or 0°C + 273) is a factor because higher temperature increases random molecular motion, disrupting order and increasing entropy. The change in free energy during a reaction is represented by $\Delta G = \Delta H - T\Delta S$. For a reaction to be spontaneous, the free energy of the system must decrease ($-\Delta G$): the system must lose energy (decrease *H*), become more disordered (increase *S*), or both. An unstable system is rich in free energy and has a tendency to change spontaneously to a more stable stage, potentially performing work in the process.

Free Energy and Equilibrium At equilibrium in a chemical reaction, the forward and backward reactions are proceeding at the same rate, and $\Delta G = 0$ because there is no net free energy change. Moving toward equilibrium is spontaneous; the ΔG of the reaction is negative. To move away from equilibrium is nonspontaneous; energy must be added to the system.

■ INTERACTIVE QUESTION 6.2

Complete the following sentences by relating free energy to stability, spontaneous change, equilibrium, and capacity to do work.

a. When a system is high in free energy,

b. When a system has low free energy,

Free Energy and Metabolism An **exergonic reaction** ($-\Delta G$) proceeds with a net release of free energy and is spontaneous. The magnitude of ΔG indicates the maximum amount of work the reaction can do. **Endergonic reactions** ($+\Delta G$) are nonspontaneous; they must absorb free energy from the surroundings. A total of 686 kcal (2870 kJ) of energy is released from the exergonic reaction that breaks a mole of glucose down to carbon dioxide and water. The reverse reaction in photosynthesis has a ΔG of +686 kcal/mol.

Metabolic disequilibrium is essential to life. Respiration and other cellular chemical reactions are reversible and could reach equilibrium if the cell did not maintain a steady supply of reactants and siphon off the products (as reactants for new processes or as waste products to be expelled).

An ecosystem's free energy comes from sunlight to photosynthetic organisms and is transferred via organic molecules to consumers.

Central to a cell's bioenergetics is **energy coupling**, using exergonic processes to power endergonic ones.

■ INTERACTIVE QUESTION 6.3

Develop a concept map on free energy and ΔG. The value in this exercise is for you to wrestle with and organize these concepts for yourself. Do not turn to the suggested concept map until you have worked on your own understanding. Remember that the concept map in the answer section is only one way of structuring these ideas—have confidence in your own organization.

■ ATP powers cellular work by coupling exergonic to endergonic reactions

A cell performs mechanical work in movement of the cell or parts of the cell, transport work in pumping molecules across membranes, and chemical work in driving endergonic reactions to synthesize cellular molecules. The immediate source of the energy to perform this work most often comes from ATP.

Structure and Hydrolysis of ATP **ATP (adenosine triphosphate)** consists of the nitrogenous base adenine bonded to the sugar ribose, which is connected to a chain of three phosphate groups. This triphosphate tail is unstable; ATP can be hydrolyzed to ADP (adenosine diphosphate) and an inorganic phosphate molecule ($\textcircled{P}_i$), releasing 7.3 kcal (31 kJ) of energy per mole of ATP when measured under standard conditions. The ΔG of the reaction in the cell is estimated to be closer to -13 kcal/mol.

How ATP Performs Work In a cell, the free energy released from the hydrolysis of ATP is used to transfer the phosphate group to another molecule, producing a **phosphorylated intermediate** that is more reactive. The phosphorylation of other molecules by ATP forms the basis for almost all cellular work.

Regeneration of ATP A cell regenerates ATP at a phenomenal rate. The formation of ATP from ADP and $\textcircled{P}_i$ is endergonic, with a ΔG of +7.3 kcal/mol (standard conditions). Cellular respiration (the catabolic processing of glucose and other organic molecules) provides the energy for the regeneration of ATP. Plants can also produce ATP using light energy.

■ INTERACTIVE QUESTION   6.4

Label the three components (**a** through **c**) of the ATP molecule shown below.

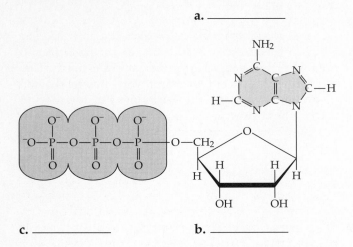

a. _____

c. _____   b. _____

d. Circle the region of instability in this molecule, and show which bonds are likely to break.

e. Explain why that part of the molecule is unstable and by what chemical mechanism the bond is broken.

f. Why is energy released?

The E_A barrier is essential to life because it prevents the energy-rich macromolecules of the cell from decomposing spontaneously. For metabolism to proceed in a cell, however, E_A must be reached. Heat, a possible source of activation energy in reactions, would be harmful to the cell and would also speed metabolic reactions indiscriminately. Enzymes are able to lower E_A for specific reactions so that metabolism can proceed at cellular temperatures. Enzymes do not change ΔG for a reaction.

■ INTERACTIVE QUESTION   6.5

In this graph of a reaction with and without an enzyme catalyst, label the parts **a** through **e**.

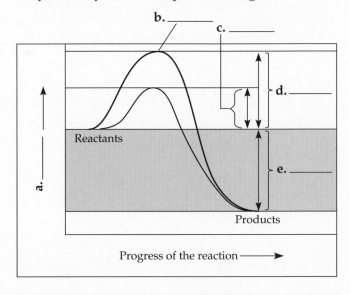

■ Enzymes speed up metabolic reactions by lowering energy barriers: *an overview*

Enzymes are biological **catalysts**—agents that change the rate of a reaction but are unchanged by the reaction.

Chemical reactions rearrange atoms by breaking and forming chemical bonds. Energy must be absorbed to break bonds and is released when bonds form. **Activation energy**, or the **free energy of activation** (E_A), is the energy that must be absorbed by reactants to reach the unstable *transition state*, in which bonds are more fragile and likely to break, and from which the reaction can proceed.

■ Enzymes are substrate-specific: *a closer look*

Most enzymes are proteins, macromolecules with characteristic three-dimensional shapes. The specificity of an enzyme for a particular **substrate** is determined by its shape. The substrate is temporarily bound to its enzyme at the **active site**, a pocket or groove found on the surface of the enzyme molecule that has a shape and charge arrangement complementary to the substrate molecule. When a substrate molecule enters the active site, the enzyme changes shape slightly, creating what is called an **induced fit** between substrate and active site and enhancing the ability of the enzyme to catalyze the chemical reaction.

The active site is an enzyme's catalytic center: *a closer look*

The substrate is held in the active site by hydrogen or ionic bonds, creating an enzyme–substrate complex. The side chains (R groups) of some of the surrounding amino acids in the active site facilitate the conversion of substrate to product. The product then leaves the active site, and the enzyme can bind with another substrate molecule.

Enzymes can catalyze reactions involving the joining of two reactants by providing active sites in which the substrates are bound closely together and properly oriented. An induced fit can stretch or bend critical bonds in the substrate molecule and make them easier to break. An active site may provide a microenvironment, such as a lower pH, that is necessary for a particular reaction. Enzymes may also actually participate in a reaction by forming brief covalent bonds with the substrate.

The rate at which an enzyme molecule works partly depends on the concentration of its substrate. The speed of a reaction will increase with increasing substrate concentration up to the point at which all enzyme molecules are saturated with substrate molecules and working at full speed.

■ INTERACTIVE QUESTION 6.6

Outline a catalytic cycle using the diagrammatic enzyme below. Sketch the appropriate substrate and products, and identify the key steps of the cycle.

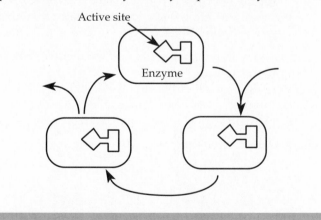

Active site

Enzyme

A cell's chemical and physical environment affects enzyme activity: *a closer look*

Effects of Temperature and pH The velocity of an enzyme-catalyzed reaction may increase with rising temperature up to the point at which increased thermal agitation begins to disrupt the hydrogen and ionic bonds and other interactions that stabilize protein conformation. A change in pH may denature an enzyme by disrupting the hydrogen bonding of the molecule. Each enzyme has a temperature and pH optimum at which it is most active.

Cofactors **Cofactors** are small molecules that bind with enzymes and are necessary for enzyme catalytic function. They may be inorganic, such as various metal atoms, or organic molecules called **coenzymes**. Most vitamins are coenzymes or precursors of coenzymes.

Enzyme Inhibitors Enzyme inhibitors selectively disrupt the action of enzymes, either reversibly by binding with the enzyme with weak bonds or irreversibly by attaching with covalent bonds. **Competitive inhibitors** compete with the substrate for the active site of the enzyme. Increasing the concentration of substrate molecules may overcome this type of inhibition. **Noncompetitive inhibitors** bind to a part of the enzyme separate from the active site and change the conformation of the enzyme, thus impeding enzyme action. Many pesticides and antibiotics are inhibitors of key enzymes and act as metabolic poisons.

Allosteric Regulation Molecules that inhibit or activate enzyme activity may bind to an **allosteric site**, a receptor site separate from the active site. Complex enzymes made of two or more polypeptide chains, each with its own active site, may have allosteric sites located where subunits join. The entire unit may oscillate between two conformational states, and the binding of an activator (or inhibitor) stabilizes the catalytically active (or inactive) conformation. Allosteric enzymes may be critical regulators of metabolic pathways.

Cooperativity Through a phenomenon called **cooperativity**, the induced-fit binding of a substrate molecule to one polypeptide subunit can change the conformation such that the active sites of all subunits are more active.

■ INTERACTIVE QUESTION 6.7

Return to your diagram in 6.6. Draw a competitive and a noncompetitive inhibitor, and indicate where they would bind to the enzyme molecule.

■ Metabolic order emerges from the cell's regulatory systems and structural organization

Feedback Inhibition   Metabolic pathways are commonly regulated by **feedback inhibition**, in which the product of a pathway acts as an inhibitor of an enzyme early in the pathway.

Structural Order and Metabolism   Specialized cellular compartments may contain high concentrations of the enzymes and substrates needed for a particular pathway. Enzymes also may be incorporated into the membranes of cellular compartments. The complex internal structure of the cell serves to order metabolic pathways in space and time.

STRUCTURE YOUR KNOWLEDGE

This long and complex chapter introduced many new concepts. See if you can step back from the details and answer the following general questions.

1. Relate the concept of free energy to metabolism.
2. What role do enzymes play in metabolism?

TEST YOUR KNOWLEDGE

MULTIPLE CHOICE: *Choose the one best answer.*

1. Catabolic and anabolic pathways are often coupled in a cell because
   a. the intermediates of a catabolic pathway are used in the anabolic pathway.
   b. both pathways use the same enzymes.
   c. the free energy released from one pathway is used to drive the other.
   d. the activation energy of the catabolic pathway can be used in the anabolic pathway.
   e. their enzymes are controlled by the same activators and inhibitors.

2. When glucose is converted to CO_2 and H_2O, changes in total energy, entropy, and free energy are as follows:
   a. $-\Delta H, -\Delta S, -\Delta G$
   b. $-\Delta H, +\Delta S, -\Delta G$
   c. $-\Delta H, +\Delta S, +\Delta G$
   d. $+\Delta H, +\Delta S, +\Delta G$
   e. $+\Delta H, -\Delta S, +\Delta G$

3. When a protein forms from amino acids, the following changes apply:
   a. $+\Delta H, -\Delta S, +\Delta G$
   b. $+\Delta H, +\Delta S, -\Delta G$
   c. $+\Delta H, +\Delta S, +\Delta G$
   d. $-\Delta H, +\Delta S, +\Delta G$
   e. $-\Delta H, -\Delta S, +\Delta G$

4. A negative ΔG means that
   a. the quantity G of energy is available to do work.
   b. the reaction is spontaneous.
   c. the reactants have more free energy than the products.
   d. the reaction is exergonic.
   e. all of the above are true.

5. According to the first law of thermodynamics,
   a. for every action there is an equal and opposite reaction.
   b. every energy transfer results in an increase in disorder or entropy.
   c. the total amount of energy in the universe is conserved or constant.
   d. energy can be transferred or transformed, but disorder always increases.
   e. potential energy is converted to kinetic energy, and kinetic energy is converted to heat.

6. According to the induced-fit hypothesis,
   a. the binding of the substrate is an energy-requiring process.
   b. a competitive inhibitor can outcompete the substrate for the active site.
   c. the binding of the substrate changes the shape of the enzyme slightly and can stress or bend substrate bonds.
   d. the active site creates a microenvironment ideal for the reaction.
   e. substrates are held in the active site by hydrogen and ionic bonds.

7. One way in which a cell maintains metabolic disequilibrium is to
   a. siphon products of a reaction off to the next step in a metabolic pathway.
   b. provide a constant supply of enzymes for critical reactions.
   c. use feedback inhibition to turn off pathways.
   d. use allosteric enzymes that can bind to activators or inhibitors.
   e. couple anabolic and catabolic pathways.

8. In an experiment, changing the pH from 7 to 6 resulted in an increase in product formation. From this we could conclude that
   a. the enzyme became saturated at pH 6.
   b. the enzyme's optimal pH is 6.
   c. this enzyme works best in a neutral pH.
   d. the temperature must have increased when the pH was changed to 6.
   e. the enzyme was in a more active conformation at pH 6.

9. When substance A was added to an enzyme reaction, product formation decreased. The addition of more substrate did not increase product formation. From this we conclude that substance A could be
   a. product molecules.
   b. a cofactor.
   c. an allosteric enzyme.
   d. a competitive inhibitor.
   e. a noncompetitive inhibitor.

10. The formation of ATP from ADP and inorganic phosphate
    a. is an exergonic process.
    b. transfers the phosphate to another intermediate that becomes more reactive.
    c. produces an unstable energy compound that can drive cellular work.
    d. has a ΔG of –7.3 kcal/mol.
    e. involves the hydrolysis of a phosphate bond.

11. At equilibrium,
    a. no enzymes are functioning.
    b. $\Delta G = 0$.
    c. the forward and backward reactions have stopped.
    d. the products and reactants have equal values of H.
    e. a reaction is no longer spontaneous.

12. In cooperativity,
    a. a cellular organelle contains all the enzymes needed for a metabolic pathway.
    b. a product of a pathway serves as a competitive inhibitor of an early enzyme in the pathway.
    c. a molecule bound to the active site of one subunit of an enzyme affects the active site of other subunits.
    d. the allosteric site is filled with an activator molecule.
    e. the product of one reaction serves as the substrate for the next in intricately ordered metabolic pathways.

FILL IN THE BLANKS

_____ 1. is the totality of an organism's chemical processes.

_____ 2. pathways require energy to combine molecules together.

_____ 3. energy is the energy of motion.

_____ 4. enzymes change between two conformations, depending on whether an activator or inhibitor is bound to them.

_____ 5. is the term for the measure of disorder or randomness.

_____ 6. is the energy that must be absorbed by molecules to reach the transition state.

_____ 7. inhibitors decrease an enzyme's activity by binding to the active site.

_____ 8. are organic molecules that bind to enzymes and are necessary for their functioning.

_____ 9. is a regulatory device in which the product of a pathway binds to an enzyme early in the pathway.

_____ 10. are more reactive molecules created by the transfer of a phosphate group from ATP.

CHAPTER 7

A TOUR OF THE CELL

FRAMEWORK

This chapter deals with the fundamental unit of life—the cell. The complexities in the processes of life are reflected in the complexities of the structure of the cell. It is easy to become overwhelmed by the number of new vocabulary terms for this array of cell organelles and membranes. The following concept map provides an organizational framework for the wealth of detail found in "a tour of the cell."

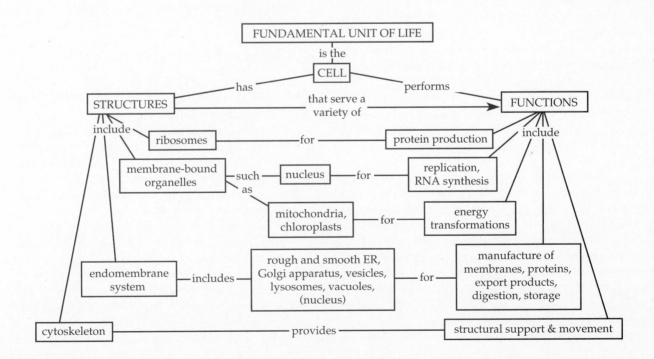

CHAPTER REVIEW

The cell is the basic structural and functional unit of all living organisms. In the hierarchy of biological organization, the capacity for life emerges from the structural order of the cell. The cell senses and responds to its environment and exchanges materials and energy with its surroundings. All cells are related through common descent, but evolution has shaped diverse adaptations.

■ Microscopes provide windows to the world of the cell

The growth of scientific knowledge and the development of new instruments and methods usually go hand in hand. The invention of the microscope in the seventeenth century led to the initial discovery and study of cells.

The glass lenses of **light microscopes (LMs)** refract (bend) the visible light passing through a specimen such that the projected image is magnified. **Resolving power** is a measure of the clarity of an image—determined by the minimum distance two points must be separated to be distinguished. The resolving power of the light microscope is limited by the wavelength of visible light, so that details finer than 0.2 μm (micrometers) cannot be resolved. Staining of specimens and using techniques such as darkfield and phase-contrast microscopy improve visibility by increasing contrast between structures that are large enough to be resolved.

Most subcellular structures, or **organelles**, are too small to be resolved by the light microscope. Cells were discovered by Robert Hooke in 1665, but their ultrastructure was largely unknown until the development of the **electron microscope (EM)** in the 1950s. The electron microscope focuses a beam of electrons through the specimen. The short wavelength of electrons allows a resolution of about 0.2 nm (nanometer), a thousand times greater than that of the light microscope.

In a **transmission electron microscope (TEM)** a beam of electrons is passed through a thin section of a specimen, and electromagnets, acting as lenses, focus and magnify the image. Contrast is increased by staining preserved cells with atoms of heavy metals.

In a **scanning electron microscope (SEM)** the electron beam scans the surface of a specimen coated with a thin gold film, exciting electrons from the specimen and collecting and focusing them onto a screen. The resulting image appears three-dimensional.

Modern cell biology integrates cytology with biochemistry to understand relationships between cellular structure and function.

■ INTERACTIVE QUESTION   7.1

a. Define cytology.

b. What do cell biologists use TEM to study?

c. What does SEM show best?

d. What advantages does LM have over TEM and SEM?

■ Cell biologists can isolate organelles to study their functions

Cell fractionation is a technique that separates major organelles of a cell so that their functions can be studied. Cells are homogenized by ultrasound or grinding, and the resulting cellular soup is separated into component fractions by differential centrifugation. The homogenate is first spun slowly, and nuclei and large particles settle to form a pellet. The remaining supernatant is centrifuged at increasing speeds, each time isolating smaller and smaller cellular components in the pellet. **Ultracentrifuges** can spin at speeds up to 80,000 rpm. Each cellular fraction contains a large quantity of the same cellular components, thus permitting the isolated study of their metabolic functions.

■ A panoramic view of the cell

Prokaryotic and Eukaryotic Cells   The kingdom Monera, the bacteria, is characterized by **prokaryotic cells**, which are cells with no nucleus. The DNA of prokaryotic cells is concentrated in a region called the **nucleoid**. **Eukaryotic cells**, which are much more structurally complex, are found in the other four kingdoms of life: protists, plants, fungi, and animals. They have a true nucleus enclosed in a nuclear membrane and numerous organelles suspended in a semifluid medium called **cytosol**. The **cytoplasm** is the entire region between the nucleus and the membrane enclosing the cell.

Cell Size   Most bacterial cells range from 1 to 10 μm in diameter, whereas eukaryotic cells are ten times larger, ranging from 10 to 100 μm.

The small size of cells is dictated by geometry and the requirements of metabolism. Area is proportional to the square of linear dimension, while volume is proportional to its cube. The **plasma membrane** surrounding every cell must provide sufficient surface area for exchange of oxygen, nutrients, and wastes relative to the volume of the cell.

■ INTERACTIVE QUESTION   7.2

a. If a eukaryotic cell has a diameter that is 10 times that of a bacterial cell, proportionally how much more surface area would the eukarotic cell have?

b. Proportionally how much more volume would it have?

The Importance of Compartmental Organization Membranes compartmentalize the eukaryotic cell, providing local environments for specific metabolic functions and participating in metabolism through membrane-bound enzymes. Membranes are composed of a bilayer of phospholipid molecules associated with diverse proteins.

■ The nucleus is a cell's genetic library

The **nucleus** is surrounded by the **nuclear envelope**, a double membrane perforated by pores that regulate the movement of large macromolecules between the nucleus and the cytoplasm. The inner membrane is lined by the **nuclear lamina**, a layer of protein filaments that helps to maintain the shape of the nucleus. There is evidence of a framework of fibers, called a nuclear matrix, extending through the nucleus.

Most of the cell's DNA is located in the nucleus, where, along with associated proteins, it is organized into **chromatin**, the substance of **chromosomes**. Each eukaryotic species has a characteristic chromosomal number. Chromosomes are visible only when coiled and condensed in a dividing cell.

The **nucleolus**, a round structure visible in the non-dividing nucleus, synthesizes ribosomal components that pass through nuclear pores for assembly in the cytoplasm.

■ INTERACTIVE QUESTION   7.3

How does the nucleus control protein synthesis in the cytoplasm?

■ Ribosomes build a cell's proteins

The prominence of nucleoli and the number of **ribosomes** are related to the rate of protein synthesis within a cell. Most of the proteins produced by free ribosomes are used within the cytosol. Bound ribosomes, attached to the endoplasmic reticulum, usually make proteins that will be included within membranes, packaged into organelles, or exported from the cell.

■ Many organelles are related through the endomembrane system

The **endomembrane system** of a cell consists of the nuclear envelope, endoplasmic reticulum, Golgi apparatus, lysosomes, vacuoles, and the plasma membrane. These membranes are all related either through direct contact or by the transfer of membrane segments by membrane-bound sacs called vesicles. Although related, the membranes differ in molecular composition and structure, depending on their functions.

■ The endoplasmic reticulum manufactures membranes and performs many other biosynthetic functions

The **endoplasmic reticulum (ER)** is the most extensive portion of the endomembrane system. It is continuous with the nuclear envelope and encloses a network of interconnected tubules or compartments called cisternae. Ribosomes are attached to the cytoplasmic surface of **rough ER**; **smooth ER** lacks ribosomes.

Functions of Smooth ER Smooth ER serves diverse functions: its enzymes are involved in fatty acid, phospholipid, steroid, and sex hormone synthesis; metabolism of carbohydrates; and detoxification of drugs and poisons. Barbiturates, alcohol, and other drugs increase a liver cell's production of smooth ER, thus leading to an increased tolerance (and thus reduced effectiveness) for these and other drugs. Smooth ER also functions in storage and release of calcium ions during muscle contraction.

Rough ER and Protein Synthesis Proteins intended for secretion are manufactured by membrane-bound ribosomes and then threaded into the cisternal space of the rough ER, where they fold into their native conformation. Many are covalently bonded to small carbohydrates to form **glycoproteins**. The secretory proteins are transported from the rough ER in tiny membrane-bound **transport vesicles** budded off from a special region called transitional ER.

Rough ER and Membrane Production   Rough ER manufactures membranes by inserting proteins formed by the attached ribosomes into the membrane. Membrane phospholipids are assembled from precursors obtained from the cytosol with the aid of enzymes built into the membrane. The newly formed membrane can be transferred to other parts of the endomembrane system by transport vesicles.

■ The Golgi apparatus finishes, sorts, and ships many products of the cell

The **Golgi apparatus** consists of a stack of flattened membranous sacs. Vesicles that bud from the ER join to the *cis* face of the Golgi apparatus, adding to it their contents and membrane. Products that travel through the Golgi apparatus are usually modified or refined as they move from one cisterna to the next. Some polysaccharides are manufactured by the Golgi. Golgi products are sorted into vesicles, which pinch off from the *trans* face of the Golgi apparatus. These vesicles may have surface molecules that help direct them to the plasma membrane or to other organelles.

■ Lysosomes are digestive compartments

Lysosomes are membrane-enclosed sacs of hydrolytic enzymes used by the cell to digest macromolecules. The cell can maintain an acidic pH for these enzymes and also protect itself from unwanted digestion by containing hydrolytic enzymes within lysosomes.

In macrophages and some protists, lysosomes fuse with food vacuoles to digest food particles ingested by **phagocytosis**. They also recycle the cell's own macromolecules by engulfing organelles or small bits of cytosol, a process known as autophagy. During development or metamorphosis, lysosomes may be programmed to destroy their cells to create a specific body form.

Storage diseases are inherited defects in which a lysosomal enzyme is missing or defective, and lysosomes become packed with indigestible substances.

■ Vacuoles have diverse functions in cell maintenance

Vacuoles are membrane-enclosed sacs that are larger than vesicles. **Food vacuoles** are formed as a result of

phagocytosis. **Contractile vacuoles** pump excess water out of freshwater protists.

A large **central vacuole** is found in mature plant cells, surrounded by a membrane called the **tonoplast**. This vacuole stores organic compounds and inorganic ions for the cell. Poisonous or unpalatable compounds, which may protect the plant from predators, and dangerous metabolic byproducts may also be contained in the vacuole. Plant cells increase in size with a minimal addition of new cytoplasm as their vacuoles absorb water and expand.

■ INTERACTIVE QUESTION   7.4

Name the components of the endomembrane system shown in this diagram and review the functions of each of these membranes.

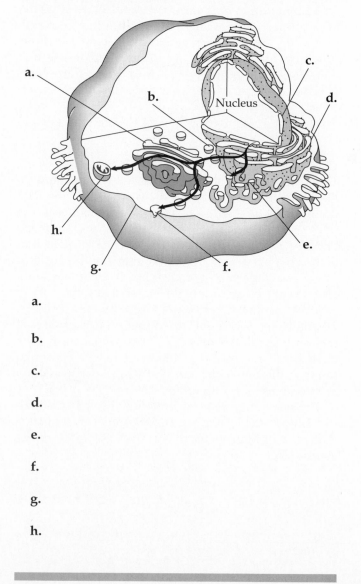

a.

b.

c.

d.

e.

f.

g.

h.

■ Peroxisomes consume oxygen in various metabolic functions

Peroxisomes are compartments enclosed by a single membrane and filled with enzymes that function in a variety of metabolic pathways, such as breaking down fatty acids to smaller molecules or detoxifying alcohol and other poisons. An enzyme that converts hydrogen peroxide (H_2O_2), a toxic byproduct of these pathways, is also packaged into peroxisomes.

Specialized peroxisomes called glyoxysomes are found in the tissues of germinating seeds and contain enzymes that convert the fatty acids stored in the seed to sugar for the developing seedling.

■ INTERACTIVE QUESTION 7.5

Why are peroxisomes not considered part of the endomembrane system?

■ Mitochondria and chloroplasts are the energy transformers of cells

Cellular respiration, the catabolic processing of fuels to produce ATP, occurs within the **mitochondria** of eukaryotic cells. Photosynthesis occurs in the **chloroplasts** of plants and eukaryotic algae, which produce organic compounds from carbon dioxide and water by absorbing solar energy. The membrane proteins of mitochondria and chloroplasts are made by ribosomes either free in the cytosol or contained within these organelles. They also contain a small amount of DNA that directs the synthesis of some of their proteins. These semiautonomous organelles are not considered part of the endomembrane system.

Mitochondria Two membranes, each a phospholipid bilayer with unique embedded proteins, enclose a mitochondrion. A narrow intermembrane space exists between the smooth outer membrane and the convoluted inner membrane. The folds of the inner membrane, called **cristae**, create a large membrane surface area and enclose the **mitochondrial matrix**. Within this matrix are enzymes that control many of the metabolic steps of cellular respiration. Other important enzymes are built into the inner membrane.

Chloroplasts **Plastids** are plant organelles that include amyloplasts, which store starch; chromoplasts, which contain pigments; and chloroplasts, which contain the green pigment chlorophyll and function in photosynthesis.

Chloroplasts, which are bounded by two membranes, are composed of a viscous fluid called the **stroma** and a membranous system of flattened sacs, called **thylakoids**, which enclose the thylakoid space. Thylakoids may be stacked together to form structures called **grana**.

■ INTERACTIVE QUESTION 7.6

Sketch a mitochondrion and a chloroplast and label their membranes and compartments.

■ The cytoskeleton provides structural support and functions in cell motility

The **cytoskeleton** is a network of fibers that function to give mechanical support; maintain or change cell shape; anchor or direct the movement of organelles and cytoplasm; and control movement of cilia, flagella, pseudopods, and even contraction of muscle cells. The cytoskeleton interacts with special proteins called motor molecules that change shape to produce cellular movements. At least three types of fibers are involved in the cytoskeleton: **microtubules**, **microfilaments**, and **intermediate filaments**.

Microtubules All eukaryotic cells have microtubules, which are hollow rods constructed of two kinds of globular proteins called α- and β-tubulins. In addition to providing the major supporting framework of the cell, microtubules serve as tracks along which organelles move with the aid of motor molecules. In many cells, microtubules radiate out from a region near the nucleus called a **centrosome**. In animal cells, a pair of **centrioles**, each composed of nine sets of triplet microtubules arranged in a ring, may help to organize microtubule assembly for cell division.

■ **INTERACTIVE QUESTION   7.7**

Fill in the following table to organize what you have learned about the components of the cytoskeleton. You may wish to refer to the textbook for additional details.

Cytoskeleton	Structure and Monomers	Functions
Microtubules	a.	b.
Microfilaments (actin filaments)	c.	d.
Intermediate filaments	e.	f.

Cilia and **flagella** are extensions of eukaryotic cells, composed of and moved by microtubules. Cilia are numerous and short; flagella occur one or two to a cell and are longer. A flagellum moves with an undulating motion, whereas a cilium generates force with a power stroke alternating with a recovery stroke (like an oar). Many protists use cilia or flagella to move through aqueous media. Cilia or flagella attached to stationary cells of a tissue sweep fluid past the cell.

Both cilia and flagella are composed of two single microtubules surrounded by a ring of nine doublets of microtubules (a nearly universal "9 + 2" arrangement), all of which are enclosed in an extension of the plasma membrane. A **basal body**, structurally identical to a centriole, anchors the tubules to the cell. The sliding of the microtubule doublets past each other occurs as arms, composed of the motor protein **dynein**, alternately attach to adjacent doublets, pull down (as the conformation of dynein changes), release, and reattach. In conjunction with the radial spokes or other anchoring structural elements, this action—driven by ATP—causes the bending of the flagella or cilia.

Microfilaments (Actin Filaments)   Microfilaments are solid rods consisting of a helix of two chains of molecules of the globular protein **actin**. In muscle cells, thousands of actin filaments interdigitate with thicker filaments made of the protein **myosin**. The sliding of actin and myosin filaments past each other, driven by ATP-powered arms extending from the myosin, causes the shortening of the cell and thus the contraction of muscles.

Microfilaments seem to be present in all eukaryotic cells. They function in support, such as in the core of microvilli; in localized contractions, such as the pinching apart of animal cells when they divide and such as the extension and retraction of **pseudopodia**; and in **cytoplasmic streaming** in plant cells.

Intermediate Filaments   Intermediate filaments are intermediate in size between microtubules and microfilaments and are more diverse in their composition—each type is constructed from different protein subunits. Intermediate fibers appear to be important in maintaining cell shape and anchoring certain organelles. The nucleus is securely held in a web of intermediate fibers, and the nuclear lamina lining the inside of the nuclear envelope is composed of intermediate filaments. The various kinds of intermediate filaments may serve as the superstructure of the entire cytoskeleton.

■ **Plant cells are encased by cell walls**

A **cell wall** is a diagnostic feature of plant cells. Plant cell walls are composed of microfibrils of cellulose embedded in a matrix of polysaccharides and protein. The exact composition of the wall varies between species and between cell types.

The **primary cell wall** secreted by a young plant cell is relatively thin and flexible. Adjacent cells are connected by the **middle lamella**, a thin layer of polysaccharides (called pectins) that glue the cells together. When they stop growing, some cells secrete a thicker and stronger **secondary cell wall** between the plasma membrane and the primary cell wall.

■ **INTERACTIVE QUESTION   7.8**

Sketch two adjacent plant cells, and show the location of the primary and secondary cell walls and the middle lamella.

■ The extracellular matrix (ECM) of animal cells functions in support, adhesion, movement, and development

Animal cells secrete an **extracellular matrix (ECM)** composed primarily of glycoproteins. **Collagen** forms strong fibers that are embedded in a network of **proteoglycans**. Cells may be attached to the ECM by **fibronectins** that bind to **integrins**, receptor proteins that span the plasma membrane and bind to and communicate with microfilaments of the cytoskeleton.

The ECM provides migratory paths for some cells in developing embryos. A cell's contact with its ECM appears to influence the activity of genes in the nucleus.

■ Intercellular junctions integrate cells into higher levels of structure and function

Plasmodesmata are channels in plant cell walls through which strands of cytoplasm connect bordering cells and water and small solutes can move. The plasma membranes of adjacent cells are continuous through the channel, linking most cells of a plant into a living continuum.

There are three main types of intercellular junctions between animal cells. **Tight junctions** are fusions between adjacent cell membranes that create an impermeable seal across a layer of epithelial cells. **Desmosomes**, reinforced by intermediate filaments, are strong connections between adjacent cells. **Gap junctions** allow for the exchange of materials between cells through protein-surrounded pores.

■ INTERACTIVE QUESTION 7.9

Return to your sketch of plant cells in 7.8 and draw in a plasmodesma.

■ The cell is a living unit greater than the sum of its parts

The compartmentalization and variety of organelles typical of cells exemplify the principle that structure correlates with function. The intricate functioning of a living cell emerges from the complex interactions of its many parts.

STRUCTURE YOUR KNOWLEDGE

1. This table lists the general functions performed by an animal cell. List the cellular structures associated with each of these functions.

Functions	Associated Organelles and Structures
Cell division	a.
Information storage and transferal	b.
Energy conversions	c.
Manufacture membranes and products	d.
Lipid synthesis, drug detoxification	e.
Digestion, recycling	f.
Conversion of H_2O_2 to water	g.
Structural integrity	h.
Movement	i.
Exchange with environment	j.
Cell–cell connection	k.

2. This table lists structures that are unique to plant
cells. Fill in the functions of these structures.

Plant Cell Structures	Functions		
Cell wall	a.		
Central vacuole	b.		
Chloroplast	c.		
Amyloplast	d.		
Plasmodesmata	e.		

3. Label the indicated structures in this diagram of a
cell.

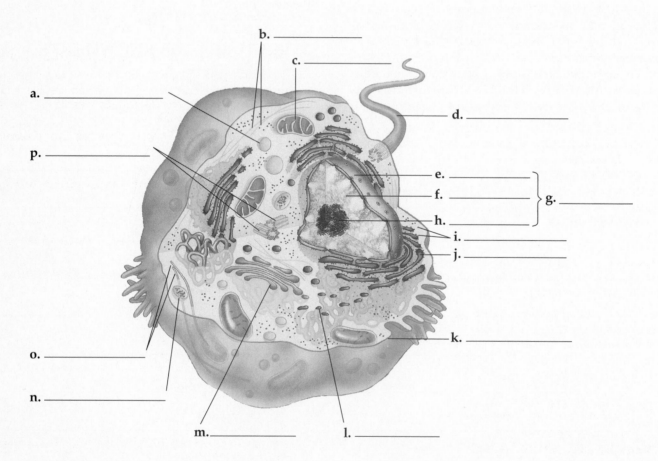

4. Create a diagram or flow chart to trace the
development of a secretory product (such as a
digestive enzyme) from the DNA code to its
export from the cell.

TEST YOUR KNOWLEDGE

MULTIPLE CHOICE: *Choose the one best answer.*

1. Which of the following is/are *not* found in a prokaryotic cell?
   a. ribosomes
   b. plasma membrane
   c. mitochondria
   d. a and c
   e. a, b, and c

2. Resolving power of a microscope is
   a. the distance between two separate points.
   b. the sharpness or clarity of an image.
   c. the degree of magnification of an image.
   d. the depth of focus on a specimen's surface.
   e. the wavelength of light.

3. Which of the following is *not* a similarity among the nucleus, chloroplasts, and mitochondria?
   a. They contain DNA.
   b. They are bounded by a double phospholipid bilayer membrane.
   c. They can divide to reproduce themselves.
   d. They are derived from the endoplasmic reticulum system.
   e. Their membranes are associated with specific proteins.

4. The pores in the nuclear envelope provide for the movement of
   a. proteins into the nucleus.
   b. ribosomal components out of the nucleus.
   c. mRNA out of the nucleus.
   d. enzymes into the nucleus.
   e. all of the above.

5. The ultrastructure of a chloroplast could be seen best using
   a. transmission electron microscopy.
   b. scanning electron microscopy.
   c. phase contrast light microscopy.
   d. cell fractionation.
   e. darkfield microscopy.

6. The largest number of bound ribosomes most likely would be found in a cell
   a. with a high metabolic rate.
   b. that produces secretory products.
   c. with many cilia.
   d. that produces steroids.
   e. that detoxifies poisons.

7. Which structure is *not* considered to be part of the endomembrane system?
   a. peroxisome
   b. smooth ER
   c. nuclear envelope
   d. lysosome
   e. Golgi apparatus

8. A growing plant cell elongates primarily by
   a. increasing the number of vacuoles.
   b. synthesizing more cytoplasm.
   c. taking up water into its central vacuole.
   d. synthesizing more cellulose.
   e. producing a secondary cell wall.

9. The innermost portion of a mature plant cell wall is the
   a. primary cell wall.
   b. secondary cell wall.
   c. middle lamella.
   d. plasma membrane.
   e. plasmodesmata.

10. Contractile elements of muscle cells are
    a. intermediate filaments.
    b. centrioles.
    c. microtubules.
    d. actin filaments.
    e. fibronectins.

11. Microtubules are components of all of the following *except*
    a. centrioles.
    b. the spindle apparatus for separating chromosomes in cell division.
    c. tracks along which organelles can move using motor molecules.
    d. flagella and cilia.
    e. the pinching apart of the cytoplasm in animal cell division.

12. Of the following, which is probably the most common route for membrane flow in the endomembrane system?
    a. rough ER → Golgi → lysosomes → vesicles → plasma membrane
    b. rough ER → transitional ER → Golgi → vesicles → plasma membrane
    c. nuclear envelope → rough ER → Golgi → smooth ER → lysosomes
    d. rough ER → vesicles → Golgi → smooth ER → plasma membrane
    e. smooth ER → vesicles → Golgi → vesicles → peroxisomes

13. Proteins to be used within the cytosol are generally synthesized
    a. by ribosomes bound to rough ER.
    b. by free ribosomes.
    c. by the nucleolus.
    d. within the Golgi apparatus.
    e. by mitochondria and chloroplasts.

14. Plasmodesmata in plant cells are similar in function to
    a. desmosomes.
    b. tight junctions.
    c. gap junctions.
    d. the extracellular matrix.
    e. integrins.

15. In a cell fractionation procedure, the first pellet formed would most likely contain
    a. the extracellular matrix.
    b. ribosomes.
    c. mitochondria.
    d. lysosomes.
    e. nuclei.

FILL IN THE BLANKS *with the appropriate cellular organelle or structure.*

_____ 1. transport membranes and products to various locations

_____ 2. infolding of mitochondrial membrane with attached enzymes

_____ 3. consists of collagen, proteglycans, and fibronectins

_____ 4. small sacs with specific enzymes for particular metabolic pathway

_____ 5. stack of flattened sacs inside chloroplasts

_____ 6. anchoring structure for cilia and flagella

_____ 7. semifluid medium between nucleus and plasma membrane

_____ 8. system of fibers that maintains cell shape, anchors organelles

_____ 9. connection between animal cells that creates impermeable layer

_____ 10. membrane surrounding central vacuole of plant cells

CHAPTER 8

MEMBRANE STRUCTURE AND FUNCTION

FRAMEWORK

This chapter presents the fluid mosaic model of membrane structure, relating the molecular structure of biological membranes to their function of regulating the passage of substances into and out of the cell.

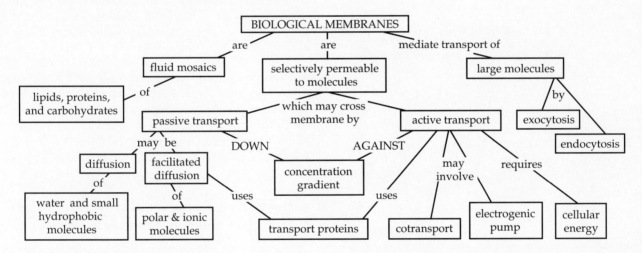

CHAPTER REVIEW

The plasma membrane is the boundary of life; this **selectively permeable** membrane allows the cell to maintain a unique internal environment and to control the movement of materials into and out of the cell.

◙ Membrane models have evolved to fit new data: *science as a process*

Membrane **phospholipids** are **amphipathic**. In 1935, Davson and Danielli proposed a molecular model of membrane structure in which a double layer of phos-

pholipids, with the hydrophobic hydrocarbon tails in the center and the hydrophilic heads facing the aqueous solution on both sides of the membrane, is covered with a coat of globular proteins. After changing to layers of protein in the pleated-sheet configuration, this sandwich model was consistent with the observed thickness of plasma membranes and with the apparent triple-layer staining results seen with electron microscopy in the 1950s. The Davson-Danielli model was widely accepted by the 1960s for all cellular membranes.

In 1972, Singer and Nicolson proposed their **fluid mosaic model** in which amphipathic membrane proteins are embedded in the phospholipid bilayer with their hydrophilic regions extending out into the aqueous environment. The phospholipid bilayer is envisioned as fluid, with a mosaic of protein molecules floating in it.

The fluid mosaic model is supported by evidence from freeze-fracture electron microscopy. This technique involves freezing a specimen, fracturing it with a cold knife, etching the fractured surface to enhance contrast, and then coating the fractured surface with platinum and carbon. The resulting replica of the surface is examined with the electron microscope.

The fluid mosaic model is currently the most accepted and useful model for organizing existing knowledge and extending further research on membrane structure.

■ **INTERACTIVE QUESTION   8.1**

Label the components in this diagram of the fluid mosaic model of membrane structure. Indicate the regions that are hydrophobic and those that are hydrophilic.

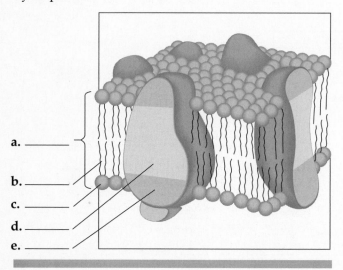

a. _____

b. _____

c. _____

d. _____

e. _____

■ **A membrane is a fluid mosaic of lipids, proteins, and carbohydrates**

The Fluid Quality of Membranes   Membranes are held together primarily by weak hydrophobic interactions that allow the lipids and some of the proteins to drift laterally.

Phospholipids with unsaturated hydrocarbon tails maintain membrane fluidity at lower temperatures. The steroid cholesterol, common in cell membranes of animals, restricts movement of phospholipids and thus reduces fluidity at warmer temperatures. Cholesterol also prevents the close packing of lipids and thus enhances fluidity at lower temperatures.

■ **INTERACTIVE QUESTION   8.2**

a. Cite some experimental evidence that shows that membrane proteins drift.

b. How might the plasma membrane of a plant cell change in response to the cold temperatures of winter?

Membranes as Mosaics of Structure and Function
Each membrane has its own unique complement of membrane proteins, which determines most of the specific functions of that membrane. **Integral proteins** extend into the hydrocarbon region of the lipids. They may extend across the membrane, with two hydrophilic ends and a hydrophobic midsection. **Peripheral proteins** are attached to the surface of the membrane, often to integral proteins. Membranes are asymmetric; they have distinct inner and outer faces related to the directional orientation of their proteins, the composition of the lipid bilayers, and, in the plasma membrane, the attachment of carbohydrates to the exterior surface.

Membrane Carbohydrates and Cell–Cell Recognition
The ability of a cell to distinguish other cells is based on recognition of membrane carbohydrates. The glycolipids and glycoproteins attached to the outside of plasma membranes vary from species to species, from individual to individual, and even among cell types.

■ **A membrane's molecular organization results in selective permeability**

The plasma membrane permits a regular exchange of nutrients, waste products, oxygen, and inorganic ions. Biological membranes are selectively permeable; the ease and rate at which small molecules pass through them differ.

Permeability of the Lipid Bilayer   Hydrophobic molecules can dissolve in and cross through a membrane.

■ **INTERACTIVE QUESTION   8.3**

What types of molecules have difficulty crossing the plasma membrane? Why?

Small polar molecules, such as H_2O and CO_2, can cross a plasma membrane rapidly.

Transport Proteins   Ions and polar molecules may move across the plasma membrane with the aid of **transport proteins**, which may provide a hydrophilic channel or may physically bind and transport a specific molecule.

Passive transport is diffusion across a membrane

Diffusion is the movement of a substance down its **concentration gradient** due to random thermal motion. This spontaneous process decreases free energy and increases entropy (disorder, randomness) in a system. The cell does not expend energy when substances diffuse across membranes down their concentration gradient; therefore, the process is called **passive transport**.

Osmosis is the passive transport of water

Osmosis is the diffusion of water across a selectively permeable membrane. Water diffuses down its own concentration gradient, which is affected by the binding of water molecules to solute particles, thus lowering the proportion of unbound water that is free to cross the membrane. Water will move from a **hypo-** tonic solution across a membrane into a **hypertonic** solution—in other words, from a region with a lesser concentration of solutes (and greater concentration of unbound water) into a region with a greater concentration of solutes. **Isotonic** solutions have equal solute concentrations, and there is no net movement of water across a membrane separating them. The direction of osmosis is determined by the difference in total solute concentration, regardless of the kinds of solute molecules involved.

Cell survival depends on balancing water uptake and loss

Water Balance of Cells Without Walls   An animal cell placed in a hypertonic environment will lose water and shrivel. If placed in a hypotonic environment, the cell will gain water, swell, and possibly lyse (burst). Cells without rigid walls must either live in an isotonic environment, such as salt water or isotonic body fluids, or have adaptations for **osmoregulation**.

■ INTERACTIVE QUESTION 8.5

What adaptations may freshwater protists have for osmoregulation?

Water Balance of Cells with Walls   The cell walls of plants, fungi, prokaryotes, and some protists play a role in water balance within hypotonic environments. Water moving into the cell causes the cell to swell against its cell wall, creating a **turgid** cell. Turgid cells provide mechanical support for nonwoody plants. Plant cells in an isotonic surrounding are **flaccid**. In a hypertonic medium, a plant cell undergoes **plasmolysis**, the pulling away of the plasma membrane from the cell wall as the cell shrivels.

■ INTERACTIVE QUESTION 8.4

A solution of 1 *M* glucose is separated by a selectively permeable membrane from a solution of 0.2 *M* fructose and 0.7 *M* sucrose. The membrane is not permeable to the sugar molecules. Indicate which side is hypertonic, which is hypotonic, and the direction of osmosis.

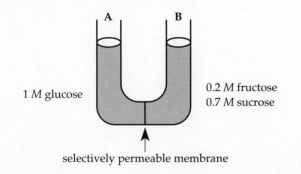

1 *M* glucose

0.2 *M* fructose
0.7 *M* sucrose

selectively permeable membrane

■ INTERACTIVE QUESTION 8.6

a. The ideal osmotic environment for animal cells is _____ .

b. The ideal environment for plant cells is _____ .

Specific proteins facilitate the passive transport of selected solutes

Facilitated diffusion involves the diffusion of polar molecules and ions across a membrane with the aid of transport proteins. Transport proteins are similar to enzymes in several ways: they are specialized with a specific binding site for the solute they transport, they can reach a maximum rate of transport when all transport proteins are saturated by a high concentration of the solute, and they can be inhibited by molecules that resemble their normal solute.

The binding of the solute to the transport protein may cause a conformational change that serves to translocate the binding site and the attached solute. A transport protein may provide a selective channel across the membrane, as in **gated channels** that open in response to an electrical or chemical stimulation of a nerve cell.

■ **INTERACTIVE QUESTION   8.7**

Why is facilitated diffusion considered passive transport?

Active transport is the pumping of solutes against their gradients

Active transport, requiring the expenditure of energy by the cell, is essential for a cell to maintain internal concentrations of small molecules that differ from environmental concentrations. The terminal phosphate group of ATP may be transferred to a transport protein, inducing it to change its conformation and translocate the bound solute across the membrane. The **sodium–potassium pump** allows the cell to exchange Na^+ and K^+ across animal cell membranes, creating a greater concentration of potassium ions and a lesser concentration of sodium ions within the cell.

Some ion pumps generate voltage across membranes

Cells have a **membrane potential**, a voltage across their plasma membrane created by the electrical potential energy resulting from the separation of opposite charges. The cytoplasm of a cell is negatively charged compared to extracellular fluid. The membrane potential favors the diffusion of cations into the cell and anions out of the cell. Diffusion of an ion is affected by both the membrane potential and its own concentration gradient; thus an ion diffuses down its **electrochemical gradient**.

Electrogenic pumps are membrane proteins that generate voltage across a membrane by active transport of ions. A **proton pump** that transports H^+ out of the cell generates membrane potentials in plants, fungi, and bacteria.

■ **INTERACTIVE QUESTION   8.8**

The Na^+–K^+ pump, the major electrogenic pump in animal cells, exchanges sodium ions for potassium ions, both of which are cations. How does this exchange generate a membrane potential?

In cotransport, a membrane protein couples the transport of one solute to another

Cotransport is a mechanism through which the active transport of a solute is indirectly driven by an ATP-powered pump that transports another substance against its gradient. As that substance diffuses back down its concentration gradient through specialized transport proteins, the solute may be cotransported against its concentration gradient across the membrane.

Exocytosis and endocytosis transport large molecules

In **exocytosis**, the cell secretes macromolecules by the fusion of vesicles with the plasma membrane. In **endocytosis**, a region of the plasma membrane sinks inward and pinches off to form a vesicle containing material that had been outside the cell. **Phagocytosis** is a form of endocytosis in which pseudopodia wrap around a food particle, creating a vacuole that then fuses with a lysosome containing hydrolytic enzymes. In **pinocytosis**, droplets of extracellular fluid are taken into the cell in small vesicles. **Receptor-mediated endocytosis** allows a cell to acquire specific substances from extracellular fluid. **Ligands**, molecules that bind specifically to receptor sites, attach to proteins usually clustered in **coated pits** on the cell surface and are carried into the cell when the coated pit buds off to form a vesicle.

■ **INTERACTIVE QUESTION 8.9**

a. How is cholesterol, which is used for the synthesis of other steroids and membranes, transported into human cells?

b. Explain why cholesterol accumulates in the blood of individuals with the disease familial hypercholesteremia.

■ **Specialized membrane proteins transmit extracellular signals to the inside of the cell**

Binding of an extracellular molecule known as a **first messenger** to a membrane receptor protein can initiate a **signal-transduction pathway**. The receptor protein activates a relay protein that then stimulates an effector membrane protein that is an enzyme. A **second messenger** is produced by the effector protein, and this cytoplasmic molecule initiates metabolic and structural responses in the cell. These membrane proteins allow a cell to receive and respond to environmental signals.

■ **INTERACTIVE QUESTION 8.10**

Label the components in this diagram of a signal-transduction pathway.

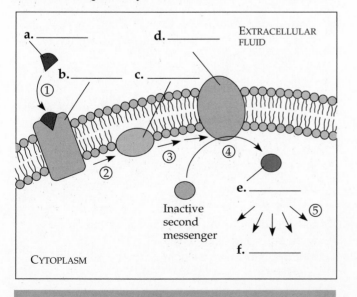

STRUCTURE YOUR KNOWLEDGE

1. Create a concept map to illustrate your understanding of osmosis. This exercise will help you practice using the words *hypotonic*, *isotonic*, and *hypertonic*, and it will help you focus on the effect of these osmotic environments on plant and animal cells. Explain your map to a friend.

2. The following diagram illustrates passive and active transport across a plasma membrane. Use it to answer questions a–d.

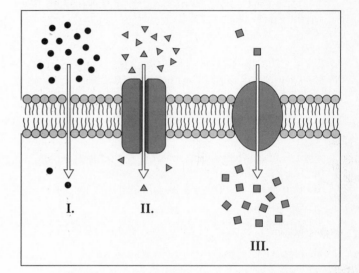

a. Which section represents facilitated diffusion?

How can you tell?

Does the cell expend energy in this transport?

Why or why not?

What types of solute molecules may be moved by this type of transport?

b. Which section shows active transport?

How can you tell?

Does the cell expend energy in this transport?

Why or why not?

c. Which section shows diffusion?

What types of solute molecules may be moved by this type of transport?

d. Which of these sections are considered passive transport?

TEST YOUR KNOWLEDGE

MULTIPLE CHOICE: *Choose the one best answer.*

1. Glycoproteins and glycolipids are important for
   a. facilitated diffusion.
   b. active transport.
   c. cell–cell recognition.
   d. cotransport.
   e. signal-transduction pathways.

2. A single layer of phospholipid molecules coats the water in a beaker. Which part of the molecules will face the air?
   a. the phosphate groups
   b. the hydrocarbon tails
   c. both head and tail because the molecules are amphipathic and will lie sideways
   d. the phospholipids would dissolve in the water and not form a membrane coat
   e. the glycolipid regions

3. Which of the following is *not* true about osmosis?
   a. It increases free energy in a system.
   b. Water moves from a hypotonic to a hypertonic solution.
   c. Solute molecules bind to water and decrease the water available to move.
   d. It increases the entropy in a system.
   e. There is no net osmosis between isotonic solutions.

4. Support for the fluid mosaic model of membrane structure comes from
   a. the freeze-fracture technique of electron microscopy.
   b. the movement of proteins in hybrid cells.
   c. the amphipathic nature of membrane proteins.
   d. both a and b.
   e. all of the above.

5. A freshwater *Paramecium* is placed into salt water. Which of the following events would occur?
   a. an increase in the action of its contractile vacuole
   b. swelling of the cell until it becomes turgid
   c. swelling of the cell until it lyses
   d. shriveling or crenation of the cell
   e. diffusion of salt ions into the cell

6. Ions diffuse across membranes down their
   a. electrochemical gradient.
   b. electrogenic gradient.
   c. electrical gradient.
   d. concentration gradient.
   e. osmotic gradient.

7. The fluidity of membranes in a plant in cold weather may be maintained by
   a. increasing the number of phospholipids with saturated hydrocarbon tails.
   b. activating a H^+ pump.
   c. increasing the concentration of cholesterol in the membrane.
   d. increasing the proportion of integral proteins.
   e. increasing the number of phospholipids with unsaturated hydrocarbon tails.

8. A plant cell placed in a hypotonic environment will
   a. plasmolyze.
   b. shrivel.
   c. become turgid.
   d. become flaccid.
   e. lyse.

9. Which of the following is *not* true of the carrier molecules involved in facilitated diffusion?
   a. They increase the speed of transport across a membrane.
   b. They can concentrate solute molecules on one side of the membrane.
   c. They have specific binding sites for the molecules they transport.
   d. They may undergo a conformational change upon binding of solute.
   e. They may be inhibited by molecules that resemble the solute to which they normally bind.

10. The membrane potential of a cell favors the
    a. movement of cations into the cell.
    b. movement of anions into the cell.
    c. action of an electrogenic pump.
    d. movement of sodium out of the cell.
    e. action of a proton pump.

11. Cotransport may involve
    a. active transport of two solutes through a transport protein.
    b. passive transport of two solutes through a transport protein.
    c. ion diffusion against the electrochemical gradient created by an electrogenic pump.
    d. first and second messengers in a signal-transduction pathway.
    e. transport of one solute against its concentration gradient in tandem with another that is diffusing down its concentration gradient.

12. Exocytosis involves all of the following *except*
    a. ligands and coated pits.
    b. the fusion of a vesicle with the plasma membrane.

c. a mechanism to transport carbohydrates to the outside of plant cells during the formation of cell walls.

d. a mechanism to rejuvenate the plasma membrane.

e. a means of exporting large molecules.

13. The proton pump in plant cells is the functional equivalent of an animal cell's
a. cotransport mechanism.
b. sodium–potassium pump.
c. contractile vacuole for osmoregulation.
d. receptor-mediated endocytosis of cholesterol.
e. signal-transduction pathway.

14. Pinocytosis involves
a. the fusion of a newly formed food vacuole with a lysosome.
b. receptor-mediated endocytosis and the formation of vesicles.
c. the pinching in of the plasma membrane around droplets of external fluid.
d. pseudopod extension as vesicles move along the cytoskeleton and fuse with the plasma membrane.
e. the accumulation of specific large molecules in a cell.

15. Watering a houseplant with too concentrated a solution of fertilizer can result in wilting because
a. the uptake of ions into plant cells makes the cells hypertonic.
b. the soil solution becomes hypertonic, causing the cells to lose water.
c. the plant will grow faster than it can transport water and maintain proper water balance.
d. diffusion down the electrochemical gradient will cause a disruption of membrane potential and accompanying loss of water.
e. the plant will suffer fertilizer burn due to a caustic soil solution.

Use the U-tube setup to answer questions 16 through 18.

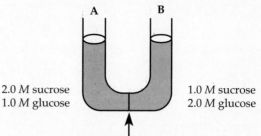

2.0 *M* sucrose
1.0 *M* glucose

1.0 *M* sucrose
2.0 *M* glucose

selectively permeable membrane

The solutions in the two arms of this U-tube are separated by a membrane that is permeable to water and glucose but not to sucrose. Side A is filled with a solution of 2.0 *M* sucrose and 1.0 *M* glucose. Side B is filled with 1.0 *M* sucrose and 2.0 *M* glucose.

16. Initially, the solution in side A, with respect to that in side B, is
a. hypotonic.
b. hypertonic.
c. isotonic.
d. lower.
e. higher.

17. After the system reaches equilibrium, what changes are observed?
a. The water level is higher in side A than in side B.
b. The water level is higher in side B than in side A.
c. The molarity of glucose is higher in side A than in side B.
d. The molarity of sucrose has increased in side A.
e. Both a and c have occurred.

18. During the period before equilibrium is reached, which molecule(s) will show net movement through the membrane?
a. water
b. glucose
c. sucrose
d. water and sucrose
e. water and glucose

CHAPTER 9

CELLULAR RESPIRATION:
HARVESTING CHEMICAL ENERGY

FRAMEWORK

The catabolic pathways of glycolysis and respiration capture the chemical energy in glucose and other fuels and store it in ATP. Glycolysis, occurring in the cytosol, produces ATP, pyruvate, and NADH; the latter two may then enter the mitochondria for respiration. A mitochondrion consists of a matrix in which the enzymes of the Krebs cycle are localized, a highly folded inner membrane in which enzymes and the molecules of the electron transport chain are embedded, and an intermembrane space between the two membranes to temporarily house H^+ that has been pumped across the inner membrane during the redox reactions of the electron transport chain. Through a chemiosmotic mechanism, a proton motive force drives oxidative phosphorylation as protons move back through ATP synthases located in the membrane.

CHAPTER REVIEW

■ Cellular respiration and fermentation are catabolic (energy-yielding) pathways

Fermentation, which occurs without oxygen, is the partial degradation of sugars to release energy. **Cellular respiration**, the most common catabolic pathway, uses oxygen to break down glucose (or other energy-rich organic compounds) and obtain energy in the usable form of ATP. This exergonic process has a free energy change of -686 kcal/mol of glucose.

■ **INTERACTIVE QUESTION 9.1**

Fill in the summary equation for cellular respiration.

$$\underline{\hspace{2cm}} + 6\,O_2 \longrightarrow \underline{\hspace{2cm}} + 6\,H_2O + \underline{\hspace{2cm}}$$

■ Cells must recycle the ATP they use for work

ATP, adenosine triphosphate, is an energy source because of the instability of the bonds between its three negatively charged phosphate groups. When enzymes shift the terminal phosphate group to another molecule, this newly phosphorylated molecule can perform work. A cell regenerates its supply of ATP from ADP and inorganic phosphate, using energy from cellular respiration.

■ Redox reactions release energy when electrons move closer to electronegative atoms

Oxidation-reduction or **redox reactions** involve the partial or complete transfer of one or more electrons from one reactant to another. The substance that loses electrons is **oxidized** and acts as a **reducing agent** (electron donor) to the substance that gains electrons. By gaining electrons, a substance is **reduced** and acts as an **oxidizing agent** (electron acceptor).

Oxygen strongly attracts electrons and is one of the most powerful oxidizing agents. As electrons shift toward a more electronegative atom, they give up potential energy. Thus, chemical energy is released in a redox reaction that passes electrons to oxygen.

Fill in the appropriate terms in this equation.

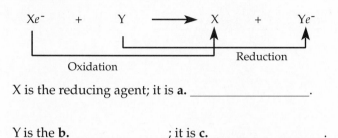

X is the reducing agent; it is **a.** _____.

Y is the **b.** _____; it is **c.** _____.

■ Electrons "fall" from organic molecules to oxygen during cellular respiration

In respiration, the energy from the oxidation of glucose and the reduction of oxygen is captured in ATP. Organic molecules with an abundance of hydrogen are rich in "hilltop" electrons that release their potential energy when they "fall" closer to oxygen.

a. In the conversion of glucose and oxygen to carbon dioxide and water, which molecule is reduced?

b. Which molecule is oxidized?

c. What happens to the energy that is released in this redox reaction?

■ The "fall" of electrons during respiration is stepwise, via NAD⁺ and an electron transport chain

At certain steps in the oxidation of glucose, two hydrogen atoms are removed by enzymes called dehydrogenases and the two electrons and one proton are passed to a coenzyme, **NAD⁺** (nicotinamide adenine dinucleotide).

Energy from respiration is slowly released in a series of small steps as electrons are passed down an **electron transport chain**, a group of carrier molecules located in the inner mitochondrial membrane, to a stable location close to a highly electronegative oxygen atom.

a. NAD⁺ is called _____.

b. Its reduced form is _____.

■ Respiration is a cumulative function of glycolysis, the Krebs cycle, and electron transport: *an overview*

Glycolysis, occurring in the cytosol, breaks glucose into two molecules of pyruvate. The **Krebs cycle**, located in the mitochondrial matrix, converts a derivative of pyruvate into carbon dioxide. In some of the steps of glycolysis and the Krebs cycle, dehydrogenase enzymes transfer electrons to NAD⁺. NADH passes electrons to the electron transport chain, from which they eventually combine with hydrogen ions and oxygen to form water. The energy released in each step of the chain is used to synthesize ATP by **oxidative phosphorylation**.

About 10% of the ATP generated for each molecule of glucose oxidized to carbon dioxide and water is produced by **substrate-level phosphorylation**, in which an enzyme transfers a phosphate group from a substrate to ADP.

■ Glycolysis harvests chemical energy by oxidizing glucose to pyruvate: *a closer look*

Glycolysis, a ten-step process occurring in the cytosol, has an energy-investment stage and an energy-payoff phase. Two molecules of ATP are consumed as glucose is split into two phosphorylated three-carbon molecules of glyceraldehyde 3-phosphate. The conversion of these molecules to pyruvate produces two NADH and four ATP by substrate-level phosphorylation. For each molecule of glucose, glycolysis yields a net gain of two ATP and two NADH.

■ **INTERACTIVE QUESTION 9.5**

Fill in the boxes with the three metabolic stages of respiration (**a–c**). Indicate whether the ATP is produced by substrate-level phosphorylation or oxidative phosphorylation (**d–f**). Label the arrows that show the flow of electrons via NADH.

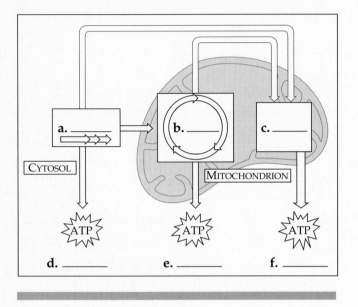

■ **INTERACTIVE QUESTION 9.6**

Fill in the blanks in the summary figure of the Krebs cycle. *Some letters are used more than once to indicate repetitive steps.*

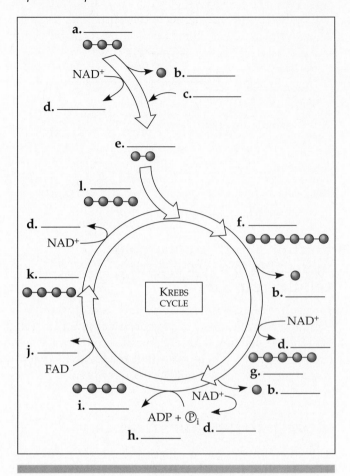

■ **The Krebs cycle completes the energy-yielding oxidation of organic molecules:** *a closer look*

Before the Krebs cycle begins, a series of steps occur within a multienzyme complex in the mitochondria: a carboxyl group is removed from the three-carbon pyruvate and released as CO_2; the remaining two-carbon group is oxidized to form acetate with the accompanying reduction of NAD^+ to NADH; and coenzyme A is attached to the acetate by an unstable bond, forming **acetyl CoA**.

The acetyl fragment of acetyl CoA is added to oxaloacetate to form citrate, which is progressively decomposed back to oxaloacetate. For each turn of the Krebs cycle, two carbons enter in the reduced form from acetyl CoA; two carbons exit completely oxidized as CO_2; three NADH and one $FADH_2$ are formed; and one ATP is made by substrate-level phosphorylation. There are two turns of the Krebs cycle for each glucose molecule oxidized.

■ **The inner mitochondrial membrane couples electron transport to ATP synthesis:** *a closer look*

Pathway of Electron Transport Most components of the chain are proteins with tightly bound, nonprotein prosthetic groups that shift between reduced and oxidized states as they accept and donate electrons. The pathway goes from a flavoprotein to an iron–sulfur protein to ubiquinone (Q) to a series of molecules called **cytochromes**, proteins with an iron-containing heme group. The last cytochrome, cyt a_3, passes its electrons to oxygen, which picks up a pair of H^+ and forms water.

FADH$_2$ adds its electrons to the chain at a lower energy level; thus one-third less energy is provided for ATP synthesis by FADH$_2$ as compared to NADH.

Chemiosmosis: The Energy-Coupling Mechanism
The electron carrier molecules are organized into three complexes. Electrons are transferred between the complexes by the mobile carriers, Q and cytochrome *c*. Some members of the chain also accept and release protons, which are pumped into the intermembrane space at three points. The resulting proton gradient stores potential energy, referred to as the **proton-motive force**. The exergonic passage of protons back through the **ATP synthase** complexes that span the membrane drives ATP synthesis. This coupling of exergonic redox reactions to produce ATP through the use of an H$^+$ gradient is called **chemiosmosis**.

Chloroplasts use chemiosmosis to make ATP during photosynthesis. Bacteria generate proton gradients across their plasma membranes and use the proton-motive force to generate ATP, pump molecules across the membrane, and even move flagella.

■ Cellular respiration generates many ATP molecules for each sugar molecule it oxidizes: *a review*

The efficiency of respiration in its energy conversions is approximately 38%. The rest of the energy is released as heat. (See Interactive Question 9.8, p. 56)

■ Fermentation enables some cells to produce ATP without the help of oxygen

Glycolysis produces a net of 2 ATP per glucose molecule, under both **aerobic** and **anaerobic** conditions. Without oxygen, glycolysis is part of fermentation—the anaerobic catabolism of organic nutrients into a waste product that regenerates NAD$^+$, the oxidizing agent for glycolysis.

In **alcohol fermentation**, pyruvate is converted into acetaldehyde, and CO$_2$ is released. Acetaldehyde is then reduced by NADH to form ethanol (ethyl alco-

■ INTERACTIVE QUESTION 9.7

Label the blanks on this diagram of the electron transport chain and oxidative phosphorylation in mitochondria.

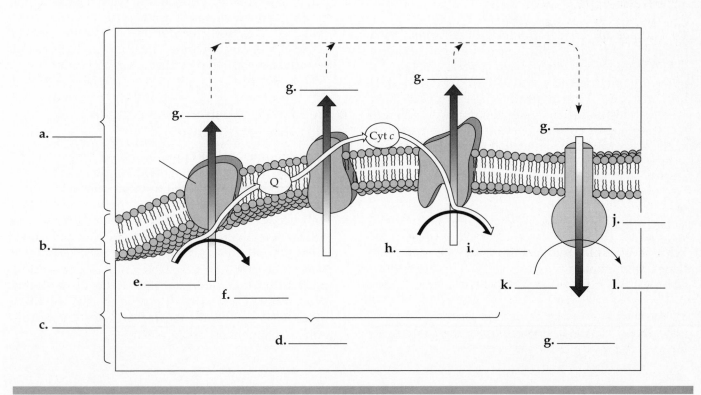

■ INTERACTIVE QUESTION   9.8

Fill in the tally for maximum ATP yield from the oxidation of one molecule of glucose to six molecules of carbon dioxide.

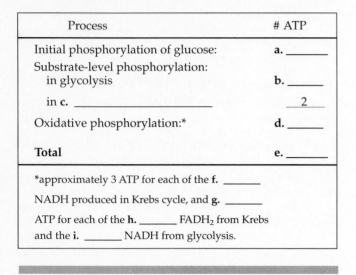

Process	# ATP
Initial phosphorylation of glucose:	a. _____
Substrate-level phosphorylation: in glycolysis	b. _____
in c. _____	2
Oxidative phosphorylation:*	d. _____
Total	e. _____

*approximately 3 ATP for each of the **f.** _____
NADH produced in Krebs cycle, and **g.** _____
ATP for each of the **h.** _____ FADH$_2$ from Krebs
and the **i.** _____ NADH from glycolysis.

hol), and NAD$^+$ is regenerated. In **lactic acid fermentation**, pyruvate is reduced to form lactate and recycle NAD$^+$; no CO$_2$ is released. Muscle cells make ATP by lactic acid fermentation when energy demand is high and oxygen supply is low.

Comparison of Fermentation and Respiration   Both fermentation and respiration use glycolysis with NAD$^+$ as the oxidizing agent to convert glucose and other organic fuels to pyruvate. To oxidize NADH back to NAD$^+$, fermentation uses pyruvate or acetaldehyde as the final electron acceptor, whereas respiration uses oxygen, via the electron transport chain. Also, with oxygen available, pyruvate can be oxidized in the Krebs cycle to produce much more ATP.

■ INTERACTIVE QUESTION   9.9

Why does respiration generate so much more ATP than fermentation?

Facultative anaerobes, such as yeast and some bacteria, can make ATP by fermentation or respiration, depending upon whether oxygen is available.

Evolutionary Significance of Glycolysis   Glycolysis is common to fermentation and respiration. This most widespread of all metabolic processes probably evolved in ancient prokaryotes before oxygen was available. The cytosolic location of glycolysis is also evidence of its antiquity.

■ Glycolysis and the Krebs cycle connect to many other metabolic pathways

The Versatility of Catabolism   Fats, proteins, and carbohydrates can all be used by cellular respiration to make ATP. Proteins are digested into amino acids, which are then deaminated and can enter into respiration at several sites. The digestion of fats yields glycerol, which is converted to an intermediate of glycolysis, and fatty acids, which are broken down by **beta oxidation** to two-carbon fragments that enter the Krebs cycle as acetyl CoA.

Biosynthesis (Anabolic Pathways)   The organic molecules of food also provide carbon skeletons for biosynthesis. Some monomers, such as amino acids, can be directly incorporated into the cell's macromolecules. Intermediates of glycolysis and the Krebs cycle serve as precursors for anabolic pathways. The molecules of carbohydrates, fats, and proteins can all be interconverted to provide for a cell's needs.

■ Feedback mechanisms control cellular respiration

Through feedback inhibition, the end product of a pathway inhibits the enzyme that initiates the pathway, thus preventing a cell from producing an excess of a particular substance. The supply of ATP in the cell regulates respiration. The allosteric enzyme that catalyzes an early step of glycolysis, phosphofructokinase, is inhibited by ATP and activated by ADP. Phosphofructokinase is also inhibited by citrate transported from the mitochondria into the cytosol, thus synchronizing the rates of glycolysis and the Krebs cycle. Other enzymes located at key intersections help to maintain metabolic balance.

STRUCTURE YOUR KNOWLEDGE

1. This chapter describes how the catabolic pathways of glycolysis and respiration capture chemical energy and store it in ATP. One of the best ways to learn the three main components of cellular respiration is to explain them to someone. Find two study partners and have each person be responsible for learning and explaining the important concepts and steps of either glycolysis, the Krebs cycle, or the electron transport chain. You may want to involve a fourth person to teach you fermentation, or learn that together as a group. Make sure your partners understand the significance of the section you are teaching. Use diagrams and sketches to help you explain the process.

2. Fill in the following table to summarize the major inputs and outputs of glycolysis, the Krebs cycle, the electron transport chain and oxidative phosphorylation, and fermentation.

3. Create a concept map to organize your understanding of oxidative phosphorylation and chemiosmosis.

TEST YOUR KNOWLEDGE

MULTIPLE CHOICE: *Choose the one best answer.*

1. In a redox reaction,
   a. the substance that is reduced loses energy.
   b. the substance that is oxidized loses energy.
   c. the oxidizing agent donates electrons.
   d. the reducing agent is the most electronegative.
   e. the electron acceptor loses energy.

2. In the reaction $C_6H_{12}O_6 + 6 O_2 \rightarrow 6 CO_2 + 6 H_2O$,
   a. oxygen becomes reduced.
   b. glucose becomes reduced.
   c. oxygen becomes oxidized.
   d. water is a reducing agent.
   e. oxygen is a reducing agent.

3. A substrate that is phosphorylated
   a. has a stable phosphate bond.
   b. has been formed by the reaction ADP + $℗_i$ → ATP.
   c. has an increased reactivity; it is primed to do work.
   d. has been oxidized.
   e. will pass its electrons to the electron transport chain.

Process	Main Function	Inputs	Outputs
Glycolysis			
Pyruvate to acetyl CoA			
Krebs cycle			
Electron transport chain and oxidative phosphorylation			
Fermentation			

4. Which of the following is *not* true of oxidative phosphorylation?
   a. It uses oxygen as the ultimate electron donor.
   b. It involves the redox reactions of the electron transport chain.
   c. It involves an ATP synthase located in the inner mitochondrial membrane.
   d. It produces approximately three ATP for every NADH that is oxidized.
   e. It depends on chemiosmosis.

5. Substrate-level phosphorylation
   a. involves the shifting of a phosphate group from ATP to a substrate.
   b. can use NADH or $FADH_2$.
   c. takes place only in the cytosol.
   d. accounts for 10% of the ATP formed by fermentation.
   e. is the energy source for facultative anaerobes under anaerobic conditions.

6. The major reason that glycolysis is not as energy-productive as respiration is that
   a. NAD^+ is regenerated by alcohol or lactate production, without the high-energy electrons passing through the electron transport chain.
   b. it is the pathway common to fermentation and respiration.
   c. it does not take place in a specialized membrane-bound organelle.
   d. pyruvate is more reduced than CO_2; it still contains much of the energy from glucose.
   e. substrate-level phosphorylation is not as energy efficient as oxidative phosphorylation.

7. The electron carrier molecules Q and cytochrome *c*
   a. are reduced as they pass electrons on to the next molecule.
   b. contain heme prosthetic groups.
   c. shuttle protons to ATP synthase.
   d. transport H^+ into the mitochondrial matrix, establishing the proton-motive force.
   e. are mobile carriers that transfer electrons between the electron carrier complexes.

8. When electrons move closer to a more electronegative atom,
   a. energy is released.
   b. energy is consumed.
   c. a proton gradient is established.
   d. water is produced.
   e. ATP is synthesized.

9. When pyruvate is converted to acetyl CoA,
   a. CO_2 and ATP are released.
   b. a multienzyme complex removes a carboxyl group and attaches a coenzyme.
   c. one turn of the Krebs cycle is completed.
   d. NAD^+ is regenerated so that glycolysis can continue.
   e. phosphofructokinase is activated and glycolysis continues.

10. How many molecules of CO_2 are generated for each molecule of acetyl CoA introduced into the Krebs cycle?
    a. 1   c. 3   e. 6
    b. 2   d. 4

11. In the chemiosmotic mechanism,
    a. ATP production is linked to the proton gradient established by the electron transport chain.
    b. the difference in pH between the intermembrane space and the cytosol drives the formation of ATP.
    c. the flow of H^+ through ATP synthases from the matrix to the intermembrane space drives the phosphorylation of ADP.
    d. the energy released by the reduction and subsequent oxidation of components of the electron transport chain is transferred as a phosphate to ADP.
    e. the production of water in the matrix by the reduction of oxygen leads to a net flow of water out of a mitochondrion.

12. Which of the following reactions is incorrectly paired with its location?
    a. ATP synthesis/inner membrane of the mitochondrion
    b. fermentation/cell cytosol
    c. glycolysis/cell cytosol
    d. substrate-level phosphorylation/cytosol and matrix
    e. Krebs cycle/cristae of mitochondrion

13. When glucose is oxidized to CO_2 and water, approximately 38% of its energy is transferred to
    a. heat.
    b. ATP.
    c. acetyl CoA.
    d. water.
    e. the Krebs cycle.

14. What do muscle cells in oxygen deprivation gain from the conversion of pyruvate?
    a. ATP and lactate
    b. ATP, NAD^+, and lactate
    c. CO_2 and lactate
    d. ATP and alcohol
    e. ATP, lactate, and CO_2

15. Glucose, made from six radioactively labeled carbon atoms, is fed to yeast cells in the absence of oxygen. How many molecules of radioactive alcohol (C_2H_5OH) are formed from each molecule of glucose?
    a. 0
    b. 1
    c. 2
    d. 3
    e. 6

16. Which of the following produces the most ATP per gram?
    a. glucose, because it is the starting place for glycolysis
    b. glycogen or starch, because they are polymers of glucose
    c. fats, because they are highly reduced compounds
    d. proteins, because of the energy stored in their tertiary structure
    e. amino acids, because they can be fed directly into the Krebs cycle

17. Fats and proteins can be used as fuel in the cell because they
    a. can be converted to glucose by enzymes.
    b. can be converted to intermediates of glycolysis or the Krebs cycle.

    c. can pass through the mitochondrial membrane to enter the Krebs cycle.
    d. contain unstable phosphate bonds.
    e. contain more energy than glucose.

18. Which is *not* true of the enzyme phosphofructokinase? It is
    a. an allosteric enzyme.
    b. inhibited by citrate.
    c. the pacemaker of glycolysis and respiration.
    d. inhibited by ADP.
    e. an early enzyme in the glycolytic pathway.

19. Substrate-level phosphorylation accounts for what percentage of ATP formation when glucose is oxidized to CO_2 and water?
    a. 0%
    b. 4%
    c. 11%
    d. 15%
    e. 20%

20. Cyanide is a poison that blocks the passage of electrons along the electron transport chain. Which of the following is a metabolic effect of this poison?
    a. The pH of the intermembrane space is much lower than normal.
    b. Electrons are passed directly to oxygen, causing cells to explode.
    c. Alcohol would build up in the cells.
    d. NADH supplies would be exhausted, and ATP synthesis would cease.
    e. No proton gradient would be produced, and ATP synthesis would cease.

CHAPTER 10

PHOTOSYNTHESIS

FRAMEWORK

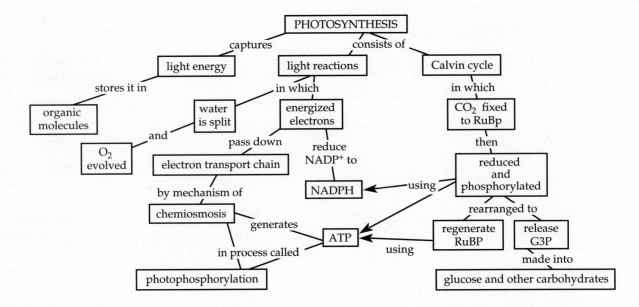

CHAPTER REVIEW

■ Plants and other autotrophs are the producers of the biosphere

In **photosynthesis**, plants convert the light energy of the sun into chemical energy stored in organic molecules. Organisms obtain the organic molecules they require for energy and carbon skeletons by **autotrophic** or **heterotrophic nutrition**. Autotrophs "feed themselves" in the sense that they make their own organic molecules from inorganic raw materials. Plants, algae, and some protists and prokaryotes are photoautotrophs. Some bacteria are chemoautotrophs, which use energy from oxidizing inorganic substances to produce organic compounds.

Heterotrophs are consumers. They may eat plants or animals or decompose organic litter, but almost all are ultimately dependent on plants for food and oxygen.

■ Chloroplasts are the sites of photosynthesis in plants

Chloroplasts, found mainly in the **mesophyll** tissue of the leaf, contain **chlorophyll**, the green pigment that absorbs the light energy that drives photosynthesis. CO_2 enters and O_2 exits the leaf through **stomata**. Veins carry water from the roots to leaves and distribute sugar to nonphotosynthetic tissue.

A chloroplast consists of a double membrane surrounding a dense fluid called the stroma and an elaborate thylakoid membrane system enclosing the thylakoid space. Thylakoid sacs may be layered to form grana. Chlorophyll is contained in the thylakoid membrane.

■ **INTERACTIVE QUESTION   10.1**

Label the indicated parts in this diagram of a chloroplast.

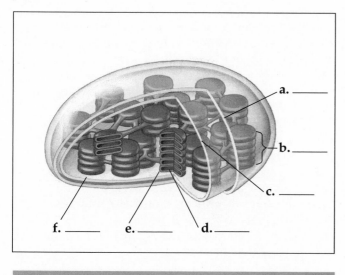

■ **Evidence that chloroplasts split water molecules enabled researchers to track atoms through photosynthesis:** *science as a process*

The process of photosynthesis is represented as follows: $6 CO_2 + 12 H_2O + light energy \longrightarrow C_6H_{12}O_6 + 6 O_2 + 6 H_2O$. If only the net consumption of water is considered, the equation for photosynthesis is the reverse of respiration. Both processes occur in plant cells.

The Splitting of Water   Using evidence from bacteria that utilize hydrogen sulfide (H_2S) for photosynthesis, van Niel hypothesized that all photosynthetic organisms need a hydrogen source and that plants split water as their hydrogen source, releasing oxygen. Scientists confirmed this hypothesis by using a heavy isotope of oxygen (^{18}O). Labeled O_2 was produced in photosynthesis only when water, rather than carbon dioxide, contained the labeled oxygen.

Photosynthesis as a Redox Process   Photosynthesis is a redox process like respiration, but differs in the direction of electron flow. The electrons increase their potential energy when they travel from water to reduce CO_2 into sugar, and light provides this energy.

■ **The light reactions and the Calvin cycle cooperate in transforming light to the chemical energy of food:** *an overview*

Solar energy is converted into chemical energy in the **light reactions**. Light energy absorbed by chlorophyll drives the transfer of electrons and hydrogen from water to the electron acceptor $NADP^+$, which is reduced to NADPH and temporarily stores energized electrons. Oxygen is released when water is split. ATP is formed during the light reactions by **photophosphorylation**.

In the **Calvin cycle**, carbon dioxide is incorporated into existing organic compounds by **carbon fixation**, and these compounds are then reduced to form carbohydrate. NADPH and ATP from the light reactions supply the reducing power and chemical energy needed for the Calvin cycle.

■ **INTERACTIVE QUESTION   10.2**

Fill in the blanks in this overview of photosynthesis in a chloroplast.

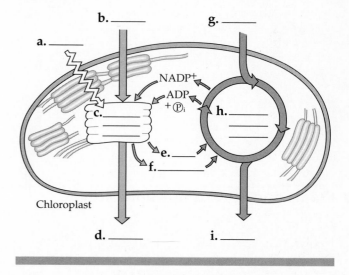

■ **The light reactions convert solar energy to the chemical energy of ATP and NADPH:** *a closer look*

The Nature of Sunlight   Electromagnetic energy, also called radiation, travels as rhythmic wave disturbances of electrical and magnetic fields. The distance between the crests of the electromagnetic waves, their **wavelength**, ranges across the **electromagnetic spectrum**. The small band of radiation from about 380 to 750 nm is called **visible light**.

Light also behaves as if it consists of discrete particles called **photons**, which have a fixed quantity of energy. The amount of energy in a photon is inversely related to its wavelength. Thus a photon of red light, with a longer wavelength, has less energy than a photon of violet light.

Photosynthetic Pigments: The Light Receptors   A **spectrophotometer** measures the ability of a pigment to absorb various wavelengths of light. The **absorption spectrum** of **chlorophyll *a***, the pigment that participates directly in the light reactions, shows that it absorbs blue and red light best.

An **action spectrum** shows the relative rates of photosynthesis under different wavelengths of light. The action spectrum for photosynthesis differs from the absorption spectrum of chlorophyll *a* because **chlorophyll *b*** and **carotenoids** absorb light of different wavelengths and transfer energy to chlorophyll *a*. Some carotenoids function in photoprotection by accepting energy from chlorophyll when it is exposed to excessive light intensity.

■ INTERACTIVE QUESTION   10.3

Label the absorption and action spectra on this graph. Why are these lines different?

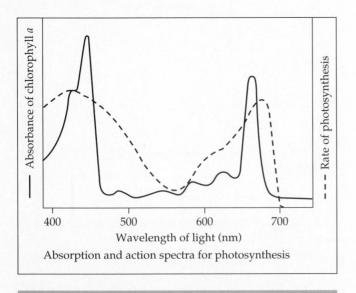

Absorption and action spectra for photosynthesis

Photoexcitation of Chlorophyll   When a pigment molecule absorbs energy from a photon, one of the molecule's electrons is elevated to an orbital where it has more potential energy. Only photons whose energy is equal to the difference between the excited state and the ground state for that molecule are absorbed.

The excited state is unstable. Energy is generally released as heat as the electron drops back to its ground-state orbital. Some pigments, including chlorophyll when it is isolated from thylakoids, may also emit photons of light, called fluorescence, as their electrons return to ground state.

Photosystems: Light-Harvesting Complexes of the Thylakoid Membrane   Chlorophyll *a* and the accessory pigments are clustered together in the thylakoid membrane in **photosystems**, consisting of an antenna complex of a few hundred pigment molecules, a **reaction-center** chlorophyll *a*, and its **primary electron acceptor**. Pigment molecules in the antenna complex absorb photons and pass the energy from molecule to molecule until it reaches the reaction-center chlorophyll *a*. In a redox reaction, a high-energy electron of the reaction-center chlorophyll is trapped by the primary electron acceptor before it can return to the ground state.

There are two types of photosystems in the thylakoid membrane. The chlorophyll molecule at the reaction center of **photosystem I** is called P700, after the wavelength of light (700 nm) it absorbs best. At the reaction center of **photosystem II** is a chlorophyll *a* molecule called P680.

■ INTERACTIVE QUESTION   10.4

What are the differences between P700, P680, and the chlorophyll *a* molecules found in the antenna complex?

Noncyclic Electron Flow   In **noncyclic electron flow**, electrons pass continuously from water to $NADP^+$. Excited electrons of P680 in photosystem II are trapped by the primary electron acceptor. Electrons are restored to oxidized P680 by an enzyme that removes electrons from water. The oxygen atom released when water is split immediately combines with another oxygen to form O_2.

The primary electron acceptor passes the photoexcited electrons to an electron transport chain consisting of plastoquinone (Pq), two cytochromes, and plastocyanin (Pc). The energy released as electrons "fall" through the electron transport chain is used to produce ATP using a chemiosmotic mechanism similar to that found in mitochondria. ATP synthesis during noncyclic electron flow is called **noncyclic photophosphorylation**.

■ INTERACTIVE QUESTION   10.5

Fill in the steps of noncyclic electron flow in the diagram below. Circle the important products that will be used to provide chemical energy and reducing power to the Calvin cycle.

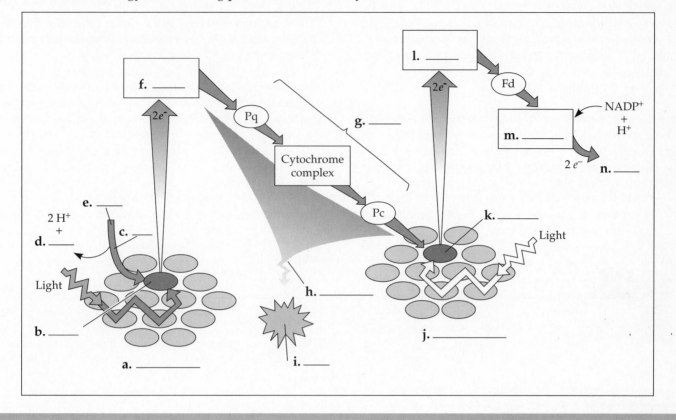

Electrons finally pass to P700 in photosystem I to replace the photoexcited electrons passed to its primary electron acceptor. This primary electron acceptor passes electrons to ferredoxin (Fd), from which the enzyme $NADP^+$ reductase transfers them to $NADP^+$.

Cyclic Electron Flow   In **cyclic electron flow**, electrons excited from P700 are passed through the electron transport chain and return to P700. Neither NADPH nor O_2 is generated. The Calvin cycle requires more ATP than NADPH, and the additional ATP may be supplied by **cyclic photophosphorylation**, perhaps in response to a buildup of NADPH.

A Comparison of Chemiosmosis in Chloroplasts and Mitochondria   Chemiosmosis in mitochondria and in chloroplasts is very similar. The key difference is that in respiration chemical energy from food is converted into ATP, whereas in chloroplasts light energy is converted to chemical energy stored in ATP.

In chloroplasts, the electron transport chain pumps

■ INTERACTIVE QUESTION   10.6

a. What path do electrons from P700 take during noncyclic electron flow?

b. What path do they take in cyclic electron flow?

c. Why is no oxygen generated by cyclic electron flow?

protons from the stroma into the thylakoid compartment. As H^+ diffuses back through ATP synthase, ATP is formed on the stroma side, where it is available for the Calvin cycle.

■ INTERACTIVE QUESTION 10.7

a. In the light, the proton gradient across the thylakoid membrane is as great as 3 pH units. On which side is the pH lowest?

b. What three factors contribute to the formation of this large difference in H^+ concentration between the thylakoid compartment and the stroma.

■ The Calvin cycle uses ATP and NADPH to convert CO_2 to sugar: *a closer look*

The Calvin cycle turns three times to fix three molecules of CO_2 into **glyceraldehyde 3-phosphate (G3P)**. The cycle can be divided into three phases:

(1) Carbon fixation: CO_2 is fixed by being added to a five-carbon sugar, ribulose bisphosphate (RuBP), in a reaction catalyzed by the enzyme RuBP carboxylase (**rubisco**). The unstable six-carbon intermediate that is

■ INTERACTIVE QUESTION 10.8

Label the three phases (**a** through **c**) and key molecules (**d** through **n**) in this diagram of the Calvin cycle.

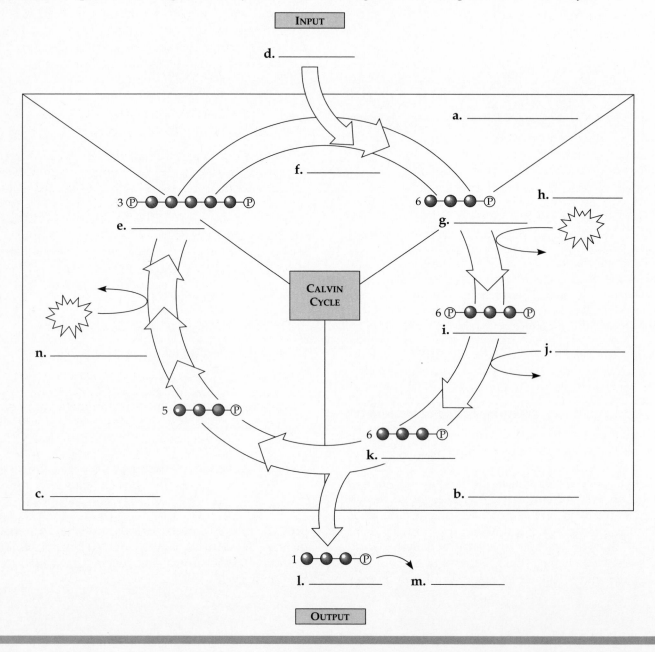

formed splits into two molecules of 3-phosphoglycerate.

(2) Reduction: Each molecule of 3-phosphoglycerate is then phosphorylated by ATP to form 1,3-bisphosphoglycerate. Two electrons from NADPH reduce this compound to create G3P. The cycle must turn three times, reducing three CO_2 molecules, to create one molecule of G3P and regenerate the three molecules of the five-carbon RuBP used in the cycle.

(3) Regeneration of starting material (RuBP): The rearrangement of five molecules of G3P into the three molecules of RuBP requires three more ATP.

Nine molecules of ATP and six of NADPH are required to synthesize one G3P.

■ Alternative mechanisms of carbon fixation have evolved in hot, arid climates

Photorespiration: An Evolutionary Relic? In most plants, CO_2 enters the Calvin cycle and the first product of carbon fixation is 3-phosphoglycerate. When these **C_3 plants** close their stomata on hot dry days to limit water loss, CO_2 concentration in the leaf air spaces falls, slowing the Calvin cycle. As more O_2 than CO_2 accumulates, rubisco adds O_2 in place of CO_2 to the Calvin cycle. The product splits and a two-carbon compound leaves the chloroplast, enters a peroxisome or mitochondrion, and is broken down to release CO_2. This seemingly wasteful process that removes fixed carbon from the Calvin cycle is called **photorespiration**.

■ INTERACTIVE QUESTION 10.9

What possible explanation is there for photorespiration, a process that can result in the loss of as much as 50% of the carbon fixed in the Calvin cycle?

C_4 Plants In C_4 plants, CO_2 is first added to a 3-carbon compound, phosphoenolpyruvate (PEP), with the aid of an enzyme (**PEP carboxylase**) that has a high affinity for CO_2. The resulting four-carbon compound formed in the **mesophyll cells** of the leaf is transported to **bundle-sheath cells** tightly packed around the veins of the leaf. The compound is broken down to release CO_2, creating concentrations high enough that rubisco will accept CO_2 and initiate the Calvin cycle.

■ INTERACTIVE QUESTION 10.10

a. Where does the Calvin cycle take place in C_4 plants?

b. Why are C_4 plants able to photosynthesize efficiently in hot, dry conditions when C_3 plants would be undergoing photorespiration?

CAM Plants Many succulent plants close their stomata during the day to prevent water loss, but open them at night to take up CO_2 and incorporate it into a variety of organic acids. These compounds are broken down to release CO_2 during daylight. Unlike the C_4 pathway, the **CAM** pathway does not structurally separate carbon fixation from the Calvin cycle; instead, the two processes are separated in time.

■ Photosynthesis is the biosphere's metabolic foundation: *a review*

About 50% of the organic material produced by photosynthesis is used as fuel for cellular respiration in the mitochondria of plant cells; the rest is used as carbon skeletons for synthesis of organic molecules (proteins, lipids, and a great deal of cellulose), stored as starch, or lost through photorespiration. About 160 billion metric tons of carbohydrate per year are produced by photosynthesis.

STRUCTURE YOUR KNOWLEDGE

1. You have already filled in the blanks in several diagrams of photosynthesis. To really understand this process, however, you should create your own representation. In a diagrammatic form, outline the key events of photosynthesis. Trace the flow of electrons through photosystem II and I, the transfer of ATP and NADPH formed by the light reactions into the Calvin cycle, and the major steps in the production of G3P. Note where these reactions occur in the chloroplast. You can compare to the sketch in the answer section. Then talk a friend through the steps in your diagram. Perhaps your study group can combine several representations into a clear, concise summary of this chapter.

2. Create a concept map to confirm your understanding of the chemiosmotic synthesis of ATP in photophosphorylation.

TEST YOUR KNOWLEDGE

MULTIPLE CHOICE: *Choose the one best answer.*

1. Which of the following is mismatched with its location?
   a. light reactions—grana
   b. electron transport chain—thylakoid membrane
   c. Calvin cycle—stroma
   d. ATP synthase—double membrane surrounding chloroplast
   e. splitting of water—thylakoid compartment

2. Photosynthesis is a redox process in which
   a. CO_2 is reduced and water is oxidized.
   b. $NADP^+$ is reduced and RuBP is oxidized.
   c. CO_2, $NADP^+$, and water are reduced.
   d. O_2 acts as an oxidizing agent and water acts as a reducing agent.
   e. G3P is reduced and the electron transport chain is oxidized.

3. Blue light has more energy than red light. Therefore, blue light
   a. has a longer wavelength than red light.
   b. has a shorter wavelength than red light.
   c. contains more photons than red light.
   d. has a broader electromagnetic spectrum than red light.
   e. is absorbed faster by chlorophyll *a*.

4. A spectrophotometer can be used to measure
   a. the absorption spectrum of a substance.
   b. the action spectrum of a substance.
   c. the amount of energy in a photon.
   d. the wavelength of visible light.
   e. the efficiency of photosynthesis.

5. Accessory pigments within chloroplasts are responsible for
   a. driving the splitting of water molecules.
   b. absorbing photons of different wavelengths of light and passing that energy to P680 or P700.
   c. extending the absorption spectrum of chlorophyll *a* so that it can absorb more light.
   d. passing electrons across the thylakoid membrane to the reaction-center molecules.
   e. anchoring chlorophyll *a* within the reaction center.

6. Below is an absorption spectrum for an unknown pigment molecule. What color would this pigment appear to you?

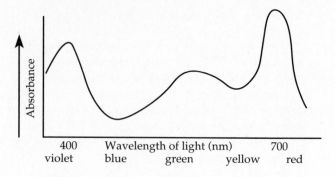

   a. violet
   b. blue
   c. green
   d. yellow
   e. red

7. Noncyclic electron flow in the chloroplast results in the production of
   a. ATP only.
   b. ATP and NADPH.
   c. ATP and G3P.
   d. RuBP.
   e. ATP, NADPH, and O_2.

8. The chlorophyll known as P680 is reduced by electrons from
   a. photosystem I.
   b. photosystem II.
   c. water.
   d. NADPH.
   e. accessory pigments.

9. CAM plants avoid photorespiration by
   a. fixing CO_2 into organic acids during the night.
   b. fixing CO_2 into four-carbon compounds in bundle-sheath cells.
   c. fixing CO_2 into four-carbon compounds in the mesophyll.
   d. using an enzyme that has a higher affinity for CO_2 than does rubisco.
   e. keeping their stomata closed during the day.

10. Electrons that flow through the two photosystems have their highest potential energy in
   a. water.
   b. P680.
   c. NADPH.
   d. the electron transport chain.
   e. photoexcited P700.

11. Chloroplasts can make carbohydrate in the dark if provided with
    a. ATP and NADPH and CO_2.
    b. an artificially induced proton gradient.
    c. organic acids or four-carbon compounds.
    d. a source of hydrogen.
    e. photons and CO_2.

12. In the chemiosmotic synthesis of ATP, H^+ diffuses through the ATP synthase
    a. from the stroma into the thylakoid compartment.
    b. from the thylakoid compartment into the stroma.
    c. from the cytoplasm into the matrix.
    d. from the cytoplasm into the stroma.
    e. from the intermembrane space into the stroma.

13. In C_4 plants, the Calvin cycle
    a. takes place at night.
    b. only occurs when the stomata are closed.
    c. takes place in the mesophyll cells.
    d. takes place in the bundle-sheath cells.
    e. uses PEP carboxylase instead of rubisco because of its greater affinity for CO_2.

14. How many "turns" of the Calvin cycle are required to produce one molecule of glucose?
    a. 1   c. 3   e. 12
    b. 2   d. 6

15. In green plants, most of the ATP for synthesis of proteins, cytoplasmic streaming, and other cellular activities comes directly from
    a. photosystem I.
    b. the Calvin cycle.
    c. oxidative phosphorylation.
    d. noncyclic photophosphorylation.
    e. cyclic photophosphorylation.

16. The six molecules of G3P formed from three turns of the Calvin cycle are converted into
    a. three molecules of glucose.
    b. three molecules of RuBP and one G3P.
    c. one molecule of glucose and four molecules of 3-phosphoglycerate.
    d. one G3P and three four-carbon intermediates.
    e. none of the above, since three molecules of G3P result from three turns of the Calvin cycle.

17. A difference between chemiosmosis in photosynthesis and respiration is that in photophosphorylation
    a. NADPH rather than NADH passes electrons to the electron transport chain.
    b. ATP synthase releases ATP into the stroma rather than into the cytosol.
    c. light provides the energy to push electrons to the top of the electron chain, rather than energy from the oxidation of foods.
    d. an H^+ concentration gradient rather than a proton-motive force drives the phosphorylation of ATP.
    e. both a and c are correct.

18-25.
Indicate if the following events occur during
    a. respiration
    b. photosynthesis
    c. both respiration and photosynthesis
    d. neither respiration nor photosynthesis

_____ 18. Chemiosmotic synthesis of ATP

_____ 19. Reduction of oxygen

_____ 20. Reduction of CO_2

_____ 21. Reduction of NAD^+

_____ 22. Oxidation of water

_____ 23. Oxidation of $NADP^+$

_____ 24. Oxidative phosphorylation

_____ 25. Electron flow along an electron transport chain

CHAPTER 11

THE REPRODUCTION OF CELLS

FRAMEWORK

This chapter details the process through which cells create new cells with identical genetic material. The following concept map summarizes cell division.

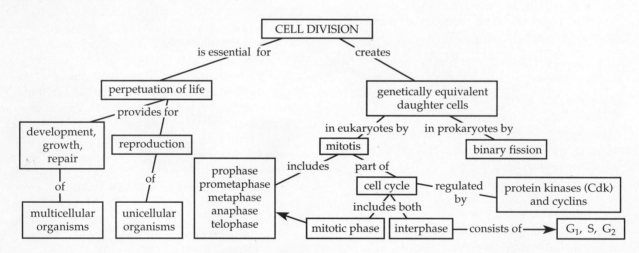

CHAPTER REVIEW

Cell division, the process through which a cell reproduces itself, is the basis for the continuity of all life.

■ Cell division functions in reproduction, growth, and repair

Cell division creates duplicate offspring in unicellular organisms and provides for growth, development, and repair in multicellular organisms. The process of re-creating a structure as intricate as a cell necessitates the exact duplication and equal division of the DNA containing the cell's genetic program.

■ **INTERACTIVE QUESTION   11.1**

A cell's complete complement of DNA is called its
_____.

■ Bacteria reproduce by binary fission

Prokaryotes, the bacteria, reproduce by a process known as **binary fission**. The single circular DNA molecule attached to the plasma membrane replicates. Growth of the membrane separates the duplicate chromosomes, the membrane pinches in between the chromosomes, and a cell wall develops—creating two identical daughter cells.

The genome of a eukaryotic cell is organized into multiple chromosomes

Copying and dividing the tens of thousands of genes in a eukaryotic cell involves a precise process. Each eukaryotic species has a characteristic number of chromosomes in each **somatic cell**; reproductive cells, or **gametes** (egg and sperm), have half that number of chromosomes.

Each chromosome is a very long DNA molecule with associated proteins that help to structure the chromosome and control the activity of the genes. This DNA–protein complex, called **chromatin**, is organized into a coiled and folded fiber. Prior to cell division, a cell copies its genome by duplicating each chromosome. Replicated chromosomes consist of two identical **sister chromatids**, joined together at a specialized region called a **centromere**. The two sister chromatids separate during **mitosis** (the division of the nucleus), and then the cytoplasm divides during **cytokinesis**, producing two separate, genetically equivalent daughter cells.

A type of cell division called **meiosis** produces daughter cells that have half the number of chromosomes of the parent cell. With the fertilization of egg and sperm, which were formed by meiosis, the chromosome number is restored in the somatic cells of the new offspring.

■ **INTERACTIVE QUESTION   11.2**

a. How many chromosomes do you have in your somatic cells?

b. How many chromosomes in your gametes?

Mitosis alternates with interphase in the cell cycle: *an overview*

The **cell cycle** includes the **mitotic (M) phase**, in which the chromosomes and cytoplasm are divided, and **interphase**, during which most of the cell's growth, metabolic activities, and chromosome replication occur. Interphase, usually lasting 90% of the cell cycle, includes the G_1 **phase**, the **S phase**, and the G_2 **phase**. Although mitosis is a dynamic continuum of chromosome separation followed by cytoplasm division, it is conventionally described in five subphases: **prophase, prometaphase, metaphase, anaphase,** and **telophase**.

■ **INTERACTIVE QUESTION   11.3**

a. How are the three subphases of interphase alike?

b. What key event happens during the S phase?

The mitotic spindle distributes chromosomes to daughter cells: *a closer look*

The **mitotic spindle** consists of fibers made of microtubules and associated proteins. The assembly of the spindle begins in the centrosome, or microtubule-organizing center. A pair of centrioles is centered in the centrosome of animal cells but is not required for normal spindle operation. Radial arrays of microtubules, called asters, extend from the centrosomes in animal cells.

The centrosome duplicates during interphase. The two centrosomes move from near the nucleus toward the poles of the cell during prophase and prometaphase, as spindle microtubules elongate from them. During prophase, the nucleoli disappear and the chromatin fibers coil and fold into visible chromosomes, consisting of sister chromatids joined at the centromere. During prometaphase, some of the spindle microtubules called **kinetochore microtubules** attach to each chromatid's **kinetochore**, a protein structure located at the centromere region. **Nonkinetochore microtubules** extend out from each centrosome and overlap at the midline. Alternate tugging on the chromosome by opposite kinetochore microtubules moves the chromosome to the midline of the cell. At metaphase, the chromosomes are aligned at the metaphase plate.

The centromeres of each chromosome separate in anaphase, and the sister chromatids (now considered chromosomes) move toward the poles. The mechanism of this movement may be motor proteins that "walk" a chromosome along the kinetochore microtubules as they shorten by depolymerizing at their kinetochore end. The extension of the spindle poles away from each other as an animal cell elongates is probably due to the sliding of nonkinetochore microtubules past each other, also using motor proteins.

During telophase, equivalent sets of chromosomes gather at the two poles of the cell. Nuclear envelopes form from fragments of the parent nuclear envelope and contributions from the endomembrane system.

■ INTERACTIVE QUESTION   11.4

The following diagrams depict interphase and the five subphases of mitosis in an animal cell. Assuming that this cell has four chromosomes, sketch the chromosomes as they would appear in each subphase. Identify the stages and label the indicated structures.

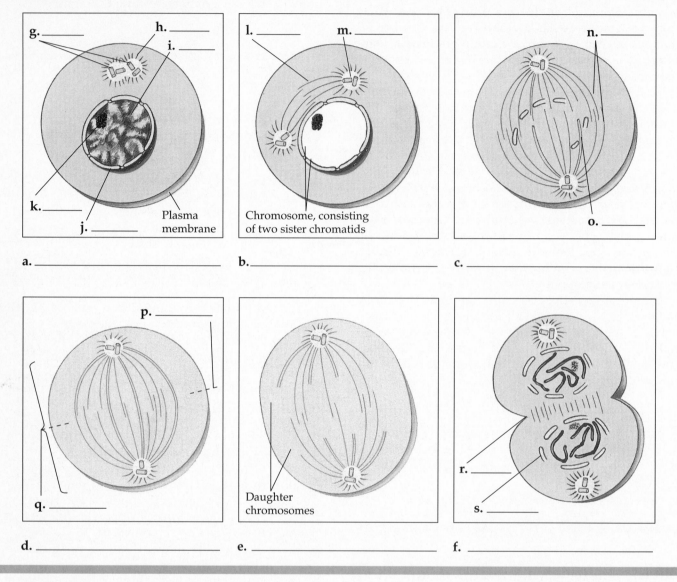

g._____   h._____   i._____
k._____   j._____   Plasma membrane
a._____

l._____   m._____
Chromosome, consisting of two sister chromatids
b._____

n._____   o._____
c._____

p._____   q._____
d._____

Daughter chromosomes
e._____

r._____   s._____
f._____

■ Cytokinesis divides the cytoplasm: *a closer look*

Cleavage is the process that separates the two daughter cells in animals. A **cleavage furrow** on the cell surface forms, as a ring of microfilaments, made of the protein actin, begins to contract on the cytoplasmic side of the membrane. The cleavage furrow deepens until the dividing cell is pinched in two.

In plant cells, a **cell plate** forms from the fusion of membrane vesicles derived from the Golgi apparatus.

The two sides of the membrane thus formed join with the plasma membrane, creating two daughter cells each with its own continuous plasma membrane. A new cell wall develops between these membranes.

■ External and internal cues control cell division

Normal growth, development, and maintenance depend on proper control of the timing and rate of cell division. Control mechanisms have been studied through the use of cell culture. Cells isolated from plants or animals are grown in glass containers supplied with a nutrient solution called a growth medium. Certain nutrients and regulatory substances called **growth factors** have been found to be essential for normal cell division. Mammalian fibroblast cells have receptors on their plasma membranes for platelet-derived growth factor (PDGF). This growth factor stimulates cell division when released from blood platelets at the site of an injury.

Density-dependent inhibition of cell division is related to diminishing supplies of essential nutrients or growth factors or to loss of anchorage to a substratum.

An important control period in cell division seems to occur during the G_1 phase of the cell cycle. Internal or external cues determine whether the cell will switch into a nondividing state (called the **G_0 phase**) or continue past a so-called **restriction point** into the S, G_2, and M phases. Cell size, determined by the cytoplasm/genome ratio, is an important cue to passing the restriction point.

■ Cyclical changes in regulatory proteins function as a mitotic clock

Protein kinases are enzymes that activate or inactivate other proteins through phosphorylation with ATP. The activity of protein kinases called **cyclin-dependent kinases (Cdks)** is controlled by the changing concentration of other regulatory proteins called **cyclins**.

A cyclin–Cdk complex called **MPF**, or maturation promoting factor, acts directly on target proteins and phosphorylates other protein kinases, leading to a cascade of activation steps during mitosis. It also activates an enzyme that destroys cyclin and thus switches itself off at the end of the cell cycle. Newly synthesized cyclin during interphase complexes with the Cdk to initiate the next cell cycle.

Control of the cell cycle occurs during G_1 when nutrients, growth factors, and cell density help determine whether the cell will pass the restriction point or become a nondividing G_0 cell. A cyclin–Cdk complex moves a cell through the restriction point when the cell reaches a threshold size. MPF appears to be the master switch that moves a cell into prophase, and protein kinases activated by cyclins sequence the steps of mitosis.

■ INTERACTIVE QUESTION   11.5

a. What is MPF?

b. Describe the concentrations of Cdks and cyclin throughout the cell cycle.

■ Cancer cells escape from the controls on cell division

Cancer cells escape from the body's normal control mechanisms. When grown in cell culture, cancer cells do not exhibit density-dependent inhibition and may continue to divide indefinitely instead of stopping after the typical 20 to 50 divisions of normal mammalian cells.

When a normal cell is transformed or converted to a cancer cell, the body's immune system usually destroys it. If it proliferates to form a **tumor**, a mass of cancer cells develops within a tissue. **Benign tumors** remain at their original site and can be removed by surgery. **Malignant tumors** cause cancer as they invade and disrupt the functions of one or more organs. Malignant tumor cells may have abnormal metabolism and unusual numbers of chromosomes. They lose their attachments to other cells and **metastasize**, entering the blood and lymph systems and spreading to surrounding tissues. Radiation and chemicals are used to treat tumors that metastasize. Much remains to be learned about the cell division processes of both normal and cancerous cells.

STRUCTURE YOUR KNOWLEDGE

1. Describe the life of one chromosome as it proceeds through an entire cell cycle starting with interphase and ending with telophase of mitosis.

2. Draw a sketch of a mitotic spindle. Idenity and list the functions of the components.

3. In this photomicrograph of cells in an onion root tip, identify the cell cycle phases for the indicated cells.

a. _____

b. _____

c. _____

d. _____

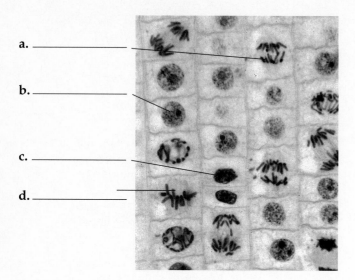

TEST YOUR KNOWLEDGE

FILL IN THE BLANK: *Identify the appropriate phase of the cell cycle.*

_____ **1.** most cells that will no longer divide are in this phase

_____ **2.** sister chromatids separate and chromosomes move apart

_____ **3.** mitotic spindle begins to form

_____ **4.** cell plate forming or cleavage furrow pinching cells apart

_____ **5.** chromosomes replicate

_____ **6.** chromosomes line up at equatorial plane

_____ **7.** phase after DNA replication

_____ **8.** chromosomes become visible

_____ **9.** kinetochore–microtubule interactions move chromosomes to midline

_____ **10.** restriction point occurs in this phase

MULTIPLE CHOICE: *Choose the one best answer.*

1. One of the major differences in the cell division of prokaryotic cells compared to eukaryotic cells is that
   a. cytokinesis does not occur in prokaryotic cells.
   b. genes are not replicated on chromosomes in prokaryotic cells.
   c. the duplicated chromosomes are attached to the nuclear membrane in prokaryotic cells and are separated from each other as the membrane grows.
   d. the chromosomes do not separate along a mitotic spindle in prokaryotic cells.
   e. the chromosome number is reduced by half in eukaryotic cells but not prokaryotic cells.

2. A plant cell has 12 chromosomes at the end of mitosis. How many chromosomes would it have in the G_2 phase of its next cell cycle?
   a. 6
   b. 9
   c. 12
   d. 24
   e. It depends on whether it is undergoing mitosis or meiosis.

3. The longest part of the cell cycle is
   a. prophase.
   b. G_1 phase.
   c. G_2 phase.
   d. mitosis.
   e. interphase.

4. In animal cells, cytokinesis involves
   a. the separation of sister chromatids.
   b. the contraction of the contractile ring of microfilaments.
   c. depolymerization of kinetochore microtubules.
   d. a protein kinase that phosphorylates other enzymes.
   e. sliding of nonkinetochore microtubules past each other.

5. Humans have 46 chromosomes. That number of chromosomes will be found in
   a. cells in anaphase.
   b. the egg and sperm cells.
   c. the somatic cells.
   d. all the cells of the body.
   e. only cells in interphase.

6. Sister chromatids
   a. have one-half the amount of genetic material as does the original chromosome.
   b. start to move along kinetochore microtubules toward opposite poles during telophase.
   c. each have their own kinetochore.
   d. are formed during prophase.

e. slide past each other along nonkinetochore microtubules.

7. Which of the following would *not* be exhibited by cancer cells?
   a. changing levels of MPF concentration
   b. passage through the restriction point
   c. density-dependent inhibition
   d. metastasis
   e. response to growth factors

8. Which of the following is *not* true of a cell plate?
   a. It forms at the site of the metaphase plate.
   b. It results from the fusion of microtubules.
   c. It fuses with the plasma membrane.
   d. A cell wall is laid down between its membranes.
   e. It forms during telophase in plant cells.

9. A cell that passes the restriction point will most likely
   a. undergo chromosome duplication.
   b. have just completed cytokinesis.
   c. continue to divide only if it is a cancer cell.
   d. show a drop in MPF concentration.
   e. move into the G_0 phase.

10. The rhythmic changes in cyclin concentration in a cell cycle are due to
    a. its increased production once the restriction point is passed.
    b. the cascade of increased production once its enzyme is phosphorylated by MPF.
    c. its degradation by an enzyme phosphorylated by MPF.
    d. the changing ratio of cytoplasm to genome.
    e. the binding of the growth factor PDGF.

UNIT **III**

THE GENE

CHAPTER **12**

MEIOSIS AND SEXUAL LIFE CYCLES

FRAMEWORK

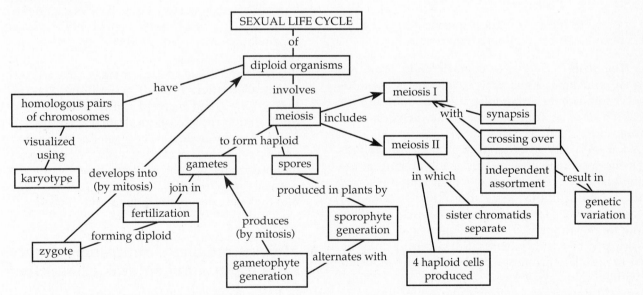

CHAPTER REVIEW

Genetics is the scientific study of the transmission of traits from parents to offspring (**heredity**) and the **variation** between and within generations.

■ Offspring acquire genes from parents by inheriting chromosomes

The inheritance of traits from parents to offspring involves the transmission of discrete units of information coded in segments of DNA known as **genes**. Specific sequences of the four nucleotides that comprise DNA translate into instructions for synthesizing proteins, such as enzymes, that then guide the development of inherited traits.

Precise copies of an organism's genes are packaged into sperm or ova. Upon fertilization, genes from both parents are contributed to the nucleus of the fertilized egg. The DNA of a eukaryotic cell is packaged into a species-specific number of chromosomes, each of which contains hundreds or thousands of genes. A gene's **locus** is its location on a chromosome.

■ Like begets like, more or less: a comparison of asexual and sexual reproduction

In **asexual reproduction**, a single parent passes all its genes on to its offspring. Occasional genetic differences may result from rare DNA changes known as mutations. A **clone** is a group of genetically identical offspring of an asexually reproducing individual.

In **sexual reproduction**, an individual receives a unique combination of genes inherited from two parents.

■ Fertilization and meiosis alternate in sexual life cycles: *an overview*

An organism's life cycle is the sequence of stages from conception to production of its own offspring.

The Human Life Cycle   In **somatic cells**, there are two chromosomes of each type, known as **homologous chromosomes** or homologs. A gene controlling a particular trait is found at the same locus on each chromosome of a homologous pair.

A **karyotype** is an ordered display of an individual's chromosomes. It is usually made by cutting the individual chromosomes out of a photograph taken of white blood cells that were stimulated to undergo mitosis, arrested in metaphase, and stained. The chromosomes are arranged into homologous pairs by size, shape, and banding pattern. Karyotyping may be used to identify chromosomal abnormalities associated with inherited disorders.

Sex chromosomes determine the sex of a person: females have two homologous *X* chromosomes; males have nonhomologous *X* and *Y* chromosomes. Chromosomes other than the sex chromosomes are called **autosomes**.

Somatic cells contain a set of chromosomes from each parent. **Gametes**, ova and sperm, contain a single set of chromosomes. The **haploid** number (*n*) of chromosomes for humans is 23. Fusion of sperm and ovum into a **zygote**, called **fertilization** or **syngamy**,

combines the paternal and maternal sets of chromosomes. The **diploid** zygote then divides by mitosis to produce the somatic cells of the body, all of which contain the diploid number (2*n*) of chromosomes.

Meiosis is a special type of cell division that halves the chromosome number in gametes to compensate for the doubling that occurs at fertilization. An alternation between diploid and haploid conditions, involving the processes of fertilization and meiosis, is characteristic of the life cycle of all sexually reproducing organisms.

■ INTERACTIVE QUESTION   12.1

If 2*n* = 14, how many chromosomes will be present in somatic cells? **a.**_____ How many chromosomes will be found in gametes? **b.** _____ If *n* = 14, how many chromosomes will be found in diploid somatic cells? **c.** _____ How many sets of homologous chromosomes will be found in gametes? **d.** _____

The Variety of Sexual Life Cycles   In most animals, meiosis occurs in the formation of gametes, which are the only haploid cells in the life cycle. In many fungi and some protists, the only diploid stage is the zygote. Meiosis occurs right after the gametes fuse, and mitosis then creates a multicellular haploid organism. Gametes are produced by mitosis in these organisms.

Plants and some species of algae have a type of life cycle called **alternation of generations** that includes both diploid and haploid multicellular stages. The multicellular diploid **sporophyte** stage produces haploid **spores** by meiosis. These spores undergo mitosis and develop into a multicellular haploid plant, the **gametophyte**, which produces gametes by mitosis. Gametes fuse to form a diploid zygote that develops into the next sporophyte generation.

■ Meiosis reduces chromosome number from diploid to haploid: *a closer look*

In meiosis, chromosome replication is followed by two consecutive cell divisions: **meiosis I** and **meiosis II**, producing four haploid daughter cells.

In interphase I, each chromosome replicates, producing two genetically identical chromatids attached at their centromeres. During prophase I, homologous chromosomes synapse, producing a tetrad, and chromatids may criss-cross, forming chiasmata that help hold the homologs together as they move toward the metaphase plate.

Complete these three diagrams of sexual life cycles with the names of processes or cells. Shade in the portions of each life cycle that is diploid.

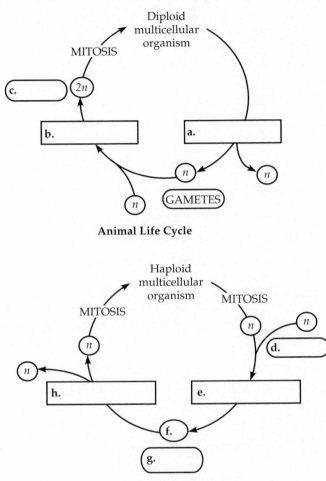

Animal Life Cycle

Some Fungi and Algae Life Cycle

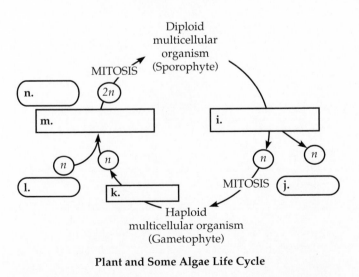

Plant and Some Algae Life Cycle

In metaphase I, the synapsed chromosomes line up on the metaphase plate with their centromeres attached to spindle fibers from opposite poles. The homologous pairs separate in anaphase I, with one homolog moving toward each pole. Sister chromatids remain attached. In telophase I, a haploid set of chromosomes, each composed of two chromatids, reaches each pole. Cytokinesis usually occurs during telophase I and may be followed by a period when nuclei re-form, or the two haploid cells may proceed directly into meiosis II. There is no replication of genetic material prior to this second division.

Meiosis II looks like a regular mitotic division, in which chromosomes line up individually on the metaphase plate, and sister chromatids separate and move apart in anaphase II. At the end of telophase II, there are four haploid daughter cells. (See Interactive Question 12.3, p. 78.)

A Comparison of Mitosis and Meiosis   Mitosis produces daughter cells that are genetically identical to the parent cell. Meiosis produces haploid cells that differ genetically from their parent cell and from each other.

The three unique events that produce this result occur during meiosis I: In prophase I, when homologous chromosomes **synapse**, genetic material is rearranged by crossing over, which is visible during this stage by the appearance of X-shaped regions called **chiasmata**. In metaphase I, chromosomes line up in pairs, not as individuals, on the metaphase plate. During anaphase I, the homologous pairs separate and one homolog goes to each pole.

■ Sexual life cycles produce genetic variation among offspring

Independent Assortment of Chromosomes   The first meiotic division results in an assortment of maternal and paternal chromosomes in the two daughter cells. Each homologous pair lines up independently at the metaphase plate; the orientation of the maternal and paternal chromosomes is random. The number of possible combinations of chromosomes is 2^n, where n is the haploid number. (See Interactive Question 12.4, p. 78.)

Crossing Over   In prophase I, segments of nonsister chromatids synapse and **cross over**, resulting in new genetic combinations of maternal and paternal genes on the same chromosome. The exact mechanism of synapsis is not known, but it involves a synaptonemal complex, a protein structure that helps to bring homologous chromosomes into precise alignment.

■ INTERACTIVE QUESTION   12.3

Label the following stages of meiosis.

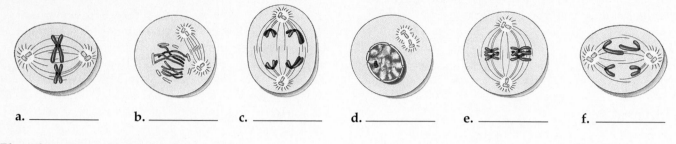

a. _____   b. _____   c. _____   d. _____   e. _____   f. _____

Place these stages in the proper sequence.

_____ _____ _____ _____ _____ _____

■ INTERACTIVE QUESTION   12.4

How many assortments of maternal and paternal chromosomes are possible in human gametes?

Random Fertilization The random nature of fertilization adds to the genetic variability established in meiosis. Human parents can produce a zygote with any of about 64 trillion (8 million × 8 million) diploid combinations.

■ Evolutionary adaptation depends on a population's genetic variation

In Darwin's theory of evolution by natural selection, genetic variations present in a population result in adaptation as the individuals best suited to an environment produce the most offspring. The process of sexual reproduction and mutation are the sources of this variation.

STRUCTURE YOUR KNOWLEDGE

1. In the table below, describe the key events of the following stages of meiosis:

Interphase I	a.
Prophase I	b.
Metaphase I	c.
Anaphase I	d.
Metaphase II	e.
Anaphase II	f.

2. Create a concept map to help you organize your understanding of the similarities and differences between mitosis and meiosis. Compare your map with those of some classmates to see different ways of organizing this material.

TEST YOUR KNOWLEDGE

MULTIPLE CHOICE: *Choose the one best answer.*

1. The restoration of the diploid chromosome number after halving in meiosis is due to
   a. synapsis.
   b. fertilization.
   c. mitosis.
   d. DNA replication.
   e. chiasmata.

2. What is a karyotype?
   a. a genotype of an individual
   b. a unique combination of chromosomes found in a gamete
   c. a blood type determination of an individual
   d. a pictorial display of an individual's chromosomes
   e. a species-specific diploid number of chromosomes

3. What are autosomes?
   a. sex chromosomes
   b. chromosomes that occur singly
   c. chromosomal abnormalities that result in genetic defects
   d. chromosomes found in mitochondria and chloroplasts
   e. none of the above

4. A synaptonemal complex would be found during
   a. prophase I of meiosis.
   b. fertilization or syngamy of gametes.
   c. metaphase II of meiosis.
   d. prophase of mitosis.
   e. anaphase I of meiosis.

5. During the first meiotic division (meiosis I),
   a. Homologous chromosomes separate.
   b. The chromosome number becomes haploid.
   c. Crossing over between nonsister chromatids occurs.
   d. Paternal and maternal chromosomes assort randomly.

   e. All of the above occur.

6. A cell with a diploid number of 6 could produce gametes with how many different combinations of maternal and paternal chromosomes?
   a. 6   c. 12   e. 128
   b. 8   d. 64

7. The DNA content of a cell is measured in the G_2 phase. After meiosis I, the DNA content of one of the two cells produced would be
   a. equal to that of the G_2 cell.
   b. twice that of the G_2 cell.
   c. one-half that of the G_2 cell.
   d. one-fourth that of the G_2 cell.
   e. impossible to estimate due to independent assortment of homologous chromosomes.

8. In many fungi and some protists,
   a. The zygote is the only haploid stage.
   b. Gametes are formed by meiosis.
   c. The multicellular organism is haploid.
   d. The gametophyte generation produces gametes by mitosis.
   e. Reproduction is exclusively asexual.

9. In the alternation of generations found in plants,
   a. the sporophyte generation produces spores by mitosis.
   b. the gametophyte generation produces gametes by mitosis.
   c. the zygote will develop into a sporophyte generation by meiosis.
   d. spores develop into the haploid sporophyte generation.
   e. the gametophyte generation produces spores by meiosis.

10. Which of the following is *not* a source of genetic variation in sexually reproducing organisms?
   a. crossing over
   b. replication of DNA during S phase before meiosis I
   c. independent assortment of chromosomes
   d. random fertilization of gametes
   e. mutation

11. Meiosis II is similar to mitosis because
   a. sister chromatids separate.
   b. homologous chromosomes separate.
   c. DNA replication precedes the division.
   d. they both take the same amount of time.
   e. haploid cells are produced.

12. Homologous chromosomes
    a. have identical genes.
    b. have genes for the same traits at the same loci.
    c. are found in gametes.
    d. separate in meiosis II.
    e. are all of the above.

13. Asexual reproduction of a diploid organism would
    a. be impossible.
    b. involve meiosis.
    c. produce identical offspring.
    d. show variation among sibling offspring.
    e. involve asexual spores produced by mitosis.

14. In a sexually reproducing species with a diploid number of 8, how many different combinations of paternal and maternal chromosomes would be possible in the offspring?
    a. 8   c. 64   e. 512
    b. 16   d. 256

15. The calculation of offspring in question 14 includes only variation resulting from
    a. crossing over.
    b. random fertilization.
    c. independent assortment of chromosomes.
    d. a, b, and c.
    e. only b and c.

MENDEL AND THE GENE IDEA

FRAMEWORK

Through his work with garden peas in the 1860s, Mendel developed the fundamental principles of inheritance and the laws of segregation and independent assortment. This chapter describes the basic monohybrid and dihybrid crosses that Mendel performed to establish that inheritance involves particulate genetic factors (genes) that segregate independently in the formation of gametes and recombine to form offspring. The laws of probability can be applied to predict the outcome of genetic crosses.

The phenotypic expression of genotype may be affected by such factors as incomplete dominance, multiple alleles, pleiotropy, epistasis, and polygenic inheritance, as well as the environment. Genetic screening and counseling for recessively inherited genetic disorders use Mendelian principles to analyze human pedigrees.

CHAPTER REVIEW

What is the genetic basis for the variation evident in the individuals of a population and how is this variation transmitted from parents to offspring? The genetic material of two parents is neither blended nor irreversibly mixed in offspring but is passed on to future generations as discrete heritable units, or genes.

■ Mendel brought an experimental and quantitative approach to genetics: *science as a process*

Gregor Mendel worked with garden peas, a good choice of study organism because they are available in many varieties, their fertilization is easily controlled, and the characteristics of their offspring can be quantified.

Mendel studied seven **characters**, or heritable features, that occurred in alternative forms called **traits**. He used **true-breeding** varieties of pea plants, which means that self-fertilizing parents always produce offspring with the parental form of the character. To follow the transmission of these well-defined traits, Mendel performed **hybridizations** in which he cross-pollinated plants that were true breeding for alternate forms of the same characteristic (a **monohybrid** cross), and then allowed the next generation to self-pollinate. The true-breeding parental plants are the **P generation** (parental); the results of the first cross are the F_1 generation (first filial); and the next generation, from the self-cross of the F_1, is known as the F_2 **generation**.

■ According to the law of segregation, the two alleles for a character are packaged into separate gametes

Mendel found that the F_1 offspring did not show a blending of the parental characteristics. Instead, only one of the parental forms of the trait was found in the hybrid offspring. In the F_2 generation, however, the missing parental form reappeared in the ratio of 3:1—three offspring with the dominant trait shown by the F_1 to one offspring with the reappearing recessive trait.

Mendel's explanation for this phenomenon contains four parts: (1) alternate forms of genes, now called **alleles**, account for variations in characters; (2) an organism has two alleles for each inherited trait, one received from each parent; (3) when two different alleles occur together, one of them, called the **dominant allele**, may be completely expressed while the other, the **recessive allele**, has no observable effect on the organism's appearance; and (4) allele pairs separate (segregate) during the formation of gametes, so an egg or sperm carries only one allele for each inherited trait. This explanation is consistent with the behavior of chromosomes during meiosis. Mendel's **law of segregation** refers to this separation of alleles in the formation of gametes.

Mendel's law of segregation explains the 3:1 ratio observed in the F_2 plants. During the segregation of allele pairs in the formation of F_1 gametes, half the gametes receive one allele while the other half receive the alternate allele. Random fertilization of gametes results in one-fourth of the plants having two dominant alleles, half having one dominant and one recessive allele, and one-fourth receiving two recessive alleles, producing a ratio of plants showing the dominant to recessive trait of 3:1. A Punnett square can be used to predict the results of simple genetic crosses. Dominant alleles are often symbolized by a capital letter, recessive alleles by a small letter.

■ **INTERACTIVE QUESTION   13.1**

Fill in this diagram of a monohybrid cross of round- and wrinkled-seeded pea plants. The round allele (*R*) is dominant and the wrinkled allele (*r*) is recessive.

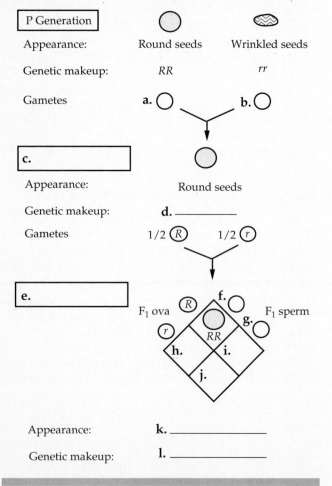

Some Useful Genetic Vocabulary   An organism that has a pair of identical alleles for a character is said to be **homozygous**. If the organism has two different alleles, it is said to be **heterozygous** for that character. Homozygotes are true breeding; heterozygotes are not, since they produce gametes with one or the other allele that can combine to produce homozygous dominant, heterozygous, and homozygous recessive offspring. **Phenotype** is an organism's expressed traits; **genotype** is its genetic makeup.

The Testcross   In a **testcross**, an organism expressing the dominant phenotype is bred with a recessive homozygote to determine the genotype of this phenotypically dominant organism.

■ **INTERACTIVE QUESTION   13.2**

A tall pea plant is crossed with a recessive dwarf pea plant. What will the phenotypic and genotypic ratio of offspring be

 a. if the tall plant was *TT*?

 b. if the tall plant was *Tt*?

■ **According to the law of independent assortment, each pair of alleles segregates into gametes independently**

Mendel used **dihybrid crosses**, involving parents differing in two traits, to determine whether the two traits were transmitted independently of each other or together in the same combination in which they were inherited from the parent plants.

If the two pairs of alleles segregate independently, then gametes from an F_1 hybrid generation (*AaBb*) should contain four combinations of alleles in equal quantities (*AB*, *Ab*, *aB*, *ab*). The random fertilization of these four classes of gametes should result in 16 (4 × 4) gamete combinations that produce four phenotypic categories in a ratio of 9:3:3:1 (nine offspring showing both dominant traits, three showing one dominant trait and one recessive, three showing the opposite dominant and recessive traits, and one showing both recessive traits). Mendel obtained these ratios when he categorized the F_2 progeny of dihybrid crosses, providing evidence for his **law of independent assortment**. This principle states that genes assort independently from each other when alleles segregate in the formation of gametes.

■ **INTERACTIVE QUESTION 13.3**

A true-breeding tall, purple-flowered pea plant (*TTPP*) is crossed with a true-breeding dwarf, white-flowered plant (*ttpp*).

   a. What is the phenotype of the F_1 generation?

_____

   b. What is the genotype of the F_1 generation?

_____

   c. What four types of gametes are formed by F_1 plants? _____   _____

_____   _____

   d. Fill in the following Punnett square to show the offspring of the F_2 generation.

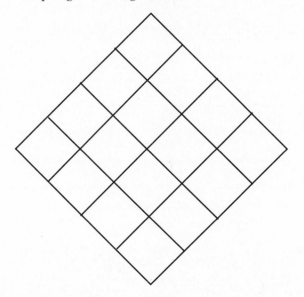

   e. List the phenotypes and ratios found in the F_2 generation.

_____   _____

_____   _____

   f. Shade each phenotype a different color in the Punnett square so you can see the 9:3:3:1 ratio of offspring.

■ **Mendelian inheritance reflects rules of probability**

General rules of probability apply to the laws of segregation and independent assortment. The probability of an event occurring is the number of times that event could occur over all the possible events; for example, the probability of drawing an ace from a deck of cards is 4/52. The outcome of independent events is not affected by previous trials.

The Rule of Multiplication   The rule of multiplication states that the probability that a certain combination of independent events will occur simultaneously is equal to the product of the separate probabilities of the independent events. The probability of a particular genotype being formed by fertilization is equal to the product of the probabilities of forming each type of gamete needed to produce that genotype.

■ **INTERACTIVE QUESTION 13.4**

Apply the rule of multiplication to a dihybrid cross. How would you determine the probability of getting an F_2 offspring that is homozygous recessive for both traits?

The Rule of Addition   If the genotype can be formed in more than one way, then the rule of addition states that its probability is equal to the sum of the separate probabilities of the different ways the event can occur. For example, a heterozygote offspring can occur if the ovum contains the dominant allele and the sperm the recessive (1/4 probability) or vice versa (1/4). Therefore, the heterozygote offspring would be the predicted result from a monohybrid cross half of the time (1/4 + 1/4 = 1/2).

Using Rules of Probability to Solve Genetics Problems   Fairly complex genetics problems can be solved by applying the rules of multiplication and addition. The probability of a particular genotype arising from a cross can be determined by considering each gene involved as a separate monohybrid cross and then multiplying the probabilities of all the independent events involved in the final genotype.

■ **INTERACTIVE QUESTION 13.5**

   a. In the following cross, what is the probability of obtaining offspring that show all three dominant traits, *A_B_C_* (_ indicates that the second allele can be either dominant or recessive without affecting the phenotype determined by the first dominant allele)?

   AaBbcc × AabbCC

   probability of offspring that are *A_B_C_* = _____

   b. What is the probability that the offspring of this *AaBbcc × AabbCC* cross will show at least two dominant traits? _____

Mendel discovered the particulate behavior of genes: *a review*

Mendel's laws of segregation and independent assortment explain the inheritance of discrete genes from generation to generation according to basic rules of probability. The larger the sample size, the more closely the results will conform to statistical predictions.

The relationship between genotype and phenotype is rarely simple

Incomplete Dominance   Some characters show **incomplete dominance**, in that the F_1 hybrids have a phenotype intermediate between that of the parents.

What Is a Dominant Allele?   In the case of an allele showing **complete dominance**, the phenotype of the heterozygote is indistinguishable from that of the homozygous dominant. At the other end of the continuum, alleles that exhibit **codominance** are both separately expressed in the phenotype, as in the case of the *MN* blood cell type. Intermediate phenotypes are characteristic of alleles showing incomplete dominance. Even when alleles exhibit dominance or incomplete dominance at the phenotypic level, both alleles may be expressed at the microscopic or molecular level.

Even though one allele may be considered "dominant," the two alleles in a cell do not interact at all. Also, whether or not an allele is dominant or recessive has no relation to how common it is in a population.

Multiple Alleles   Most genes exist in more than two allelic forms. The gene that determines human blood groups has three alleles. The alleles I^A and I^B are codominant with each other; each codes for a carbohydrate found on the surface of red blood cells. The allele *i* does not code for a blood protein and is recessive to I^A and I^B. Blood type is critical in transfusions because, if the carbohydrate coating on the donor's blood cells is foreign to the recipient, the recipient's antibodies will cause agglutination, or clumping.

■ INTERACTIVE QUESTION   13.6

List the possible genotypes for the following blood groups.

  a. A _____   **b.** B _____

  c. AB _____   **d.** O _____

Pleiotropy   **Pleiotropy** is the characteristic of a single gene having multiple phenotypic effects in an individual. Many hereditary diseases with complex sets of symptoms are caused by a single allele.

Epistasis   In **epistasis**, a gene at one locus may affect the expression of another gene. F_2 ratios that differ from the typical 9:3:3:1 often indicate epistasis.

■ INTERACTIVE QUESTION   13.7

A dominant allele *M* is necessary for the production of the black pigment melanin; mm individuals are white. A dominant allele *B* results in the deposition of a lot of pigment in an animal's hair, producing a black color. The genotype *bb* produces brown hair. Two black animals heterozygous for both genes are bred. Fill in the following table for the offspring of this $MmBb \times MmBb$ cross.

Phenotype	Genotype	Ratio
Black		
	M_bb	

Polygenic Inheritance   **Quantitative characters**, such as height or skin color, vary along a continuum in a population. Such variation is usually due to **polygenic inheritance**, in which two or more genes have an additive effect on a single phenotypic trait. Each dominant allele contributes one "unit" to the expressed trait. A polygenic trait may result in a normal distribution (forming a bell-shaped curve) of the trait within a population.

■ INTERACTIVE QUESTION   13.8

The height of spike weed is a result of polygenic inheritance involving three genes, each of which can contribute an additional 5 cm to the base height of the plant. The base height of the weed is 10 cm, and the tallest plant can reach 40 cm.

  a. If a tall plant (*AABBCC*) is crossed with a base-height plant (*aabbcc*), what is the height of the F_1 plants?

  b. How many phenotypic classes will there be in the F_2?

Nature Versus Nurture: Environmental Impact on Phenotype The phenotype of an individual is the result of complex interactions between its genotype and the environment. Genotypes have a phenotypic range called a **norm of reaction** within which the environment influences phenotypic expression. Polygenic characters are often **multifactorial**, meaning that a combination of genetic and environmental factors influences phenotype.

Integrating a Mendelian View of Heredity and Variation The phenotypic expression of most genes is influenced by other genes and by the environment. The theory of particulate inheritance and Mendel's principles of segregation and independent assortment, however, still form the basis for modern genetics.

■ Pedigree analysis reveals Mendelian patterns in human inheritance

A family **pedigree** is a family tree with the history of a particular trait shown across the generations. By convention, circles represent females, squares are used for males, and solid symbols indicate individuals that express the phenotype in question. Parents are joined by a horizontal line, and offspring are listed below parents from left to right in order of birth. The genotype of individuals in the pedigree can often be deduced by following the patterns of inheritance. (See Interactive Question 13.9.)

■ Many human disorders follow Mendelian patterns of inheritance

Recessively Inherited Disorders Only homozygous recessive individuals express the phenotype for the several thousand genetic disorders that are inherited as simple recessive traits. Carriers of the disorder are heterozygotes who are phenotypically normal but may transmit the recessive allele to their offspring.

Cystic fibrosis is the most common lethal genetic disease in the United States; it is found more frequently in Caucasians than in other races. This recessive allele results in defective chloride channels in cell membranes. Accumulating extracellular chloride leads to the build-up of thickened mucus in various organs and a predisposition to pneumonia and other infections.

Tay-Sachs disease is a lethal disorder in which the brain cells are unable to metabolize a type of lipid that then accumulates and damages the brain, resulting in an early death.

■ **INTERACTIVE QUESTION 13.9**

Consider this pedigree for the trait albinism (lack of skin pigmentation) in three generations of a family. (Solid symbols represent individuals who are albinos.) From your knowledge of Mendelian inheritance, answer the following questions.

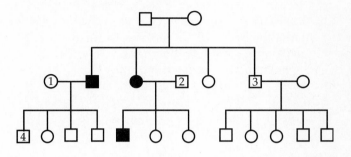

a. Is this trait caused by a dominant or recessive allele? How can you tell?

b. Determine the probable genotypes of the parents in the first generation. (Let *AA* and *Aa* represent normal pigmentation and *aa* be the albino genotype.) Genotype of father_____; of mother_____.

c. Determine the probable genotypes of the mates of the albino offspring in the second generation and the grandson 4 in the third generation. Genotypes: mate 1 _____ mate 2 _____ grandson 4 _____.

d. Can you determine the genotype of son 3 in the second generation? Why or why not?

■ **INTERACTIVE QUESTION 13.10**

a. What is the probability that a mating between two carriers will produce an offspring with a recessively inherited disorder?

b. What is the probability that a phenotypically normal child produced by a mating of two heterozygotes will be a carrier?

Sickle-cell disease is the most common inherited disease among African-Americans. Due to a single amino acid substitution in the hemoglobin protein, low oxygen concentration in the blood causes red blood cells to deform into a sickle shape, triggering blood clotting and other pleiotropic effects. Heterozygous individuals are said to have sickle cell trait but are usually healthy. The resistance to malaria that accompanies the sickle cell trait may explain why this lethal recessive allele remains in relatively high frequency in areas where malaria is common.

The likelihood of two mating individuals carrying the same rare deleterious allele increases when the individuals have common ancestors. Such consanguineous matings, between siblings or close relatives, are indicated on pedigrees by double lines.

Dominantly Inherited Disorders   A few human disorders are due to dominant genes. In **achondroplasia**, dwarfism is due to a single copy of a mutant allele. Homozygosity for this dominant allele is lethal.

Dominant lethal alleles are more rare than recessive lethals because the harmful allele cannot be masked in the heterozygote. A late-acting lethal dominant allele can be passed on if the symptoms do not develop until after reproductive age. Molecular geneticists have developed a method to detect the lethal gene for **Huntington's disease**, a degenerative disease of the nervous system that does not develop until later in life.

Multifactorial Disorders   Many diseases have genetic (usually polygenic) and environmental components. These multifactorial disorders include heart disease, diabetes, cancer, and others.

■ Technology is providing new tools for genetic testing and counseling

The risk of a genetic disorder being transmitted to offspring can sometimes be determined through genetic counseling and testing before a child is conceived or in the early stages of pregnancy. The probability of a child having a genetic defect can be determined by considering the family history of the disease.

Carrier Recognition   Determining whether parents are carriers is important for determining the risk of a genetic disorder. Biochemical tests that permit carrier recognition have been developed for a number of heritable disorders.

Fetal Testing   **Amniocentesis** is a procedure done between the fourteenth and sixteenth weeks of pregnancy by extracting a small amount of amniotic fluid from the sac surrounding the fetus. The fluid is analyzed biochemically, and fetal cells present in the fluid are cultured for several weeks and then tested and karyotyped to check for certain chromosomal defects.

Chorionic villus sampling (CVS) is a technique in which a small amount of the fetal tissue is suctioned from the placenta. These rapidly growing cells can be karyotyped immediately. This procedure can be performed at only 8 to 10 weeks of pregnancy.

Ultrasound is a simple, noninvasive procedure that can reveal major abnormalities. Fetoscopy, the insertion of a needle-thin viewing scope and light into the uterus, allows the fetus to be checked for anatomical problems.

Newborn Screening   Some genetic disorders can be detected at birth. Routine screening is now used to test for the recessively inherited disorder phenylketonuria (PKU) in newborns. If detected, dietary adjustments can lead to normal development.

STRUCTURE YOUR KNOWLEDGE

1. Relate Mendel's two laws of inheritance to the behavior of chromosomes in meiosis that you studied in Chapter 12.

2. Mendel worked with characters that exhibited two alternate forms: smooth or wrinkled, green or yellow, tall or short. In his F_1 generation, the dominant allele was always expressed, with the recessive trait reappearing in the F_2 offspring. But not all allele pairs operate by complete dominance; some show incomplete dominance or codominance. And some that show dominance on the phenotypic level are actually incompletely dominant or codominant when looked at on the microscopic or molecular level. Taking into account the mechanisms by which genotype becomes expressed as phenotype, explain how this range in dominance can occur.

▓ INTERACTIVE QUESTION   13.11

If two prospective parents both have siblings who had a recessive genetic disorder, what is the chance that they would have a child who inherits the disorder?

GENETICS PROBLEMS

*One of the best ways to learn genetics is to work problems. You can't memorize genetic knowledge; you have to practice using it. Work through the problems presented below, methodically setting down the information you are given and what you are to determine. Write down the symbols used for the alleles and genotypes, and the phenotypes resulting from those genotypes. Avoid using Punnett squares; while useful for learning basic concepts, they are much too laborious and mistake-prone. Instead, break complex crosses into their monohybrid components and rely on the rules of multiplication and addition. **Always** look at the answers you get, to see if they make logical sense. And remember that study groups are great for going over your problems and helping each other "see the light." Answers and explanations are provided at the end of the book.*

1. Pattipan squash are either white or yellow. You start growing pattipans and find out that if you want to get white pattipans then at least one of the parents must be white. Which color is dominant?

2. For the following crosses, determine the probability of obtaining the indicated genotype in an offspring.

Cross	Offspring	Probability
AAbb × AaBb	*AAbb*	**a.**
AaBB × AaBb	*aaBB*	**b.**
AABbcc × aabbCC	*AaBbCc*	**c.**
AaBbCc × AaBbcc	*aabbcc*	**d.**

3. Tall red-flowered plants are crossed with dwarf white-flowered plants. The resulting F$_1$ generation consists of all tall pink-flowered plants. Assuming that height and flower color are each determined by a single gene locus, predict the results of an F$_1$ cross of *TtRr* plants. List the phenotypes and predicted ratios for the F$_2$ generation.

4. Blood typing has often been used as evidence in paternity cases, when the blood type of the mother and child may indicate that a man alleged to be the father could not possibly have fathered the child. For the following mother and child combinations, indicate which blood groups of potential fathers would be exonerated.

Blood Group of Mother	Blood Group of Child	Man Exonerated If He Belongs to Blood Group(s)
AB	A	**a.**
O	B	**b.**
A	AB	**c.**
O	O	**d.**
B	A	**e.**

5. In rabbits, the homozygous *CC* is normal, *Cc* results in rabbits with deformed legs, and *cc* is lethal. For a gene for coat color, the genotype *BB* produces black, *Bb* brown, and *bb* a white coat. Give the phenotypic proportions of offspring from a cross of a deformed-leg, brown rabbit with a deformed-leg, white rabbit.

6. Polydactyly (extra fingers and toes) is due to a dominant gene. A father is polydactyl, the mother has the normal phenotype, and they have had one normal child. What is the genotype of the father? Of the mother? What is the probability that a second child will have the normal number of digits?

7. In dogs, black (*B*) is dominant to chestnut (*b*), and solid color (*S*) is dominant to spotted (*s*). What are the genotypes of the parents that would produce a cross with 3/8 black solid, 3/8 black spotted, 1/8 chestnut solid, and 1/8 chestnut spotted puppies? (Hint: first determine what genotypes the offspring must have before you deal with the fractions.)

8. When hairless hamsters are mated with normal-haired hamsters, about one-half the offspring are hairless and one-half are normal. When hairless hamsters are crossed with each other, the ratio of normal-haired to hairless is 1:2. How do you account for the results of the first cross? How would you explain the unusual ratio obtained in the second cross?

9. Two pigs whose tails are exactly 25 cm in length are bred over 10 years and they produce 120 piglets with the following tail lengths: 7 piglets at 15 cm, 31 at 20 cm, 44 at 25 cm, 29 at 30 cm, and 9 at 35 cm.
   a. How many pairs of genes are regulating the tail length character?
   b. What offspring phenotypes would you expect from a mating between a 15-cm and a 30-cm pig?

10. Fur color in rabbits is determined by a single gene locus for which there are four alleles. Four phenotypes are possible: black, Chinchilla (gray color caused by white hairs with black tips), Himalayan (white with black patches on extremities), and white. The black allele (C) is dominant over all other alleles, the Chinchilla allele (C^{ch}) is dominant over Himalayan (C^h), and the white allele (c) is recessive to all others.
   a. A black rabbit is crossed with a Himalayan, and the F_1 consists of a ratio of 2 black to 2 Chinchilla. Can you determine the genotypes of the parents?
   b. A second cross was done between a black rabbit and a Chinchilla. The F_1 contained a ratio of 2 black to 1 Chinchilla to 1 Himalayan. Can you determine the genotypes of the parents of this cross?

11. In Labrador retriever dogs, the dominant gene B determines black coat color and bb produces brown. A separate gene E, however, shows dominant epistasis over the B and b alleles, resulting in a "golden" coat color. The recessive e allows expression of B and b. A breeder wants to know the genotypes of her three dogs, so she breeds them and makes note of the offspring of several litters. Determine the genotypes of the three dogs.
   a. golden female (Dog 1) × golden male (Dog 2) offspring: 7 golden, 1 black, 1 brown
   b. black female (Dog 3) × golden male (Dog 2) offspring: 8 golden, 5 black, 2 brown

12. The ability to taste phenylthiocarbamide (PTC) is controlled in humans by a single dominant allele (T). A woman nontaster married a man taster and they had three children, two boy tasters and a girl nontaster. All the grandparents were tasters. Create a pedigree for this family for this trait. (Solid symbols should signify nontasters (tt).) Where possible, indicate whether tasters are TT or Tt.

TEST YOUR KNOWLEDGE

MATCHING: *Match the definition with the correct term.*

_____ 1. codominance   A. true breeding variety

_____ 2. homozygous   B. cross between two hybrids

_____ 3. heterozygous   C. cross that involves two different gene pairs

_____ 4. phenotype   D. an allele that is not expressed in the phenotype

_____ 5. polygenic trait   E. the physical characteristics of an individual

_____ 6. pleiotropy   F. genotype with two different alleles for same locus

_____ 7. epistasis   G. genotype with multiple allele for same locus

_____ 8. testcross   H. one gene influences the expression of another gene

_____ 9. dihybrid cross   I. both alleles are fully expressed in heterozygote

_____ 10. incomplete dominance   J. single gene with multiple phenotypic effects

K. heterozygote is intermediate between homozygous phenotypes

L. two or more genes with additive effect on phenotype

M. cross with homozygous recessive to determine genotype of unknown

MULTIPLE CHOICE: *Choose the one best answer.*

1. According to Mendel's law of segregation,
   a. There is a 50% probability that a gamete will get a dominant allele.
   b. Gene pairs segregate independently of other genes in gamete formation.
   c. Allele pairs separate in gamete formation.
   d. The laws of probability determine gamete formation.
   e. There is a 3:1 ratio in the F_2 generation.

2. After obtaining two heads from two tosses of a coin, the probability of tossing the coin and obtaining a head is
   a. 1/2.   c. 1/6.   e. 1/16.
   b. 1/4.   d. 1/8.

3. The probability of tossing three coins simultaneously and obtaining two heads and one tail is
   a. 1/2.   c. 1/8.   e. 3/8.
   b. 3/4.   d. 1/16.

4. A multifactorial disease
   a. can usually be traced to consanguineous matings.
   b. is caused by recessively inherited lethal genes.
   c. has both genetic and environmental causes.
   d. has a collection of symptoms traceable to an epistatic gene.
   e. is usually associated with quantitative traits.

5. The F_2 generation
   a. has a phenotypic ratio of 3:1.
   b. is the result of the self-fertilization or crossing of F_1 individuals.
   c. can be used to determine the genotype of individuals with the dominant phenotype.
   d. has a phenotypic ratio that equals its genotypic ratio.
   e. has 16 different genotypic possibilities.

6. The base height of the dingdong plant is 10 cm. Four genes contribute to the height of the plant, and each dominant allele contributes 3 cm to height. If you cross a 10-cm plant (quadruply homozygous recessive) with a 34-cm plant, how many phenotypic classes will there be in the F_2?
   a. 4   c. 8   e. 64
   b. 5   d. 9

7. A 1:1 phenotypic ratio in a testcross indicates that
   a. The alleles are dominant.
   b. One parent must have been homozygous recessive.
   c. The dominant phenotype parent was a heterozygote.
   d. The alleles segregated independently.
   e. The alleles are codominant.

8. Carriers of a genetic disorder
   a. are indicated by solid symbols on a family pedigree.
   b. are involved in consanguineous matings.
   c. will produce children with the disease.
   d. are heterozygotes for the gene that can cause the disorder.
   e. have a homozygous recessive genotype.

9. If both parents are carriers of a lethal recessive gene, the probability that their child will inherit and express the disorder is
   a. 1/8.
   b. 1/4.
   c. 1/2.
   d. $1/2 \times 1/2 \times 1/4$, or 1/16.
   e. $2/3 \times 2/3 \times 1/4$, or 1/9.

10. Two true-breeding varieties of garden peas are crossed. One parent had red, axial flowers, and the other had white, terminal flowers. All F_1 individuals had red, terminal flowers. If 100 F_2 offspring were counted, how many of them would you expect to have red, axial flowers?
    a. 6   c. 25   e. 75
    b. 19   d. 56

THE CHROMOSOMAL
BASIS OF INHERITANCE

FRAMEWORK

```
                          CHROMOSOMES
              may be          are            may have
                          location of
    sex                                                    alterations in
 chromosomes    autosomes  GENES                         number and structure
                on different    may be    on same
 can carry               unlinked      linked              such as
 sex-linked      show        do not      may      deletions,   aneuploidy   polyploidy
   traits                    show                duplications,
                                   cross         translocations,  such as     seen in
             independent          over          and inversions
             assortment                                   trisomy,      plants
                                  gives   provide        monosomy
             results in
                                                can cause
   genetic          recombination   cytological
 recombination       frequency       evidence            genetic
                         can use to                      disorders
                      map location
                        of genes
```

CHAPTER REVIEW

By 1900, three botanists had independently reproduced the plant-breeding results that Mendel had
explained 35 years previously.

■ Mendelian inheritance has its physical basis in the behavior of chromosomes during sexual life cycles

Mendel's principles, combined with cytological evidence of the processes of mitosis and meiosis gathered
in the late 1800s, led to the **chromosome theory of
inheritance** that Mendelian genes have loci on chromosomes that undergo segregation and independent
assortment in the process of gamete formation.

■ Morgan traced a gene to a specific chromosome: *science as a process*

Morgan, working with the fruit fly, *Drosophila
melanogaster*, was the first to associate a specific gene
with a specific chromosome. Fruit flies are prolific
breeders and have only four pairs of chromosomes,
which are easily distinguishable with a microscope.
The sex chromosomes occur as *XX* in females and *XY*
in male flies.

The normal phenotype for a character found most
commonly in nature is called the **wild type**, whereas
alternative traits, assumed to have arisen as mutations, are called **mutant phenotypes**.

A Note on Genetic Symbols The notation commonly
used by geneticists employs a small letter to signify a
recessive mutant allele and the small letter with a

superscripted plus sign for the wild-type allele. A capital letter is used to signify dominant mutant alleles.

Discovery of a Sex-Linked Gene Morgan discovered a mutant white-eyed male fly that he mated with a wild-type red-eyed female. The F$_1$ were all red-eyed. In the F$_2$, however, all female flies were red-eyed, whereas half of the males were red-eyed and half were white-eyed. Morgan deduced that the gene for eye color was **sex-linked**, occurring only on the X sex chromosome. Males have only one X, so their phenotype is determined by the eye-color allele they inherit from their mother. This association of a specific gene with a chromosome provided evidence for the chromosome theory of inheritance.

■ **INTERACTIVE QUESTION 14.1**

Complete this summary of Morgan's crosses with the mutant white-eyed fly by filling in the Punnett square and showing the phenotypes and genotypes of the F$_2$ generation.

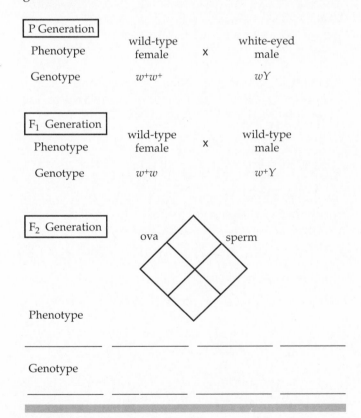

■ Linked genes tend to be inherited together because they are located on the same chromosome

Genes that are located on the same chromosome are **linked**. Morgan performed a testcross of heterozygote

wild-type flies with flies that were homozygous recessive for black bodies and vestigial wings and found that the offspring were not in the predicted 1:1:1:1 phenotypic classes. Rather, most of the offspring were the same phenotypes as the parents—either wild type (gray, normal wings) or double mutant (black, vestigial). Morgan deduced that these traits were inherited together because their genes were located on the same chromosome.

■ Independent assortment of chromosomes and crossing over cause genetic recombination

Genetic recombination results in offspring with new combinations of traits inherited from their parents.

The Recombination of Unlinked Genes: Independent Assortment of Chromosomes In a dihybrid cross between a heterozygote and a homozygous recessive, one-half of the offspring will be **parental types** and have phenotypes like one or the other parent, and one-half of the offspring, called **recombinants**, will have combinations of the two traits that are unlike the parents. This 50% frequency of recombination is observed when two genes are located on different chromosomes, and it results from the random alignment of homologous chromosomes at metaphase I and the resulting independent assortment of alleles.

■ **INTERACTIVE QUESTION 14.2**

In a testcross between a heterozygote tall, purple-flowered pea plant and a dwarf, white-flowered plant,

  a. what are the phenotypes of offspring that are parental types?

  b. what are the phenotypes of offspring that are recombinants?

The Recombination of Linked Genes: Crossing Over Linked genes do not assort independently, and one would not expect to see recombination of parental traits in the offspring. Recombination of linked genes does occur, however, due to crossing over, the reciprocal trade between non-sister chromatids of synapsed homologous chromosomes during prophase of meiosis I.

■ INTERACTIVE QUESTION   14.3

With unlinked genes, an equal number of parental and recombinant offspring are produced. With linked genes, (more/fewer) parentals than recombinants are produced. (Circle and then explain your answer.)

■ Geneticists can use recombination data to map a chromosome's genetic loci

Morgan's group was the first to work out methods for mapping genes on particular chromosomes. Sturtevant suggested that recombination frequencies reflect the relative distance between genes; genes located farther apart have a greater probability that a crossover event will occur between them. Sturtevant used recombination data to locate genes on a chromosomal map, defining one map unit (or centimorgan) as equal to a 1% recombination frequency.

The sequence of genes on a chromosome can be determined by finding the recombination frequency between different pairs of genes. Linkage cannot be determined if genes are 50 or more map units apart because they would have the maximum of 50% recombination frequency typical of unlinked genes. Distant genes on the same chromosome may be mapped by adding the recombination frequencies determined for intermediate genes.

Sturtevant and his co-workers found that the genes for the various known mutations of *Drosophila* clustered into four groups of linked genes, providing

■ INTERACTIVE QUESTION   14.4

Recombination frequency is given below for several gene pairs. Create a linkage map for these genes, showing the map unit distance between loci.

j, k 12%   k, l 6%

j, m 9%   l, m 15%

additional evidence that genes are located on chromosomes.

The frequency of crossing over may vary along the length of a chromosome, and a **linkage map** provides the sequence but not the exact location of genes on chromosomes. **Cytological mapping** pinpoints the exact locations of gene loci, often using a technique that associates a mutant phenotype with a visible chromosomal defect or other feature.

■ The chromosomal basis of sex produces unique patterns of inheritance

The Chromosomal Basis of Sex in Humans   Sex is a phenotypic character usually determined by the presence or absence of special chromosomes. Females, who are *XX*, produce ova that each contain an *X* chromosome. Males, who are *XY*, produce two kinds of sperm, each with either an *X* or a *Y* chromosome. Whether or not the gonads of an embryo develop into testes or ovaries depends on the presence or absence of the gene *Sry*, found on the *Y* chromosome. Other sex-determination systems include *X-O* (in grasshoppers and some other insects), *Z-W* (in birds and some fishes and insects), and haplo-diploid (in bees and ants).

Sex-Linked Disorders in Humans   Sex chromosomes may carry genes for traits that are not related to sex. Males inherit their *X*-linked alleles (coding for what

■ INTERACTIVE QUESTION   14.5

Two normal color-sighted individuals produce the following children and grandchildren. Fill in the probable genotype of the indicated individuals in this pedigree. *Squares are males, circles are females, and solid symbols represent color blindness. Choose an appropriate notation for the genotypes.*

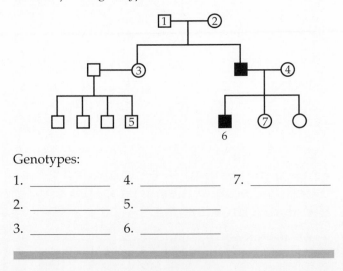

Genotypes:

1. _____   4. _____   7. _____

2. _____   5. _____

3. _____   6. _____

are usually called sex-linked traits) only from their mothers; daughters inherit sex-linked alleles from both parents.

Recessive sex-linked traits are seen more often in males, since they are hemizygous for sex-linked genes. **Duchenne muscular dystrophy** is a sex-linked disorder resulting from the lack of a key muscle protein. **Hemophilia** is a sex-linked recessive trait characterized by excessive bleeding due to the absence of a blood-clotting protein.

X-Inactivation in Females   Only one of the X chromosomes is fully active in most mammalian female somatic cells. The other X chromosome is inactive and contracted into a **Barr body** located inside the nuclear membrane. Lyon demonstrated that the selection of which X chromosome is inactivated is a random event occurring independently in embryonic cells. A female who is heterozygous for a sex-linked trait is a mosaic and will express one allele in approximately half her cells and the alternate in the other cells.

As a result of X-inactivation, both males and females have an equal dosage of X-linked genes. A gene called *XIST* is active on the X chromosome that forms the Barr body. Its RNA product may trigger DNA methylation and X-inactivation.

■ Alterations of chromosome number or structure cause some genetic disorders

Alterations of Chromosome Number: Aneuploidy and Polyploidy   **Nondisjunction** occurs when a pair of homologous chromosomes does not separate properly in meiosis I or sister chromatids do not separate in meiosis II. As a result, a gamete receives either two or no copies of that chromosome. A zygote formed with one of these aberrant gametes has a chromosomal alteration known as **aneuploidy**, a nontypical number of chromosomes. The zygote will be either **trisomic** for that chromosome (chromosome number is $2n + 1$) or **monosomic** ($2n - 1$). Aneuploid organisms usually have a set of symptoms caused by the abnormal dosage of genes. A mitotic nondisjunction early in embryonic development is likely to be harmful.

Polyploidy is a chromosomal alteration in which an organism has more than two complete chromosomal sets, as in triploid ($3n$) or tetraploid ($4n$) organisms. Polyploidy is common in the plant kingdom and has played an important role in the evolution of plants.

■ INTERACTIVE QUESTION   14.6

a. What is the difference between a trisomic and a triploid organism?

b. Which of these is likely to show the most deleterious effects of its chromosomal imbalance?

Alterations of Chromosome Structure   Chromosome breakage can result in chromosome fragments that are lost, called **deletion**; that join to the homologous chromosome, called **duplication**; that join a nonhomologous chromosome, called **translocation**; or that rejoin the original chromosome in the reverse orientation, called **inversion**. Errors in crossing over can result in a deletion and duplication in non-sister chromatids, caused by nonequal exchange of chromatids.

A homozygous deletion is usually lethal. Duplications and translocations also are typically harmful. Even though all the genes are present in proper quantities in inversions and translocations, the phenotype may be altered due to the influence of neighboring genes on the expression of the relocated genes.

■ INTERACTIVE QUESTION   14.7

Two nonhomologous chromosomes have gene orders, respectively, of A-B-C-D-E-F-G-H-I-J and M-N-O-P-Q-R-S-T. What types of chromosome alterations would have occurred if daughter cells were found to have a gene sequence on the first chromosome of A-B-C-O-P-Q-G-J-I-H?

Human Disorders Due to Chromosomal Alterations   The frequency of aneuploid zygotes may be fairly high in humans, but development is usually so disrupted that the embryos spontaneously abort. Some genetic disorders, expressed as syndromes of characteristic traits, are the result of aneuploidy. Fetal testing can detect such disorders before birth.

Down syndrome, caused by trisomy of chromosome 21, results in characteristic facial features, short stature, heart defects, and mental retardation. The incidence of Down syndrome increases for older mothers, perhaps because older women are less likely to spontaneously abort trisomic embryos.

XXY males exhibit Klinefelter syndrome, a condition in which the individual has abnormally small testes, is sterile, may have feminine body contours, and is usually of normal intelligence.

Males with an extra *Y* chromosome may be somewhat taller than average males, but they do not exhibit any well-defined syndrome. Trisomy *X* results in females who are healthy and distinguishable only by karyotype. Monosomy *X* individuals (*XO*) exhibit Turner syndrome and are phenotypically female, sterile individuals with short stature and usually normal intelligence.

Structural alterations of chromosomes, such as deletions or translocations, may be associated with specific human disorders. The *cri du chat* syndrome is caused by a deletion in chromosome 5; chronic myelogenous leukemia is a cancer that is associated with a reciprocal chromosomal translocation; and some individuals with Down syndrome have an extra part of a third chromosome 21 attached to another chromosome.

▒ INTERACTIVE QUESTION   14.8

What is an explanation for the observation that most sex-chromosome aneuploidies have less deleterious effects than do autosomal aneuploidies?

■ **The phenotypic effects of some genes depend on whether they were inherited from the mother or the father**

Genomic Imprinting   Some traits, including some genetic disorders, seem to depend on which parent supplied the alleles for the trait. A child with a deletion of a segment of chromosome 15 will exhibit Prader-Willi syndrome if the abnormal chromosome came from the father, or Angelman syndrome if the chromosome came from the mother. These phenomena provide evidence for *genomic imprinting*, a process in which certain genes are imprinted differently, per-

haps by the addition of methyl groups, depending on the individual's sex. These genes then produce different effects in the offspring. In the formation of gametes, chromosomes are re-imprinted according to the sex of the individual.

Fragile X and Triplet Repeats   Genomic imprinting may help to explain **fragile X syndrome**, a mental retardation more prevalent in males than females and seemingly linked to whether the fragile X chromosome was inherited from the mother. The repetitious CGG sequence near the tip of an X chromosome may be elongated from generation to generation, eventually creating a fragile X chromosome. This region is imprinted in the female by addition of methyl groups, which may prevent neighboring genes from functioning normally in the offspring.

Such **triplet repeats** appear in many places in the human genome, but progressive addition to such repeats may lead to disorders such as fragile X syndrome and Huntington's disease.

■ **Extranuclear genes exhibit a non-Mendelian pattern of inheritance**

Exceptions to Mendelian inheritance are found in the case of extranuclear genes located in cytoplasmic organelles, such as mitochondria and plant plastids, which are transmitted to offspring in the cytoplasm of the ovum.

STRUCTURE YOUR KNOWLEDGE

1. Mendel's law of independent assortment applies only to genes that are on different chromosomes. However, two of the genes Mendel studied were actually located on the same chromosome. Explain why genes located more than 50 map units apart behave as though they are not linked. How can one determine whether these genes are linked and what the relative distance is between them?

2. You have found a new mutant phenotype in fruit flies that you suspect is recessive and sex-linked. What is the single, best cross you could make to confirm your predictions?

3. Various human disorders or syndromes are related to chromosomal abnormalities. What explanation can you give for the adverse phenotypic effects associated with these chromosomal alterations?

GENETICS PROBLEMS

1. The following pedigree traces the inheritance of a genetic trait.

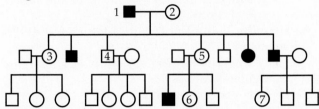

a. What type of inheritance does this trait show?

b. Give the predicted genotype for the following individuals:

1. _____   5. _____

2. _____   6. _____

3. _____   7. _____

4. _____

c. What is the probability that a child of individual #6 and a phenotypically normal male will have this trait?

2. The following recombination frequencies were found. Determine the order of these genes on the chromosome.

a, c 10%	b, c 4%	c, d 20%
a, d 30%	b, d 16%	
a, e 6%	b, e 20%	

3. In guinea pigs, black (*B*) is dominant to brown (*b*), and solid color (*S*) is dominant to spotted (*s*). A heterozygous black, solid-colored pig is mated with a brown, spotted pig. The total offspring for several litters are black solid = 16, black spotted = 5, brown solid = 5, and brown spotted = 14. Are these genes linked or nonlinked? If they are linked, how many map units are they apart?

4. A woman is a carrier for a sex-linked lethal gene that causes spontaneous abortions. She has nine children. How many of these children do you expect to be boys?

5. A dominant sex-linked gene *B* produces white bars on black chickens, as seen in the Barred Plymouth Rock breed. A clutch of chicks has equal numbers of black and barred chicks. (Remember that sex is determined by the *Z-W* system in birds: *ZZ* are males, *ZW* are females.)

a. If only the females are found to be black, what were the genotypes of the parents?

b. If males and females are evenly represented in the black and barred chicks, what were the genotypes of the parents?

TEST YOUR KNOWLEDGE

MULTIPLE CHOICE: *Choose the one best answer.*

1. The chromosomal theory of inheritance states that
   a. Genes are located on chromosomes.
   b. Chromosomes and their associated genes undergo segregation during meiosis.
   c. Chromosomes and their associated genes undergo independent assortment in gamete formation.
   d. Mendel's laws of inheritance relate to the behavior of chromosomes in meiosis.
   e. All of the above are correct.

2. A wild type is
   a. the phenotype found most commonly in nature.
   b. the dominant allele.
   c. designated by a small letter if it is recessive or a capital letter if it is dominant.
   d. your basic party animal.
   e. a trait found on the X chromosome.

3. Sex-linked traits
   a. are carried on an autosome but expressed only in males.
   b. are coded for by genes located on a sex chromosome.
   c. are found in only one or the other sex, depending on the sex-determination system of the species.
   d. are always inherited from the mother in mammals and fruit flies.
   e. depend on whether the gene was inherited from the mother or the father.

4. Linkage and cytological maps for the same chromosome
   a. are both based on mutant phenotypes and recombination data.
   b. may have different sequences of genes.
   c. have both the same sequence of genes and intergenic distances.
   d. have the same sequence of genes but different intergenic distances.
   e. are created using chromosomal abnormalities.

5. The genetic event that results in Turner Syndrome (*XO*) is probably
   a. nondisjunction.
   b. deletion.
   c. parental imprinting.
   d. monoploidy.
   e. independent assortment.

6. A 1:1:1:1 ratio of offspring from a dihybrid test-cross indicates that
   a. The genes are linked.
   b. The dominant organism was homozygous.
   c. Crossing over has occurred.
   d. The genes are 25 map units apart.
   e. The genes are not linked.

7. Genes *A* and *B* are linked 12 map units apart. A heterozygous individual, whose parents were *AAbb* and *aaBB*, would be expected to produce gametes in the following frequencies:
   a. 44% *AB*   6% *Ab*   6% *aB*   44% *ab*
   b. 6% *AB*   44% *Ab*   44% *aB*   6% *ab*
   c. 12% *AB*   38% *Ab*   38% *aB*   12% *ab*
   d. 6% *AB*   6% *Ab*   44% *aB*   44% *ab*
   e. 38% *AB*   12% *Ab*   12% *aB*   38% *ab*

8. A female tortoise-shell cat is heterozygous for the gene that determines black or orange coat color, which is located on the *X* chromosome. A male tortoise-shell cat
   a. is hemizygous at this loci.
   b. must have had a tortoise-shell mother.
   c. must have resulted from a nondisjunction and has an extra Barr body in his cells.
   d. must have three alleles for coat color, one from his father and two from his mother.
   e. would be hermaphroditic.

9. A son inherits color blindness from his
   a. mother.
   b. father.
   c. mother only if she is color-blind.
   d. father only if he is color-blind.
   e. mother only if she is not color-blind.

10. Genomic imprinting
    a. explains cases in which the gender of the parent from whom an allele is inherited affects the expression of that allele.
    b. is greatest in females because of the larger maternal contribution of cytoplasm.
    c. is more likely to occur in offspring of older mothers.
    d. may explain the transmission of Duchenne muscular dystrophy.
    e. involves both a and b.

11. A cross of a wild-type red-eyed female *Drosophila* with a violet-eyed male produces all red-eyed offspring. If the gene is sex-linked, what should the reciprocal cross (violet-eyed female × red-eyed male) produce? (Assume that the red allele is dominant to the violet allele.)
    a. all violet-eyed flies
    b. 3 red-eyed flies to 1 violet-eyed
    c. a 1:1 ratio of red and violet eyes in both males and females
    d. red-eyed females and violet-eyed males
    e. all red-eyed flies

12. Which of the following chromosomal alterations does not alter genic balance but may alter phenotype because of differences in gene expression?
    a. deletion
    b. inversion
    c. duplication
    d. nondisjunction
    e. genomic imprinting

CHAPTER 15

THE MOLECULAR BASIS
OF INHERITANCE

FRAMEWORK

This chapter outlines the key evidence that was gathered to establish DNA as the molecular basis of inheritance. Watson and Crick's double helix, with its rungs of specifically paired nitrogenous bases and twisting side ropes of phosphate and sugar groups, provided the three-dimensional model that explained DNA's ability to encode a great variety of information and produce exact copies of itself through semiconservative replication. The replication of DNA is an extremely fast and accurate process involving many enzymes and proteins.

CHAPTER REVIEW

Deoxyribonucleic acid, DNA, is the genetic material, the substance of genes, the basis of heredity. Nucleic acids' unique ability to direct their own replication allows for the precise copying and transmission of DNA to all the cells in the body and from one generation to the next. DNA encodes the blueprints that direct and control the biochemical, anatomical, physiological, and behavioral traits of organisms.

■ **The search for the genetic material led to DNA:** *science as a process*

The role of DNA in heredity was first established through work with microorganisms—bacteria and viruses. By the 1940s, chromosomes were known to carry hereditary information and to consist of proteins and DNA. Most scientists believed that the proteins carried the genetic program because of their specificity and heterogeneity.

Evidence That DNA Can Transform Bacteria The work of Griffith in 1928 provided the first evidence that the genetic material was some sort of specific molecule. Griffith worked with two strains of *Streptococcus pneumoniae*—a smooth strain (S) that synthesized a polysaccharide capsule and a rough strain (R) that did not form such a coat. Only live S cells caused pneumonia when injected in mice. However, when Griffith injected a mixture of heat-killed S cells and live R cells, the mice died. Moreover, live S cells could be isolated from the blood of these mice. These bacteria had incorporated external genetic material in a process now called **transformation**.

Avery worked for a decade to identify the transforming agent by purifying chemicals from heat-killed S cells. In 1944, Avery, McCarty, and MacLeod announced that DNA was the molecule that transformed the bacteria.

Evidence That Viral DNA Can Program Cells Viruses consist of little more than DNA, or sometimes RNA, contained in a protein coat. They reproduce by infecting a cell and commandeering that cell's metabolic machinery. Bacteriophages, or **phages**, are viruses that infect bacteria. In 1952, Hershey and Chase showed that DNA was the genetic material of a phage known as T2 which infects the bacterium *Escherichia coli (E. coli)*.

Hershey and Chase devised an experiment using radioactive isotopes to determine whether the phage's DNA or protein was transferred to the bacteria. One batch of T2 was grown with radioactive sulfur that became incorporated into protein; another was grown with radioactive phosphorus that labeled the DNA.

Separate samples of *E. coli* were infected with the labeled T2 cells, then blended and centrifuged to isolate the bacterial cells from the lighter viral particles. In the samples with the labeled proteins, radioactivity was found in the supernatant, indicating that the phage protein did not enter the cells. In the samples with the labeled DNA, most of the radioactivity was found in the bacterial cell fraction. And when these *E. coli* cells were returned to culture, they released phages containing radioactive phosphorus. Hershey and Chase concluded that viral DNA is injected into the host cell and serves as the hereditary material.

Additional Evidence That DNA Is the Genetic Material of Cells   Circumstantial evidence that DNA is the genetic material came from the observation that a eukaryotic cell doubles its DNA content prior to mitosis and that diploid cells have twice as much DNA as haploid gametes of the same organism.

Chargaff, in 1947, reported that the ratio of nitrogenous bases in the DNA from various organisms was species specific. Chargaff also determined that the number of adenines and thymines were approximately equal, and the number of guanines and cytosines were also equal in the DNA from all the organisms he studied. The A=T and G=C properties of DNA became known as Chargaff's Rules.

■ **INTERACTIVE QUESTION   15.1**

Take a minute to review the key findings of the scientists who helped to establish DNA as the genetic material. Complete the table.

Investigator	Experimental Organisms and Conclusions
Griffith	a.
Avery	b.
Hershey & Chase	c.
Chargaff	d.

■ **Watson and Crick discovered the double helix by building models to conform to X-ray data:** *science as a process*

By the early 1950s, the arrangement of covalent bonds in a nucleic acid polymer was established, and the race was on to determine the three-dimensional structure of DNA.

Crick was studying protein structure using a technique called X-ray crystallography. An X-ray beam passed through a crystal can expose photographic film to produce a pattern of spots that a crystallographer interprets into information about three-dimensional atomic structure. Watson saw an X-ray photo produced by Franklin that indicated the basic shape of DNA was a helix. He and Crick deduced that the helix had a width of 2 nm, suggesting that the helix consisted of two strands, thus the term **double helix**.

Watson and Crick constructed wire models to build a double helix that would conform to the X-ray measurements and the known chemistry of DNA. They finally arrived at a model that paired the nitrogenous bases on the inside of the helix with the sugar–phosphate chains on the outside. The helix makes one full turn every 3.4 nm; thus ten layers of nucleotide pairs, stacked 0.34 nm apart, are present in each turn of the helix.

To produce the molecule's uniform 2-nm width, a purine base must pair with a pyrimidine. The molecular arrangements of the side groups of the bases permit two hydrogen bonds to form between adenine and thymine and three hydrogen bonds between guanine and cytosine. This complementary pairing explains Chargaff's rules.

In 1953, Watson and Crick published a paper in *Nature* reporting the double helix as the molecular model for DNA. (See Interactive Question 15.2, p. 99.)

■ **During DNA replication, base pairing enables existing DNA strands to serve as templates for new complementary strands**

Watson and Crick noted that the base-pairing rule of DNA sets up a mechanism for its replication. Each side of the double helix is an exact complement to the other. When the two sides of a DNA molecule separate in the replication process, each strand serves as a template for rebuilding a double-stranded molecule.

The **semiconservative model** of gene replication predicts that the two daughter DNA molecules each have one old strand from the parent DNA and one newly formed strand. In contrast, a conservative model predicts that the parent strand remains intact and the duplicated molecule is totally new, whereas a dispersive model predicts that all four strands of the two DNA molecules are a mixture of parent and new DNA.

■ INTERACTIVE QUESTION 15.2

Review the structure of DNA by labeling the following diagrams.

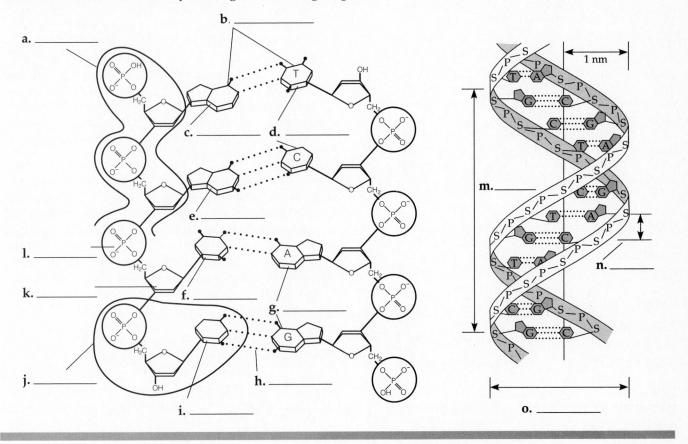

Meselson and Stahl tested these models by growing *E. coli* in a medium with ^{15}N, a heavy isotope that the bacteria incorporated into their nitrogenous bases. Cells with labeled DNA were transferred to a medium with the lighter isotope, ^{14}N. After one generation of bacterial growth, the DNA extracted from the culture was all of intermediate density. A second replication produced both light and hybrid DNA. These results confirmed the semiconservative model of DNA replication.

■ A team of enzymes and other proteins functions in DNA replication

Getting Started: Origins of Replication   Replication begins at special sites, called **origins of replication**, where proteins that initiate replication bind to a specific sequence of nucleotides and separate the two strands to form a replication "bubble." Replication proceeds in both directions in the Y-shaped **replication forks**.

Elongating a New DNA Strand   Enzymes called **DNA polymerases** connect nucleotides to the grow-

■ INTERACTIVE QUESTION 15.3

Using different colors for light and heavy strands of DNA, sketch the results of the replication cycles of heavy DNA when *E. coli* were moved to ^{14}N medium for two generations. Show the predicted density bands in the centrifuge tubes.

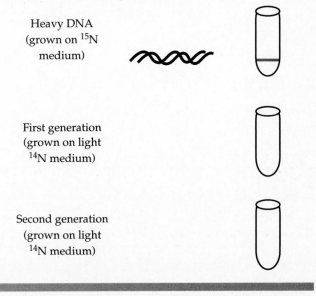

Heavy DNA (grown on ^{15}N medium)

First generation (grown on light ^{14}N medium)

Second generation (grown on light ^{14}N medium)

ing end of the new DNA strand. A nucleoside triphosphate lines up with its complementary base on the template strand; the hydrolysis of its two tail phosphate groups provides the energy for polymerization.

The Problem of Antiparallel DNA Strands　The two strands of a DNA molecule are antiparallel; their sugar-phosphate backbones run in opposite directions. The deoxyribose sugar of each nucleotide is connected to its own phosphate group at its 5' carbon and connects to the phosphate group of the adjacent nucleotide by its 3' carbon. Thus a strand of DNA has polarity, with a 5' end at the final nucleotide's phosphate group, and a 3' end where a hydroxyl group is attached to the 3' carbon of the end nucleotide.

■ INTERACTIVE QUESTION　15.4

Look back to question 15.2 and label the 5' and 3' ends of both strands of the DNA molecule.

DNA polymerase adds nucleotides to the 3' end of a growing strand; DNA is replicated in a 5'→3' direction. Because the DNA strands run in an antiparallel direction, the simultaneous synthesis of both strands presents a problem. The **leading strand** is the new 5'→3' strand being formed as DNA polymerase moves along the template as the replication fork progresses. The **lagging strand** is created as a series of short segments, called **Okazaki fragments**, that are formed in the 5'→3' direction away from the replication fork. An enzyme called **DNA ligase** joins the fragments. In both cases, DNA polymerase moves along the template in a 3'→5' direction.

Priming DNA Synthesis　DNA polymerase cannot initiate synthesis of a DNA strand; it can only add nucleotides to an existing chain. An enzyme called **primase** pairs about 10 RNA nucleotides to the DNA strand to form the **primer** of RNA needed to start the chain. Each fragment on the lagging strand requires a primer. A continuous strand of DNA is produced after an enzyme replaces the RNA primer with DNA nucleotides and ligase joins the fragments.

■ INTERACTIVE QUESTION　15.5

In this diagram showing the replication of DNA, label the following items: leading and lagging strands, Okazaki fragment, DNA polymerase, DNA ligase, helicase, primase, single-strand binding proteins, RNA primer, replication fork, and 5' and 3' ends of parental DNA.

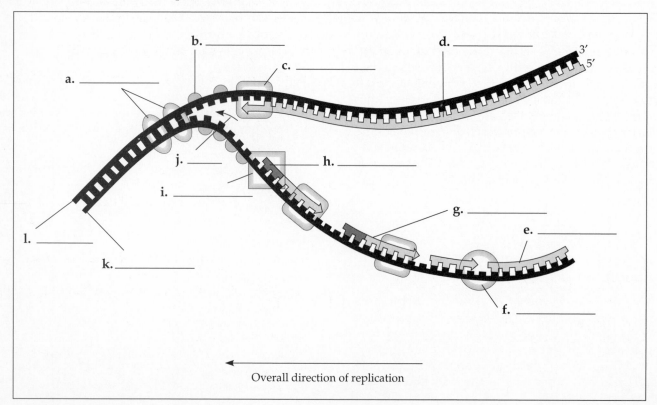

Other Proteins Assisting DNA Replication Many other proteins are involved in DNA replication. Enzymes called **helicases** unwind the helix and separate the parent strands, and **single-strand binding proteins** keep the separated strands apart.

■ Enzymes proofread DNA during its replication and repair damage to existing DNA

Initial pairing errors in nucleotide placement may occur as often as 1 per 10,000 bases. The amazing accuracy of DNA replication is achieved by proofreading and correcting pairing errors. In bacteria, DNA polymerase checks each new nucleotide against its template and backs up and replaces incorrect nucleotides. In eukaryotic cells, several proteins in addition to polymerase are involved in **mismatch repair**.

DNA molecules may be altered by reactive chemicals, radioactive emissions, X-rays, and ultraviolet (UV) light. These changes, or mutations, may be corrected by many types of DNA repair enzymes. In **excision repair**, the damaged strand is cut out by a repair enzyme and the gap is correctly filled through the action of DNA polymerase and DNA ligase. In skin cells, excision repair frequently corrects thymine dimers caused by ultraviolet rays of sunlight.

STRUCTURE YOUR KNOWLEDGE

1. Summarize the evidence and techniques Watson and Crick used to deduce the double helix structure of DNA.

2. Review your understanding of DNA replication by describing the key enzymes and proteins (in the order of their functioning) that direct replication.

TEST YOUR KNOWLEDGE

MULTIPLE CHOICE: *Choose the one best answer.*

1. One of the reasons most scientists believed proteins were the carriers of genetic information was that
   a. Proteins were more heat stable than nucleic acids.
   b. The protein content of duplicating cells always doubled prior to division.
   c. Proteins were much more complex molecules than nucleic acids.
   d. Early experimental evidence pointed to proteins as the hereditary material.
   e. Proteins were found in DNA.

2. Transformation involves
   a. the transfer of genetic material, often from one bacterial strain to another.
   b. the creation of a strand of RNA from a DNA molecule.
   c. the infection of bacterial cells by phage.
   d. the type of semiconservative replication shown by DNA.
   e. the replication of DNA along the lagging strand.

3. The DNA of an organism has thymine as 20% of its bases. What percentage of its bases would be guanine?
   a. 20%   c. 40%   e. 80%
   b. 30%   d. 60%

4. In his work with pneumonia-causing bacteria and mice, Griffith found that
   a. DNA was the transforming agent.
   b. The R and S strains mated.
   c. Heat-killed S cells could cause pneumonia when mixed with heat-killed R cells.
   d. Some heat-stable chemical was transferred to R cells to transform them into S cells.
   e. A T2 phage transformed R cells to S cells.

5. When T2 phages are grown with radioactive sulfur,
   a. Their DNA is tagged.
   b. Their proteins are tagged.
   c. Their DNA is found to be of medium density in a centrifuge tube.
   d. They transfer their radioactivity to *E. coli* chromosomes when they infect the bacteria.
   e. Their excision enzymes repair the damage caused by the radiation.

6. Meselson and Stahl
   a. provided evidence for the semiconservative model of DNA replication.
   b. were able to separate phage protein coats from *E. coli* by using a blender.
   c. found that DNA labeled with ^{15}N was of intermediate density.
   d. grew *E. coli* on labeled phosphorus and sulfur.
   e. found that DNA composition was species specific.

7. Watson and Crick concluded that each base could not pair with itself because
   a. There would not be room for the helix to make a full turn every 3.4 nm.
   b. The width of 2 nm would not permit two purines to pair together.
   c. The bases could not be stacked 0.34 nm apart.
   d. Identical bases could not hydrogen-bond together.
   e. They would be on antiparallel strands.

8. The joining of nucleotides in the polymerization of DNA requires energy from
   a. DNA polymerase.
   b. the hydrolysis of the terminal phosphate group of ATP.
   c. RNA nucleotides.
   d. the hydrolysis of GTP.
   e. the hydrolysis of phosphates from nucleoside triphosphates.

9. Continuous elongation of a new DNA molecule along one strand of DNA
   a. requires the action of DNA ligase as well as polymerase.
   b. occurs because DNA ligase can only elongate in the 5'→ 3' direction.
   c. makes a single Okazaki fragment.
   d. occurs on the leading strand.
   e. occurs on the lagging strand.

10. Which of the following statements about DNA polymerase is *incorrect*?
    a. It joins complementary base pairs to each other.
    b. It is able to proofread and correct for errors in its base-pairing in bacteria.
    c. It is unable to join free nucleotides unless an RNA primer is present.
    d. It only works in the 5' → 3' direction.
    e. It is found in eukaryotes and prokaryotes.

11. Thymine dimers, covalent links between adjacent thymine bases in DNA, may be induced by UV light. When they occur, they are repaired by
    a. excision enzymes.
    b. DNA polymerase.
    c. ligase.

d. primase.
e. a, b, and c are all needed.

Use the following diagram to answer questions 12 through 15.

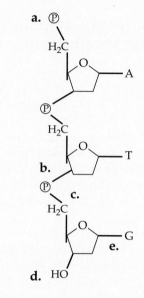

12. Which letter indicates the 5' end of this single DNA strand?
    a.   b.   c.   d.   e.

13. At which letter would the next nucleotide be added?
    a.   b.   c.   d.   e.

14. Which letter indicates a phosphodiester bond formed by DNA polymerase?
    a.   b.   c.   d.   e.

15. The base sequence of the DNA strand made from this template would be (from top to bottom)
    a. A T C.
    b. C G A.
    c. T A C.
    d. U A C.
    e. A T G.

CHAPTER 16

FROM GENE TO PROTEIN

FRAMEWORK

This chapter deals with the pathway from DNA to RNA to proteins. The triplet code instructions of DNA are transcribed to a sequence of codons in mRNA. In eukaryotes, mRNA is processed before it leaves the nucleus. Complexed with ribosomes, mRNA is translated into a sequence of amino acids in a protein as tRNAs match their anticodons to the mRNA codons.

CHAPTER REVIEW

The hereditary information of DNA is contained in specific sequences of nucleotides. These sequences are translated into proteins, which are responsible for the specific traits of an organism and thus form the link between genotype and phenotype.

▧ The study of metabolic defects provided evidence that genes specify proteins: *science as a process*

In 1909, Garrod first suggested that genes determine phenotype through the action of enzymes. Garrod reasoned that inherited diseases were caused by an inability to make certain enzymes.

How Genes Control Metabolism   In the 1930s, Beadle and Ephrussi speculated that the mutations causing various eye colors in *Drosophila* resulted from a nonfunctioning enzyme at some point in the metabolic pathway leading to pigment formation.

Beadle and Tatum, working with mutants of a bread mold, *Neurospora crassa*, demonstrated the relationship between genes and enzymes. They studied several nutritional mutants, called **auxotrophs**, that could not grow on the minimal medium that sufficed for wild-type mold. By growing an auxotroph on complete growth medium and then transferring samples to various combinations of minimal medium and an added nutrient, Beadle and Tatum were able to identify the specific metabolic defect for each auxotroph.

Three classes of *Neurospora* mutants that were unable to synthesize arginine were identified by supplementing different precursors of the pathway. Beadle and Tatum reasoned that the metabolic pathway of each class was blocked at a different enzymatic step. They formulated the one gene–one enzyme hypothesis that the function of a gene is to control production of a specific enzyme.

One Gene–One Polypeptide   Molecular biologists revised this idea to one gene–one protein, because not all proteins are enzymes. Because many proteins consist of more than one polypeptide chain, each of which is codified by its own gene, the axiom is now the **one gene–one polypeptide hypothesis**.

■ Transcription and translation are the two main steps from gene to protein: *an overview*

RNA is the link between a gene and the protein for which it codes. RNA differs from DNA in two ways: the sugar component of its nucleotides is **ribose**, rather than **deoxyribose**, and **uracil (U)** replaces thymine as one of its nitrogenous bases. All three molecules are composed of specific sequences of monomers: nucleotides in DNA and RNA and amino acids in proteins.

Transcription is the transfer of information from DNA to **messenger RNA (mRNA)**, using the "language" of nucleic acids. **Translation** transfers information from mRNA to a polypeptide, changing from the language of nucleotides to that of amino acids.

In prokaryotes, which lack a nucleus, transcription of DNA to mRNA and translation of mRNA by ribosomes into protein occur almost simultaneously. In eukaryotes, the mRNA must exit the nucleus before translation can occur. **RNA processing**, the modification of mRNA within the nucleus, occurs only in eukaryotes.

■ INTERACTIVE QUESTION   16.1

Fill in the sequence in the synthesis of proteins. Put the name of the process above each arrow.

_____ → _____ → _____

■ In the genetic code, a particular triplet of nucleotides specifies a certain amino acid: *a closer look*

A sequence of three nucleotides provides 4^3 or 64 possible unique sequences of nucleotides, more than enough to code for the 20 amino acids. The translation of nucleotides into amino acids uses a **triplet code** to specify each amino acid.

The base triplets along the **template strand** of a gene are transcribed into mRNA **codons**. The same strand of a long DNA molecule can be the template strand for one gene and the complementary strand for another. The mRNA is complementary to the DNA template since its bases follow the same base-pairing rules, with the exception that uracil substitutes for thymine in RNA. Thus the base triplet ATG is transcribed as the mRNA codon UAC. Each mRNA codon specifies one of the 20 amino acids to be sequenced in the polypeptide chain.

Cracking the Genetic Code   In the early 1960s, Nirenberg synthesized artificial mRNA by linking uracil RNA nucleotides. Adding this "poly-U" to a test tube containing all the biochemical ingredients necessary for protein synthesis, he obtained a polypeptide containing the single amino acid phenylalanine. The codons AAA, GGG, and CCC were deciphered in the same manner.

Using more elaborate techniques to decode the triplets with mixed bases, scientists had deciphered all 64 codons by the mid-1960s. Three codons function as stop signals, or termination codons. The codon AUG both codes for methionine and functions as an initiation codon, a start signal for translation; thus all polypeptides are synthesized with methionine as their first amino acid.

The code is often redundant, meaning that more than one codon may specify a single amino acid. The code is never ambiguous; no codon specifies two different amino acids.

In general, a particular nucleotide sequence on mRNA is read in only one **reading frame**, starting at a start codon and reading each triplet sequentially.

Evolutionary Significance of a Common Genetic Language   The genetic code of codons and their corresponding amino acids is almost universal. A bacterial cell can translate the genetic messages of human cells. Recently, exceptions to the constancy of the genetic code have been found in several single-celled ciliates and in the DNA of mitochondria and chloroplasts. The near universality of a common genetic language, however, lends compelling evidence to the antiquity of the code and the evolutionary connection of all living organisms.

■ INTERACTIVE QUESTION   16.2

Practice using the dictionary of the genetic code in your textbook. Determine the amino acid sequence for a polypeptide coded for by the following mRNA transcript: AUGCCUGACUUUAAGUAG

■ Transcription is the DNA-directed synthesis of RNA: *a closer look*

Transcription begins when **RNA polymerases** separate the two strands of DNA and link RNA nucleotides that base-pair along the template strand to the 3' end of the growing RNA polymer. A **transcription unit** is the entire sequence of DNA, including the initiation and termination sites, that is transcribed into one mRNA molecule. In eukaryotes, this unit represents a single gene; in prokaryotes, a transcription unit may include a few genes that code for related functions.

Bacteria have one type of RNA polymerase. Eukaryotes have three types; the one that synthesizes mRNA is called RNA polymerase II.

RNA Polymerase Binding and Initiation of Transcription   RNA polymerases bind to **promoters**, which include an initiation site for RNA synthesis and recognition sequences such as the *TATA box* in eukaryotes, a stretch of DNA rich in thymine and adenine bases which is upstream from the initiation site. **Transcription factors** are proteins that first bind to the promoter and aid RNA polymerase II in locating and binding to this region.

Elongation of the RNA Strand RNA polymerase II untwists and separates the double helix, exposing DNA nucleotides for base pairing with RNA nucleotides, and joins the nucleotides to the 3' end of the growing polymer. The growing mRNA peels away from the DNA template. Several molecules of RNA polymerase may be transcribing simultaneously from a single gene, enabling a cell to produce large quantities of a protein.

Termination of Transcription Transcription ends when RNA polymerase recognizes a termination site on the coding strand, often the sequence AATAAA, and releases the newly made RNA strand. In bacteria, this mRNA can be translated immediately. In eukaryotes, RNA is processed before it leaves the nucleus.

■ **INTERACTIVE QUESTION 16.3**

Review the key steps of transcription:

a.

b.

c.

■ Translation is the RNA-directed synthesis of a polypeptide: *a closer look*

Transfer RNA (tRNA) molecules are specific for the amino acid they carry to the ribosomes. They each have a specific base triplet, called an **anticodon**, that binds to a complementary codon on mRNA, thus assuring that amino acids are arranged in the sequence prescribed by the transcription from DNA.

The Structure and Function of Transfer RNA As with other RNAs, transfer RNA is transcribed in the nucleus of a eukaryote and moves into the cytoplasm where it can be used repeatedly. These single-stranded, short RNA molecules are arranged into a cloverleaf shape by hydrogen bonding between complementary base sequences and then folded into a three-dimensional, roughly L-shaped structure. The anticodon is at one end of the L; the 3' end is the attachment site for its amino acid.

Sixty-one codons for amino acids can be read from mRNA, but there are only about 45 different tRNA

molecules. A phenomenon known as *wobble* enables the third nucleotide of some tRNA anticodons to pair with more than one kind of base in the codon. Thus, one tRNA can recognize more than one mRNA codon, all of which code for the same amino acid carried by that tRNA. The modified base inosine (I) is in the third position on several tRNAs and can pair with U, C, or A.

■ **INTERACTIVE QUESTION 16.4**

Using some of the codons and the amino acids you identified in Interactive Question 16.2, fill in the following table.

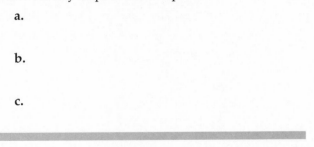

DNA Triplet	mRNA Codon	Anticodon	Amino Acid
			methionine
		GGA	
TTC			
	UAG		

Aminoacyl-tRNA Synthetases Each amino acid has a specific **aminoacyl-tRNA synthetase** that attaches it to its appropriate tRNA molecule to create an amino acid-tRNA complex. The hydrolysis of ATP drives this process.

Ribosomes Ribosomes facilitate the specific coupling of tRNA anticodons with mRNA codons during protein synthesis. They consist of a large and a small subunit, each composed of proteins and a specialized form of RNA called **ribosomal RNA (rRNA)**. Subunits are constructed in the nucleolus in eukaryotes. Prokaryotic ribosomes are smaller and differ enough in molecular composition that some antibiotics can inhibit them without affecting eukaryotic ribosomes.

A large and small subunit join to form a ribosome when they attach to an mRNA molecule. Ribosomes have a binding site for mRNA, a **P site** that holds the tRNA carrying the growing polypeptide chain, and an **A site** that binds to the tRNA carrying the next amino acid. The transfer of this amino acid to the carboxyl end of the growing polypeptide chain is catalyzed by the ribosome.

Building a Polypeptide The three stages of protein synthesis—chain initiation, chain elongation, and chain termination—all require proteins (mostly enzymes). The first two stages also require energy, which is provided by GTP (guanosine triphosphate).

The initiation stage begins as the mRNA and an initiator tRNA bind to the small subunit of the ribosome. The initiator tRNA, carrying methionine, attaches to the start codon AUG on the mRNA. With the aid of proteins called *initiation factors* and the expenditure of one GTP, the large subunit of the ribosome attaches to the small one. The initiator tRNA fits into the P site of the now-functional ribosome.

The addition of amino acids in the elongation stage involves several proteins called *elongation factors* and occurs in a three-step cycle: codon recognition, peptide bond formation, and translocation. In the codon recognition step, an elongation factor brings the correct tRNA into the A binding site of the ribosome, where the anticodon hydrogen-bonds with the mRNA codon. This step requires energy from the hydrolysis of GTP. In the second step, a component of the large ribosomal subunit catalyzes the formation of a peptide bond between the polypeptide held in the P site and the amino acid in the A site. The polypeptide is now held by the amino acid in the A site. The tRNA carrying the growing polypeptide is now translocated to the P site, a process requiring energy from the hydrolysis of another GTP molecule. The next mRNA codon moves into the A site as the mRNA ratchets through the ribosome in the 5'→3' direction.

Termination occurs when a termination codon—UAA, UAG, or UGA—reaches the A site of the ribosome and binds with a *release factor*. The ribosome now attaches a water molecule to the polypeptide chain, freeing the completed polypeptide from the tRNA in the P site. The ribosome then separates into its small and large subunits.

Polyribosomes   An mRNA may be translated simultaneously by several ribosomes in clusters called **polyribosomes**.

From Polypeptide to Functional Protein   During and following translation, a polypeptide folds spontaneously into its secondary and tertiary structure. The protein may need to undergo posttranslational modifications: amino acids may be chemically modified; one or more amino acids at the beginning of the chain may be enzymatically removed; segments of the polypeptide may be excised; the chain may be cleaved into several pieces; or several polypeptides may associate into a quaternary structure.

■ Some polypeptides have signal sequences that target them to specific destinations in the cell

All ribosomes are identical, whether they are free ribosomes that synthesize cytosolic proteins or ER-bound ribosomes that make membrane and secretory

■ INTERACTIVE QUESTION   16.5

Identify the structures in the following diagrams of protein synthesis. Name the stages (**1–4**), identify the components (**a–k**), and then briefly describe what is happening in each stage.

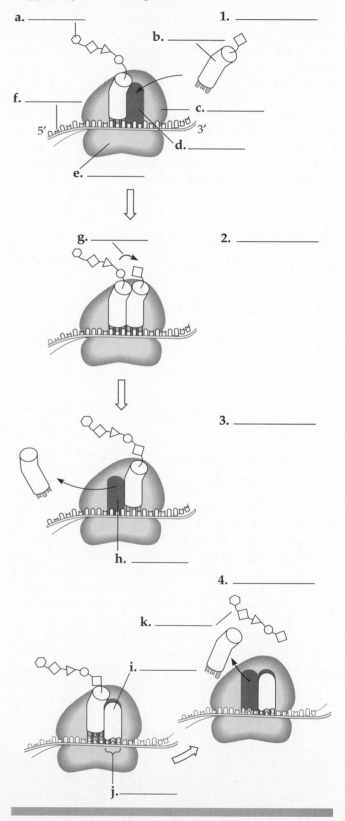

proteins. All protein synthesis begins in the cytoplasm. If a protein is destined to be a secretory protein, its polypeptide chain will begin with a **signal sequence** of amino acids that is recognized by a *signal recognition particle*, which attaches the ribosome to a receptor site on the ER membrane. As the growing polypeptide threads into the cisternal space of the ER, the signal sequence is enzymatically removed. Other signal sequences direct newly made proteins to specific sites such as mitochondria or chloroplasts.

■ **INTERACTIVE QUESTION   16.6**

What determines if a ribosome remains free in the cytosol?

■ **Comparing protein synthesis in prokaryotes and eukaryotes:** *a review*

The basic processes of transcription and translation are the same in bacteria and eukaryotes, although the RNA polymerases and ribosomes differ. In bacteria, these processes occur almost simultaneously, whereas in eukaryotes, transcription is physically separated from translation by the nuclear envelope, allowing RNA processing to occur before RNA leaves the nucleus.

■ **Eukaryotic cells modify RNA after transcription**

Alteration of the Ends of mRNA   A modified guanine nucleotide is attached to the 5' end of an mRNA, and a string of adenine nucleotides, called a **poly-A tail**, is added to the 3' end. The **5' cap** serves as a recognition site for the small ribosomal subunit, and both the 5' cap and poly-A tail protect the ends of the mRNA from hydrolytic enzymes.

Split Genes and RNA Splicing   Long segments of noncoding base sequences, known as **introns** or intervening sequences, occur within the boundaries of eukaryotic genes. The remaining coding regions are called **exons**, since they are expressed in protein synthesis. An entire transcript is made of the gene, forming an oversized RNA called *heterogeneous nuclear RNA (hnRNA)*. Enzymes remove the introns and join the exons before the RNA leaves the nucleus. This **RNA splicing** also occurs in the production of tRNA and rRNA.

Signals for RNA splicing are sets of a few nucleotides at either end of each intron. *Small nuclear ribonucleoproteins (snRNPs)*, composed of proteins and *small nuclear RNA (snRNA)*, are components of a molecular complex called a **spliceosome** that is involved

in RNA splicing. The spliceosome snips an intron out of the RNA transcript and connects the adjoining exons.

Ribozymes   Several other schemes of RNA processing have been identified for tRNA and rRNA. In the protozoan *Tetrahymena*, splicing occurs with the intron RNA acting as an enzyme and catalyzing the process itself. RNA molecules that act as enzymes are called **ribozymes**. Ribosomal RNA appears to have enzymatic functions during translation.

■ **INTERACTIVE QUESTION   16.7**

RNA is a versatile molecule with many different shapes and functions. What two commonly accepted statements have been made obsolete by recent discoveries concerning RNA? Explain why.

The Functional and Evolutionary Importance of Introns   Several hypotheses attempt to explain the functions of introns. One is that they are somehow involved in regulating gene activity or in the flow of mRNA into the cytoplasm. The splicing of the RNA transcript may allow different cells to use the same gene to synthesize different proteins. Exons may code for polypeptide **domains**, functional segments of a protein, such as binding and active sites. Introns may facilitate recombination of exons between different alleles to create novel proteins.

■ **A point mutation can affect the function of a protein**

Mutations, changes in a cell's genetic information, may involve large portions of a chromosome or affect just one nucleotide, as in a **point mutation**. If the mutation is in a gamete, it may be passed on to offspring.

Types of Point Mutations   A **base-pair substitution** replaces one nucleotide and its complementary partner with another pair of nucleotides. Due to the redundancy of the genetic code, some base-pair substitutions in the third nucleotide of a codon, called *silent mutations*, do not affect gene translation. A substitution may result in the insertion of a different amino acid without altering the character of the protein if the new amino acid has similar properties or is not located in a region crucial to that protein's function.

A base-pair substitution that results in a different amino acid in a critical portion of a protein, such as the active site of an enzyme, may significantly impair protein function.

A substitution that results in an incorrectly coded amino acid is called a **missense mutation**. **Nonsense mutations** occur when the point mutation changes an amino acid codon into a stop codon, prematurely halting the translation of the polypeptide chain and usually creating a nonfunctional protein.

Base-pair **insertions** or **deletions** that are not in multiples of three nucleotides alter the reading frame. All nucleotides downstream from the mutation will be improperly grouped into codons, creating extensive missense and usually prematurely ending in nonsense. These **frameshift mutations** almost always produce nonfunctional proteins.

Mutagens   Mutations can occur in a number of ways. *Spontaneous mutations* include base-pair substitutions, insertions, and deletions that occur during DNA replication, repair, or recombination. Physical agents, such as X-rays and UV light, and various chemical agents that cause mutations are called **mutagens**. Chemical mutagens include base analogs that substitute for normal bases in DNA synthesis and result in mispairing and base-pair substitutions. The Ames test measures the mutagenic strength of compounds by testing their ability to cause mutations in bacteria.

■ What is a gene?

Our definition of a gene has evolved from Mendel's heritable factors, to Morgan's loci along chromosomes, to the one gene–one polypeptide axiom.

■ **INTERACTIVE QUESTION   16.8**

Define the following, and explain what type of point mutation could cause each of these mutations.

   a. silent mutation

   b. missense mutation

   c. nonsense mutation

   d. frameshift mutation

Research continually refines our understanding of the structural and functional aspects of genes, which now include introns and promoter regions. The best working definition of a gene is that it is a region of DNA that is required to produce an RNA molecule.

STRUCTURE YOUR KNOWLEDGE

1. Make sure you understand and can explain the processes of transcription and translation. You may find that filling in the following table in a study group helps you to review these processes.

	Transcription	Translation
Template		
Location		
Molecules involved		
Enzymes involved		
Control—start and stop		
Product		
Product processing		
Energy source		

2. What is the genetic code? Explain redundancy and the wobble phenomenon. What is meant by saying that the genetic code is almost universal?

3. Prepare a concept map showing the types and consequences of point mutations.

TEST YOUR KNOWLEDGE

MULTIPLE CHOICE: *Choose the one best answer.*

1. In Beadle and Tatum's study of *Neurospora*, they were able to identify three classes of arginine auxotrophs, mutants that needed arginine added to minimal media in order to grow. The pathway in the production of arginine includes the following steps: precursor → ornithine → citrulline → arginine. What nutrient(s) had to be added to the minimal medium in order for the class of mutant with a defective enzyme for the precursor → ornithine step to grow?
   a. precursor only
   b. ornithine only
   c. citrulline only
   d. ornithine or citrulline
   e. precursor, ornithine, and citrulline

2. Transcription involves the transfer of information from
   a. DNA to RNA.
   b. RNA to DNA.
   c. mRNA to an amino acid sequence.
   d. DNA to an amino acid sequence.
   e. the nucleus to the cytoplasm.

3. Which of the following is *not* true of an anticodon?
   a. It consists of three nucleotides.
   b. It is the basic unit of the genetic code.
   c. It extends from one end of a tRNA molecule.
   d. It may pair with more than one codon, especially if it has the base inosine in its third position.
   e. Its base uracil base-pairs with adenine.

4. RNA polymerase
   a. is the protein responsible for the production of ribonucleotides.
   b. is the enzyme that creates hydrogen bonds between nucleotides on the DNA template strand and their complementary RNA nucleotides.
   c. is the enzyme that transcribes exons, but does not transcribe introns.
   d. is a ribozyme composed of snRNPs.
   e. begins transcription at a promoter sequence and moves along the template strand of DNA in a 5'→ 3' direction.

5. Transfer RNA
   a. forms hydrogen bonds with the anticodon in the A site of a ribosome.
   b. binds to its specific amino acid in the active site of an aminoacyl-tRNA synthetase.
   c. uses GTP as the energy source to bind its amino acid.
   d. is translated from mRNA.
   e. is formed in the nucleolus.

6. Translocation involves
   a. the hydrolysis of a GTP molecule.
   b. the movement of the tRNA in the A site to the P site.
   c. the movement of the mRNA strand one triplet length in the A site.
   d. the release of the tRNA in the P site.
   e. all of the above.

7. Changes in a polypeptide following translation may involve
   a. the addition of sugars or lipids to certain amino acids.
   b. the action of enzymes to add amino acids at the beginning of the chain.
   c. the removal of poly-A from the end of the chain.
   d. the addition of a 5' cap of a modified guanosine residue.
   e. all of the above.

8. Several proteins may be produced at the same time from a single mRNA by
   a. the action of several ribosomes in a cluster called a polyribosome.
   b. several RNA polymerase molecules working sequentially.
   c. signal sequences that associate ribosomes with rough ER.
   d. containing several promoter regions.
   e. the involvement of multiple spliceosomes.

9. Which enzyme is responsible for the synthesis of tRNA?
   a. RNA replicase
   b. RNA polymerase
   c. aminoacyl-tRNA synthetase
   d. ribosomal enzymes
   e. ribozymes

10. In RNA processing,
    a. Exons are excised before the mRNA is translated.
    b. Assemblies of protein and snRNPs, called spliceosomes, may catalyze splicing.
    c. The RNA transcript that leaves the nucleus may be much longer than the original transcript.
    d. Large quantities of rRNA are assembled into ribosomes.
    e. Signal sequences are added to the 5' end of the transcript.

11. Base-pair substitutions may have little effect on the resulting protein for all of the following reasons *except* which one?
    a. The redundancy of the code may result in a silent mutation.
    b. As long as the substitution is three nucleotides, the reading frame is not altered.
    c. The missense mutation may not occur in a critical part of the protein.
    d. The new amino acid may have similar properties to the replaced one.
    e. The wobble phenomenon would result in no change in translation.

12. A ribozyme is
    a. an exception to the one gene–one RNA molecule axiom.
    b. an enzyme that adds the 5' cap and poly-A tail to mRNA.
    c. an example of rearrangement of protein domains caused by RNA splicing.
    d. an RNA molecule that functions as an enzyme.

    e. an enzyme that produces both small and large ribosomal subunits.

13. A signal sequence
    a. is most likely to be found on proteins produced by bacterial cells.
    b. directs an mRNA molecule into the cisternal space of ER.
    c. is a sign to help bind the small ribosomal unit at the initiation codon.
    d. would be the first 20 or so amino acids of a protein destined for secretion from the cell.
    e. is part of the 5' cap.

14. A base deletion early in the coding sequence of a gene may result in
    a. a nonsense mutation.
    b. a frameshift mutation.
    c. multiple missense mutations.
    d. a nonfunctional protein.
    e. all of the above.

15. The bonds between the anticodon of a tRNA molecule and the complementary codon of mRNA are
    a. formed by the input of energy from GTP.
    b. formed by the input of energy from ATP.
    c. hydrogen bonds.
    d. catalyzed by aminoacyl-tRNA synthetase.
    e. formed in the P site of the ribosome.

16. A prokaryotic gene 600 nucleotides long can code for a polypeptide chain of about how many amino acids?
    a. 100   c. 300   e. 1800
    b. 200   d. 600

CHAPTER 17

MICROBIAL MODELS: THE GENETICS OF VIRUSES AND BACTERIA

FRAMEWORK

A virus is an infectious particle consisting of a genome of single- or double-stranded DNA or RNA enclosed in a protein capsid. Viruses replicate using the metabolic machinery of their bacterial, animal, or plant host. Viral infections may destroy the host cell and cause diseases within the host organism. Viruses may have evolved from plasmids or transposons.

Bacteria have a circular chromosome and may have additional genes carried on plasmids. In conjunction with the genetic diversity generated by mutation, genetic recombination, and transposons, the short generation time of bacteria facilitates their adaptation to changing environments. Individual cells may alter their metabolic response to changing environmental conditions through feedback inhibition of enzyme activity and the regulation of gene expression.

CHAPTER REVIEW

Molecular biology originated in the laboratories of microbiologists. The study of viruses and bacteria has provided information about the molecular genetics of all organisms, an appreciation for the special genetic features of microbes, and an understanding of how viruses and bacteria cause disease, and it has led to the development of new powerful techniques of manipulating genes that have had an immense impact on basic research and biotechnology.

■ **Researchers discovered viruses by studying a plant disease:** *science as a process*

The search for the cause of tobacco mosaic disease led to the discovery of viruses. In 1883, Mayer found that,

although he could not find any microbe, he could transmit the disease by spraying sap from an infected plant onto a healthy plant. Ivanowsky found that infected sap still transmitted the disease after it had passed through a filter designed to remove bacteria. Beijerinck determined that the disease could not be caused by a bacterial toxin because the infectious agent was able to reproduce in a plant sprayed with filtered sap and then infect other plants. Unlike bacteria, the infectious agent could not be cultivated on nutrient media nor was it inactivated by alcohol. In 1935, Stanley crystallized the infectious particle, now known as tobacco mosaic virus (TMV). Since that time, many viruses have been seen with the electron microscope.

■ **Most viruses consist of a genome enclosed in a protein shell**

Viral Genomes   Viral genomes may be single- or double-stranded DNA or single- or double-stranded RNA. Viral genes are contained on a single linear or circular nucleic acid molecule.

Capsids and Envelopes   The **capsid**, or protein shell, is built from a large number of often identical protein subunits (capsomeres) and may be rod-shaped (helical), polyhedral, or more complex in shape. Membranous **viral envelopes**, derived from membranes of the host cell but also including viral proteins and glycoproteins, may cloak the capsids of viruses found in animals.

Complex capsids are found among bacteriophages, or **phages**, the viruses that infect bacteria. Of the seven phages (T1–T7) that infect the bacterium *E. coli*, T2, T4, and T6 have similar capsid structures consisting of an icosahedral (20-sided) head and a protein tailpiece with tail fibers for attaching to a bacterium.

■ Viruses can only reproduce within a host cell

Viruses are obligate intracellular parasites that lack metabolic enzymes and other equipment needed to express their genes and reproduce. Each virus type has a limited **host range** due to proteins on the outside of the virus that recognize only specific receptor molecules on the host cell surface.

Once the viral genome enters the host cell, the cell's enzymes, nucleotides, amino acids, ribosomes, ATP, and other resources are used to make copies of the viral genome and capsid proteins. DNA viruses use host DNA polymerases to copy their genome, whereas RNA viruses bring their own enzymes for replicating their RNA genome.

After replication, viral nucleic acid and capsid proteins spontaneously assemble to form new virus particles within the host cell, a process called self-assembly. Hundreds or thousands of newly formed viruses are released, often destroying the host cell in the process.

■ Phages exhibit two reproductive cycles: the lytic and lysogenic cycles

The Lytic Cycle   A replication cycle of a virus that culminates in death of the host cell is known as a **lytic cycle**. **Virulent viruses** cause their host cells to lyse and release newly produced phages.

The T4 phage uses its tail fibers to stick to a receptor site on the surface of an *E. coli* cell. The sheath of the tail contracts using ATP stored in the tailpiece and thrusts its viral DNA into the cell, leaving the empty capsid behind. The *E. coli* cell begins to transcribe and translate phage genes, one of which codes for an enzyme that chops up host cell DNA. Nucleotides from the degraded bacterial DNA are used to create viral DNA. Capsid proteins are made and assembled into phage tails, tail fibers, and polyhedral heads. The viral components assemble into 100 to 200 phage particles that are released after a lysozyme is manufactured that digests the bacterial cell wall.

Bacteria defend against viral infection by mutations that change their receptor sites and by produc-

■ INTERACTIVE QUESTION   17.1

In this diagram of a lytic and lysogenic cycle, describe the steps **1–8** and label structures **a–e**.

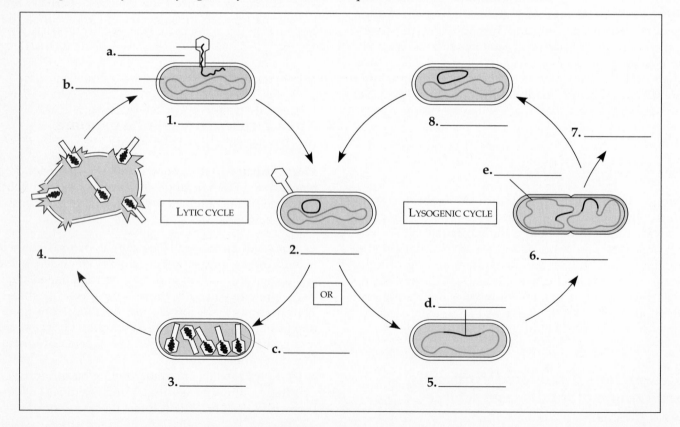

ing restriction enzymes that chop up viral DNA once it enters the cell.

The Lysogenic Cycle

In a **lysogenic cycle**, a virus reproduces its genome without killing its host. **Temperate viruses** can reproduce by the lytic and lysogenic cycles.

When the phage lambda (λ) injects its DNA into an *E. coli* cell, it can begin a lytic cycle, or its DNA may be incorporated into the host cell's chromosome and begin a lysogenic cycle. Most of the genes of the inserted phage genome, known as a **prophage**, are repressed by a protein coded for by a prophage gene. Reproduction of the host cell replicates the prophage along with the bacterial DNA. The prophage can become virulent if it is excised from the bacterial chromosome, usually in response to environmental stimuli, and starts the lytic cycle.

Expression of prophage genes may cause a change in the bacterial phenotype. Several disease-causing bacteria would be harmless except for the expression of prophage genes that code for toxins.

■ Animal viruses are diverse in their modes of infection and mechanisms of replication

Reproductive Cycles of Animal Viruses

A membranous envelope surrounding the capsid is present in several groups of animal viruses. Glycoproteins extending from the viral membrane attach to receptor sites on the host cell plasma membrane, and the two membranes fuse, transporting the capsid into the cell. The viral genome replicates and uses host cell ribosomes for protein synthesis. Glycoproteins are deposited in patches in the plasma membrane, where new viruses bud off within an envelope derived from the host's plasma membrane.

The envelopes of herpesviruses come from the host nuclear membrane. The herpesvirus' double-stranded DNA can integrate into the cell's genome as a **provirus**, similar to a bacterial prophage, and remain latent until it initiates herpes infections in times of stress.

RNA viruses infect animal hosts as well as plants and some bacteria. In the complicated reproductive cycle of **retroviruses**, the viral RNA genome is transcribed into DNA by a viral enzyme, **reverse transcriptase**. This viral DNA is then integrated into a chromosome, where it is transcribed by the host cell into viral RNA, which acts both as new viral genome and template for viral proteins. **HIV (human immunodeficiency virus)** is a retrovirus that causes **AIDS (acquired immunodeficiency syndrome)**.

Important Viral Diseases in Animals

The symptoms of a viral infection may be caused by toxins produced by infected cells, toxic components of the viruses themselves, cells killed or damaged by the virus, or the body's defense mechanisms fighting the infection.

■ INTERACTIVE QUESTION 17.2

Summarize the flow of genetic information during replication of an RNA virus. Indicate the enzymes that catalyze this flow.

Enzymes_____ _____

Vaccines are variants or derivatives of pathogens that induce the immune system to react against the actual disease agent. In 1796, Jenner used the cowpox virus to vaccinate against smallpox. Vaccinations have greatly reduced the incidence of many viral diseases.

Unlike bacteria, viruses use the host's cellular machinery to replicate, and few drugs have been found to treat or cure viral infections. Some antiviral drugs interfere with viral nucleic acid synthesis.

Emerging Viruses

The emergence of "new" viral diseases may be linked to the evolution of an existing virus (as in influenza viruses), the spread from one host species to another (as in the monkeypox virus), or the dissemination of an existing virus to a more widespread population (as in hantavirus and HIV).

Viruses and Cancer

Viruses that can cause cancer in animals are called tumor viruses. When tumor viruses infect cells growing in tissue culture, the cells are transformed—they assume rounded shapes and lose the contact inhibition regulation of growth. Viral nucleic acid becomes permanently integrated into host cell DNA. There is strong evidence linking some viruses to certain types of human cancer.

Oncogenes are genes responsible for triggering cancerous transformation in cells. Surprisingly, these genes have been found not only in tumor viruses, but also within the genomes of normal cells of many species. Some tumor viruses lack oncogenes but trans-

---

■ INTERACTIVE QUESTION 17.3

What do oncogenes code for?

form cells simply by turning on the cells' oncogenes. **Carcinogens**, nonviral cancer-causing agents, most likely affect the activity of cellular oncogenes.

Plant viruses are serious agricultural pests

Most plant viruses are RNA viruses. Plant viral diseases may spread through vertical transmission from a parent plant or through horizontal transmission from an external source. Plant injuries increase susceptibility to viral infections, and insects can act as carriers of viruses.

Viral particles spread easily through the plasmodesmata, the cytoplasmic connections between plant cells. Reducing the spread of disease and breeding resistant varieties are the best preventions for plant viral infections.

Viroids and prions are infectious agents even simpler than viruses

Viroids are very small molecules of naked RNA that can disrupt the metabolism of a plant cell and severely stunt plant growth. Viroid RNA has been found to be very similar to sequences of self-excising introns in some eukaryotic rRNA genes. It is likely that viroids disrupt the regulatory control of cellular genes.

Prions are protein infectious agents that may be linked to some diseases of the human nervous system. Although proteins cannot reproduce, prions may spread disease by converting normal cellular proteins into the defective form of the prion.

Viruses may have evolved from other mobile genetic elements

Viruses inhabit a gray area between life and nonlife—they are inert molecules containing a genetic program that can be expressed only within living cells.

The genomes of viruses are more similar to those of their host cells than to the genomes of viruses infecting other hosts. Viruses may have evolved from fragments of cellular nucleic acids that moved from one cell to another and eventually evolved special packaging. Sources of viral genomes may have been plasmids, self-replicating circles of DNA found in bacteria and yeast, and transposons, segments of DNA that can change location on a chromosome.

The short generation span of bacteria facilitates their evolutionary adaptation to changing environments

The circular bacterial chromosome is simpler in structure than eukaryotic chromosomes. This double-

---

■ INTERACTIVE QUESTION   17.4

Complete the following concept map that summarizes these sections on viruses.

stranded DNA molecule is tightly packed into a region of the cell called the **nucleoid**. Many bacteria also have plasmids, small rings of DNA with a few genes.

Replication of the bacterial chromosome proceeds bidirectionally from a single origin prior to binary fission. Most bacteria in a colony are genetically identical. Mutations, although statistically rare, produce a great deal of genetic diversity because bacteria can reproduce as often as once every 20 minutes. Genetically well-adapted bacteria clone themselves rapidly.

■ Genetic recombination and transposition produce new bacterial strains

Genetic recombination, the combining of genetic material from two individuals, also adds to the genetic diversity of bacterial populations. Evidence that genetic recombination occurs in bacteria is provided by growing two mutant strains of *E. coli*, each of which is unable to produce a particular amino acid, together on a complete culture medium. When later transferred to minimal media, numerous colonies grow that are now able to synthesize both amino acids. Genetic recombination in bacteria involves the mechanisms of transformation, transduction, and conjugation rather than the eukaryotic mechanisms of meiosis and fertilization.

Transformation Some bacteria can take up segments of naked DNA in a process called **transformation**. When the foreign DNA is integrated into the bacterial chromosome by an exchange similar to crossing over, a new combination of genes is produced. Many bacteria have surface proteins specialized for uptake of DNA. *E. coli* can be artificially induced to take up foreign DNA, a procedure important to biotechnology.

Transduction Bacteriophages can transfer genes from one bacterium to another by *generalized* transduction, when a random piece of host DNA is accidentally packaged within a phage capsid and introduced into a new bacterium, or by *specialized* transduction, when bacterial genes adjacent to a prophage insertion site are excised with the prophage from the bacterial chromosome. By either method, recombination occurs when the newly introduced bacterial genetic material replaces the homologous region of a bacterial chromosome.

Conjugation and Plasmids In **conjugation**, two cells temporarily join by appendages called sex pili and transfer DNA from one to the other. The ability to

■ **INTERACTIVE QUESTION 17.5**

a. What type of phage and reproductive cycle would most likely cause generalized transduction?

b. What type of phage would cause specialized transduction?

form pili and donate DNA requires the presence of an F plasmid.

Plasmids, small circular DNA molecules, replicate independently, either in concert with the bacterial chromosome or on their own schedule. Some plasmids, called **episomes**, can reversibly incorporate into the cell's chromosome. Temperate viruses are also considered to be episomes. Plasmid genes are not required for reproduction and survival under normal conditions, but may confer advantages to bacteria in stressful environments.

Bacterial cells containing the **F plasmid** are called F^+. The F plasmid replicates in synchrony with the bacterial chromosome, and F^+ cells pass the trait to daughter cells. The plasmid also replicates before conjugation and is transferred to the recipient cell, changing it from an F^- to an F^+ cell.

The F plasmid is an episome. Cells in which it is inserted into the bacterial chromosome are called Hfr cells, for "high frequency of recombination." When these cells undergo conjugation, the F plasmid transfers an attached copy of the bacterial chromosome to the recipient cell. Movement may disrupt the mating, resulting in a partial transfer of genes. Recombination between the new DNA and the recipient cell's chromosome produces new genetic combinations in this bacterium and its offspring.

Hfr bacteria of a given strain always transfer chromosomal genes in the same order, beginning with the F plasmid. In experiments that interrupt mating at different time intervals, the loci of many *E. coli* genes have been determined by genetic analysis of the recombinants cultured from different time samples.

R plasmids carry genes that code for antibiotic-destroying enzymes. R plasmids can be transferred to nonresistant cells during conjugation, creating the medical problem of antibiotic-resistant pathogens.

Transposons Transposable genetic elements, or **transposons**, are mobile segments of DNA that may move within a chromosome and to and from plasmids. In conservative transposition, the transposon simply changes location; in replicative transposition, the transposon first replicates, thus remaining in its original position and also inserting in a new location. Unlike other forms of genetic recombination where alleles exchange between homologous regions, transposons can move genes to totally new areas.

Insertion sequences are the simplest transposons, consisting of only a transposase gene and inverted repeats, which serve as recognition sites for transposase. Transposase is the enzyme responsible for the cutting and ligating of DNA required for transposition. Other enzymes are also required, such as DNA polymerase, which forms direct repeats at both ends of a transposon in its new site.

Insertion sequences cause mutations that often inactivate a gene when they insert in coding regions. They can also change transcription rates when they insert in regulatory regions. Even though insertion sequences transpose rarely in *E. coli*, they are a significant source of genetic variation due to the rapid proliferation of bacteria.

Complex transposons contain other genes between two insertion sequences. They have been shown to confer selective advantage by moving beneficial genes, such as those for antibiotic resistance, about in the bacterial genome.

Wandering genes were first described by McClintock from her corn breeding experiments in the 1940s and 50s. Transposons have since been found to be important components of eukaryotic genomes.

■ The control of gene expression enables individual bacteria to adjust their metabolism to environmental change

Feedback inhibition allows regulation of enzyme activity in response to changing environmental conditions, and regulated genes can be switched on and off as metabolic needs change. The operon model for gene regulation was first described by Jacob and Monod in 1961.

Operons: The Basic Concept **Structural genes** (genes that code for polypeptides) for the different enzymes of a single metabolic pathway may be grouped together on the chromosome and served by a single promoter. An **operator** is a segment of DNA within the promoter region or between it and the structural genes that controls the access of RNA polymerase to the structural genes. An **operon** is the DNA segment that includes the clustered structural genes, the promoter, and the operator.

Operators are normally "on." A **repressor** is a protein that binds to a specific operator, blocking attachment of RNA polymerase and thus turning the operator "off." **Regulatory genes** code for repressor proteins and usually produce them at a slow rate. The activity of the repressor protein is determined by the presence or absence of a **corepressor**. In the *trp* operon, tryptophan is the corepressor that binds to the

---

■ INTERACTIVE QUESTION   17.6

Complete the following concept map that summarizes the genetic characteristics of bacteria.

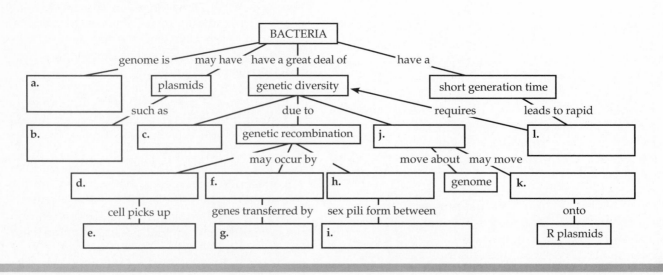

■ **INTERACTIVE QUESTION 17.7**

In the following diagram of an operon for inducible enzymes located on a stretch of a bacterial chromosome, identify components **a** through **i**.

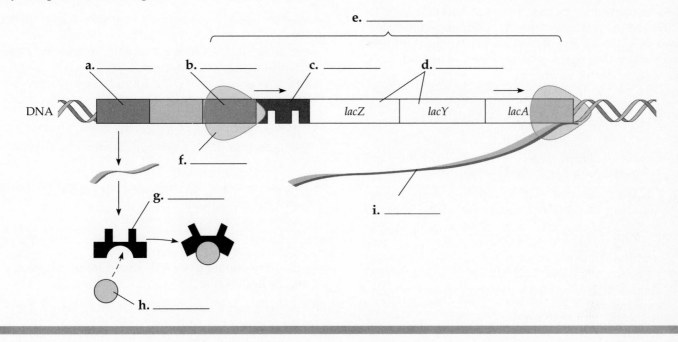

repressor protein, changing its conformation into its active state, which has a high affinity for the operator and switches off the *trp* operon. Should the tryptophan concentration of the cell fall, repressor proteins are no longer bound with tryptophan, the operator is no longer repressed, RNA polymerase attaches to the promoter, and mRNA for tryptophan synthesis is produced.

Repressible vs. Inducible Enzymes: Two Types of Negative Regulation   The synthesis of *repressible enzymes*, such as the enzymes of the tryptophan pathway, is inhibited by a metabolic end product. The synthesis of *inducible enzymes* is stimulated by specific small molecules. The *lac* operon, controlling lactose metabolism in *E. coli*, is an inducible operon that contains three structural genes. The *lac* repressor is innately active, binding to the *lac* operator and switching off the operon. Allolactose, an isomer of lactose, acts as an **inducer**, a small molecule that binds to and inactivates the repressor protein, so that the operon can be transcribed.

An Example of Positive Gene Regulation   With a regulatory system that uses positive control, an activator molecule interacts directly with the genome to turn on transcription.

E. coli cells preferentially use glucose as their source of energy and carbon skeletons. Should glu-

■ **INTERACTIVE QUESTION 17.8**

a. Repressible enzymes usually function in _____ pathways. The pathway's end product serves as a _____ to activate the repressor and turn off enzyme synthesis and prevent overproduction of the end product of the pathway. Genes for repressible enzymes are usually switched _____ and the repressor is synthesized in an_____ form.

b. Inducible enzymes usually function in _____ pathways. Nutrient molecules serve as _____ to stimulate production of the enzymes necessary for their breakdown. Genes for inducible enzymes are usually switched _____ and the repressor is synthesized in an _____ form.

cose levels fall, transcription of operons for other catabolic pathways can be increased through the action of **catabolite activating protein (CAP). Cyclic AMP (cAMP)**, derived from ATP, accumulates in the cell when glucose is absent and binds with CAP. The cAMP–CAP complex attaches to the promoter region and stimulates transcription by increasing the promoter's ability to associate with RNA polymerase.

The regulation of the *lac* operon includes negative control by the repressor protein that is inactivated by the presence of lactose, and positive control by CAP that functions when complexed with cAMP. *E. coli* is able to control its gene expression and conserve its RNA and protein synthesis depending upon the presence of glucose and the availability of catabolites (such as lactose) when glucose supply is limited.

STRUCTURE YOUR KNOWLEDGE

1. Create a concept map that describes the lytic and lysogenic cycles of a bacteriophage.

2. Create a concept map to develop your understanding of the mechanisms by which bacteria regulate their gene expression in response to varying metabolic needs. Distinguish repressible and inducible enzymes, which are both examples of negative control, and catabolite activator protein, which illustrates positive control of gene expression.

TEST YOUR KNOWLEDGE

MULTIPLE CHOICE: *Choose the one best answer.*

1. The study of the genetics of viruses and bacteria has done all the following *except*
   a. provide information on the molecular biology of all organisms.
   b. illuminate the sexual reproductive cycles of viruses.
   c. develop new techniques for manipulating genes.
   d. develop an understanding of the causes of cancer.
   e. show that genetic recombination occurs even in asexual bacteria.

2. Beijerinck concluded that the cause of tobacco mosaic disease was not a filterable toxin because
   a. The infectious agent could not be cultivated on nutrient media.
   b. A plant sprayed with filtered sap would develop the disease.
   c. The infectious agent could be crystallized.
   d. The infectious agent reproduced and could be passed on from a plant infected with filtered sap.
   e. The filtered sap was infectious even though microbes could not be found in it.

3. Viral genomes may be any of the following *except*
   a. several molecules of single-stranded DNA.
   b. double-stranded RNA.
   c. a circular DNA molecule.
   d. a linear single-stranded RNA molecule.
   e. double-stranded DNA.

4. Retroviruses have a gene for reverse transcriptase that
   a. uses viral RNA as a template for making complementary RNA strands.
   b. protects viral DNA from degradation by restriction enzymes.
   c. destroys the host cell DNA.
   d. translates RNA into proteins.
   e. uses viral RNA as a template for DNA synthesis.

5. Virus particles are formed from capsid proteins and nucleic acid molecules
   a. by spontaneous self-assembly.
   b. at the direction of viral enzymes.
   c. by using host cell enzymes.
   d. using energy from ATP stored in the tailpiece.
   e. by both b and d.

6. A virus has a base ratio of $(A + G)/(U + C) = 1$. What type of virus is this?
   a. a single-stranded DNA virus
   b. a single-stranded RNA virus
   c. a double-stranded DNA virus
   d. a double-stranded RNA virus
   e. a retrovirus

7. Vertical transmission of a plant viral disease may involve
   a. the movement of viral particles through the plasmodesmata.
   b. the inheritance of an infection from a parent plant.
   c. a bacteriophage transmitting viral particles.
   d. insects carrying viral particles between plants.
   e. the transfer of filtered sap.

8. Bacteria defend against viral infection through the action of
   a. antibiotics that they produce.
   b. restriction enzymes that chop up foreign DNA.
   c. their R plasmids.
   d. reverse transcriptase.
   e. episomes that incorporate viral DNA into the bacterial chromosome.

9. Drugs that are effective in treating viral infections
   a. induce the body to produce antibodies.
   b. inhibit the action of viral ribosomes.
   c. interfere with the synthesis of viral nucleic acid.
   d. change the cell-recognition sites on the host cell.
   e. produce vaccines that stimulate the immune system.

10. Which of the following is *not* true of tumor viruses?
    a. They can integrate viral nucleic acid into the host cell genome.
    b. They may transform cells growing in tissue culture into rounded cells that lose their contact inhibition.
    c. They may turn on the host cell's oncogenes.
    d. They may carry oncogenes.
    e. They may be transferred by bacteriophage.

11. The herpesvirus
    a. acts as a provirus when its DNA becomes incorporated into the host cell's genome.
    b. is a retrovirus that uses restriction enzymes to transcribe DNA from its RNA genome.
    c. has an envelope derived from the host cell's plasma membrane.
    d. is the retrovirus that has been linked to HIV, the virus that causes AIDS.
    e. can be used to vaccinate against hepatitis B.

12. Which of the following would never be an episome?
    a. an F plasmid
    b. a prophage
    c. a provirus
    d. a retrovirus
    e. a virulent phage

13. Tiny molecules of naked RNA that may act as infectious agents are
    a. retroviruses.
    b. transposons.
    c. viroids.
    d. episomes.
    e. prions.

14. Regulatory genes are genes that
    a. code for repressor proteins.
    b. are transcribed continuously.
    c. are not contained in the operon they control.
    d. may code for active or inactive repressors.
    e. are all of the above.

15. Inducible enzymes
    a. are usually involved in anabolic pathways.
    b. are produced when a catabolite inactivates the repressor protein.
    c. are produced when an activator molecule enhances the attachment of RNA polymerase with the operator.
    d. are regulated by inherently inactive repressor molecules.
    e. are regulated by feedback inhibition.

16. In *E. coli*, tryptophan switches off the *trp* operon by
    a. inactivating the repressor protein.
    b. inactivating the first enzyme in the pathway by feedback inhibition.
    c. binding to the repressor and increasing the latter's affinity for the operator.
    d. binding to the operator.
    e. binding to the promoter.

17. A mutation that renders nonfunctional the product of a regulatory gene for an inducible operon would result in
    a. continuous transcription of the structural genes of the operon.
    b. complete blocking of the attachment of RNA polymerase to the promoter.
    c. irreversible binding of the repressor to the operator.
    d. no difference in transcription rate when an activator protein was present.
    e. negative control of transcription.

18. A complex transposon
    a. carries a gene only for transposase between two inverted repeats.
    b. may transpose genes for antibiotic resistance to R plasmids.
    c. involves the exchange of homologous regions of DNA when it is a replicative transposition.
    d. is necessary for the F plasmid to incorporate into the bacterial chromosome to form an Hfr cell.
    e. transports genes in a distinct order between bacteria during conjugation.

MATCHING: *Match these components of the* lac *operon with their functions:*

____ 1. β-galactosidase   **A.** is inactivated when attached to lactose

____ 2. cAMP–CAP complex   **B.** codes for synthesis of repressor

____ 3. lactose   **C.** hydrolyzes lactose

____ 4. operator   **D.** stimulates gene expression

____ 5. promoter   **E.** repressor attaches here

____ 6. regulator gene   **F.** RNA polymerase attaches here

____ 7. repressor   **G.** acts as inducer that inactivates repressor

____ 8. structural gene   **H.** codes for an enzyme

GENOME ORGANIZATION AND EXPRESSION IN EUKARYOTES

FRAMEWORK

```
                    ┌─────────────────────────────────┐
                    │ GENE EXPRESSION IN EUKARYOTES   │
                    └─────────────────────────────────┘
```

GENE EXPRESSION IN EUKARYOTES

depends on

when abnormal
may lead to

physical availability of genes	structural or chemical changes	transcriptional control	posttranscriptional control	translational control	posttranslational control	cancer

due to — DNA packing — consists of — nucleosomes, 30-nm chromatin fibers, looped domains

such as — methylation, amplification, rearrangements

may include — transcription factors, enhancer and promoter, hormones and other chemical signals

such as — RNA processing

such as — initiation factors or repressors

such as — cleavage, chemical additions, targeting, degradation

caused by — oncogenes and tumor-suppressor genes

CHAPTER REVIEW

■ Each cell of a multicellular eukaryote expresses only a small fraction of its genome

Eukaryotic cells in multicellular organisms control the expression of their genes in response to their external and internal environments and also to direct the **cellular differentiation** necessary to create specialized cells. In both prokaryotes and eukaryotes, gene activity is regulated by DNA-binding proteins. In eukaryotes, however, chromosome structure, gene organization, and cellular structures are more complex, providing additional challenges for control of gene activity.

■ The structural organization of chromatin sets coarse controls on gene expression

Each chromosome consists of a single, extremely long DNA molecule precisely complexed with a large amount of protein. During interphase, chromatin is extended and intertwined, whereas discrete, condensed chromosomes appear during mitosis.

Nucleosomes, or "Beads on a String"   **Histones** are small, positively charged proteins that bind tightly to the negatively charged DNA to make up chromatin. Partially unfolded chromatin appears as a string of beads, each bead a **nucleosome** consisting of the DNA helix wound around a protein core of four pairs of different histone molecules. A fifth histone, H1, may attach to the outside of the "bead." Nucleosomes are the basic unit of DNA packing and may limit or direct the access of transcription proteins to DNA.

Higher Levels of DNA Packing   The *30-nm chromatin fiber* is a tightly coiled cylinder of nucleosomes organized with the aid of histone H1. The **looped domain** is a loop of the 30-nm chromatin fiber attached to a non-histone protein scaffold. In the mitotic chromosome, looped domains may also coil and fold, further compacting the chromatin.

Interphase chromatin may be attached to discrete locations of the nuclear envelope by protein scaffolding, which helps to organize areas of transcription. Chromatin visible with the light microscope during interphase is called **heterochromatin**, a highly compacted DNA that is not actively transcribed. The more open form of interphase chromatin is called **euchromatin**.

■ INTERACTIVE QUESTION 18.1

a. List the multiple levels of packing in a metaphase chromosome in order of increasing complexity.

b. Give an example of heterochromatin common in mammalian cells.

■ Noncoding sequences and gene duplications account for much of a eukaryotic genome

Unlike the prokaryote genome, in which most of the DNA contains uninterrupted codes for proteins (or RNA), the eukaryote genome has multiple copies of some DNA sequences and contains mostly noncoding DNA, long stretches of which interrupt coding sequences.

Repetitive Sequences Highly repetitive, short sequences of nucleotides make up 10–25% of the DNA in multicellular eukaryotes. This so-called **satellite DNA** is located primarily at the centromeres and **telomeres**, or chromosome tips. Repeating telomere sequences are periodically added by a special protein–RNA complex, telomerase, perhaps preventing the loss of a DNA tip during DNA duplication on the lagging strand.

Many highly repetitive sequences are transposons, some of which have been associated with diseases and some forms of cancer. Mutations that extend the number of repetitive sequences within genes may lead to gene malfunction.

Multigene Families Most genes are present in single copies as unique sequences. Multigene families are collections (often clustered but occasionally scattered throughout the genome) of similar or identical genes that probably evolved from a single ancestral gene. Identical genes usually code for RNA products. The identical genes coding for the major rRNA molecules are arranged in huge tandem arrays that enable cells to produce the millions of ribosomes needed for protein synthesis.

Examples of multigene families of nonidentical genes are the two families of genes that code for the α and β polypeptide chains of hemoglobin. Different versions of each chain are clustered together on two different chromosomes and are expressed at the appropriate time during development. Both types appear to have evolved from a common ancestral globin.

Families of identical genes most likely arose by repeated gene duplication due to mistakes in DNA replication and recombination. Nonidentical gene families probably arose from mutations in duplicated genes. **Pseudogenes**, sequences of DNA similar to real genes but lacking sites needed for gene expression, may be present within gene families.

A large amount of noncoding DNA is found in introns. A few examples have been found, however, where introns code for RNA molecules that regulate expression of some genes.

■ INTERACTIVE QUESTION 18.2

Name the two human genetic diseases that have been linked to elongations of base triplet repeats.

■ The control of gene expression can occur at any step in the pathway from gene to functional protein

DNA packing and the location of genes relative to the protein scaffold and nucleosomes may help control which genes are available for transcription.

Organization of a Typical Eukaryotic Gene A typical protein-encoding gene consists of a promoter sequence, where RNA polymerase attaches, and a sequence of introns interspersed among the coding exons. After transcription, RNA processing removes the introns and adds the guanosine cap at the 5' end and a poly-A tail at the 3' end. **Enhancers** are noncoding sequences that influence transcription and may be located at some distance from the promoter.

Transcriptional Control Both prokaryotic and eukaryotic cells have transcription factors that bind with RNA polymerase to the promoter region, forming an initiation complex so that polymerase begins

transcription. In addition, eukaryotic genomes have transcription factors that stimulate transcription of specific genes when they bind to enhancer sites. A loop in the DNA may bring the enhancer and its attached transcription factor into contact with the transcription factors and RNA polymerase at the promoter. The hundred or so transcription factors that have been identified have one of three basic DNA-binding domains as well as variable regions that permit recognition of different DNA sequences and proteins.

Posttranscriptional Control   Gene expression, measured by the amount and types of functional proteins that are made, can be blocked or stimulated at any posttranscriptional step. In the nucleus, RNA processing involves the removal of the coded intron DNA segments and the splicing together of exons, as well as the addition of a 5' cap and a poly-A tail. Processed mRNA, perhaps with attached proteins, exits from the nucleus through nuclear pores. In the cytoplasm it may interact with proteins and associate with ribosomes for translation. Much remains to be learned about how these steps are regulated.

In contrast to prokaryotic mRNA that is degraded after a few minutes, eukaryotic mRNA can last hours or even weeks. The length of time before degradation by cellular enzymes influences the amount of protein synthesis. Some proteins that bind to mRNA and block translation may also block degradation when mRNA is to be stored in the cell.

Translational and Posttranslational Control   The translation of certain eukaryotic mRNA can be delayed by the binding of repressor proteins to the 5' ends of messages that prevent ribosome binding or by the inactivation of initiation factors. A great deal of mRNA is synthesized and stored in egg cells; translation is begun when initiation factors are activated following fertilization.

Following translation, polypeptides are often cleaved or chemical groups added to yield an active protein. Selective degradation of proteins may serve as a control mechanism in the cell.

Arrangement of Coordinately Controlled Genes Unlike prokaryotic genes, which are often organized into operons that are controlled by the same regulatory sites and transcribed together, eukaryotic genes coding for enzymes in the same metabolic pathway are often scattered throughout the chromosomes and separately transcribed. The integrated control of these scattered genes may involve a specific regulatory element or enhancer associated with each related gene that serves as a recognition signal for a single type of transcription factor.

Gene Expression and Differentiation   Cellular differentiation during development is determined by chemical signals from within the cytoplasm or received from neighboring cells. These signals activate transcription factors that then set off a cascade of regulatory proteins that control expression of tissue-specific genes.

■ **INTERACTIVE QUESTION   18.3**

Label the components of this diagram of how enhancers and transcription factors may control eukaryotic gene expression.

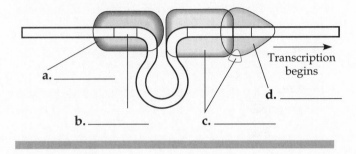

■ **Chemical signals that help control gene expression include hormones**

As in prokaryotes, small organic molecules combine with regulatory proteins to influence transcription in eukaryotes.

Chromosome Puffs: Evidence for the Regulatory Role of Steroid Hormones in Insects   Chromosome puffs appear along the polytene chromosomes found in the salivary glands of certain insect larvae. Autoradiography has shown that the puffs correspond to regions of active RNA synthesis. The shifting location of the puffs, especially during critical times in development, indicates that genes are turned on and off selectively. These changes in puff patterns can be induced by ecdysone, the steroid hormone that initiates molting, showing that gene regulation is responsive to this steroid chemical signal.

Action of Steroid Hormones in Vertebrates   Lipid-soluble steroid hormones diffuse across the plasma membrane. A hormone molecule binds with a specific receptor protein, either before or after it enters the nucleus. The inhibitory protein that was associated with the receptor is released, and the now-activated receptor protein can bind to an enhancer or other regulatory region and initiate transcription. Nonsteroid

hormones bind to a receptor protein on the outside of the cell and initiate an intracellular pathway that may activate specific transcription factors.

■ **INTERACTIVE QUESTION 18.4**

Complete the following five key points that summarize the control of gene expression in eukaryotes:

a. Different cell types of a multicellular organism _____.

b. Physical state or organization of the genome makes certain genes _____.

c. _____ may occur at each step from gene to functional protein.

d. Control of transcription is especially important. The _____ activates transcription of specific genes.

e. Some of these DNA-binding proteins are responsive to _____.

■ **Chemical modification or relocation of DNA within a genome can alter gene expression**

Gene Amplification and Selective Gene Loss In an amphibian ovum, a million or more extra copies of the rRNA genes are synthesized and contained in nucleoli, enabling the developing egg to make the huge numbers of ribosomes needed for protein synthesis. Selective replication of a gene is called **gene amplification**.

In certain insects, genes may be selectively lost, and whole or parts of chromosomes may be eliminated from some cells early in development.

Cancer cells resistant to chemotherapeutic drugs contain amplified genes conferring drug resistance. Selective gene amplification may confer drug resistance on some parasites.

Rearrangements in the Genome Transposons may affect gene expression if they move into the middle of a coding sequence of a gene or insert into a sequence that regulates transcription. If a transposon inserts downstream from a promoter, a gene carried on the transposon may be activated.

B lymphocytes are white blood cells that produce highly specific antibodies, or **immunoglobulins**, that recognize and bind to viruses, bacteria, and other invading molecules. DNA segments of the genes for the light and heavy polypeptide chains of an antibody molecule are permanently rearranged during cellular differentiation. An immunoglobulin gene is pieced together from hundreds of different variable regions that are widely separated from several constant regions. Each differentiated B lymphocyte and its descendants possess a distinctive genome that will produce one specific antibody.

DNA Methylation In a process called **DNA methylation**, methyl groups ($-CH_3$) are added to DNA bases (usually cytosine) after DNA synthesis. Genes are generally more heavily methylated in cells in which they are not expressed; drugs that inhibit methylation can induce gene reactivation, even in Barr bodies. DNA methylation may reinforce gene regulatory decisions of early development.

■ **INTERACTIVE QUESTION 18.5**

What accounts for the huge variety and specificity of immunoglobulins?

■ **Cancer can result from the abnormal expression of genes that regulate cell growth and division**

A change in a regulatory mechanism that leads to cancer is most often caused by carcinogens, physical mutagens, or viruses. **Oncogenes**, or cancer-causing genes, were first found in certain RNA viruses and later recognized in the genomes of humans and other animals. A cellular **proto-oncogene**, which may control an essential function in normal cells such as cell growth, division, and adhesion, may become an oncogene by several mechanisms. A mutation may result in more copies present than normal (amplification), transposition or chromosomal translocation (both of which may separate it from its normal control regions), or a nucleotide sequence that creates a more active or resilient protein. Mutations in **tumor-suppressor genes** can contribute to the onset of cancer if they result in a decrease in the activity of proteins that prevent uncontrolled cell growth.

More than one somatic mutation appears to be needed to produce a cancerous cell. Mutation of a single proto-oncogene can stimulate cell growth and division, but both tumor-suppressor alleles at a partic-

ular locus must be defective to allow defective growth. A genetic predisposition to certain cancers may involve the inheritance of a recessive mutant allele for a suppressor gene.

Breast cancer cases linked to family history have been traced to mutant alleles for either BRCA1 on chromosome #17 or BRCA2 on chromosome #13. Noninherited breast cancer also seems to involve somatic mutations for these tumor-suppressor genes. In most cases of colorectal cancer, at least four DNA changes need to occur: the activation of a cellular oncogene and the inactivation of three tumor-suppressor genes. An inherited defective DNA repair mechanism is involved in approximately 15% of colorectal tumors.

Virus-associated cancers are thought to account for 15% of human cancer cases. The virus might add an oncogene or affect a proto-oncogene or tumor-suppressor gene.

The ability to diagnose inherited predispositions for cancer has improved, and, hopefully, better cancer treatments will come from increased knowledge of eukaryotic genomes.

STRUCTURE YOUR KNOWLEDGE

1. Fill in the following table to help you organize the major mechanisms that can regulate the expression of eukaryotic genes.

2. **a.** What are proto-oncogenes? How do they become oncogenes?

   b. What is the role of tumor-suppressor genes in the development of cancer?

TEST YOUR KNOWLEDGE

MULTIPLE CHOICE: *Choose the one best answer.*

1. The control of gene expression is more complex in eukaryotic cells because
   a. Chromosomes are contained in a nucleus.
   b. Gene expression differentiates specialized cells.
   c. The chromosomes are linear and more numerous.
   d. Operons are controlled by more than one promoter region.
   e. Inhibitory or activating molecules may help regulate transcription.

2. Histones are
   a. small, positively charged proteins that bind tightly to DNA.
   b. small bodies in the nucleus involved in rRNA synthesis.
   c. basic units of DNA packing consisting of DNA wound around a protein core.
   d. repeating arrays of six nucleosomes organized around an H1 molecule.
   e. proteins responsible for producing repeating sequences at telomeres.

Level of Control		Examples
Gene availability (physical and chemical changes)	a.	
Transcriptional control	b.	
Posttranscriptional control	c.	
Translational control	d.	
Posttranslational control	e.	

3. Heterochromatin
   a. has a higher degree of packing than does euchromatin.
   b. is visible with the light microscope during interphase.
   c. is not actively involved in transcription.
   d. makes up Barr bodies.
   e. is all of the above.

4. DNA methylation of cytosine residues
   a. can be induced by drugs that reactivate genes.
   b. may contribute to long-term gene inactivation.
   c. produces the promoter regions that specifically bind RNA polymerase.
   d. makes satellite DNA a different density so it can be separated by ultracentrifugation.
   e. may be related to the transformation of proto-oncogenes to oncogenes.

5. Chromosome puffs along the polytene chromosomes of *Drosophila*
   a. appear at specific sites during developmental stages and give visual evidence of selective gene activation.
   b. are associated with the presence of nonsteroid hormones that act via a second messenger.
   c. are regions of DNA methylation causing highly compacted and inactive areas on the chromosome.
   d. contain collections of multigene families.
   e. contain multiple copies of rRNA genes.

6. Which of the following is *not* true of enhancers?
   a. They may be located thousands of nucleotides upstream from the genes they affect.
   b. When bound with transcription factors, they interact with the promoter region and other transcription factors to increase the activity of a gene.
   c. They may complex with steroid-activated receptor proteins and thus selectively activate specific genes at appropriate stages in development.
   d. They may coordinate the transcription of enzymes in the same metabolic pathway.
   e. They are located within the promoter, and, when complexed with a steroid or other small molecule, they release an inhibitory protein and thus make DNA more accessible to RNA polymerase.

7. Which of the following is *not* an example of the control of gene expression that occurs after transcription?
   a. mRNA stored in the cytoplasm needing a control signal to initiate translation
   b. the length of time mRNA lasts before it is degraded

c. rRNA genes amplified in tandem arrays
d. RNA processing before mRNA exits from the nucleus
e. splicing or modification of a polypeptide

8. Pseudogenes are
   a. tandem arrays of rRNA genes that enable actively synthesizing cells to create enough ribosomes.
   b. genes that can become oncogenes when induced by carcinogens.
   c. genes of multigene families that are expressed at different times during development.
   d. sequences of DNA that are similar to real genes but lack signals for gene expression.
   e. both c and d.

9. Which of the following might a proto-oncogene code for?
   a. DNA polymerase
   b. reverse transcriptase
   c. receptor proteins for growth factors
   d. immunoglobulins
   e. transcription factors that inhibit cell-division genes

10. A gene can develop into an oncogene when it
    a. is present in more copies than normal.
    b. undergoes a translocation that removes it from its normal control region.
    c. develops a mutation that creates a more active or resistant protein.
    d. is transposed to a new location where its expression is enhanced.
    e. does any of the above.

11. Gene amplification involves
    a. the production of stereotypic loud genes.
    b. the enlargement of chromosomal areas undergoing transcription.
    c. the synthesis of extra copies of genes to meet particular metabolic needs.
    d. the transformation of genes into tumor-suppressor genes.
    e. an enhancer or promoter that is activated and increases transcription rate.

12. A tumor-suppressor gene could cause the onset of cancer
    a. if both alleles have mutations that decrease the activity of the gene product.
    b. if only one allele has a mutation that alters the gene product.
    c. if it is inherited in mutated form from a parent.
    d. if one allele of a proto-oncogene has changed to an oncogene.
    e. if both a and d have happened.

CHAPTER 19

DNA TECHNOLOGY

FRAMEWORK

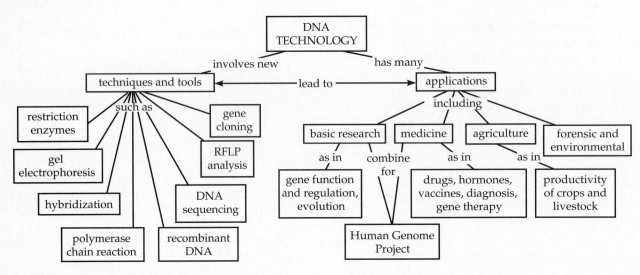

CHAPTER REVIEW

Genetic engineering, the manipulation of genetic material for practical purposes, has begun an industrial revolution in **biotechnology**. This use of living organisms to manufacture desirable products dates back centuries, but advances in recombinant DNA technology, along with new methods for manipulating and analyzing DNA, have resulted in hundreds of new products. Genes from different organisms and species are now routinely combined to form recombinant DNA and inserted into living cells in which these genes are expressed and their products studied or harvested. Genetic engineering has led to major advances in our knowledge of the organization and regulation of the eukaryotic genome and in the development of research techniques.

■ DNA technology makes it possible to clone genes for basic research and commercial applications: *an overview*

A general approach to genetic engineering takes advantage of the plasmids of bacterial cells. **Recombinant DNA** may be made by inserting foreign DNA into plasmids. These plasmids are put back into bacterial cells where they will replicate as the bacteria reproduce. Such cloned DNA may also be used to produce a protein coded for by the foreign DNA, transfer a metabolic capability into another organism, or produce multiple copies of the gene itself for biological research.

The toolkit for DNA technology includes restriction enzymes, DNA vectors, and host organisms

Restriction Enzymes **Restriction enzymes** protect bacteria from the DNA of viruses or other bacteria by cutting up foreign DNA in a process called *restriction*. Most restriction enzymes recognize short nucleotide sequences and cut at specific points within them. The cell protects its own DNA from restriction by methylating nucleotide bases within its own recognition sequences.

Recognition sequences are usually symmetrical sequences of four to eight nucleotides running in opposite directions on the two strands. The restriction enzyme usually cuts phosphodiester bonds in a staggered way, between the same adjacent nucleotides on both strands, leaving **sticky ends** of short single-stranded sequences on both sides of the resulting **restriction fragment**.

DNA from different sources can be combined in the laboratory when the DNA is cut by the same restriction enzyme and the complementary bases on the sticky ends of the restriction fragments pair by hydrogen bonding. DNA ligase is used to seal the strands together.

Vectors **Cloning vectors** are used to move recombinant DNA from the test tube into a cell. Plasmids are easily returned to bacterial cells where they will replicate cloned DNA. Bacteriophages can also serve as vectors. When a recombinant phage infects a bacterial cell, phage DNA replicates to form new phage particles, simultaneously cloning the inserted genes. These phages can then infect other bacterial cells and transmit the foreign DNA.

Yeast cells have plasmids that enable recombinant DNA to be cloned in eukaryotic cells. Retroviruses are used to deliver recombinant DNA directly to chromosomes in animal cells.

Host Organisms The ease with which DNA can be isolated from and returned to bacterial cells make bacteria a common host in genetic engineering. Eukaryotic hosts are better able to transcribe and translate eukaryotic genes and are preferable when a protein must be modified following translation. In addition to yeast cells, animal and plant cells in culture can serve as host cells.

■ **INTERACTIVE QUESTION** 19.1

a. Which of these DNA sequences would function best as a recognition sequence for a restriction enzyme? Why?

··CAGCAG·· ··GTGCTG·· ··GAATTC··

··GTCGTC·· ··CACGAC·· ··CTTAAG··

b. List the four most common vectors for recombinant DNA.

c. Why do bacteria make excellent hosts for genetic engineering?

Recombinant DNA technology provides a means to transplant genes from one species into the genome of another

Steps for Using Bacteria and Plasmids to Clone Genes The plasmid method of gene cloning involves treating antibiotic-resistant plasmids with a restriction enzyme that cuts the DNA ring at a single **restriction site** and disrupts a gene whose activity is easily determined, such as *lacZ*, the gene for β-galactosidase. The clipped plasmids are mixed with foreign DNA that has been treated with the same restriction enzyme and has complementary sticky ends. The sticky ends form hydrogen bonds with each other, and DNA ligase seals the recombinant molecules. The plasmids are introduced into bacterial cells by transformation and reproduce as the bacteria develop clones of cells. Clones carrying the recombinant DNA plasmids are identified by plating onto a medium containing the appropriate antibiotic and X-gal, a compound that is cleaved by β-galactosidase and yields a blue product. Colonies that are able to grow on the medium and are not blue (because of the foreign DNA inserted in the middle of the β-galactosidase gene) are carrying a recombinant plasmid.

Sources of Genes for Cloning The genes used for cloning can be isolated from an organism by cutting its DNA into thousands of pieces with restriction enzymes and inserting them into plasmids or viral DNA. The recombinant DNA is then used to transform bacterial cells or is inserted into phage. The collection of the thousands of clones of bacteria or phage derived from this shotgun approach is called a **genomic library**.

■ INTERACTIVE QUESTION 19.2

This schematic diagram shows the steps in plasmid cloning of a gene. Identify components **a–j**. Briefly describe the five steps of the process. How are bacterial clones that have picked up the recombinant plasmid identified?

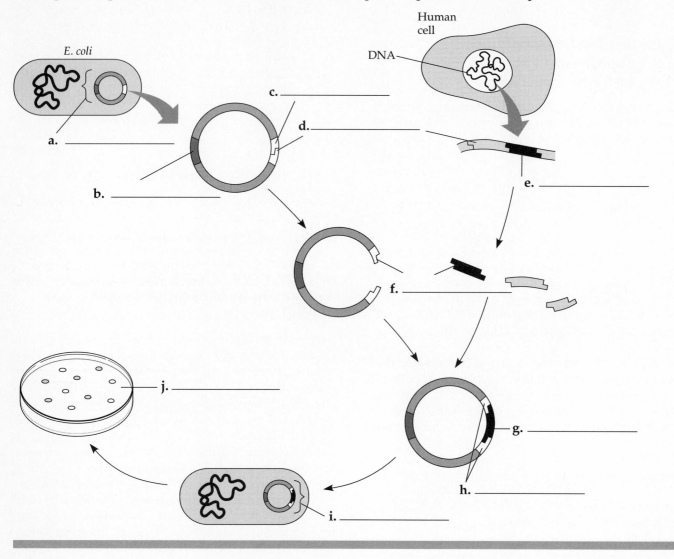

Creating artificial genes eliminates the problem caused by large introns that can make bacterial cells, which lack RNA-processing enzymes, unable to express eukaryotic genes. Using reverse transcriptase from retroviruses, mRNA—which has no introns—is used as a template to produce **complementary DNA**, or **cDNA**. Control sequences for transcription and translation must be provided within the vector DNA. A partial genomic library can be produced from the mRNA molecules found in a cell, and it contains only the genes that are expressed (transcribed) within the cell.

Inserting DNA into Cells   Recombinant DNA is usually inserted into bacterial cells through transforma-

tion or infection with phage vectors. Yeast cells can also be transformed by plasmids. Yeasts and other eukaryotic cells can take up linear DNA that then becomes incorporated into a chromosome through recombination.

Other means for introducing DNA into eukaryotic cells include **electroporation**, in which an electric pulse briefly opens holes in the plasma membrane through which DNA can enter. DNA can be injected into cells using microscopically thin needles or fired into plant cells on microscopic metal particles using a gene gun.

Selection: Finding a Gene of Interest   Finding a bacterial clone that contains the gene of interest within a huge genomic library is a difficult challenge. If the gene is expressed, the presence of the protein product can be determined by its activity or structure (using antibodies).

The gene itself can be detected using a nucleic acid sequence called a **probe**, which has complementary sequences to segments of the gene and can hybridize with the denatured (single-stranded) DNA of the gene. The probe is located by its radioactively labeled molecules or fluorescent tag. Once the desired clone is identified, it can be grown in culture and the gene of interest isolated in large quantities.

Achieving Expression of Cloned Genes Differences between prokaryotic and eukaryotic mechanisms for transcription and translation can be overcome by engineering signal sequences into the cloning vector so that the host cell will express the foreign gene. Expression can be maximized by attaching the eukaryotic gene to a bacterial gene produced normally in large quantities. When bacterial cells are engineered to secrete the protein product, the task of purification is simplified.

■ INTERACTIVE QUESTION 19.3

What steps can be taken to facilitate the expression of a eukaryotic gene in bacteria?

■ Additional methods for analyzing and cloning nucleotide sequences increase the power of DNA technology

Gel Electrophoresis Nucleic acids and proteins can be separated on the basis of their size and electrical charge by **gel electrophoresis**. Due to the negative charge of their phosphate groups, restriction fragments of DNA migrate through the electric field produced in a thin slab of gel toward the positive electrode. Fragments move at a rate inversely proportional to their size, producing band patterns in the gel of fragments of decreasing size. Individual fragments can be isolated from the gel and still retain their biologic activity.

DNA Synthesis and Sequencing Gene sequencing is made possible with base-specific chemical or enzymatic reactions and high-resolution gel electrophoresis, which can separate fragments that differ by as little as one nucleotide.

In the Sanger method of DNA sequencing, samples of a single-stranded restriction fragment are incubat-

ed with modified nucleotides (dideoxyribonucleotides) that randomly block further synthesis when they are incorporated into a growing DNA strand. The sets of radioactive or fluorescently tagged strands of varying lengths are separated by gel electrophoresis, and the nucleotide sequence is read from the sequence of bands.

Thousands of DNA sequences are being collected in computer data banks, where they can be analyzed for genetic control elements and similarities among genes of different organisms, and can be automatically translated into amino acid sequences.

■ INTERACTIVE QUESTION 19.4

The following gel was produced from four samples of a single-stranded DNA fragment that were incubated with radioactively labeled primer, DNA polymerase, A, T, C, and G nucleotides, and a different one of the four dideoxy nucleotides. Starting from the bottom of the gel, what is the sequence of nucleotides in the original single-stranded DNA fragment?

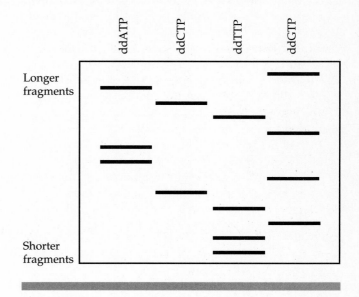

The Polymerase Chain Reaction (PCR) A technique developed in 1985, known as **polymerase chain reaction (PCR)**, has revolutionized molecular biology by being able rapidly to produce millions of copies of a section of DNA *in vitro*. A solution of DNA containing the region of interest is 1) heated to separate the strands, 2) incubated with the four nucleotides, DNA polymerase, and specially synthesized primers that bind upstream from the target sequence, and 3) heated again to repeat the process. The desired DNA segment does not need to be purified from the starting material, and very small samples can be used. Some limits on the number of accurate copies exist due to the accumulation of relatively rare copying errors.

Hybridization   Labeled probes consisting of sequences of DNA complementary to a gene of interest can search for homologous DNA in other organisms. The steps of Southern hybridization include restriction enzyme treatment of DNA, gel electrophoresis, Southern blotting (transfer of DNA from the gel to nitrocellulose or nylon membranes), addition of labeled DNA probes, and autoradiography. This technique will locate complementary sequences among the restriction fragments on a gel.

Northern blotting is used to hybridize mRNA with probes in order to determine whether a particular gene is being expressed, how much mRNA is present, and whether the abundance changes during developmental stages.

RFLP Analysis   **Restriction fragment length polymorphisms (RFLPs)** provide valuable markers for mapping chromosomes, diagnosing genetic diseases, and solving crimes. Restriction fragment length refers to the length of DNA fragments, shown by the pattern of bands produced by gel electrophoresis, that result when DNA is cut by a specific restriction enzyme. Fragment length variations or polymorphisms occur as a result of differences in base sequences that may delete or add restriction sites to homologous chromosomes. When these naturally occurring variations in DNA occur frequently in a population, geneticists can use them as reference points along a chromosome.

In RFLP analysis, DNA extracted usually from white blood cells is mixed with a restriction enzyme, and the resulting restriction fragments are separated by gel electrophoresis. Using Southern blotting, an RFLP marker is used as the probe to show the location of DNA bands or fragments that hybridize with the probe.

■ DNA technology is catalyzing progress in many fields of biology

New techniques in DNA research have contributed to research advances in almost all fields of biology and have facilitated the study of gene structure, function, and regulation on a molecular level. Gene product function has been studied using *in vitro* **mutagenesis**, changing specific sequences of a cloned gene, returning the gene to the cell, and examining changes in cell physiology or developmental pattern.

Cloned genes can be used as labeled probes to locate similar DNA segments in the same or other genomes, providing information on evolutionary relationships. Such cDNA probes allow location and

■ INTERACTIVE QUESTION   19.5

A bloody crime has occurred. Police have collected blood samples from the victim, two suspects, and blood found at the scene. Briefly list the steps the lab went through to produce the following autoradiograph.

a.

b.

c.

d.

e.

Which suspect would you charge with the crime?

study of the natural form of the gene, including its regulatory and noncoding sequences.

DNA probes can map genes on eukaryotic chromosomes with an *in situ* **hybridization** technique. Labeled DNA probes base-pair with intact chromosomes (that have been treated to separate DNA strands) on a microscope slide, and autoradiography and chromosome staining show the location of the gene. Similarly, probes can be used to bind with mRNA to identify cells that are expressing a particular gene.

Many molecules controlling cell metabolism and development are produced in such small quantities that they cannot be purified and characterized by standard biochemical techniques. The production of cDNA from mRNA molecules and its cloning in bacteria provides large enough quantities of these proteins to investigate their structures and functions.

■ The Human Genome Project is an enormous collaborative effort to map and sequence DNA

There are four complementary components to the *Human Genome Project*, the current effort to determine the nucleotide sequence of the human genome. Locating genetic markers, including RFLPs, throughout the chromosomes will facilitate mapping other genes by testing for genetic linkages to known markers. Physical mapping involves determining the order of identifiable fragments of each chromosome. This sequencing has been facilitated with the technique of **chromosome walking**, in which overlapping fragments produced from a known starting point indicate the order of genes and markers on a chromosome. The genes for cystic fibrosis and Huntington's disease have been isolated in this way. Sequencing the 3 billion nucleotide pairs of the haploid human genome may be the most time-intensive part of the project. The analysis of the genomes of other species will help develop strategies and new technologies for the project, as well as allow comparisons with the human genome.

Information from the Human Genome Project should contribute to the diagnosis, treatment, and prevention of genetic diseases, and provide insight into basic questions of molecular genetics and evolutionary biology.

The combination of the relatively new molecular techniques with classical genetic analysis will contribute to this huge project. For example, using polymerase chain reaction amplification of DNA from single sperm cells, the products of meiotic recombination can be directly analyzed, and linkage maps can be developed based on the frequency of crossovers between genes. Advances in both DNA technology and computer software will contribute to the success of the Human Genome Project.

■ DNA technology is reshaping the medical and pharmaceutical industries

Diagnosis of Diseases PCR and labeled DNA probes are being used to identify difficult pathogens and to diagnose infectious diseases. HIV probes prepared from its base sequence can be used with small blood samples amplified by PCR to detect HIV infection.

DNA technology has led to the diagnosis of over 200 human genetic disorders. Genes have been cloned for a number of genetic diseases, making it possible to produce a probe that can locate the corresponding gene in DNA from individuals who are being tested. Hybridization can be used to detect mutant alleles. Even if the gene has not yet been cloned, a disease gene may be diagnosed when closely linked with an RFLP marker. Comparisons of blood samples within a family must be used to determine which variant of the RFLP marker is linked to the abnormal allele.

Human Gene Therapy Genetic engineering may provide the means for correcting genetic disorders in individuals by replacing or supplementing defective genes. New genes would be introduced into somatic cells of types that actively reproduce within the body. Thus the gene would be replicated when the cells are reinserted into the individual, and the protein product of the introduced normal gene could then correct the biochemical defect. This type of gene therapy is currently being used successfully in children with an immunodeficiency disease caused by a lack of the enzyme adenosine deaminase. The patient's own T lymphocytes are engineered to contain a normal ADA allele and then are returned to the body, providing temporary correction of the disease. Treating the reproducing bone marrow cells should result in a more long lasting treatment.

Some of the technical problems involved with human gene therapy include how to get the proper control mechanisms to operate on the transferred gene, and when in development and into which tissues or cells the gene should be introduced. The huge expense of gene therapy presents ethical and social questions concerning the availability of such treatments. The most difficult ethical question is whether germ cells should be treated to correct defects in future generations. Opponents fear that tampering with human genes will eventually lead to the practice

■ INTERACTIVE QUESTION 19.6

List the four components of the Human Genome Project.

a.

b.

c.

d.

of eugenics, the deliberate effort to control the genetic makeup of human populations.

Vaccines and Other Pharmaceutical Products Recombinant DNA techniques have been used to make large amounts of protein molecules from disease-causing viruses, bacteria, and other microbes, which can be used as vaccines if the protein subunit triggers an immune response against the pathogen. Genetic engineering methods can also be used to directly modify the genome of a pathogen so as to attenuate it (make it nonpathogenic). The vaccinia virus, which is the basis of the smallpox vaccine, has been modified to carry genes that induce immunity to other diseases.

Gene splicing has been used to produce hormones and proteins in bacteria. Insulin and human growth hormone were the first two polypeptide hormones made by recombinant DNA techniques to be approved for use in the United States. Genetically engineered human growth hormone is used to treat children born with hypopituitarism and may have other medical uses. The hormone erythropoietin (EPO) is now being produced by biotechnology and used to stimulate red blood cell production in cases of anemia.

Two new approaches to fighting diseases are under development. **Antisense nucleic acid** is single-stranded DNA or RNA molecules that would base-pair with and block the translation of mRNA. Interfering with viral mRNA or the transformation of cells to a cancerous state could prevent diseases from spreading. Surface-receptor blockers or mimics are drugs that interfere with the binding of viral particles to receptors on cell membranes or bind with the virus to keep it from infecting cells.

■ INTERACTIVE QUESTION   19.7

a. Why is it easier to perform a test for Huntington's disease now that the gene has been cloned?

b. What are some of the practical and ethical considerations in human gene therapy?

■ DNA technology offers forensic, environmental, and agricultural applications

Forensic Uses of DNA Technology   RFLP analysis can be used in criminal cases to compare the **DNA fingerprint**, or specific pattern of RFLP bands, of a victim, suspect, and crime sample. Variations in the number of tandem repeated base sequences (**variable number tandem repeats** or **VNTRs**) found in satellite DNA are now commonly used in forensic DNA fingerprint analysis. Forensic tests require only five or ten small regions of the genome that are known to be highly variable from one person to the next in order to provide a high statistical probability that matching DNA fingerprints come from the same individual.

Environmental Uses of DNA Technology Microorganisms that are able to extract heavy metals, such as copper, lead, and nickel, may become important in mining and cleaning up mining waste. Microbes are used in sewage treatment plants to degrade many organic compounds into nontoxic form. Engineering organisms to degrade chlorinated hydrocarbons and other toxic compounds is an active area of research. Environmental disasters such as oil spills and waste dumps are other areas for which detoxifying microbes are being developed.

Agricultural Uses of DNA Technology   Genetic engineering is providing means to improve the productivity of agricultural plants and animals. Vaccines, growth hormones, and antibodies are produced with recombinant DNA technology. **Transgenic organisms** containing genes from other species are being developed for potential agricultural use. Rainbow trout and salmon engineered with a foreign gene for growth hormone grow at least twice as fast.

The ability to regenerate plants from single cells growing in tissue culture has made plant cells easier to genetically manipulate than mammalian cells. The bacterium *Agrobacterium tumefaciens* produces crown gall tumors in the plants it infects when its **Ti plasmid** (tumor inducing) integrates a segment of DNA into the plant chromosomes. Using DNA technology, foreign genes are inserted into Ti plasmids whose disease-causing properties have been eliminated. The recombinant plasmid can be either directly introduced into plant cells growing in culture or put back into *Agrobacterium* which then is used to infect cells. When these cells regenerate whole plants, the foreign gene is included in the plant genome.

Only dicotyledons are susceptible to infection by *Agrobacterium*. Techniques such as electroporation and

DNA guns are being used to transfer foreign genes into important monocots such as corn and wheat. Challenges still remain in identifying beneficial genes that will improve plant traits, particularly those traits that are polygenic.

Early results of genetic engineering in plants have been positive. Plant strains that carry a bacterial gene for resistance to herbicides have been developed, which will enable crops to be grown while weeds in their midst are killed. Crop plants are being engineered to be resistant to infectious pathogens and insect pests. Crop yields may be improved through the selective enlargement of plant parts or improvements to a plant's food value.

The production of bacteria with increased nitrogen-fixing potential and the engineering of plants that can fix nitrogen themselves are examples of a valuable potential use of recombinant DNA technology for improving plant productivity.

■ INTERACTIVE QUESTION 19.8

An antisense gene has been used to retard spoilage in tomatoes. Explain how this gene would work.

■ DNA technology raises important safety and ethical questions

Scientists have worried that there might be dangerous consequences to DNA technology, in particular, the production of new pathogens and their release into the environment. Several federal agencies set policy and regulate new developments in genetic engineering.

Safety issues include concerns about potential harmful side effects of medical products. With genetically engineered agricultural products, there are potential dangers in the impact of such organisms on native species. Transgenic species may become "superweeds" or introduce some of their genes for resistance to herbicides, insect pests, and natural diseases into wild plants. Researchers are trying to develop mechanisms to prevent the escape of engineered plant genes.

Ethical questions about human genetic information include who should have access to information about a person's genome and how that information should be used. Potential ethical, environmental, and health issues must be considered in the development of these powerful genetic techniques and remarkable products of biotechnology.

STRUCTURE YOUR KNOWLEDGE

1. Fill in this table on the basic tools of gene manipulation used in DNA technology.

Technique or Tool	Brief Description	Some Uses in DNA Technology
Restriction enzymes	a.	
Gel electrophoresis	b.	
cDNA	c.	
Labeled probe and autoradiography	d.	
Southern hybridization	e.	
DNA sequencing, Sanger method	f.	
Polymerase chain reaction (PCR)	g.	
RFLP analysis	h.	

2. Describe several examples of the many possible applications for DNA technology in agriculture and in medicine.

TEST YOUR KNOWLEDGE

MULTIPLE CHOICE: *Choose the one best answer.*

1. The role of restriction enzymes in DNA technology is to
   a. provide a vector for the transfer of recombinant DNA.
   b. produce cDNA from mRNA.
   c. produce a cut (usually staggered) at specific recognition sequences on DNA.
   d. reseal "sticky ends" after base-pairing of complementary bases.
   e. digest DNA into single strands that can hybridize with complementary sequences.

2. This segment of DNA has restriction sites I and II which create restriction fragments a, b, and c. Which of the following gel(s) produced by electrophoresis would represent the separation and identity of these fragments?

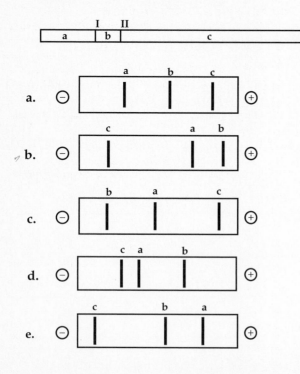

3. Yeast has become important in genetic engineering because it
   a. has RNA splicing machinery.
   b. has plasmids that can be genetically engineered.
   c. allows the study of eukaryotic gene regulation and expression.

   d. grows readily and rapidly in the laboratory.
   e. does all of the above.

4. Which of the following DNA sequences would most likely be a restriction site?
   a. AACCGG
      TTGGCC
   b. GGTTGG
      CCAACC
   c. AAGG
      TTCC
   d. AATTCCGG
      TTAAGGCC
   e. CTGCAG
      GACGTC

5. Genes for cloning may be chemically synthesized
   a. when the exact sequence of nucleotides is known.
   b. through the use of restriction enzymes and gel electrophoresis to separate restriction fragments.
   c. by the Sanger method.
   d. by making complementary DNA from genes without introns.
   e. with Southern hybridization.

6. Petroleum-lysing bacteria are being engineered for the removal of oil spills. What is the most realistic danger of these bacteria to the environment?
   a. mutations leading to the production of a strain pathogenic to humans
   b. extinction of natural microbes due to the competitive advantage of the "petro-bacterium"
   c. destruction of natural oil deposits
   d. poisoning of the food chain
   e. none of the above

7. You are attempting to introduce a gene that imparts larval moth resistance to bean plants. Which of the following vectors are you most likely to use?
   a. bacteriophage
   b. plasmids
   c. Ti plasmids
   d. particle bombardment
   e. either c or d

8. An attenuated virus
   a. is a virus that is nonpathogenic.
   b. in an elongated viral particle.
   c. can transfer recombinant DNA to other viruses.
   d. will not produce an immune response.
   e. is made with antisense DNA.

9. Difficulties in getting prokaryotic cells to express eukaryotic genes include the fact that
   a. The signals that control gene expression are different and prokaryotic promoter regions must be added to the vector or recombinant DNA.

b. The genetic code differs between the two because prokaryotes substitute the base uracil for thymine.

c. Prokaryotic cells cannot transcribe introns because their genes do not have them.

d. The ribosomes of prokaryotes are not large enough to handle long eukaryotic genes.

e. The RNA splicing enzymes of bacteria are different from those of eukaryotes.

10. Complementary DNA does not create as complete a gene library as the shotgun approach because
a. It has eliminated introns from the genes.
b. A cell produces mRNA for only a small portion of its genes at any one time.
c. The shotgun approach produces more restriction fragments.
d. cDNA is not as easily integrated into plasmids or phage genomes.
e. Northern blotting does not do as complete a job as Southern blotting.

11. Which of the following is *not* true of recognition sequences?
a. Modification by methylation of bases within them prevents restriction of bacterial DNA.
b. They are usually symmetrical sequences of four to eight nucleotides.
c. They signal the attachment of RNA polymerase.
d. Each recognition sequence is cut by a specific restriction enzyme.
e. Cutting a recognition sequence in the middle of a functional and identifiable gene is used to screen clones that have taken up foreign DNA.

Use the following choices to answer questions 12–14.
a. restriction enzyme
b. reverse transcriptase
c. ligase
d. DNA polymerase
e. RNA replicase

12. Which enzyme is used to make cDNA? b
13. Which enzyme is used in the polymerase chain reaction? d
14. Which enzyme is used to produce DNA fragments for DNA fingerprinting? a

15. A plasmid has two antibiotic-resistance genes, one for ampicillin and one for tetracycline. It is treated with a restriction enzyme that cuts in the middle of the ampicillin gene. DNA fragments containing a human globin gene were cut with the same enzyme. The plasmids and fragments are mixed, treated with ligase, and used to transform bacterial cells. Clones that have taken up the recombinant DNA are the ones that
a. can grow on plates with both antibiotics.
b. can grow on plates with ampicillin but not with tetracycline.
c. can grow on plates with tetracycline but not with ampicillin.
d. cannot grow with any antibiotics.
e. can grow on plates with tetracycline and are not blue.

16. VNTRs are a valuable tool for
a. forming overlapping sections in chromosome walking.
b. infecting dichotomous plant cells with recombinant DNA.
c. acting as probes in Northern blots.
d. DNA fingerprinting.
e. PCR to produce multiple copies of a DNA segment.

MECHANISMS OF EVOLUTION

CHAPTER 20

DESCENT WITH MODIFICATION: A DARWINIAN VIEW OF LIFE

FRAMEWORK

This chapter describes Darwin's formulation of evolution—descent from a common ancestor modified by the mechanism of natural selection, resulting in the evolution of species adapted to their environments. The scientific and philosophical climate of Darwin's day was quite inhospitable to the implications of evolution, but most biologists accepted the theory of evolution quite rapidly. Only later was natural selection recognized as a mechanism of evolution. Evidence for evolution is drawn from biogeography, the fossil record, comparative anatomy, comparative embryology, and molecular biology.

CHAPTER REVIEW

Evolution refers to the processes that have changed life on Earth from its early beginnings to the present diversity of organisms. Darwin presented the first convincing case for evolution in his book *On the Origin of Species by Means of Natural Selection*, published in 1859. Darwin made two major claims: Species were not specially created in their present forms but evolved from ancestral species, and **natural selection** provides the mechanism for evolution.

■ Western culture resisted evolutionary views of life

Darwin's theory was truly radical, for it challenged both the prevailing scientific views and the world view that had been held for centuries in Western culture.

The Scale of Life and Natural Theology The Greek philosopher Plato believed in two worlds, an ideal and eternal real world and the illusory world perceived by the senses. According to this philosophy of idealism, or **essentialism**, the variations in plant and animal populations were simply imperfect representatives of ideal forms. Plato's student Aristotle believed that all living forms could be arranged on a "scale of nature" of increasing complexity in which each group of organisms was fixed, was permanent, and did not change.

The Judeo-Christian account of creation embedded the idea of the fixity of species in Western thought. Biology during Darwin's time was dominated by **natural theology**, the study of nature to reveal the Creator's plan. One of the goals of natural theology was to classify species in order to reveal the rungs on the scale of life that God had created. Linnaeus, working in the eighteenth century, developed both a binomial system for naming organisms according to their genus and species, and a hierarchy of classifications for grouping species. **Taxonomy**, the branch of biology that names and classifies organisms, originated in the work of Linnaeus.

Cuvier, Fossils, and Catastrophism   **Fossils** are remnants or impressions of organisms laid down in rock, usually **sedimentary rocks**, such as sandstone and shale, that form through the compression of layers of sand and mud into superimposed layers called strata. Fossils reveal a succession of flora and fauna (plant and animal life).

Cuvier, the father of **paleontology**, the study of fossils, observed the differences between older fossils and modern life forms and the occurrences of extinctions. Adopting the view of history known as **catastrophism**, he speculated that the differences in fossil strata were the result of local catastrophic events such as floods or drought and were not indicative of evolution.

■ Theories of geological gradualism helped clear the path for evolutionary biologists

Gradualism, the idea that immense change is the cumulative result of slow but continuous processes, was used by Hutton in 1795 to explain the geological state of the Earth. Lyell extended gradualism to a theory of **uniformitarianism**, proposing that uniform rates and effects of geological processes balance out through time.

Darwin took two ideas from the observations of Hutton and Lyell: the Earth must be very old if geological change is slow and gradual, and very slow processes can produce substantial change.

■ Lamarck placed fossils in an evolutionary context

Lamarck, in 1809, was the first to publish a theory of evolution that explained how life evolves. Lamarck believed that evolution was driven by the tendency toward greater complexity and that, as organisms evolved, they became better adapted to their environment. Lamarck explained the mechanism of evolution with two principles: The use or disuse of body parts leads to their development or deterioration, and acquired characteristics can be inherited. In the creationist–essentialist climate of his time, Lamarck's views were dismissed; under present genetic knowledge, his ideas are sometimes ridiculed. Lamarck's theory, however, presented many key evolutionary ideas: that evolution is the best explanation for the fossil record and the current diversity of life, that Earth is very old, and that adaptation to the environment is the main result of evolution.

■ INTERACTIVE QUESTION   20.1

a. Match the theory or philosophy and its proponent(s) with the following description.

A. catastrophism	a. Aristotle
B. essentialism	b. Cuvier
C. inheritance of acquired characteristics	c. Hutton
	d. Lamarck
D. gradualism	e. Linnaeus
E. natural selection	f. Lyell
F. natural theology	g. Plato
G. scale of nature	h. Darwin
H. uniformitarianism	

Theory   Proponent

1. ____   ____   Discovery of the Creator's plan through the study of nature.

2. ____   ____   History of Earth marked by floods or droughts that resulted in extinctions.

3. ____   ____   Ideal world with perfect forms of which world of senses is imperfect representation.

4. ____   ____   Early explanation of mechanism of evolution.

5. ____   ____   Profound change is the cumulative product of slow but continuous processes.

6. ____   ____   Fixed species on a continuum from simple to complex.

7. ____   ____   Different reproductive success leads to adaptation to environment and evolution.

8. ____   ____   Geological processes have constant rates and balance out through time.

b. Now place 1 through 8 in chronological order.

—   —   —   —

—   —   —   —

Field research helped Darwin frame his view of life: *science as a process*

Natural theology was the prevailing view when Darwin was born in 1809. Growing up with a strong interest in nature, Darwin was sent to medical school but ended up attending college to become a naturalist and a clergyman.

The Voyage of the Beagle Darwin was 22 years old when he sailed from Great Britain as the naturalist on the *H.M.S. Beagle.* He spent the voyage collecting thousands of specimens of the fauna and flora of South America, observing the various adaptations of organisms living in very diverse habitats, and making special note of the geographic distribution of the taxonomically related species of South America. He was particularly struck by the uniqueness of the fauna of the Galapagos Islands. Most of the animal species on the islands were unique to the Galapagos, although they resembled species from the nearby mainland. Darwin also read and was influenced by Lyell's *Principles of Geology.*

Darwin Focuses on Adaptation Upon learning that the 13 types of finches he had collected on the Galapagos were indeed separate species, Darwin began, in 1837, the first of several notebooks on the origin of species. He began to link the origin of new species to the process of adaptation to different environments.

In 1844, Darwin wrote a long essay on the origin of species and natural selection but was reluctant to introduce his theory publicly. In 1858, Darwin received Wallace's manuscript describing a theory of natural selection identical to Darwin's. Wallace's paper and extracts of Darwin's unpublished essay of 1844 were jointly presented to the Linnaean Society. Darwin published *On The Origin of Species* the next year. Within a decade, Darwin's book and its defenders had convinced the majority of biologists that evolution was the best explanation for the diversity of life.

The Origin of Species developed two main points: the occurrence of evolution and natural selection as its mechanism

Descent with Modification Darwin's concept of **descent with modification** included the notion that all organisms were related through descent from some unknown ancient prototype and had developed increasing modifications as they adapted to various habitats. Darwin's view of the history of life is analogous to a tree with a common ancestor at the fork of each new branch and modern species at the tips of the

living twigs. Most branches are dead ends; about 99% of all species that have lived are extinct.

The taxonomy developed by Linnaeus provided a hierarchical organization of groups that suggested to Darwin the branching genealogy of the tree of life. The major taxonomic groups are kingdom → phylum → class → order → family → genus → species. Genetic analysis now shows that taxonomy reflects evolutionary relatedness.

Natural Selection and Adaptation Darwin's book focused on how populations of individual species become adapted to their environments through natural selection. Ernst Mayr described Darwin's theory of natural selection as follows:

Observation 1: Species have the potential for their population size to increase exponentially.

Observation 2: Most population sizes are stable.

Observation 3: Natural resources are limited.

Inference 1: Since only a fraction of offspring survive, there is a struggle for limited resources.

Observation 4: Individuals vary within a population.

Observation 5: Much of this variation is inherited.

Inference 2: Individuals whose inherited characteristics fit them best to the environment are likely to leave more offspring.

Inference 3: Unequal reproduction leads to gradual change in a population and the accumulation of favorable characteristics.

Variation arises through chance events of mutation and genetic recombination, but natural selection is the result of definite environmental criteria for reproductive success.

■ INTERACTIVE QUESTION 20.2

Summarize in your own words Darwin's theory of natural selection as the mechanism of evolution.

Artificial selection used in the breeding of domesticated plants and animals provided Darwin with evidence that selection among the variations present in a population can lead to substantial changes. He reasoned that natural selection, working over hundreds or thousands of generations, could gradually create the modifications essential for the present diversity of life. Gradualism is basic to the Darwinian view of evolution.

Natural selection results in the evolution of populations, groups of interbreeding individuals of the same species in a common geographic area. Evolution is measured only as change in the relative proportions of variations in a population over time. Natural selection affects only those traits that are heritable—acquired characteristics cannot evolve. And natural selection is a local and temporal phenomenon, depending on the specific environmental factors present in a region at a given time.

P. and R. Grant documented the correlation between variations in the depth of finch beaks and the availability of small and large seeds during wet and dry years on a tiny islet of the Galapagos. During dry periods, birds with stronger beaks (greater depth) more successfully feed on large seeds and thus leave more offspring. During wet periods, the smaller-beaked finches more efficiently feed on the abundant small seeds and have a reproductive advantage.

In another example of selection operating in natural populations, Singer and Parmesan documented a switch in plant preference for egg laying in a butterfly population. As a new weed invaded its meadow, the egg-laying pattern of these butterflies, which is genetically determined, adapted to changing vegetation over a period of 10 years. This study and many others of natural populations, as well as laboratory experimental studies, have provided evidence for natural selection as the mechanism of evolution.

■ Evidence from many fields of biology validates the evolutionary view of life

Biogeography   The geographic distribution of species, or **biogeography**, first suggested common descent to Darwin. Islands have endemic species that are related to species on the nearest island or mainland. Widely separated areas having similar environments are more likely to have species taxonomically related to those of their region, regardless of environment, than to each other. Biogeographical distribution patterns are explained by evolution; modern species are found where they are because they evolved from ancestors who inhabited those regions.

The Fossil Record   The succession of fossil forms supports the existence of the major branches of descent that were established with evidence from anatomy, biochemistry, molecular biology, and other sources. Paleontologists have discovered many transitional fossils linking modern species to their ancestral forms.

Comparative Anatomy   The anatomical similarities among species grouped in the same taxonomic category, known as **homology**, provide evidence of evolution. The same skeletal elements make up the forelimbs of all mammals regardless of function or external shape. These forelimbs are **homologous structures**, similar because of their common ancestry. Comparative anatomy illustrates that evolution is a remodeling process in which ancestral structures are modified for new functions.

Vestigial organs are rudimentary structures, of little or no value to the organism, that are historical remnants of ancestral structures.

Comparative Embryology   Comparative embryology shows that closely related organisms have similar stages in their embryonic development. Early embryos of all vertebrates pass through a stage in which they have gill pouches, which develop into gills in fish but into different structures in other vertebrates.

In the late nineteenth century, many embryologists adopted the view that "ontogeny recapitulates phylogeny," stating that the embryonic development (**ontogeny**) of an organism is a replay of its evolutionary history (**phylogeny**). Although the recapitulation theory is an overstatement, ontogeny can provide evidence of homology between structures that are very different in their adult forms.

Molecular Biology   An organism's DNA reflects its ancestry; closely related species should have a larger proportion of DNA and proteins in common than do more distantly related species.

Darwin's hypothesis—that all forms of life descended from the earliest organisms and are thus related—is supported through molecular biology. For example, all aerobic species have the respiratory protein cytochrome *c* (with some variations in the amino acid sequences but with the same essential structure and function). A common genetic code, passed along through all branches of evolution, is also important evidence that all life is related.

■ What is theoretical about the Darwinian view of life?

The evolution of modern species from ancestral forms is supported by historical facts such as fossils, biogeography, and molecular biology. The second of Darwin's claims, that natural selection is the main mechanism of evolution, is a theory that explains the historical facts of evolution. A scientific "theory" is a

■ **INTERACTIVE QUESTION   20.3**

Complete the following concept map that summarizes the five main sources of evidence for evolution.

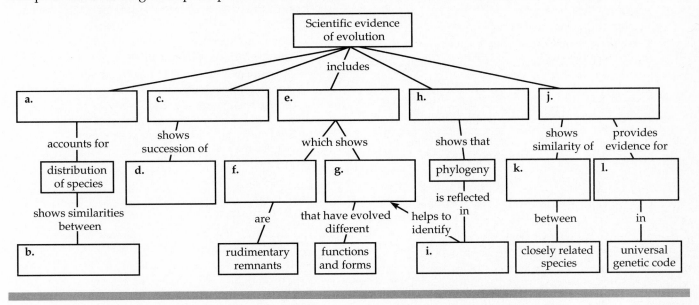

unifying concept with broad explanatory power and predictions that have been tested by experiments and observations.

STRUCTURE YOUR KNOWLEDGE

1. Briefly state the main components of Darwin's theory of evolution.

TEST YOUR KNOWLEDGE

MULTIPLE CHOICE: *Choose the one best answer.*

1. The classification of organisms into hierarchical groups is called
   a. the scale of nature.
   b. taxonomy.
   c. natural theology.
   d. ontogeny.
   e. phylogeny.

2. The study of fossils is called
   a. phylogeny.
   b. gradualism.
   c. paleontology.
   d. fossilogy.

   e. biogeography.

3. To Cuvier, the differences in fossils from different strata were evidence for
   a. changes occurring as a result of cumulative but gradual processes.
   b. divine creation.
   c. evolution by natural selection.
   d. continental drift.
   e. local catastrophic events such as droughts or floods.

4. Darwin proposed that new species evolve from ancestral forms by
   a. the gradual accumulation of adaptations to changing environments.
   b. the inheritance of acquired adaptations to the environment.
   c. the struggle for limited resources.
   d. the accumulation of mutations.
   e. the exponential growth of populations.

5. The best description of natural selection is
   a. the survival of the fittest.
   b. the struggle for existence.
   c. the reproductive success of the members of a population best adapted to the environment.
   d. the overproduction of offspring in environments with limited natural resources.
   e. a change in the proportion of variations within a population.

6. The remnants of pelvic and leg bones in a snake
    a. are vestigial structures.
    b. provide support for the fact that ontogeny recapitulates phylogeny.
    c. are homologous structures.
    d. provide evidence for inheritance of acquired characteristics.
    e. resulted from artificial selection.

7. Which of the following would provide the best information for distinguishing phylogenetic relationships between several very similar-appearing organisms?
    a. the fossil record
    b. homologous structures
    c. comparative anatomy
    d. taxonomy

    e. genetic analyses and protein comparisons

8. Darwin's claim that all of life descended from a common ancestor is best supported with evidence from
    a. the fossil record.
    b. comparative embryology.
    c. taxonomy.
    d. molecular biology.
    e. comparative anatomy.

9. The smallest unit that can evolve is
    a. a genome.
    b. an individual.
    c. a species.
    d. a population.
    e. a community.

CHAPTER 21

THE EVOLUTION OF POPULATIONS

FRAMEWORK

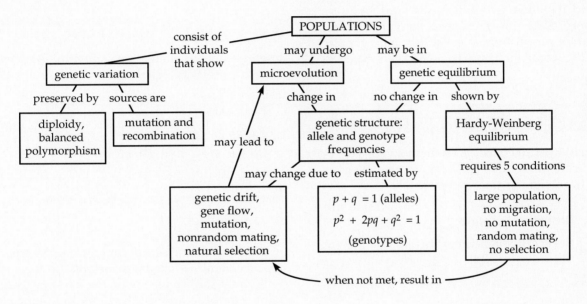

CHAPTER REVIEW

Although it is individuals that are selected for or against by natural selection, it is populations that actually evolve. The accumulated effect of differential reproductive success over generations leads to microevolution, changes in the genetic makeup of a population.

◪ The modern evolutionary synthesis integrated Darwinism and Mendelism: *science as a process*

Most biologists rapidly accepted evolution but not Darwin's proposal of natural selection as the mechanism of evolution. Without a theory of genetics, the transmission of chance variations from parents to off-spring could not be explained. When Mendel's work was rediscovered in the early 1900s, many geneticists believed that Darwin's focus on the inheritance of quantitative traits that vary on a continuum could not be explained by the inheritance of discrete Mendelian traits.

During the 1920s, research focused on rapid phenotypic change caused by mutations as an alternative mechanism to explain evolution. Also, many scientists supported the idea of goal-oriented evolution, or *orthogenesis*, as opposed to Darwin's mechanistic explanation based on natural selection.

The emergence of **population genetics** in the 1930s, with its emphasis on quantitative inheritance and genetic variation within populations, reconciled Mendelism with Darwinism.

In the early 1940s, a comprehensive theory of evolution, known as the **modern synthesis**, was developed that emphasizes the importance of populations as the units of evolution, the essential role of natural selection, and the gradualness of evolution.

A population has a genetic structure defined by its gene pool's allele and genotype frequencies

A **population** is a localized group of individuals of the same species. A **species** is a group of populations that have the ability to interbreed in nature. Within the geographic range of a species, populations may be totally isolated or contiguous.

The **gene pool** is the term for all the genes present in a population at any given time. For individuals of a diploid species, the pool includes two alleles for each gene locus. If all individuals are homozygous for the same allele, the allele is said to be *fixed*. More often, two or more alleles are present in the gene pool in some relative proportion or frequency. The **genetic structure** of a population is its allele and genotype frequencies.

■ INTERACTIVE QUESTION   21.1

In a population of 200 mice, 98 are homozygous dominant for brown coat color (*BB*), 84 are heterozygous (*Bb*), and 18 are homozygous recessive (*bb*).

a. The allele frequencies of this population are
_____ *B* allele   *b* allele_____

b. The genotype frequencies of this population are
_____ *BB*   _____ *Bb*   _____ *bb*

The Hardy-Weinberg theorem describes a nonevolving population

In the absence of selection pressure and other agents of change, the allele frequencies within a population will remain constant from one generation to the next, in spite of the shuffling of alleles by meiosis and random fertilization. This stasis is formulated as the **Hardy-Weinberg theorem**, named for its originators.

The allele frequency within a population determines the proportion of gametes that will contain an allele. The random combination of gametes will yield offspring with genotypes that reflect and reconstitute the allele frequencies. The frequencies of both alleles and genotypes will remain stable in a population that is in Hardy-Weinberg equilibrium.

With the Hardy-Weinberg equation, the frequencies of alleles within a population can be estimated from the genotype frequencies and vice versa. In a simple case of having only two alleles at a particular gene locus, the letters p and q represent the proportions of the two alleles within the population. The combined frequencies of the alleles must equal 100 percent of the genes for that locus: $p + q = 1$. The frequencies of the genotypes in the offspring reflect the frequencies of the alleles and the probability of each combination. According to the rule of multiplication, the probability that two gametes containing the same allele will come together in a zygote is equal to ($p \times p$) or p^2, or ($q \times q$) or q^2. A p and q allele can combine in two different ways, depending on which parent contributes which allele; therefore, the frequency of a heterozygous offspring is equal to $2pq$. The sum of the frequencies of all possible genotypes in the population adds up to one: $p^2 + 2pq + q^2 = 1$.

■ INTERACTIVE QUESTION   21.2

Use the allele frequencies you determined in question 21.1 to predict the genotype frequencies of the next generation.

Frequencies of

B (p) = _____ b (q) = _____

$BB = p^2$ = _____ $Bb = 2pq$ = _____

$bb = q^2$ = _____

Allele frequencies can be determined from genotype frequencies. The frequency of p in the gene pool will equal the frequency of the homozygous dominant genotype and one-half the frequency of heterozygous genotypes. If the frequency of homozygous recessive individuals is known (q^2), then the frequency of q may be determined as the square root of q^2 (assuming the population is in Hardy-Weinberg equilibrium for that gene). From our example above,

$q = \sqrt{bb} = 0.3$.

Microevolution is a generation-to-generation change in a population's allele or genotype frequencies: *an overview*

Hardy-Weinberg equilibrium serves as a baseline against which allele and genotype frequencies of a population can be compared to determine whether evolution is occurring. **Microevolution** is the generation-to-generation change in a population's genetic structure.

■ INTERACTIVE QUESTION 21.3

Practice using the Hardy-Weinberg equation so that you can easily determine genotype frequencies from allele frequencies and vice versa.

a. The allele frequencies in a population are $A = 0.6$ and $a = 0.4$. Predict the genotype frequencies for the next generation.

AA_____   Aa _____   aa _____

b. What would the allele frequencies be for the generation you predicted above in **a**?

A_____   a _____

c. Suppose that one gene locus determines stripe pattern in skunks. *SS* skunks have two broad stripes; *Ss* skunks have two narrow stripes; *ss* skunks have white speckles down their backs. A sampling of a population of skunks found 65 broad-striped skunks, 14 narrow-striped, and 1 speckled skunk. What are the allele frequencies for *S* and *s*?

S_____   s_____

The Hardy-Weinberg equilibrium is maintained only if all of the following five conditions are met: a very large population, isolation from other populations, no net changes in the gene pool due to mutation, random mating, and no natural selection.

Five sources of microevolution arise from departures from one or more of the five conditions of Hardy-Weinberg equilibrium. Natural selection, leading to the unequal reproductive success of different genotypes, tends to increase the fitness of a population to its environment; the other four agents are chance events and usually nonadaptive.

■ Genetic drift can cause evolution via chance fluctuation in a small population's gene pool: *a closer look*

Chance deviations from expected results are more likely to occur in a small sample due to sampling error. **Genetic drift** is a chance change in the gene pool of a small population due to sampling error and is likely to play a role in the microevolution of populations of less than 100 individuals.

The Bottleneck Effect The **bottleneck effect** occurs when some disaster or other factor reduces the population size dramatically, and the few surviving individuals are unlikely to represent the genetic makeup of the original population. Genetic drift will remain a factor in the population until it is large enough for chance events to be less significant. A bottleneck usually reduces variability because some alleles are lost from the gene pool.

The Founder Effect Genetic drift that occurs when only a few individuals colonize a new area is known as the **founder effect**. Allele frequencies in the small sample are unlikely to be representative of the parent population, and genetic drift will affect the gene pool of the new population until it is larger.

■ Gene flow can cause evolution by transferring alleles between populations: *a closer look*

Gene flow, the migration of individuals or the transfer of gametes between populations, may result in the gain or loss of alleles. Differences in allele frequencies between populations, which may have developed by natural selection or genetic drift, tend to be reduced by gene flow.

■ Mutations can cause evolution by substituting one allele for another in a gene pool: *a closer look*

The altering of allele frequency due to mutation is probably of little importance in microevolution due to very low mutation rates for most gene loci (one mutation in every 10^5 to 10^6 gametes). Mutation is central to evolution, however, because it is the original source of genetic variation.

■ Nonrandom mating can cause evolution by shifting the frequencies of genotypes in a gene pool: *a closer look*

Individuals tend to mate more often with close neighbors than with more distant population members. The effect of **inbreeding** between closely related partners is that genotype frequencies may vary from those predicted from Hardy-Weinberg equilibrium. The proportions of homozygotes will increase, and more individuals will express the recessive phenotype. Although inbreeding may change the ratios of genotypes and phenotypes, allele frequencies will remain the same.

In **assortative mating**, the choice of mates reflects a preference for like individuals. As with inbreeding, this nonrandom mating results in fewer heterozygous individuals but does not change allele frequencies in the gene pool. The shift in genotype frequencies, however, is a change in a population's genetic structure and is considered to be evolution.

Natural selection can cause evolution via differential reproductive success among varying members of a population: *a closer look*

For the Hardy-Weinberg equilibrium to be maintained, there must be no differential success in survival and reproduction, a condition that is probably never met. Individuals that are more successful in producing viable, fertile offspring pass their alleles to the next generation in disproportionate number. Natural selection is likely to be adaptive; favorable genotypes are increased and maintained in a population. As the environment changes, selection favors genotypes that are adapted to the new conditions.

Genetic variation is the substrate for natural selection

How Extensive Is Genetic Variation Within and Between Populations?   Individual variation, the slight differences between individuals as a result of their unique genomes, is the raw material for natural selection. Polygenic traits, those that are additively affected by two or more gene loci, provide much of the heritable variation within a population. Some traits vary categorically as distinct phenotypes and may be determined by a single gene locus. **Polymorphism** occurs when two or more discrete forms, or *morphs*, are evident in a population.

Quantitative measures of genetic variation include the percent of gene loci that have two or more alleles in a population and the average percent of loci that are heterozygous in individuals of a population. The extent of genetic variation is evident in the molecular differences found by using biochemical methods such as electrophoresis to compare the protein products of specific gene loci among individuals in a population.

Geographic variations are regional differences in genetic structure among populations of a species. These variations may be due to differing environmental selection factors or simply to genetic drift. A **cline**, or graded variation within a species along a geographic axis, may parallel an environmental gradient.

How Is Genetic Variation Generated?   New alleles originate by mutations, most of which occur in somatic cells and cannot be passed on to the next generation. Point mutations that alter a protein enough to

■ INTERACTIVE QUESTION   21.4

Fill in the following concept map that summarizes the five sources of microevolution. Do not worry if your word choice differs from the answer key. Better still, create your own concept map to help you review microevolution.

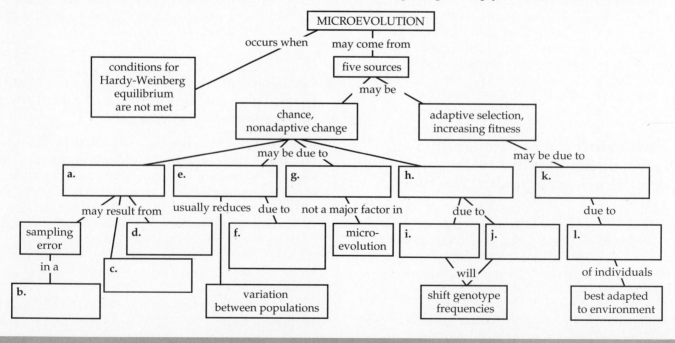

affect its function are more often harmful than beneficial. Rarely, however, a mutation may result in an individual better adapted to the environment and more reproductively fit, or a mutation already present in the population may be selected for when the environment changes.

Chromosomal mutations are most often deleterious. Occasionally, a translocation of a chromosomal piece may bring alleles together that are beneficial in combination. Duplication of chromosomal segments is usually harmful, but those that do not upset the genetic balance within cells may provide an expanded genome with extra loci that could eventually take on new functions by mutation. The shuffling of exons may also result in new genes.

■ **INTERACTIVE QUESTION   21.5**

a. What is a major source of genetic variation for bacteria and microorganisms?

b. What is the major source of genetic variation for plants and animals?

c. Explain why your answers to *a* and *b* are different.

How Is Genetic Variation Preserved?   Natural selection selects for favorable genotypes and tends to eliminate others, setting a trend toward genetic uniformity. Several mechanisms help to preserve variation.

Diploidy in most eukaryotes maintains genetic variation by hiding recessive alleles in heterozygotes, enabling them to perpetuate and be selected for, should the environment change.

Balanced polymorphism occurs when natural selection maintains variation at some gene loci. When individuals heterozygous at a certain gene locus are reproductively more fit, this **heterozygote advantage** tends to maintain two or more alleles at this locus.

Hybrid vigor may be seen in plants when two highly inbred varieties are crossbred. The crossbreeding may mask harmful recessive alleles that were homozygous in the inbred varieties and produce a heterozygote advantage at other loci.

Balanced polymorphism may be maintained in a population that ranges across a patchy environment in which different phenotypes are beneficial in different subregions. In **frequency-dependent selection**, a morph's reproductive success declines if it becomes too common, often due to increased predation or, in the case of "left- and right-mouthed" cichlid fish, decreased feeding success.

Does All Genetic Variation Affect Survival and Reproductive Success?   Some of the diversity seen in populations may be **neutral variations** that do not confer a selective advantage or disadvantage. According to the *theory of neutral evolution*, many such neutral alleles will not be affected by natural selection and will change randomly by genetic drift. There is no consensus on how much variation, if any, is truly neutral.

■ **Natural selection is the mechanism of adaptive evolution**

Adaptive evolution is a combination of chance and sorting: the chance occurrence of new genetic variation by mutation and sexual recombination, and the sorting or selecting of those variations most fit for the environment.

Fitness   **Darwinian fitness** is a measure of an individual's relative contribution to the gene pool of the next generation. Population geneticists speak in terms of the **relative fitness** of a genotype as its contribution to the next generation as compared to the contribution of other genotypes for the same locus. The most fecund variants are said to have a relative fitness of 1, whereas the fitness of another genotype is the percentage of offspring it produces in comparison.

■ **INTERACTIVE QUESTION   21.6**

A gene locus has two alleles, *B* and *b*. The genotype *BB* has a relative fitness of 0.5 and *bb* has a relative fitness of 0.25.

a. What is the relative fitness of the genotype *Bb*?

b. If a new generation has 100 *BB* individuals, how many *Bb* and *bb* would you expect to find?

Bb _____   *bb* _____

c. What might account for the different relative fitness values for these genotypes?

What Does Selection Act On? Selection acts on phenotype, the physical, biochemical, and behavioral traits of an organism, and it indirectly adapts a population to its environment by selecting for and maintaining favorable genotypes in the gene pool.

Some of the effects of pleiotropic genes may be positive, whereas others may be negative, and the fitness of a genotype depends on the balance of these effects on survival and reproduction. Most variations that natural selection acts upon are polygenic, quantitative characters, influenced by several gene loci. Selection acts on an organism that is an integrated composite of many phenotypic features. The relative fitness of any genotype depends on the entire genetic context of the individual.

Modes of Natural Selection The frequency of a trait, especially a quantitative trait, may be affected by three modes of natural selection. **Stabilizing selection** acts against extreme phenotypes and favors more intermediate forms, tending to reduce phenotypic variation. **Directional selection** occurs most frequently during periods of environmental change when individuals deviating in one direction from the average for some phenotypic character may be favored. **Diversifying selection** occurs when the environment favors individuals on both extremes of a phenotypic range. Balanced polymorphism may result from diversifying selection.

Sexual Selection **Sexual dimorphism** is the distinction between males and females on the basis of secondary sexual characteristics. In vertebrates, the male is usually the showier sex, and these characteristics may serve to attract females or compete with other males for females. **Sexual selection** is the selection for traits that may not be adaptive to the environment but do enhance reproductive success by increasing an individual's success in attracting a mate. Females, through their choice of mates, play an important role in the evolution of such traits.

■ Does evolution fashion perfect organisms?

There are at least four reasons why evolution does not produce perfect organisms. First, each species has evolved from a long line of ancestral forms, many of whose structures have been co-opted for new situations. Second, adaptations are often compromises between the need to do several different things, such as swim and walk, be agile and strong. Third, the evolution that occurs as the result of chance events, such

as genetic drift, is not adaptive. Fourth, natural selection can act on only those variations that are available; new alleles do not arise when needed.

STRUCTURE YOUR KNOWLEDGE

1. a. What is the Hardy-Weinberg theorem?
    b. Define the variables of the Hardy-Weinberg equation. Make sure you can use this equation to determine allele frequencies and predict genotype frequencies.

2. It seems that natural selection would work toward genetic unity; the genotypes that are most fit produce the most offspring, increasing the frequency of adaptive alleles and eliminating less beneficial alleles from the population. Yet there remains a great deal of variability within the populations of a species. Describe some of the factors that contribute to this genetic variability.

3. You collect 100 samples from a large butterfly population. Fifty specimens are dark brown, 20 are speckled, and 30 are white. Coloration in this species of butterfly is controlled by one gene locus: *BB* individuals are brown, *Bb* are speckled, and *bb* are white. What are the allele frequencies for the coloration gene in this population? Is this population in Hardy-Weinberg equilibrium? Explain your answer.

TEST YOUR KNOWLEDGE

MULTIPLE CHOICE: *Choose the one best answer.*

1. Darwinian fitness is a measure of
    a. survival.
    b. number of matings.
    c. adaptation to the environment.
    d. successful competition for resources.
    e. number of viable offspring.

2. According to the Hardy-Weinberg theorem,
    a. The genetic structure of a population should remain constant from one generation to the next if five conditions are met.
    b. Only natural selection, resulting in unequal reproductive success, will cause evolution.
    c. The square root of the frequency of individuals showing the recessive trait will always equal the frequency of *q*.

    d. Genetic drift, gene flow, mutations, and non-random mating are nonadaptive causes of microevolution; natural selection is the only adaptive cause.

    e. All of the above are correct.

3. If a population has the following genotype frequencies, *AA* = 0.42, *Aa* = 0.46, *aa* = 0.12, what are the allele frequencies?

    a. *A* = 0.42 *a* = 0.12

    b. *A* = 0.6   *a* = 0.4

    c. *A* = 0.65 a = 0.35

    d. *A* = 0.76 a = 0.24

    e. *A* = 0.88 a = 0.12

4. In a population with two alleles, *B* and *b*, the allele frequency of *b* is 0.4. What would be the frequency of heterozygotes if the population is in Hardy-Weinberg equilibrium?

    a. 0.16

    b. 0.24

    c. 0.48

    d. 0.6

    e. You cannot tell from this information.

5. In a population that is in Hardy-Weinberg equilibrium for two alleles, *C* and *c*, 16 percent of the population show a recessive trait. Assuming *C* is dominant to *c*, what percent show the dominant trait?

    a. 36 percent

    b. 48 percent

    c. 60 percent

    d. 84 percent

    e. 96 percent

6. Genetic drift is likely to be seen in a population

    a. that has a high migration rate.

    b. that has a low mutation rate.

    c. in which there is assortative mating.

    d. that is very small.

    e. for which environmental conditions are changing.

7. Gene flow often results in

    a. populations that are better adapted to the environment.

    b. an increase in sampling error in the formation of the next generation.

    c. adaptive microevolution.

    d. nonassortative matings.

    e. a reduction of the allele frequency differences between populations.

8. The existence of two distinct phenotypic forms in a species is known as

    a. geographic variation.

    b. stabilizing selection.

    c. heterozygote advantage.

    d. polymorphism.

    e. directional selection.

9. Assortative mating will most likely result in

    a. a change in allele frequency.

    b. an increase in the gene loci that are homozygous.

    c. sexual selection.

    d. stabilizing selection.

    e. neutral variations.

10. Mutations are rarely the cause of microevolution because

    a. They are most often harmful and do not get passed on.

    b. They may be masked in diploid individuals and are not able to be selected for.

    c. They occur very rarely.

    d. They are only passed on when they occur in gametes.

    e. of all of the above.

11. In a study of a population of field mice, you find that 48 percent of the mice have a coat color that indicates that they are heterozygous for a particular gene. What would be the frequency of the dominant allele in this population?

    a. 0.24

    b. 0.48

    c. 0.50

    d. 0.60

    e. You cannot estimate allele frequency from this information.

12. In a random sample of a population of shorthorn cattle, 73 animals were red ($C^R C^R$), 63 were roan ($C^R C^r$—a mixture of red and white), and 13 were white ($C^r C^r$). Estimate the allele frequencies of C^R and C^r, and determine whether the population is in Hardy-Weinberg equilibrium.

    a. C^R = 0.64, C^r = 0.36; because the population is large and a random sample was chosen, the population is in equilibrium.

    b. C^R = 0.7, C^r = 0.3; the genotype ratio is not what would be predicted from these frequencies and the population is not in equilibrium.

    c. C^R = 0.7, C^r = 0.3; the genotype ratio is what would be predicted from these frequencies and the population is in equilibrium.

    d. C^R = 1.04, C^r = 0.44; the allele frequencies add up to greater than 1 and the population is not in equilibrium.

    e. You cannot estimate allele frequency from this information.

13. A scientist observes that the height of a certain species of asters decreases as the altitude on a mountainside increases. She gathers seeds from samples at various altitudes, plants them in a uniform environment, and measures the height of the new plants. All of her experimental asters grow to approximately the same height. From this she concludes that
    a. Height is not a quantitative trait.
    b. The cline she observed was due to genetic variations.
    c. The differences in the parent plant's heights were due to directional selection.
    d. The height variation she initially observed was influenced more by environmental factors than by genetic ones.
    e. Stabilizing selection was responsible for height differences in the parent plants.

14. Sexual selection will
    a. select for traits that enhance an individual's chance of mating.
    b. increase assortative mating.
    c. result in individuals better adapted to the environment.
    d. result in stabilizing selection.
    e. result in a relative fitness of more than 1.

15. The greatest source of genetic variation in plant and animal populations is from
    a. mutations.
    b. recombination.
    c. selection.
    d. polymorphism.
    e. recessive masking in heterozygotes.

16. A plant population is found in an area that is becoming more arid. The average surface area of leaves has been decreasing over the generations. This is an example of
    a. nonrandom mating.
    b. directional selection.
    c. disruptive selection.
    d. gene flow.
    e. genetic drift.

CHAPTER 22

THE ORIGIN OF SPECIES

FRAMEWORK

A concept map showing: SPECIES — may be defined by / may originate by.

Under "may be defined by":
- **morphological concept** — based on — **shared similar morphology**
- **biological species concept** — must be — **reproductively isolated** — resulting from — **prezygotic barriers**, **postzygotic barriers**
- **recognition concept** — based on traits that — **maximize recognition and successful mating**
- **cohesion concept** — based on — **species integrity** — resulting from — **integrated complex of genes and adaptations**

Under "may originate by":
- **allopatric speciation** — may result in — **adaptive radiation**; evolve at — **tempo** — may be — **gradual**, **punctuated equilibrium**
- **sympatric speciation** — seen in — **polyploidy in plants**

CHAPTER REVIEW

Evolutionary theory attempts to determine the mechanisms of speciation, the origin of species. **Anagenesis**, or **phyletic evolution**, involves the transformation of an entire species into a new species. In **cladogenesis**, or **branching evolution**, new species arise from parent species that continue to exist. Cladogenesis is both the more common pattern of evolution and the process that increases biological diversity.

■ The biological species concept emphasizes reproductive isolation

Taxonomists often find that their classification of local species corresponds to the folk taxonomy of a region.

Species are most often characterized by their physical form or morphology, although taxonomists now consider physiology, biochemistry, behavior, and genetics to distinguish species.

According to the **biological species concept**, developed by Mayr in 1942, a species is a population or group of populations of individuals that have the potential to interbreed in nature and produce viable, fertile offspring, but which do not interbreed with other species in nature. Members of a species are said to be **conspecific**.

The biological species concept does not work for species that are completely asexual, such as prokaryotes and some protists and fungi. Extinct species also cannot be grouped based on the criterion of interbreeding.

Taxonomists must infer whether populations that are geographically isolated are able to interbreed in nature. Populations that differ slightly and live in separate areas within the species' geographic range are sometimes called subspecies. Genetic exchange among subspecies may be so circuitous or slight that it is difficult to decide whether some subspecies should be designated as separate species. These groups may be in the midst of evolution into new species. Even if not universally applicable, the biological species concept is a practical taxonomic tool and usually agrees with species identified by morphological criteria.

■ **INTERACTIVE QUESTION 22.1**

List four areas in which the biological species concept is difficult to apply.

a.

b.

c.

d.

■ Reproductive barriers separate species

Any intrinsic mechanism that prevents two species from producing viable, fertile hybrids is a reproductive barrier serving to preserve the genetic integrity of a species.

Prezygotic Barriers **Prezygotic barriers** function before the formation of a zygote by preventing mating between species or the successful formation of a zygote. Prezygotic barriers include *habitat isolation*, in which two species may live in the same area but occupy different habitats; *temporal isolation*, in which two species breed at different times; *behavioral isolation*, in which courtship rituals and behavioral signals attract only conspecific mates; *mechanical isolation*, in which anatomical incompatibility or mechanical barriers prevent mating with nonconspecifics; and *gametic isolation*, in which the gametes of different species fail to fuse due to an inhospitable female reproductive tract environment or lack of specific recognition molecules on the surfaces of gametes.

Postzygotic Barriers Should a hybrid zygote form, **postzygotic barriers** prevent it from developing into a viable, fertile adult. Postzygotic barriers include *reduced hybrid viability*, in which a hybrid zygote fails to survive embryonic development due to genetic incompatibility; *reduced hybrid fertility*, in which a viable hybrid individual is sterile, often due to the inability to produce normal gametes in meiosis; or

hybrid breakdown, in which the hybrids are viable and fertile, but their offspring are feeble or sterile.

Introgression When fertile hybrids do successfully mate with one of their parent species, genes may pass between species in a process called **introgression**. This small amount of gene transplantation increases the reservoir of genetic variation present in a species without threatening its integrity as a distinct species.

■ **INTERACTIVE QUESTION 22.2**

Identify the type of reproductive barrier illustrated by the following examples and indicate whether they are pre- or postzygotic barriers.

Type of Barrier	Pre- or Postzygotic	Example
a.	b.	Two species of frogs are mated in the lab and produce viable, but sterile, offspring.
c.	d.	Two species of sea urchin release their gametes at the same time, but cross-specific fertilization does not occur.
e.	f.	Two species of orchid have different length nectar tubes and are pollinated by different species of moths.
g.	h.	Two species of mayflies emerge during different weeks in springtime.
i.	j.	Two species of salamanders will mate in the lab and produce viable, fertile offspring, but offspring of these hybrids are sterile.
k.	l.	Two similar species of birds have different mating rituals.
m.	n.	When two species of mice are bred in the lab, embryos usually abort.
o.	p.	Peepers breed in woodland ponds, whereas leopard frogs breed in swamps.

Geographical isolation can lead to the origin of species: allopatric speciation

When the gene pool of a population becomes separated from other populations, the isolated population can follow its own evolutionary course as a result of selection, genetic drift, and mutation. Speciation mechanisms can be grouped by biogeographical factors: **Allopatric speciation** occurs when the gene pool of a population becomes segregated geographically from other populations; **sympatric speciation** occurs when a subpopulation becomes reproductively isolated while still surrounded by its parent population.

Geographical Barriers   Geological change can isolate populations. The extent of the geographical barrier necessary to maintain genetic separation depends on the ability of the organisms to disperse. Colonization of a new area may also geographically isolate populations.

Conditions Favoring Allopatric Speciation Geographic isolation of a small population at the fringe of the parent population's range is most likely to result in speciation because of three factors: First, the gene pool of the *peripheral isolate* probably represents the extremes of any clines in the original population. Also, the founder effect may result in a population whose gene pool differs from the parent pool. Second, genetic drift in the small gene pool may cause divergence from the parent population by chance. Third, selection pressures are likely to be different, and possibly more severe, on the fringe of a population's range. These factors do not guarantee successful speciation, and most pioneer populations probably become extinct.

Adaptive Radiation on Island Chains   Allopatric speciation may occur on island chains when small founding populations evolve in isolation and under somewhat differing environmental conditions. A single ancestral species of finch probably gave rise to the 13 species of finches now found on the Galapagos through the process of isolation and recolonization of islands. **Adaptive radiation** is the evolution of numerous, diversely adapted species from an ancestral species.

A new species can originate in the geographical midst of the parent species: sympatric speciation

A radical genetic change may result in the reproductive isolation of these mutants in the midst of the parent population. Mistakes during cell division may lead to **polyploidy** in plants. An **autopolyploid** has

more than two sets of chromosomes that have all come from a single species. Tetraploids ($4N$) can fertilize themselves or mate with other tetraploids but cannot mate with diploids from the parent population, resulting in reproductive isolation in just one generation.

Polyploid species arise more commonly through **allopolyploidy**, which is interspecific hybridization. Such hybrids are usually sterile due to difficulties in the meiotic production of gametes, but future mitotic or meiotic nondisjunctions may result in the production of a fertile polyploid. The chromosome number of an allopolyploid is equal to the sum of the diploid chromosome numbers of the parent species.

Speciation of polyploids, especially allopolyploids, has been frequent and important in plant evolution. Many of our agricultural plants are polyploids.

Sympatric speciation in animals may involve isolation within the geographic range of the parent population based on different resource usage. Assortative mating within a population that had a balanced polymorphism could also lead to sympatric speciation.

Population genetics can account for speciation

Speciation by Adaptive Divergence   Reproductive barriers may arise coincidentally as two populations accumulate differences in genotype and allele frequencies as they adapt to different environments. Postzygotic reproductive barriers that affect the viability of hybrids may result from pleiotropic effects of important genes controlling development. As populations diverge in their adaptation to specific environments, prezygotic habitat barriers can also evolve.

Sexual selection, which enhances an organism's reproductive success with its own species, may lead indirectly to reproductive barriers. The **recognition concept of species** emphasizes that natural selection acts on characteristics that maximize successful mating within a species, as opposed to driving the development of reproductive barriers in an isolated population that may be speciating.

Speciation by Shifts in Adaptive Peaks According to Wright's adaptive landscape metaphor, **adaptive peaks** occur where a population's gene pool is at an equilibrium that maximizes fitness in that environment. Valleys separating the peaks represent low average fitness of individuals. Natural selection will tend to push the population back to the former peak when there has been some slight nonadaptive change in allele frequency. When environmental conditions change, the adaptive landscape itself changes, creating new peaks and valleys.

Peak shifts can be initiated by a bottleneck or by other nonadaptive changes in a population's genetic structure. Genetic drift can knock a small population off its adaptive peak, and natural selection can then push it to a new adaptive peak that was present but formerly unreachable by a genetically stable population. Following a founder effect and the resulting genetic drift, adaptive evolution can push a destabilized gene pool to a new adaptive peak in a new adaptive landscape.

Hybrid Zones and the Cohesion Concept of Species When two formerly allopatric populations come back into contact, the two populations may freely interbreed, or they may have become reproductively isolated and thus separate species. A third option is the establishment of a **hybrid zone** where the two diverged populations interbreed. Introgression of alleles may not extend far, and the two parent populations may remain distinct on either side of a stable hybrid zone and be considered separate species.

The **cohesion concept of species** emphasizes cohesive forces that hold a species together, such as an integrated set of genes and adaptations, which have been selected for in a species' evolutionary history.

How Much Genetic Change Is Required for Speciation? Reproductive isolation may occur by a change at just a few gene loci or by cumulative divergence at many gene loci.

■ **INTERACTIVE QUESTION   22.5**

Fill in the following table to review the four ways in which a species can be conceptualized.

Concept	Explanation
a.	Emphasizes anatomical differences used by taxonomists to identify species
b.	Population(s) with potential to interbreed, reproductively isolated from other species
c.	Mating adaptations that result in recognition and sucessful mating of conspecifics
d.	Cohesion of phenotype, species integrity maintained by integrated complex of genes and adaptations

■ The theory of punctuated equilibrium has stimulated research on the tempo of speciation

The traditional evolutionary concept of the origin of species involves the gradual divergence of populations by microevolution, with each newly formed species continuing to evolve over long periods of time. However, in the fossil record, new forms appear rather suddenly, persist unchanged for a long time, and then disappear.

In 1972, Eldredge and Gould proposed the theory of **punctuated equilibrium**, in which long periods of stasis are punctuated by episodes of relatively rapid speciation and change. Speciation by polyploidy in plants and by small allopatric populations may occur fairly rapidly. In geological time, a few thousand years for a species to evolve is small compared with the few millions of years a successful species may remain in existence. Long periods of stasis may result from the tendency of stabilizing selection to maintain a population at an adaptive peak.

Some gradualists maintain, however, that fossils show stasis only in external anatomy and that changes in internal anatomy, physiology, and behavior go unrecorded. Even reports of long periods of morphological stasis are debatable: After analyzing a particularly complete set of trilobite fossils, Sheldon found a gradual change in aspects of their morphology, which challenged earlier categorizations of the youngest and oldest fossils in each evolutionary lineage as different species. Cheetham's extensive analysis of fossilized bryozoans, however, supports punctuated equilibrium. Additional studies of fossil lineages will be needed to determine the roles of gradual and punctuated tempos in the origin of species.

■ INTERACTIVE QUESTION   22.6

a. Compare the gradual and punctuated equilibrium theories of evolution.

b. What evidence is used to support each theory?

STRUCTURE YOUR KNOWLEDGE

1. How are speciation and microevolution different?

2. When the environment changes, a population may move to a new adaptive peak. How is this different from a peak shift?

TEST YOUR KNOWLEDGE

MULTIPLE CHOICE: *Choose the one best answer.*

1. Most new species probably have arisen by
   a. anagenesis.
   b. cladogenesis.
   c. phyletic evolution.
   d. branching evolution.
   e. both b and d.

2. Which of the following is *not* a type of intrinsic reproductive isolation?
   a. mechanical isolation
   b. behavioral isolation
   c. geographical isolation
   d. gametic isolation
   e. temporal isolation

3. The biological species concept may not apply to
   a. organisms that look different but are able to interbreed.
   b. organisms that do not breed in nature but may do so in captivity.
   c. organisms that do not breed due to recognition adaptations.
   d. populations that are geographically separated.
   e. subspecies in which there is substantial genetic exchange.

4. For which of the following is the biological species concept least appropriate?
   a. plants
   b. animals
   c. bacteria
   d. mules
   e. field biologists

5. A horse ($2n = 64$) and a donkey ($2n = 62$) can mate and produce a mule. How many chromosomes would there be in a mule's cells?
   a. 31   c. 63   e. 126
   b. 62   d. 64

6. What prevents horses and donkeys from hybridizing to form a new species?
   a. gametic isolation
   b. behavioral isolation
   c. mechanical isolation
   d. reduced hybrid viability
   e. reduced hybrid fertility

7. Introgression occurs when
   a. A hybrid successfully breeds with an individual of a parent species.
   b. Hybrids successfully breed with each other.
   c. An allopolyploid is produced.
   d. Subspecies breed.
   e. The hybrid zone remains distinct and separate.

8. Allopatric speciation is most likely to occur when
   a. The splinter population is small.
   b. A peripheral isolate is exposed to more severe selection pressures at the boundary of the population's range.
   c. A small subset of the population becomes adapted to a new food source.
   d. Both a and b are true.
   e. a, b, and c are true.

9. A tetraploid plant species is probably the result of
   a. allopolyploidy.
   b. autopolyploidy.
   c. hybridization and nondisjunction.
   d. allopatric speciation.
   e. a and c.

10. Sexual selection may lead indirectly to reproductive barriers because
    a. Isolated populations are exposed to different selection pressures.
    b. Natural selection acts on characteristics that maximize successful mating within a species.
    c. Hybrids may be reproductively less fit.
    d. Prezygotic barriers are more likely to evolve before postzygotic barriers.
    e. A cohesive set of phenotypic characters results from a complex of adapted genes.

11. There are 28 morphologically diverse species of a group of sunflowers called silverswords found on the Hawaiian archipelago. These species are an example of
    a. a geographical cline.
    b. adaptive radiation.
    c. allopolyploidy.
    d. the bottleneck effect.
    e. sympatric speciation.

12. According to Wright's evolutionary metaphor,
    a. A population has only one adaptive peak and is maintained there by natural selection.
    b. A small population that drifts into a valley surrounding an adaptive peak will either become extinct or evolve into a new species.
    c. An environmental change may create a new adaptive landscape, and natural selection can push a destabilized gene pool to a new adaptive peak.
    d. The valleys between peaks represent changes in environmental conditions and populations must move through these valleys to become a new species.
    e. Peak shifts are caused by natural selection operating on a small population.

13. The punctuated equilibrium theory maintains that
    a. Long periods of stasis are punctuated by episodes of relatively rapid speciation and change.
    b. Microevolution is the driving force of speciation.
    c. Most rapid speciation events involve polyploidy in plants.
    d. Evolution occurs rapidly when the adaptive landscape changes.
    e. Hybrid zones are the location of most rapid speciation, whereas the populations on both sides remain in equilibrium.

14. A new plant species C formed from hybridization of species A ($2n = 18$) with species B ($2n = 12$) would probably have a gamete chromosome number of
    a. 12.   c. 18.   e. 60.
    b. 15.   d. 30.

CHAPTER 23

TRACING PHYLOGENY: MACROEVOLUTION, THE FOSSIL RECORD, AND SYSTEMATICS

FRAMEWORK

This chapter considers the major events and evolutionary trends that led to the biological diversity of today. Mechanisms for macroevolution include the gradual modification of preadapted structures for new functions; evolutionary trends, perhaps as a result of species selection; and adaptive radiations as new adaptive zones appear when evolutionary novelties develop or mass extinctions occur. Continental drift and other major geological events affect the course of macroevolution.

A goal of systematics is to determine the phylogenetic history of species. The branch of systematics called taxonomy names and classifies species according to their presumed evolutionary relationships. Phylogenetic trees are constructed from evidence gathered from the fossil record and from anatomical and molecular homologies.

The relationship between microevolution and macroevolution and the roles of natural selection and chance in macroevolution are controversial topics in contemporary evolutionary theory.

CHAPTER REVIEW

Macroevolution is the origin of taxonomic groups higher than species. When biologists study macroevolution, they consider the major events revealed by the fossil record: large diversifications of taxonomic groups, the origin of novel biological designs, major extinctions, and the mechanisms that may have produced these evolutionary developments.

■ The fossil record documents macroevolution

Paleontologists reconstruct evolutionary history by collecting and interpreting the succession of organisms found in the fossil record.

How Fossils Form Fossils are preserved impressions or remnants of past organisms. Sedimentary rocks formed from the compression of sand and mud deposits are the richest sources of fossils. Hard parts of animals, such as bones, teeth, or shells, may remain as fossils. Petrification, the replacement of organic tissue with dissolved minerals, may turn the fossil to stone. Sometimes enough organic material is retained in the fossil that biochemical analysis and electron microscopy on cells can be done.

Molds of organisms, left when their bodies decay and minerals dissolved in water crystallize in the cast, are a common type of fossil. Trace fossils formed in footprints, burrows, or other impressions of animal activities are like fossils, giving clues to how animals moved or lived. Some fossils are the entire body of an organism preserved in a substance, such as ice, bogs, or amber, that retards decomposition.

Limitations of the Fossil Record The formation of a fossil is an unlikely occurrence. The incompleteness of the fossil record is understandable, considering that a large number of species that lived probably left no fossils, most fossils are destroyed, and only a fraction of existing fossils have been found.

■ **INTERACTIVE QUESTION   23.1**

What types of organisms are most likely to appear in the fossil record?

■ Paleontologists use a variety of methods to date fossils

Relative Dating The order in which fossils appear in the strata of sedimentary rocks indicates their relative age. *Index fossils*, such as shells of widespread animals, are used to correlate strata from different locations. Gaps may appear in the sequence, indicating that an area was above sea level or underwent erosion during some period of geological time.

Geologists have developed a **geological time scale** with four eras—the Precambrian, Paleozoic, Mesozoic, and Cenozoic—that are delineated by major transitions, such as mass extinction and extensive radiation. The eras are subdivided into periods and epochs, which are associated with particular evolutionary developments shown in the fossils found in the strata of those times.

Absolute Dating   **Radiometric dating** is used to determine the ages of rocks and fossils. During an organism's lifetime, it accumulates radioactive isotopes in proportions equal to the relative abundance of the isotopes in the environment. After the organism dies, the isotopes decay at a fixed rate, known as the **half-life**, or the number of years it takes for one-half of the radioactive isotopes in the specimen to decay. Carbon-14, with a half-life of 5600 years, is used for determining the age of relatively young fossils; uranium-238, with a half-life of 4.5 billion years, can be used to date rocks much older.

During an organism's life, only L-amino acids are synthesized, but following death L-amino acids are slowly converted to D-amino acids. The proportion of L- and D-amino acids and the rate of this racemization can be used to date some fossils.

■ **INTERACTIVE QUESTION   23.2**

a. A fossil has one-eighth of the normal ratio of C-14 to C-12. Estimate the age of this fossil.

b. Potassium-40 has a half-life of 1.3 billion years. If an organism had 1 mg of potassium-40 when it died and its fossil now has 0.25 mg, how old is this fossil?

■ **What are the major questions about macroevolution?** *an overview*

Key questions about the mechanisms of macroevolution include how novel features develop, what causes evolutionary trends, what are the effects of global geological changes, and how have mass extinctions and major adaptive radiations influenced macroevolution.

■ **Some evolutionary novelties are modified versions of older structures**

Evolutionary novelties that define higher taxa may evolve by the gradual modification of existing structures for new functions. **Preadaptation** is the term for structures that evolved and functioned in one setting and were then co-opted for a new function.

■ **INTERACTIVE QUESTION   23.3**

Give examples of reptilian structures that were preadapted for flight in birds.

■ **Genes that control development play a major role in evolutionary novelty**

Slight changes in the function of genes that control development may create major changes in adult morphology. **Allometric growth**, the different rates of growth in various parts of the body, leads to the final shape of the organism. A minor genetic alteration that affects allometric growth can have a major morphological effect.

Novel organisms can also be produced by small genetic changes that affect the timing of developmental events. **Paedomorphosis** is the retention in the adult of juvenile traits of ancestral organisms.

Both **heterochrony**, evolutionary changes in the rate or timing of development, and **homeosis**, changes that alter the spatial arrangement of body parts, have been important in the macroevolution of new forms of organisms.

■ **INTERACTIVE QUESTION   23.4**

How does the growth of chimpanzee brains differ from that of human brains, and what is the impact of this difference?

Recognizing trends in the fossil record does not mean that macroevolution is goal-oriented

Equus, the modern horse, descended from its much smaller ancestor, *Hyracotherium*, through a series of speciation episodes that produced many different species documented in the fossil record. *Equus* did not phyletically evolve as the direct result of trends of increasing size, reduction in the number of toes, and changes in dentition.

Evolutionary trends that are evident in the fossil record are most often the result of punctuated equilibrium, in which new species change by increments as they branch from ancestral ones.

According to Stanley's model of **species selection**, an evolutionary trend is analogous to a trend in a population produced by natural selection. The successful species that last the longest before extinction and speciate the most will determine the direction of the trend. Evolutionary trends are ultimately dictated by environmental conditions; if conditions change, an evolutionary trend may end or change.

■ INTERACTIVE QUESTION 23.5

Even when the fossil record shows a particular trend, such as increasing size or decreasing number of digits, paleontologists may interpret the trend as developing not through gradual phyletic evolution of individual populations but through species selection. Explain this distinction.

Macroevolution has a biogeographical basis in continental drift

Biogeography, the geographical distribution of species, correlates with the geological history of the Earth. The continents rest on great plates of crust that float on the molten mantle. Earthquakes, volcanoes, and mountains are formed in regions where these shifting plates abut.

Large-scale continental drift brought all the land masses together into a supercontinent named **Pangaea** about 250 million years ago, near the end of the Paleozoic era. This tremendous change undoubtedly had a great environmental impact as shorelines were eliminated, oceans deepened, shallow coastal areas were drained, and the majority of the land had a drier and more erratic climate. Species that had evolved in isolation came together and competed, many species became extinct, and new opportunities for remaining species became available.

About 180 million years ago, during the early Mesozoic era, Pangaea broke up and the continents drifted apart, creating a huge geographical isolation event. Current biogeography reflects this separation.

■ INTERACTIVE QUESTION 23.6

Marsupials evolved in what is now North America, yet their greatest diversity is found in Australia. How can you account for this biogeographical distribution?

The history of life is punctuated by mass extinctions followed by adaptive radiations of the survivors

Examples of Major Adaptive Radiations Major adaptive radiations occurred early in the history of some taxa, when the evolution of a novel characteristic opened a new **adaptive zone**, or way of life with unexploited opportunities. Nearly all the animal phyla that exist today evolved in 5 to 10 million years during the middle Cambrian, the first period of the Paleozoic era. The origin of shells and skeletons in a few taxa opened an adaptive zone by making more complex body designs possible and changing predator–prey relationships.

An empty adaptive zone can be exploited only if appropriate evolutionary novelties arise, and novelties that do arise cannot enable organisms to move into adaptive zones that do not exist or are filled. Mass extinctions are often followed by new adaptive radiations.

Examples of Mass Extinctions A species may become extinct due to a change in its physical or biological environment. Extinction, inevitable in a changing world, usually occurs at a rate of between 2.0 and 4.6 families per million years. During periods of major environmental change, mass extinctions may occur.

The Permian extinctions, occurring at the boundary between the Paleozoic and Mesozoic eras about 250 million years ago, claimed over 90 percent of marine and terrestrial species. These extinctions coincided with the formation of Pangaea and may be related to environmental changes associated with that event. Massive volcanic eruptions in Siberia may have lowered temperatures and contributed to the Permian extinctions.

The Cretaceous extinctions, which mark the boundary between the Mesozoic and Cenozoic eras about 65 million years ago, claimed over one-half the marine species, many families of terrestrial plants and animals, and the dinosaurs. The climate was cooling during that time, and shallow seas were receding from continental lowlands.

There is evidence that an asteroid or comet collided with Earth around the time of the Cretaceous extinctions. The thin layer of clay enriched in iridium that separates Mesozoic from Cenozoic sediments may have been the fallout from a huge cloud of dust created when an asteroid hit Earth. This cloud would have blocked sunlight and severely affected weather. The huge Chicxulub crater on the Yucatan coast of Mexico may have been the site of the impact. Acid precipitation and global fires may also have contributed to the massive extinctions explained by this **impact hypothesis**. Determining the rate of extinctions during that period may help scientists distinguish between the impact hypothesis and the more gradual causes of climate and habitat changes associated with continental drift and increased vulcanism.

■ INTERACTIVE QUESTION   23.7

Why do extensive adaptive radiations often follow mass extinctions?

■ Systematics connects biological diversity to phylogeny

Phylogeny is the evolutionary history of a species. A phylogenetic tree is a diagram of proposed evolutionary relationships of various groups. **Systematics** is the branch of biology concerned with the diversity of life and its phylogenetic history.

Taxonomy and Hierarchical Classification   Taxonomy involves the identification and classification of species. Taxonomy assigns names, describes diagnostic characteristics for new species, and groups species into **taxa**, the various categories. Linnaeus developed our present classification hierarchy, which organizes

similar groups into more general categories. Related species are grouped into genera, which are then grouped into **families, orders, classes, phyla**, and finally into **kingdoms**.

Linnaeus also developed the system that assigns each species a two-part Latin name—a **binomial**—consisting of the **genus** and the **specific epithet,** or species name. Both words of the binomial are italicized, and the names for all taxa at the genus level and higher are capitalized.

The Relationship of Classification to Phylogeny   A goal of systematics is that classification reflects the evolutionary relationships between species. Each taxon should be **monophyletic**, containing only species that are derived from a single ancestor. A **polyphyletic** taxon includes groups having different ancestry, and a **paraphyletic** taxon excludes some species that share a common ancestor with other species in the taxon.

Sorting Homology from Analogy   **Homology** is likeness based on shared ancestry; **analogy** is similarity due to evolutionary convergence. In **convergent evolution**, unrelated species develop similar features because they have similar ecological roles and natural selection has led to similar adaptations.

Phylogenetic trees are built on homologous similarities. In general, the more homology between two species, the more closely they are related. When two similar structures are very complex, they are more likely to be homologous and have a shared origin.

■ INTERACTIVE QUESTION   23.8

What two complications may make it difficult to determine phylogenetic relationships based on homology?

■ Molecular biology provides powerful new tools for systematics

Protein Comparison   Comparisons of amino acid sequences, which are genetically determined, provide evidence of shared ancestry. Differences between the amino acid sequences increase as organisms are more phylogenetically distant. Phylogeny based on cytochrome c is consistent with that derived from comparative anatomy and fossil evidence.

DNA Comparison Genome comparison is a direct measure of shared ancestry. **DNA–DNA hybridization** is a technique that measures the degree of base pairing between single strands of DNA from two species. The more extensive the pairing, the tighter the hybridized strands are bound and the higher the temperature needed to separate them, indicating a greater homology between the DNA sequences and a closer phylogenetic relationship. Evidence from this technique has been used to settle some old taxonomic disputes.

Restriction mapping of DNA compares fragments of DNA from different species cut by restriction enzymes. If two species have diverged greatly, their collections of restriction fragments will not match closely. Restriction mapping of mitochondrial DNA (mtDNA), which mutates about ten times faster than nuclear DNA, provides phylogenetic information for closely related species or populations.

DNA sequencing, determining the actual order of nucleotides in DNA, provides the most accurate information on genetic divergence between species. The sequencing of ribosomal RNA, whose genes change slowly, can be used to determine early phylogenetic relationships.

Analysis of Fossilized DNA Molecular systematics has now been applied to DNA extracted from fossils. Small quantities of DNA are amplified using polymerase chain reaction and tested for homology with modern species.

Molecular Clocks Some proteins seem to change or evolve at consistent rates. The number of amino acid substitutions in homologous proteins is proportional to the time elapsed since two species diverged. DNA comparisons also have been used as molecular clocks to date phylogenetic branching. Dates obtained by these methods are generally consistent with the fossil record.

The consistent rate of protein change and the rate of DNA divergence in groups of related species indicate that the accumulation of neutral mutations may change the genome as a whole more than do changes brought about by natural selection. Disagreement about the extent of neutral mutations has caused some evolutionists to rely on molecular clocks to determine the sequence of branches in phylogeny but not the actual dates for the origin of taxa.

■ **INTERACTIVE QUESTION 23.9**

How can the tools of molecular biology contribute valuable information to the fields of systematics, paleontology, and anthropology?

■ Cladistics highlights the phylogenetic significance of systematics

Phylogenetic trees indicate the relative time of origin of different taxa and the degree of divergence that develops between branches.

Phenetics **Phenetics** is a school of systematics that determines taxa strictly on the basis of phenotypic similarities and differences for as many characteristics as possible. This numerical analysis is particularly useful in analyzing DNA data.

Cladistics **Cladistics** is a school that classifies organisms according to the order in which clades, or evolutionary branches, arise. A **cladogram** represents these relationships as a series of dichotomous forks, each of which is defined by novel homologies for the species on that branch. *Shared primitive characters* are common for all the taxa to be grouped, and *shared derived characters*, or **synapomorphies**, are identified as homologies that evolved in an ancestor common to the taxa on a branch.

In the cladistic approach, birds are closer relatives of crocodiles than crocodiles are of lizards and snakes. Birds and crocodiles have synapomorphies not found in snakes and lizards. According to cladistics, the class Aves does not exist because birds are found within the clade of reptiles.

Classical Evolutionary Systematics Predating both phenetics and cladistics, **classical evolutionary systematics** considers both the divergence of structures and the sequence of branching in developing phylogenetic trees. In cases of conflict between these two considerations, a subjective decision is made. According to classical taxonomy, even though birds share a close phylogenetic branch with crocodiles, they are assigned to their own class on the basis of the major adaptive divergence that has resulted from the ability to fly.

■ Is a new evolutionary synthesis necessary?

The modern synthesis, which has dominated evolutionary theory for 50 years, combines contributions from paleontology, biogeography, systematics, population genetics, and molecular biology. This paradigm maintains that the gradual accumulation of many small changes occurring over vast periods of time can result in large-scale evolutionary changes. Microevolution is sufficient to explain most macroevolution, and natural selection is seen as the major cause of evolution at all levels.

■ INTERACTIVE QUESTION 23.10

Indicate whether the following statements are descriptive of phenetics (P), cladistics (C), or classical evolutionary (CE) systematics.

_____ **a.** method for reconstructing phylogeny

_____ **b.** determines taxa based only on quantifying multiple phenotypic comparisons

_____ **c.** considers only the order of branching of taxa

_____ **d.** uses synapomorphies to determine taxa

_____ **e.** classification scheme used in this book

_____ **f.** places humans in a family with chimpanzees and gorillas, separate from orangutans

_____ **g.** places birds in a class separate from reptiles

A current evolutionary debate concerns both the rate of evolution (gradualism versus punctuated equilibrium) and the relative contribution of microevolution to macroevolution. Some scientists favor a hierarchical theory, which maintains that events that lead to speciation and episodes of macroevolution may have little to do with adaptation and a lot to do with **historical contingency**, the occurrence of chance events. Evolutionary biologists agree that natural selection is the mechanism that adapts a population to its environment and refines the unique adaptations of new taxa.

STRUCTURE YOUR KNOWLEDGE

1. Briefly describe how each of the following mechanisms may contribute to macroevolution.

Mechanism	Effect on Macroevolution
Preadaptation	a.
Genes that control development	b.
Species selection	c.
Continental drift	d.
Extinctions	e.
Adaptive radiations	f.

2. Develop a concept map to organize your understanding of systematics, the objectives of taxonomy, and the three schools of systematics. Compare your map with those of your classmates or the suggested one in the answer section, but remember that the value of this exercise is in the process of creation.

3. Compare microevolution and macroevolution. What mechanisms are involved in both and at what level of evolution do they operate?

TEST YOUR KNOWLEDGE

MULTIPLE CHOICE: *Choose the one best answer.*

1. The richest source of fossils is found
   a. in coal and peat moss.
   b. along gorges.
   c. within sedimentary rock strata.
   d. encased in volcanic rocks.
   e. in amber.

2. Which of the following is least likely to leave a fossil?
   a. a soft-bodied land organism such as a slug
   b. a marine organism with a shell such as a mussel
   c. a vascular plant embedded in layers of mud
   d. a fresh-water snake
   e. a human

3. Index fossils are fossils of
   a. unique organisms that are used to determine the relative rates of evolution in different areas.
   b. widespread organisms that allow geologists to correlate strata of rocks from different locations.
   c. transitional forms that link major evolutionary groups.
   d. extinct organisms that indicate the separation of different eras.
   e. organisms that had the preadaptations that led to new taxa.

4. The half-life of carbon-14 is 5600 years. A fossil that is 22,400 years old would have what amount of the normal proportion of C-14 to C-12?
   a. 1/2   c. 1/6   e. 1/16
   b. 1/4   d. 1/8

5. The Permian extinctions between the Paleozoic and Mesozoic eras
   a. resulted in mass extinction of many marine and terrestrial species.
   b. coincided with the breaking apart of Pangaea.
   c. made way for the adaptive radiation of mammals, birds, and pollinating insects.
   d. appear to have been caused by a large asteroid striking Earth.
   e. did all of the above.

6. Related families are grouped into the next highest taxon called a
   a. class.
   b. order.
   c. phylum.
   d. genus.
   e. kingdom.

7. Species selection
   a. can cause evolutionary trends when species following the trend exist for long periods and speciate often.
   b. is not a gradual process because most species form, change little, and then become extinct.
   c. is the result of historical contingency, not natural selection.
   d. can explain the rapid adaptive radiation of species into adaptive zones following mass extinctions.
   e. is the basis for the classification into clades.

8. Convergent evolution may result
   a. when older structures are preadapted for new functions.
   b. when homologous structures are adapted for different functions.
   c. as a result of adaptive radiation.
   d. when species are widely separated geographically.
   e. when species have similar ecological roles.

9. *Lystrasaurus,* a fossil dinosaur, is found on most continents. This distribution is best explained by
   a. the relative ease of finding dinosaur bones.
   b. convergent evolution.
   c. divergent evolution.
   d. dispersal capabilities of ancient reptiles.
   e. biogeography.

10. The development of feathers in the ancestors of birds is an example of
    a. convergent evolution.
    b. allometric growth.
    c. analogous structures.
    d. preadaptation.
    e. adaptive radiation.

11. Synapomorphies are
    a. used to characterize all the species on a branch of a cladogram.
    b. determined through a computer comparison of all anatomical characteristics of a group of species.
    c. homologous structures that develop during adaptive radiation.
    d. used to indicate the degree of morphological differences between clades.
    e. placed at the base of a cladogram.

12. Allometric growth
    a. is the uneven growth of different body parts.
    b. involves paedomorphosis.

c. results in an evolutionary trend of increasing body size.
d. is an example of homeotic alteration.
e. is a common preadaptation that develops into an evolutionary novelty.

13. When human DNA is hybridized with both chimpanzee and orangutan DNA, which hybrid DNA would you expect to have the lowest melting temperature?
    a. human–chimp
    b. human–orangutan
    c. chimp–orangutan
    d. human DNA
    e. They would all be the same.

14. According to the hierarchical theory, macroevolution is primarily the result of
    a. natural selection.
    b. microevolution.
    c. geographic isolation.
    d. continental drift.
    e. historical contingency.

TRUE OR FALSE: *Indicate T or F, and then correct the false statements.*

_____ 1. A monophyletic taxon includes only species that share a common ancestor.

_____ 2. The more the sequences of amino acids in homologous proteins vary, the more recently the two species have diverged.

_____ 3. Phylogenetic trees determined on the basis of similar structures may be inaccurate when adaptive radiations have created large differences or when convergent evolution has created misleading analogies.

_____ 4. Phenetics is the school of taxonomy that is concerned with the order in which new groups branch over time.

_____ 5. The retention in the adult of juvenile traits of ancestors is called paedomorphosis.

_____ 6 Carbon-14 dating is not a good technique for sequencing fossils from the Carboniferous Period because carbon is too abundant in that period to allow accurate dating.

_____ 7. The layer of clay enriched in iridium between the sediments of the Mesozoic and Cenozoic eras is indicative of volcanic activity.

_____ 8. According to the modern synthesis, natural selection is seen as the major cause of evolution at all levels.

THE EVOLUTIONARY HISTORY OF BIOLOGICAL DIVERSITY

CHAPTER 24

EARLY EARTH AND THE ORIGIN OF LIFE

FRAMEWORK

This chapter describes the formation of Earth and presents a possible scenario for the chemical evolution of life between 4.0 and 3.5 billion years ago. Conditions on the primitive Earth are thought to have favored the spontaneous formation of organic monomers, the linking of these monomers into polymers, the grouping of aggregates of organic molecules into droplets (called protobionts) that became capable of metabolism and reproduction, and the development of self-replicating genetic information capable of directing metabolism and reproduction.

From this proposed beginning, an incredible biological diversity has evolved. This heterogeneous group of organisms is now classified into five kingdoms: Monera, Protista, Plantae, Fungi, and Animalia.

CHAPTER REVIEW

Life on Earth has been evolving for 3.5 billion years. Geological events have altered the course of biological evolution, and life has changed the Earth. The fossil record documents, however incompletely, the episodic development of much of the diversity of life. The origin of life on the young planet Earth, however, remains unrecorded and a matter of scientific speculation.

■ Life on Earth originated between 3.5 and 4.0 billion years ago

Precambrian fossils provide evidence of animals dating back 700 million years and prokaryotes spanning nearly 3 billion years before that. Fossils the size of bacteria have been discovered in a rock formation called the Fig Tree Chert, which is 3.4 billion years old. Fossils resembling spherical and filamentous prokaryotes have been found in 3.5 billion-year-old rocks called **stromatolites**, which are banded domes of sediment that form around the jellylike coats of motile microbes. It is possible that the earliest life forms may have emerged as long ago as 4 billion years, only 0.6 billion years after the origin of Earth. (See Interactive Question 24.1, p. 166.)

■ The first cells may have originated by chemical evolution on a young Earth: *an overview*

Between 4.0 billion years ago, when Earth's crust began to solidify, and 3.5 billion years ago, when records of stromatolites were deposited, life began.

■ INTERACTIVE QUESTION   24.1

On this time line, indicate the age of the oldest fossils of prokaryotes and animals that have been found. Show the approximate time when life on Earth is thought to have originated. Continue to fill in key events on this time line as you work through the chapters in this unit.

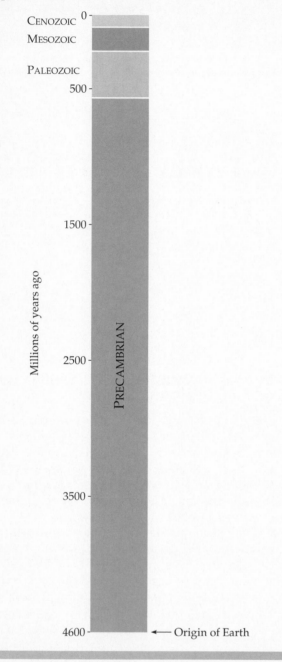

Most biologists believe that life developed on Earth from nonliving materials that became ordered into molecular aggregates capable of self-replication and metabolism.

■ Abiotic synthesis of organic monomers is a testable hypothesis: *science as a process*

In the 1920s, Oparin and Haldane independently hypothesized that conditions on the primitive Earth, in particular the reducing atmosphere, lightning, and intense ultraviolet radiation, favored the synthesis of organic compounds from inorganic precursors available in the atmosphere and seas.

In 1953, Miller and Urey tested the Oparin–Haldane hypothesis with an apparatus that simulated early Earth conditions. After a week of applying sparks (lightning) to a warmed flask of water (the primeval sea) in an atmosphere of H_2O, H_2, CH_4, and NH_3, Miller and Urey found a variety of amino acids and other organic compounds in the flask.

Various laboratory replications of the early Earth, even with the addition of low concentrations of O_2, have been able to produce all 20 amino acids, sugars, lipids, nitrogenous bases, and even ATP. The abiotic synthesis and accumulation of organic monomers could have been a natural process on primeval Earth.

■ Laboratory simulations of early Earth conditions have produced organic polymers

The abiotic synthesis of polymers may have occurred when dissolved organic monomers splashed onto hot rocks. Fox has created polypeptides he calls proteinoids by dripping dilute solutions of organic monomers onto hot sand or rock in the laboratory.

Clay may have been an important substratum for polymerization because of its ability to concentrate organic monomers, which bind to charged particles of clay. Metal atoms, such as iron and zinc, at some of the binding sites may have functioned as catalysts, facilitating the dehydration reactions that link monomers. Iron pyrite has also been proposed as a substratum for organic synthesis.

■ Protobionts can form by self-assembly

Protobionts, aggregates of abiotically produced organic molecules, probably preceded living cells. Protobionts may have developed such capabilities as metabolism, excitability, and the ability to maintain an internal chemical environment different from their surroundings.

Laboratory experiments indicate that protobionts could have formed spontaneously. Proteinoids, when mixed with cool water, self-assemble into tiny droplets called microspheres, which are coated by a selectively permeable protein membrane. An energy-storing membrane potential may develop across this membrane and discharge in nervelike fashion.

When the organic ingredients include lipids, liposomes, which are surrounded by a lipid bilayer, may spontaneously form. Coacervates are colloidal droplets that have been experimentally synthesized by shaking a solution of polypeptides, nucleic acids, and polysaccharides. If enzymes are included in the mix, the coacervates are capable of absorbing substrates, catalyzing reactions, and releasing the products.

Some abiotically produced molecules with weak catalytic capabilities could have enabled protobionts to develop a simple metabolism.

■ RNA was probably the first genetic material

The last step necessary for the evolution of life was the origin of genetic information. Successful protobionts would need not only to grow and divide, but also to develop a mechanism for replicating their successful characteristics, for creating instructions for making their key molecules. One hypothesis maintains that the first genes were short strands of RNA, which have been shown in the laboratory to be capable of self-replication. Cech's recent discovery of RNA catalysts, called **ribozymes**, indicates that RNA molecules may have been self-replicating in the prebiotic world.

Early populations of RNA molecules may have undergone natural selection as those molecules with the most stable three-dimensional conformations and greatest autocatalytic activity within a particular environment successfully competed for monomers and generated families of similar sequences.

RNA-directed protein synthesis may have begun with the weak binding of specific amino acids to bases along RNA molecules and their linkage to form a short polypeptide. This polypeptide may have then behaved as an enzyme to help the RNA molecule replicate.

When these early RNA and polypeptide molecules became packaged into protobionts, molecular cooperation could become more efficient due to the concentration of components and the potential for the protobiont to evolve as a unit. If primitive enzymes contained within a membrane-bound protobiont developed the ability to extract energy from an organic fuel, that energy would be available for the other reactions within the protobiont, and a simple metabolism could evolve.

■ The origin of hereditary information made Darwinian evolution possible

This four-step scenario results in a hypothetical antecedent of a cell. As protobionts grew, split, and distributed copies of their genes to offspring, errors in the copying of RNA would have led to variations, some of which could have had metabolic or genetic advantages. Differential reproductive success of these offspring would have accumulated metabolic and hereditary improvements. DNA, a more stable genetic molecule, eventually replaced RNA as the carrier of genetic information.

■ **INTERACTIVE QUESTION 24.2**

Describe the four steps of chemical evolution that are proposed as a possible route to the first organisms.

a.

b.

c.

d.

■ Debate about the origin of life abounds

Laboratory simulations cannot establish that this sequence actually occurred in the evolution of life, but they have shown that some of the key steps are possible. Alternative explanations include *panspermia*, the idea that some organic compounds may have reached the early Earth in meteorites and comets.

Some biologists suggest that RNA is too complicated to have formed the first genetic system and that simpler self-replicating organic molecules evolved first. Because the surface of early Earth was such an inhospitable environment, bombarded by asteroids and other cosmic debris, some scientists speculate that life began on the sea floor, perhaps at deep-sea vents.

Via whatever route, at some point membrane-enclosed compartments capable of metabolism and genetic replication moved past the gray border separating them from true cells so that, by 3.5 billion years ago, prokaryotes had evolved.

■ Arranging the diversity of life into kingdoms is a work in progress

The kingdom is the most inclusive taxonomic category. Intuitively and historically, we have divided the diversity of life into two kingdoms—plants and animals. Whittaker proposed a five-kingdom system in 1969. The prokaryotes are set apart from the eukaryotes and placed in the kingdom Monera. The kingdoms Plantae, Fungi, and Animalia consist of multicellular eukaryotes, defined by characteristics of nutrition, structure, and life cycle. Plants are autotrophic organisms; animals are heterotrophic organisms that ingest and digest their food; and fungi are absorptive heterotrophic organisms. The kingdom Protista includes unicellular eukaryotes or simple multicellular organisms that are believed to have descended from them.

The currently accepted five-kingdom system may be replaced as we increase our understanding of the evolutionary relationships among organisms.

STRUCTURE YOUR KNOWLEDGE

1. What do scientists think the primitive Earth was like? How could life possibly evolve in such an inhospitable environment?
2. Develop a simple concept map of the five kingdoms of life, indicating the key characteristics that are used to differentiate the kingdoms.

TEST YOUR KNOWLEDGE

MULTIPLE CHOICE: *Choose the one best answer.*

1. The primitive atmosphere of Earth may have favored the synthesis of organic molecules because
   a. It was highly oxidative.
   b. It was reducing and had energy sources in the form of lightning and UV radiation.
   c. It had a great deal of methane and organic fuels.
   d. It had plenty of water vapor, carbon, and nitrogen, providing the C, H, O, and N needed for organic molecules.
   e. It had no oxygen.

2. Life on Earth is thought to have begun
   a. 540 million years ago at the beginning of the Paleozoic era.
   b. 700 million years ago during the Precambrian era.
   c. 2.5 billion years ago when oxygen accumulated in the atmosphere.
   d. between 3.5 and 4.0 billion years ago.
   e. 4.6 billion years ago when the Earth formed.

3. Stromatolites are
   a. synthetic molecules consisting of amino adenosine and an ester that are able to self-replicate.
   b. a type of protobiont that forms when lipids assemble into a bilayer surrounding organic molecules.
   c. members of the kingdom Protista that are mobile and photosynthetic.
   d. fossils appearing to be prokaryotes that are the size of bacteria.
   e. banded domes of sediment that contain the earliest fossils.

4. Coacervates are
   a. droplets of molecules covered with a selectively permeable lipid membrane.
   b. droplets that self-assemble when polypeptides, nucleic acids, and polysaccharides are mixed.
   c. polypeptides formed in the laboratory by dripping organic monomers onto hot rocks.
   d. organic monomers associated with clay crystals.
   e. RNA self-replicating molecules.

5. A proposed hypothesis for the origin of genetic information is that
   a. Early DNA molecules coded for RNA, which then catalyzed the production of proteins.
   b. Early polypeptides became associated with RNA bases, and a catalyst linked the bases into RNA molecules.
   c. Short RNA strands were capable of self-replication and evolved by the natural selection of molecules that were most stable and autocatalytic.
   d. As protobionts grew and split, they distributed copies of their molecules to their offspring.
   e. Early RNA molecules coded for DNA, which then dictated the order of amino acids in a polypeptide.

6. Evidence that protobionts may have formed spontaneously comes from
   a. the discovery of ribozymes, showing that prebiotic RNA molecules may have been autocatalytic.
   b. the laboratory synthesis of microspheres, liposomes, and coacervates.

c. the fossil record found in the Fig Tree Chert.

d. the abiotic synthesis of polymers from monomers dripped onto hot rocks and clay.

e. the production of organic compounds in the Miller-Urey experimental apparatus.

7. Polymer formation from organic monomers may have been facilitated by
   a. the electrification of monomers by lightning.
   b. proteinoid formation by pH changes.
   c. amino acid concentration on clay particles.
   d. sedimentary accumulation on stromatolites.
   e. condensation of water into small pools.

8. According to the theory of panspermia,
   a. Organic molecules, abiotically synthesized in outer space, may have reached Earth carried on meteorites.
   b. RNA was the first genetic material.
   c. Life on Earth originated in shallow basins near volcanoes.
   d. Clay and zinc served as the first catalysts for abiotic synthesis of polymers.
   e. The reducing atmosphere of early Earth assisted the abiotic synthesis of organic molecules, and the oxidizing atmosphere of Earth prevents such abiotic evolution now.

9. The hypothesis that life originated in the sea is based on
   a. that fact that most species are marine.
   b. the discovery of deep-sea vents.
   c. the inhospitable asteroid-crashing environment of early Earth.
   d. the discovery of early fossils in sediments.
   e. both b and c.

10. In the five-kingdom system of classification,
    a. Prokaryotes are placed in the kingdom Protista.
    b. The fungi include absorptive heterotrophs.

c. Single-celled or simple multicellular eukaryotes are placed in the kingdom Monera.

d. The kingdoms Animalia, Plantae, and Fungi contain the only multicellular organisms.

e. Autotrophic multicellular prokaryotes are placed in the kingdom Plantae.

11. In order to place a newly identified species into one of the five kingdoms, the first piece of information you would need to acquire is
    a. the way in which it acquires food.
    b. its life cycle.
    c. the absence or presence of chloroplasts.
    d. the absence or presence of a nuclear membrane.
    e. whether the organism is multicellular.

MATCHING: *Match the scientist with the theory or experiment.*

_____ 1. produced organic compounds in a lab apparatus

_____ 2. discovered RNA catalysts called ribozymes

_____ 3. developed the five-kingdomsystem

_____ 4. claimed that conditions of primitive Earth favored synthesis of organic compounds

_____ 5. produced proteinoids in the laboratory by dripping monomers onto hot sand or clay

A. Fox

B. Oparin and Haldane

C. Crick

D. Cech

E. Whittaker

F. Miller and Urey

CHAPTER 25

PROKARYOTES AND THE ORIGINS OF METABOLIC DIVERSITY

FRAMEWORK

Many systematists place the prokaryotes of the kingdom Monera in the separate domains eubacteria and archaebacteria due to their early evolutionary divergence. This chapter presents the morphology and reproduction of these generally single-celled organisms. Every mode of nutrition and most metabolic pathways evolved within the prokaryotes. They exist in all habitats and in various symbiotic relationships with other organisms. These most numerous and diverse of all organisms are essential to the chemical cycles necessary to maintain life on Earth.

CHAPTER REVIEW

The history of prokaryotes, collectively called bacteria, spans at least 3.5 billion years, and they evolved alone for the first 2 billion years.

■ They're (almost) everywhere! *an overview of prokaryotic life*

Prokaryotes outnumber all eukaryotes combined and flourish in all habitats, including ones that are too harsh for any other forms of life. These small organisms lack membrane-enclosed organelles, and most have cell walls which differ in composition from those of plants, protists, and fungi. Their genomes are simpler than those of eukaryotes and differ in some details of replication, expression, and recombination.

The collective impact of these microscopic organisms is huge. The chemical cycles necessary for life depend on prokaryotes. Many prokaryotes live in symbiotic relationships. Mitochondria, chloroplasts, and the other four kingdoms probably evolved from symbiotic associations of prokaryotes.

■ Archaebacteria and eubacteria are the two main branches of prokaryotic evolution

The **archaebacteria** and **eubacteria** differ in many key characteristics. Archaebacteria, of ancient origin, are found in extreme habitats reminiscent of early Earth. Most prokaryotes are eubacteria, differing from archaebacteria in structure, biochemistry, and physiology. Woese and other researchers propose a three-**domain** system—**domain Archaea**, **domain Bacteria**, and **domain Eucarya**—to account for the more ancient divergence of Archaea and Bacteria than of Archaea and Eucarya.

■ INTERACTIVE QUESTION 25.1

Fill in the domains on this cladogram of the proposed origin of these three groups.

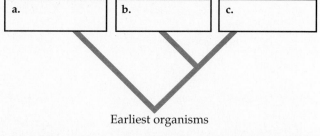

Earliest organisms

■ The success of prokaryotic life is based on diverse adaptations of form and function

Morphological Diversity of Prokaryotes Prokaryotes exist primarily as single cells, although some species may form permanent aggregates of identical cells called colonies and some exhibit a simple multicellular form. Prokaryotic cells are usually 1–5 μm in diameter, one-tenth the size of eukaryotic cells. The three most common shapes of prokaryotes are spheres (cocci), rods (bacilli), and helices (spirilla and spirochetes).

The Cell Surface of Prokaryotes The cell walls of eubacteria contain **peptidoglycan**, a matrix composed of polymers of sugars cross-linked by short polypeptides. Archaebacterial cell walls lack peptidoglycan.

The **Gram stain** is an important tool for identifying eubacteria as **gram-positive** (bacteria with walls containing a thick layer of peptidoglycan) or **gram-negative** (bacteria with more complex walls including an outer lipopolysaccharide membrane). Pathogenic gram-negative bacteria are often more harmful than gram-positive strains because their lipopolysaccharides may be toxic. This outer membrane also protects them from the defenses of their hosts and from antibiotics. Many antibiotics, such as penicillin, inhibit the synthesis of cross-links in peptidoglycan and prevent wall formation.

Many prokaryotes secrete a sticky **capsule** outside the cell wall that serves as protection and a glue for adhering to a substratum or other prokaryotes. Bacteria may also attach by means of surface appendages called **pili**, which may be specialized to hold bacteria together during conjugation.

Motility of Prokaryotes Many bacteria are equipped with flagella, either scattered over the cell surface or concentrated at one or both ends of the cell. These flagella differ from eukaryotic flagella in size, lack of a plasma membrane cover, structure, and function. Spirochetes have filaments attached at both ends of the cell that spiral around the cell under the cell wall and move past each other, causing the cell to move in a corkscrew fashion. Other motile bacteria glide on a slime that they secrete.

Many motile bacteria exhibit **taxis**, an oriented movement in response to chemical, light, magnetic, or other stimuli.

Internal Membranous Organization Extensive internal membrane systems are not found in prokaryotic cells, although there may be some infoldings of the plasma membrane that function in respiration. Thylakoid membranes in cyanobacteria function in photosynthesis.

Prokaryotic Genomes The circular double-stranded DNA chromosome is concentrated in a **nucleoid region**. It contains one one-thousandth as much DNA as a eukaryotic genome and has very little protein associated with it. Smaller rings of DNA, called plasmids, may carry genes for antibiotic resistance, metabolism of unusual nutrients, or other functions.

They replicate independently of the genophore and may be transferred between bacteria during conjugation.

Bacterial ribosomes are smaller than eukaryotic ribosomes and differ in their protein and RNA content. Some antibiotics work by binding specifically to bacterial ribosomes and blocking protein synthesis.

Growth, Reproduction, and Gene Exchange Prokaryotes reproduce asexually by **binary fission**, producing a colony of progeny. Some bacteria form **endospores**, which are tough-walled cells that can resist even boiling water; thus microbiologists must use autoclaves to sterilize laboratory equipment and media.

Bacterial growth rate in an optimal environment is geometric, with generation times from 20 minutes to 3 hours. The growth of bacterial colonies usually stops due to the exhaustion of nutrients or the toxic accumulation of wastes. Some bacteria, as well as certain protists and fungi, release **antibiotics**, which inhibit the growth of competitors.

Despite the lack of meiosis and sexual cycles, there are three mechanisms of genetic recombination in bacteria: **transformation**, the uptake of genes from the environment; **conjugation**, the direct transfer of genetic material from one bacterium to another; and **transduction**, the injection of genes by a viral bacteriophage. Mutations are the major source of genetic variation in prokaryotes. Due to short generation times, favorable mutations and beneficial genomes resulting from recombination are rapidly spread.

■ INTERACTIVE QUESTION 25.2

Fill in the following table with a brief description of the characteristics of prokaryotic cells.

Property	Description
Cell shape	a.
Cell size	b.
Cell surface	c.
Motility	d.
Internal membranes	e.
Genome	f.
Growth and reproduction	g.

■ All major types of nutrition and metabolism evolved among prokaryotes

Major Modes of Nutrition Nutrition refers to how an organism obtains energy (photo- or chemotrophs) and the carbon it uses for synthesizing organic compounds (auto- or heterotrophs).

Thus, there are four major nutritional categories: (1) **photoautotrophs**, cyanobacteria and other photosynthetic prokaryotes that use light energy and CO_2 to synthesize organic compounds; (2) **chemoautotrophs**, which obtain energy by oxidizing inorganic substances (such as H_2S, NH_3, or Fe^{2+}) and need only CO_2 as a carbon source; (3) **photoheterotrophs**, which use light energy to generate ATP but obtain carbon in organic form; and (4) **chemoheterotrophs**, which use organic molecules as both an energy and carbon source.

Nutritional Diversity Among Chemoheterotrophs The majority of bacteria are either **saprobes**, which decompose and absorb nutrients from dead organic matter, or **parasites**, which absorb nutrients from living hosts.

Some fastidious bacteria require all 20 amino acids, vitamins, and other organic compounds in order to grow, whereas others may flourish with glucose as their only organic nutrient. The few classes of synthetic organic compounds that cannot be broken down by any bacteria are called nonbiodegradable.

Nitrogen Metabolism Nitrogen, an essential component of proteins and nucleic acids, is cycled through the ecosystem with the aid of bacteria that are able to metabolize various nitrogenous compounds. Through **nitrogen fixation**, some species of bacteria convert atmospheric N_2 to ammonia.

Metabolic Relationships to Oxygen **Obligate aerobes** need oxygen for cellular respiration; **facultative anaerobes** can use oxygen but also can grow in anaerobic conditions using fermentation; **obligate anaerobes** cannot use—and are poisoned by—oxygen. Some obligate anaerobes use anaerobic respiration to break down nutrients, with inorganic molecules other than oxygen serving as the final electron acceptor.

■ The evolution of prokaryotic metabolism was both cause and effect of changing environments on Earth

All major metabolic pathways evolved in prokaryotes before there were eukaryotes, probably within the first billion years of life on Earth. Hypotheses about metabolic evolution, including the following proposed sequence of events, are based on geological, molecular, and metabolic evidence.

The Origin of Glycolysis The first prokaryotes probably were chemoheterotrophs that absorbed abiotically synthesized organic compounds. Based on its universal role in energy exchange in all modern organisms, ATP must have been an early important energy molecule. Glycolysis may have gradually evolved in cells that had enzymes capable of breaking down organic compounds and regenerating ATP from ADP

■ **INTERACTIVE QUESTION   25.3**

Complete the following concept map that organizes the various modes of nutrition found in bacteria.

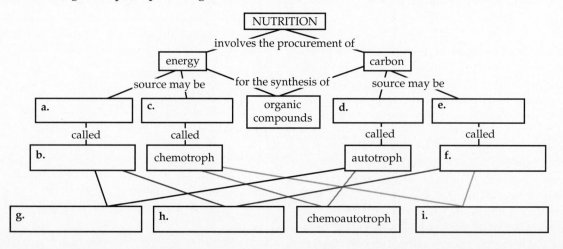

by substrate phosphorylation. Glycolysis is the only metabolic pathway common to nearly all modern organisms.

Glycolysis and fermentation do not require molecular oxygen. The archaebacteria and other obligate anaerobes that use fermentation are believed to have forms of nutrition most like that of the original prokaryotes that developed on anaerobic Earth.

The Origin of Electron Transport Chains and Chemiosmosis The chemiosmotic synthesis of ATP is common to all domains, again implying an early origin. Transmembrane proton pumps, driven by ATP, may have originally functioned to regulate the internal pH of early prokaryotes. An electron transport chain that coupled the oxidation of organic acids with the transport of H^+ out of the cell would have conserved ATP. Excess H^+ could then diffuse back, reverse the proton pump, and thus generate ATP. This type of anaerobic respiration persists in some modern bacteria.

The Origin of Photosynthesis Light-absorbing pigments may have protected cells from excess light and then become coupled with electron transport systems to drive ATP synthesis. A simple mechanism of photophosphorylation has been found in some archaebacteria called extreme halophiles in which bacteriorhodopsin, a membrane-bound pigment molecule, uses light energy to pump protons out of the cell. The resulting gradient is used for ATP synthesis.

The development of pigments and photosystems co-opting components of electron transport chains may have allowed some prokaryotes to use light energy to drive electrons from H_2S to $NADP^+$ and thus generate reducing power to fix CO_2 and make their own organic molecules. The nutrition of the green and purple sulfur bacteria probably is most like that of the early photosynthetic prokaryotes.

Cyanobacteria, the Oxygen Revolution, and the Origins of Cellular Respiration The first **cyanobacteria** evolved a mechanism that reduced CO_2 using water as a source of electrons and hydrogen. They began to release O_2 as a by-product of their photosynthesis.

Cyanobacteria evolved at least 2.5 billion years ago. Banded iron formations are found in marine sediments from that time, probably the result of the oxygen evolved by the cyanobacteria precipitating dissolved iron ions as iron oxide. When the dissolved iron was exhausted, oxygen began to accumulate in the seas and bubble out into the atmosphere. Oxidized iron is found in terrestrial rocks from about 2 billion years ago.

The change to a more oxidizing atmosphere probably caused the extinction of many bacteria, while others survived in anaerobic habitats where their descendants are found today. Some bacteria evolved antioxidant mechanisms that protected them from rising oxygen levels. Some photosynthetic prokaryotes, however, began using the oxidizing power of O_2 to pull electrons from organic molecules down existing transport chains, leading to the development of aerobic respiration by the co-opting of transport chains from photosynthesis. Purple nonsulfur bacteria are photoheterotrophs that still use a hybrid photosynthetic and respiratory electron transport system. Several bacterial lines gave up photosynthesis and their electron transport chains became adapted to function exclusively in aerobic respiration.

■ **INTERACTIVE QUESTION   25.4**

Place in order the following steps in the proposed evolution of prokaryotic metabolism and briefly describe how they may have come about: proton pumps, photosynthesis using H_2S, glycolysis, electron transport chain and chemiosmosis, aerobic respiration, photosynthesis evolving oxygen.

a.

b.

c.

d.

e.

f.

■ Molecular systematics is leading to a phylogenetic classification of prokaryotes

Comparisons of base sequences in ribosomal RNA and other nucleic acids revealed unique **signature sequences** and indicated that prokaryotes split into two divergent lineages very early in the history of life.

Domain Archaea (Archaebacteria) The archaebacteria, which live in extreme environments, consist of three main groups.

Methanogens have a unique energy metabolism in which H_2 is used to reduce CO_2 to methane (CH_4). These strict anaerobes often live in swamps and marshes, are important decomposers in sewage treatment, and are gut inhabitants that contribute to the nutrition of cattle and other herbivores.

The **extreme halophiles** (salt-lovers) live in saline places such as the Great Salt Lake and the Dead Sea. Their photosynthetic pigment bacteriorhodopsin gives them a purple-red color.

Extreme thermophiles may be found oxidizing sulfur for energy in hot sulfur springs and near deep-sea hydrothermal vents. Extreme thermophiles appear to be the prokaryotes most closely related to eukaryotes.

Domain Bacteria (Eubacteria)   Eubacteria show great diversity in their modes of nutrition and metabolism. Molecular systematics has helped to establish about 12 phylogenetically related groups, five of which are described in the text.

Proteobacteria are the most diverse group of bacteria and include the photoautotrophic or photoheterotrophic anaerobic purple bacteria, chemoautotrophic bacteria that are important in the nitrogen cycle, and the chemoheterotrophic enteric bacteria that inhabit intestinal tracts.

Most of the *gram-positive eubacteria* are chemoheterotrophs, some of which form resistant endospores. Mycoplasmas, the smallest of all cells, are the only eubacteria that lack cell walls. Actinomycetes are branching soil bacteria, many of which produce antibiotics.

The *cyanobacteria* have plantlike photosynthesis. Some filamentous cyanobacteria have heterocysts, cells specialized for nitrogen fixation.

Spirochetes are helical chemoheterotrophs that move in a corkscrew fashion. Spirochetes cause syphilis and Lyme disease.

Chlamydias are obligate intracellular animal parasites. One of these gram-negative species is a common cause of blindness and nongonococcal urethritis, a common sexually transmitted disease.

■ Prokaryotes continue to have an enormous ecological impact

Prokaryotes and Chemical Cycles   Prokaryotes are indispensable in recycling chemical elements between the biological and physical worlds. Bacteria are **decomposers** that return carbon, nitrogen, and other elements to the environment for assimilation into new living forms. Autotrophic bacteria bring carbon from CO_2 into the food chain and release O_2 to the atmosphere, and cyanobacteria fix atmospheric nitrogen

■ INTERACTIVE QUESTION   25.5

a. Identify the large structure inside this *Bacillus* cell. What is its significance?

Bacillus

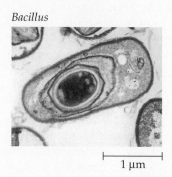

|—— 1 µm ——|

b. Identify the spherical cell in this filamentous cyanobacteria, *Anabaena*. What is the function of this cell?

Anabaena

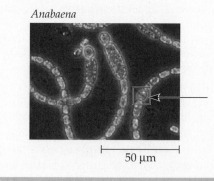

|—— 50 µm ——|

and supply plants with the nitrogen they need to make proteins.

Symbiotic Bacteria   **Symbiosis**, an ecological relationship involving direct contact between organisms of two different species, probably played a major role in the evolution of prokaryotes and the origin of eukaryotes. The organisms are called **symbionts** and, if one is much larger than the other, it is called the **host**. In **mutualism**, both symbionts benefit. Nodules on the roots of legumes house symbiotic bacteria that fix nitrogen for the host and, in return, receive nutrients. In **commensalism**, one symbiont benefits while the other is neither harmed nor helped. Many of the bacteria found in and on the human body are commensal, although some may be mutualistic. In **parasitism**, the symbiont, now called a parasite, benefits at the expense of the host. Parasitic bacteria that cause disease are called pathogens.

Bacteria and Disease   About one-half of all human diseases are caused by pathogenic bacteria that manage to invade the body, resist internal defenses, and reproduce enough to harm the host. **Opportunistic** bacteria are normal inhabitants of the human body that cause illness when the body's defenses are weakened.

Koch's postulates include four criteria for establishing a pathogen as the cause of a disease: (1) find the same pathogen in each diseased individual; (2) isolate and grow the pathogen in a pure culture; (3) induce the disease in experimental animals using the cultured pathogen; and (4) isolate the same pathogen from the experimental animal after it develops the disease.

Some bacteria, such as the actinomycetes that cause tuberculosis and leprosy, cause disease when they invade host tissues. Pathogenic bacteria more commonly cause disease by producing toxins. **Exotoxins**, among the most potent poisons known, are proteins secreted by bacteria that cause such diseases as botulism and cholera. **Endotoxins**, which are components of the outer membrane of certain gram-negative bacteria, all produce fever and aches in the host. Examples are typhoid fever and *Salmonella* food poisoning.

In the past century, improved hygiene and sanitation and the development of antibiotics have decreased the incidence and severity of bacterial disease, reduced infant mortality, and extended life expectancy in developed countries. More than half of our antibiotics come from the soil bacteria genus *Streptomyces*, an actinomycete. The evolution of antibiotic-resistant strains of pathogenic bacteria poses a serious health threat.

Putting Bacteria to Work The diverse metabolic capabilities of prokaryotes have been used to digest organic wastes, produce chemical products, make vitamins and antibiotics, and produce food products such as yogurt and cheese. Research using *E. coli* and other bacteria has expanded our understanding of molecular biology, and recombinant DNA techniques using bacteria develop both basic knowledge and practical applications.

STRUCTURE YOUR KNOWLEDGE

1. What is the rationale for separating the archaebacteria, the eubacteria, and all the eukaryotes into three domains?

2. Describe four positive ways in which prokaryotes have an impact on our lives and on the world around us.

3. Add to the time line on page 166 in Chapter 24 the approximate time when oxygen from photosynthetic cyanobacteria began to accumulate in the atmosphere.

TEST YOUR KNOWLEDGE

MULTIPLE CHOICE: *Choose the one best answer.*

1. Which of the following is not true of plasmids?
   a. They replicate independently of the main chromosome.
   b. They may carry genes for antibiotic resistance.
   c. They are essential for the existence of bacterial cells.
   d. They may be transferred between bacteria during conjugation.
   e. They may carry genes for special metabolic pathways.

2. Gram-positive bacteria
   a. have peptidoglycan in their cell walls, whereas gram-negative bacteria do not.
   b. have an outer membrane around their cell walls.
   c. have lipopolysaccharides in their cell walls and thus may be more pathogenic than gram-negative bacteria.
   d. are members of the archaebacteria.
   e. have simpler, thick peptidoglycan cell walls.

3. Many prokaryotes secrete a sticky capsule outside the cell wall that
   a. allows them to glide through a slime layer.
   b. serves as protection from host defenses and glue for adherence.
   c. reacts with the Gram stain.
   d. is used for attaching cells during conjugation.
   e. is composed of peptidoglycan, polymers of modified sugars cross-linked by short polypeptides.

4. Chemoautotrophs
   a. are photosynthetic.
   b. use organic molecules for an energy and carbon source.
   c. oxidize inorganic substances for energy and use CO_2 as a carbon source.
   d. use light to generate ATP but need organic molecules for a carbon source.
   e. use light energy to extract electrons from H_2S.

5. The major source of genetic variation in prokaryotes is
   a. binary fission.
   b. mutation.
   c. conjugation.
   d. plasmid exchange.
   e. transformation.

6. A symbiotic relationship in which the symbiont benefits and the host is neither harmed nor helped is called
   a. saprocism.
   b. opportunism.
   c. mutualism.
   d. commensalism.
   e. parasitism.

7. Facultative anaerobes
   a. can survive with or without oxygen.
   b. are poisoned by oxygen.
   c. are anaerobic chemoautotrophs.
   d. are all able to fix atmospheric nitrogen to make NH_3, which is then available for plant growth.
   e. include the methanogens that reduce CO_2 to CH_4.

8. Banded iron formations in marine sediments provide evidence of
   a. the first prokaryotes around 3.5 billion years ago.
   b. the evolution of chemoautotrophs that oxidized ferrous ions for energy.
   c. oxidized iron layers in terrestrial rocks.
   d. the release of O_2 by cyanobacteria.
   e. the early evolution of extreme thermophiles in hydrothermal vents.

9. Glycolysis
   a. is the only metabolic pathway common to nearly all modern organisms and may have been the first metabolic pathway to evolve.
   b. breaks down organic molecules and generates ATP when transmembrane proton pumps are reversed.
   c. uses the electron transport chain to create a proton gradient to produce ATP.
   d. is an aerobic process that generates organic acids that lower the pH of cells.
   e. is or does all of the above.

10. Aerobic respiration may have originated when
    a. Some bacteria evolved antioxidant mechanisms to protect them from rising oxygen levels.
    b. Photosynthetic bacteria produced an excess of organic molecules.
    c. Some bacteria used modified electron transport chains from photosynthesis to pass electrons from organic molecules to O_2.
    d. Some bacteria gave up photosynthesis and reverted to chemoheterotrophic nutrition.
    e. The supply of free ATP dwindled in early environments.

11. The archaebacteria
    a. are believed to be more closely related to eukaryotes than to eubacteria.
    b. have cell walls that lack peptidoglycan.
    c. are found in harsh habitats, reminiscent of the environment of early Earth.
    d. include the methanogens, extreme halophiles, and extreme thermophiles.
    e. are all of the above.

FILL IN THE BLANKS

_____ 1. the name for spherical bacteria

_____ 2. region in which the prokaryotic chromosome is found

_____ 3. common laboratory technique for identifying bacteria

_____ 4. surface appendages of bacteria used for adherence to substrate

_____ 5. an oriented movement in response to light or chemical stimuli

_____ 6. nutrition type in which organic molecules are used for both energy and carbon source

_____ 7. resistant cell that can survive harsh conditions

_____ 8. organism that must grow anaerobically

_____ 9. organisms that decompose and absorb nutrients from dead organic matter

_____ 10. proteins that are secreted by bacteria and are potent poisons

_____ 11. thickenings on roots that house nitrogen-fixing bacteria

_____ 12. type of bacteria with plantlike photosynthesis

CHAPTER 26

THE ORIGINS OF EUKARYOTIC DIVERSITY

FRAMEWORK

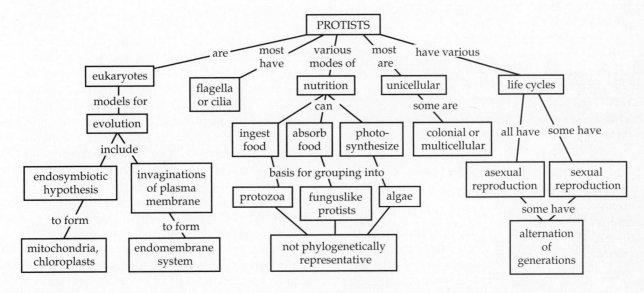

CHAPTER REVIEW

Protists are primarily eukaryotic unicellular organisms, with a few colonial and multicellular organisms. Ancient protists, probably the first eukaryotes to evolve from prokaryotes, were the ancestors of modern protists, plants, fungi, and animals.

■ Eukaryotes originated by symbiosis among prokaryotes

How the complex eukaryotic cell evolved from simpler prokaryotic cells is a fundamental question in biology. The increase in complexity and organization of prokaryotes followed three separate trends: the move toward multicellular prokaryotes, such as filamentous cyanobacteria with specialized cells; the evolution of bacterial communities in which species bene-

fited from the metabolic specialties of other species; and the trend to compartmentalize different functions within a cell that led to eukaryotic cells. Invaginations of the plasma membrane may have created the endomembrane system, and endosymbiosis may have led to mitochondria and chloroplasts.

According to the **endosymbiotic theory** developed by Margulis, eukaryotic cells evolved from symbiotic combinations of prokaryotic cells. Chloroplasts are descendants of photosynthetic prokaryotes that became endosymbionts within larger cells. Mitochondria developed from aerobic heterotrophic bacteria that entered larger cells as prey or parasites.

Similarities exist between modern eubacteria and the chloroplasts and mitochondria of eukaryotes, including size, membrane enzymes and transport systems, circular DNA molecules not associated with proteins, and the process of division. Chloroplasts and mitochondria contain ribosomes and other equipment needed to transcribe and translate their DNA into proteins. These ribosomes more closely resemble prokaryotic ribosomes than they do cytoplasmic ribosomes. These similarities, as well as the present-day existence of endosymbiotic relationships, are cited as evidence of the endosymbiotic theory. Molecular systematics, comparing base sequences of ribosomal RNA, also suggest eubacterial origins for chloroplasts and mitochondria.

A theory for the development of the eukaryotic cell must also explain the evolution of the 9 + 2 microtubule arrangement in flagella and cilia and the processes of mitosis and meiosis, which also rely on microtubules.

■ INTERACTIVE QUESTION   26.1

a. How is the endomembrane system of eukaryotic cells thought to have evolved?

b. Review the evidence for the endosymbiotic model of eukaryotic cell evolution.

■ Archezoans provide clues to the early evolution of eukaryotes

Diplomonads such as *Giardia* are included with a few other phyla in **archezoa**, an ancient lineage that branched from the eukaryotic tree perhaps more than 2 billion years ago. These organisms have relatively simple cytoskeletons, ribosomes that more closely resemble those of prokaryotes, and no mitochondria or plastids. Some systematists advocate a separate kingdom for archezoans.

The two haploid nuclei of diplomonads may be remnants of an early evolutionary stage in the development of diploidy in eukaryotes, which branched off before the endosymbiotic evolution of mitochondria. Serial endosymbiosis represents a merger of evolutionary lineages in the creation of a unique life form, a mechanism sometimes referred to as reticulate evolution.

■ The diversity of protists represents "experiments" in the evolution of eukaryotic organization

The oldest fossils of protists, found in rocks 2.1 billion years old, are **acritarchs**, remnants that resemble rup-

tured coats of cysts. Due to their early divergence, modern protists vary in cellular anatomy, ecological roles, and life histories more than any other group.

Protists abound almost anywhere there is water—they occupy freshwater, marine, and moist terrestrial habitats or live symbiotically within the bodies of hosts. They form an important part of **plankton**, the drifting community of tiny organisms found in bodies of water.

Nearly all protists are aerobic and use mitochondria for their respiration. Nutritionally, protists can be photoautotrophs, heterotrophs, or mixotrophs, which are both photosynthetic and heterotrophic. Based on their mode of nutrition and not phylogeny, protists are often grouped for convenience into three categories: photosynthetic, plantlike protists (**algae**); ingestive, animal-like protists (**protozoa**); and absorptive, funguslike protists.

Most protists have flagella or cilia at some point during their lifetime. These structures are extensions of the plasma membrane that encase cytoplasm and a 9 + 2 arrangement of microtubules.

Mitosis occurs in most phyla of protists, but details of the process vary. All protists can reproduce asexually; some also have sexual reproduction, and others may use meiosis and **syngamy** (the union of gametes) to exchange genes between two otherwise asexually reproducing individuals. All three basic types of sexual life cycles, and variations on those types, are found among protists. Many life histories include the formation of resistant **cysts**.

These unicellular organisms are exceedingly complex at the cellular level. A single cell must be capable of performing all the basic functions that are divided among specialized cells in plants and animals.

■ INTERACTIVE QUESTION   26.2

Fill in this summary table of protist characteristics.

Characteristic	Brief Description
Fossil evidence	a.
Habitats	b.
Nutrition	c.
Locomotion	d.
Reproduction, life cycles	e.

Protistan taxonomy is in a state of flux

The kingdom Protista, originally created for unicellular eukaryotes, has been expanded to include multicellular forms whose cell ultrastructure and life cycles suggested they were more closely related to certain protists than to plants or fungi. Molecular comparisons, especially of ribosomal RNA, are resulting in three taxonomic trends: (1) reassessment of protistan phyla; (2) arrangement of phyla into a cladogram, and (3) reevaluation of the five-kingdom system.

Diverse modes of locomotion and feeding evolved among protozoa

Protists that ingest food, seeking out and consuming bacteria, other protists, or detritus, or living protists as symbionts are informally grouped as protozoans, which are subdivided into phyla on the basis of how they feed and move.

Rhizopoda (Amoebas) The **amoebas** are naked or shelled unicellular protists without flagella that move and feed with **pseudopodia**. Microtubules and microfilaments of the cytoskeleton function in the formation of these cellular extensions. Their asexual reproduction occurs by nontypical forms of mitosis; there is no known sexual reproduction in this group. Most amoebas are found free-living in freshwater, marine, or soil habitats.

Actinopoda (Heliozoans and Radiozoans) The slender projections called axopodia help these planktonic organisms to remain buoyant and to phagocytize microscopic food organisms. **Heliozoans** are freshwater actinopods. **Radiozoans** are primarily marine and have delicate silica shells.

Foraminifera (Forams) These marine organisms are known for their porous, calcareous shells. Cytoplasmic strands extending through the pores function in swimming, shell formation, and feeding. Many **forams** derive nourishment from symbiotic algae. Foram fossils are useful as index fossils for dating sedimentary rocks.

Apicomplexa (Sporozoans) These animal parasites have complex life cycles that often include sexual and asexual stages and several host species. Infections are spread by tiny infectious cells called **sporozoites**. The control of malaria, caused by the apicomplexan *Plasmodium*, is complicated by the development of insecticide resistance in *Anopheles* mosquitoes, which spread the disease, and drug resistance in *Plasmodium*, as well as by the sequestering of the parasite within human liver and blood cells and the ability of the parasite to change the surface proteins presented to the host's immune system.

Zoomastigophora (Zooflagellates) These protozoa are characterized by many whiplike flagella. **Zooflagellates** are heterotrophic organisms that may be free-living or symbiotic. Mutualistic flagellates that live in the gut of termites digest the cellulose eaten by the host. *Trypanosoma*, a parasitic zooflagellate, causes African sleeping sickness.

Ciliophora (Ciliates) These protists are characterized by the cilia they use to move and feed. Their numerous cilia, coordinated by a submembrane system of microtubules, may be widespread or clumped. Most **ciliates** are complex unicellular organisms found in fresh water.

Ciliates have two types of nuclei—a large macronucleus, which controls everyday functions of the cell, has 50 or more copies of the genome, and splits during binary fission (asexual reproduction); and many small micronuclei, which are exchanged in a sexual process called **conjugation**.

▓ INTERACTIVE QUESTION 26.3

Label the indicated structures in this diagram of a *Paramecium*. To which phylum does this organism belong?

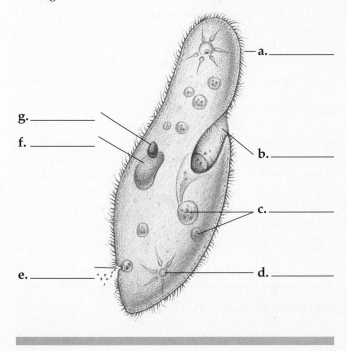

Funguslike protists have morphological adaptations and life cycles that enhance their ecological role as decomposers

Although the slime molds resemble fungi, their cellular organization, reproduction, and life cycles suggest that they are more closely related to amoeboid protists. Molecular comparisons indicate that water molds are related to certain algae.

Myxomycota (Plasmodial Slime Molds) **Plasmodial slime molds** are heterotrophic, engulfing food particles by phagocytosis as they grow through leaf litter or rotting logs. A coenocytic (multinucleate) amoeboid mass called a **plasmodium** is the feeding stage. During the sexual stage, sporangia on erect stalks produce resistant spores by meiosis. Spores germinate, the amoeboid or flagellated haploid cells fuse, and the diploid nucleus repeatedly divides without cytoplasmic divisions to form a new plasmodium.

Acrasiomycota (Cellular Slime Molds) During the feeding stage of the life cycle, **cellular slime molds** consist of solitary amoeboid cells. In the absence of food, the individual cells congregate into a mass resembling a plasmodial slime mold. Cellular slime molds are haploid and the fruiting bodies produce resistant spores asexually by mitosis. In sexual reproduction, two amoebas fuse to form a zygote that develops into a giant cell enclosed in a protective wall. Following meiosis and mitosis, haploid amoebas are released.

Oomycota (Water Molds) The water molds, white rusts, and downy mildews of this phylum resemble fungi in appearance and nutrition, but they have cell walls made of cellulose, a predominant diploid stage in their life cycles, and flagellated cells—all characteristics unlike those of true fungi.

The large egg cell of a water mold is fertilized by a smaller sperm nucleus. Zygotes, following dormancy within resistant walls, germinate to form coenocytic hyphae, which are tipped by zoosporangia that asexually produce flagellated zoospores. The saprophytic water molds are important decomposers in aquatic ecosystems. White rusts and downy mildews can be destructive parasites of land plants.

Eukaryotic algae are key producers in most aquatic ecosystems

Eukaryotic algae are relatively simple, aquatic, and mostly photosynthetic organisms that most taxonomists place in kingdom Protista in the five-kingdom

INTERACTIVE QUESTION   26.4

Compare the following aspects of the life cycles of plasmodial slime molds, cellular slime molds, and water molds.

Phylum	Haploid or Diploid	Common Body Form	Gametes or Sexual Cells
Plasmodial slime molds	a.	b.	c.
Cellular slime molds	d.	e.	f.
Water molds	g.	h.	i.

system. An important ecological group, algae generate about half of all photosynthetically produced organic material and form the base of aquatic food webs.

All algae have chlorophyll *a*, as do plants. Accessory pigments vary and are used—along with cell wall chemistry, stored food products, number and position of flagella, and chloroplast structure—as classification characteristics.

Dinoflagellata (Dinoflagellates) **Dinoflagellates** make up a large proportion of the phytoplankton at the base of most marine food chains. Blooms of dinoflagellates are responsible for the often harmful red tides, the brownish-red color coming from their xanthophyll accessory pigments.

The beating of two flagella in perpendicular grooves between the internal cellulose plates of the cell produces a characteristic spinning movement. The process of dinoflagellate nuclear division during asexual reproduction is unusual. Photosynthesis by symbiotic dinoflagellates, living in small coral cnidarians, provides the main food source for coral reef communities. Some dinoflagellates are parasitic or carnivorous heterotrophs.

Bacillariophyta (Diatoms) **Diatoms** are common unicellular marine and freshwater plankton with brown plastids and unique glasslike shells. Diatoms usually reproduce asexually, although eggs and flagellated sperm may be produced. Food reserves in the form of an oil contribute to buoyancy. Massive quantities of fossilized shells make up diatomaceous earth.

Chrysophyta (Golden Algae) The color of **golden algae** results from yellow and brown carotenoid and xanthophyll accessory pigments. There are many colonial forms of these biflagellated, freshwater plankton. Golden algae may form resistant cysts during environmental stress. The fossil acritarchs resemble ruptured cysts of chrysophytes and other algae.

Phaeophyta (Brown Algae) The mostly marine **brown algae** are the largest and most complex algae. Based on their accessory pigments and chloroplast structure, brown algae are related to golden algae and diatoms.

Large marine brown (Phaeophyta), red (Rhodophyta), and green (Chlorophyta) algae are called seaweeds. Unique anatomical and biochemical adaptations allow seaweeds to survive in the challenging intertidal zone. Some seaweeds have tissues and organs that are analogous to those found in plants. Seaweeds may have a plantlike **thallus**, or body, consisting of a rootlike **holdfast** and a stemlike **stipe** that supports leaflike **blades**. Some brown algae have floats, which keep the blades near the surface. The giant, fast-growing brown algal kelps live in deeper waters.

Biochemical adaptations to intertidal habitats include cell walls containing gel-forming polysaccharides that protect from abrasion and drying, and calcium carbonate-encrusted cell walls that protect from invertebrate grazers. Humans use some seaweeds for food. Commercial uses of seaweeds include thickeners for processed foods, lubricants, and the culture media, agar.

Most brown algae have an **alternation of generations**. The multicellular, haploid **gametophyte** produces gametes. After syngamy, the diploid zygote grows into the multicellular **sporophyte**, which then produces spores by meiosis. The gametophyte and sporophyte may be similar in appearance (**isomorphic**) or distinct (**heteromorphic**).

Rhodophyta (Red Algae) The color of **red algae** is due to the accessory pigment phycoerythrin belonging to the family of pigments called phycobilins, found only in cyanobacteria and red algae. Phycoerythrin is able to absorb the wavelengths of light that penetrate into deep water. Most red algae are marine and multicellular, often with filamentous, delicately branched thalli.

Alternation of generations is common, although life cycles are diverse. Red algae have no flagellated cells in their life cycles.

Chlorophyta (Green Algae) The chloroplasts of **green algae** resemble those of plants, and most botanists believe that plants evolved from a group of chlorophytes. Most species live in fresh water, although many are marine. Unicellular forms may be planktonic, inhabitants of damp soil, symbionts in

invertebrates and protozoa, or mutualistic partners with fungi—forming associations known as **lichens**. Colonial species may be filamentous. Some large multicellular forms are considered seaweeds.

Three separate evolutionary trends within the green algae have led from a unicellular, flagellated ancestor to increasingly complex flagellated colonies, multinucleated filaments, and true multicellular forms.

The life cycles of most green algae include sexual (with biflagellated gametes) and asexual stages. Conjugating algae, such as *Spirogyra*, produce amoeboid gametes.

In the life cycle of the unicellular flagellated *Chlamydomonas*, the haploid mature organism reproduces asexually by dividing mitotically to form four zoospores, which grow into mature haploid cells. Under harsh conditions, the cell may produce many haploid gametes, which pair off and fuse with similar-looking gametes (**isogamy**) of opposite mating strains. The resulting diploid zygote secretes a protective coat. Following dormancy, the zygote undergoes meiosis, forming four haploid cells that grow into mature individuals.

Some species of green algae produce gametes that differ in size (**anisogamy**) and morphology and may exhibit **oogamy**, in which a flagellated sperm fertilizes a larger, nonmotile egg. Some multicellular green algae have an alternation of generations, as in the isomorphic *Ulva*.

■ **INTERACTIVE QUESTION   26.5**

Define the following terms:

  a. sporophyte

  b. gametophyte

  c. heteromorphic

  d. isogamy

  e. oogamy

■ **Systematists continue to refine their hypotheses about eukaryotic phylogeny**

New fossils, comparisons of cell structure and function, and molecular biology contribute to the efforts of systematists to produce phylogenetic classifications. Several new systematic changes are being suggested, including an eight-kingdom system.

The eight-kingdom system separates Monera into the kingdoms Archaebacteria and Eubacteria. In addition, pre-mitochondrial, early archezoans, such as *Giardia*, would be assigned to **kingdom Archezoa**. The protozoa and some of the algae would stay in the kingdom Protista (perhaps with a name change to Protozoa). Brown algae and some related phyla would be placed in a new **kingdom Chromista**. Chromistans have unusual chloroplasts that appear to have descended from eukaryotic, red algal endosymbionts. In the eight-kingdom system, green algae and red algae would move to the plant kingdom.

■ **INTERACTIVE QUESTION 26.6**

List the kingdoms that would be included in the eight-kingdom system.

Multicellularity originated independently many times

Eukaryotic organization allowed for the development of more complex structures. The step from unicellular protists to multicellular forms opened new opportunities for specialization and adaptation. Multicellularity evolved in the kingdom Protista several times, creating the ancestors of multicellular algae, plants, fungi, and animals. Colonial aggregations of cells probably became more interdependent, leading to specialization of cells and division of labor.

STRUCTURE YOUR KNOWLEDGE

1. List the general characteristics of the organisms placed in the kingdom Protista.

2. Describe some of the current issues over the phylogenetic relationships of groups currently placed in this kingdom.

3. Indicate on the time line in Chapter 24, page 166, the age and name of the oldest known eukaryotic fossils.

TEST YOUR KNOWLEDGE

MATCHING: *Match the protistan phyla with their descriptions.*

_____ 1. naked or shelled amoebas

_____ 2. cellular slime molds, haploid

_____ 3. heliozoa, radiolarians, siliceous skeletons

_____ 4. plasmodial slime molds, coenocytic

_____ 5. marine, with calcareous porous shells

_____ 6. green algae, ancestors of plants

_____ 7. brown algae, large, complex seaweeds

_____ 8. golden algae, freshwater plankton

_____ 9. diatoms, two-piece shells of silica

_____10. red algae, pigment phycoerythrin

_____11. unicellular, ciliated, macro- and micronuclei

_____12. water molds, white rusts, downy mildews

_____13. many flagella; free-living, mutualistic, parasitic

_____14. parasites, sporozoites, complex life cycles

_____15. whirling movement, cause red tides

A. Acrasiomycota

B. Actinopoda

C. Apicomplexa

D. Bacillariophyta

E. Chlorophyta

F. Chrysophyta

G. Ciliophora

H. Dinoflagellata

I. Foraminifera

J. Myxomycota

K. Oomycota

L. Phaeophyta

M. Rhizopoda

N. Rhodophyta

O. Zoomastigophora

MULTIPLE CHOICE: *Choose the one best answer.*

1. A coenocytic organism
   a. uses an amoeboid type of movement.
   b. consists of a thallus with no true roots, stems, or leaves.
   c. is multinucleate.
   d. produces asexual fruiting bodies.
   e. has many haploid cells.

2. The slime molds and multicellular algae are presently included in the kingdom Protista because
   a. They appear to be more closely related to unicellular eukaryotes.
   b. They lack important characteristics of the fungi and plants.
   c. Kingdom Protista includes eukaryotic organisms that do not clearly belong in the other three kingdoms.
   d. The kingdom Protista is polyphyletic.
   e. of all of the above.

3. Genetic variation is generated in the ciliates when
   a. A micronucleus replicates its genome many times and becomes a macronucleus.
   b. Isogametes, in the form of zoospores, fuse.
   c. Micronuclei are exchanged in conjugation.
   d. Oogamous sexual reproduction occurs.
   e. Plasmids are exchanged in conjugation.

4. Phytoplankton
   a. form the basis of most marine food chains.
   b. include the multicellular green, red, and brown algae.
   c. are mutualistic symbionts that provide food for coral reef communities.
   d. are unicellular protozoans.
   e. Both a and d are correct.

5. The chlorophyta are believed to be the ancestors of plants because
   a. They are the only multicellular algal protists.
   b. They do not have flagellated gametes.
   c. They are oogamous and heteromorphic.
   d. They are similar in chloroplasts and pigment composition.
   e. They exhibit alternation of generations.

6. According to the endosymbiont theory,
   a. Multicellularity evolved when primitive cells incorporated prokaryotic cells that then took on specialized functions.
   b. The symbiotic associations found in lichens resulted from the incorporation of algal protists into the ancestors of fungi.
   c. The chloroplasts and mitochondria of eukaryotic cells began as prokaryotic endosymbionts.
   d. The infoldings and specializations of the plasma membrane led to the evolution of the endomembrane system.
   e. The nuclear membrane evolved first, then chloroplasts, and then mitochondria.

7. Examples of mutualistic symbiotic relationships include
   a. dinoflagellates producing red tides.
   b. termites and zooflagellates.
   c. sporozoans in their multiple hosts.
   d. amoeboid cellular slime molds that congregate to form a fruiting body.
   e. isomorphic sporophyte and gametophyte generations.

8. The proposed kingdom Archezoa would include
   a. the diatoms, golden algae, and brown algae.
   b. the green algal ancestors of plants.
   c. the slime molds and water molds that are the ancestors of fungi.
   d. the diplomonads that lack mitochondria and have dual haploid nuclei.
   e. the ancient acritarchs.

CHAPTER 27

PLANTS AND THE COLONIZATION OF LAND

FRAMEWORK

This chapter details the evolution of plants and their adaptations to terrestrial habitats. The kingdom Plantae includes multicellular, photoautotrophic eukaryotes that develop from embryos retained in a protective jacket of cells. Plants exhibit an alternation of generations in which the diploid sporophyte is the more conspicuous stage in all divisions except the bryophytes. The development of a cuticle and jacketed reproductive organs, vascular tissue, seeds and pollen, and flowers are linked to the major periods of plant evolution in which mosses, ferns, conifers, and flowering plants appeared and radiated.

CHAPTER REVIEW

For the first 3 billion years, life was confined to marine and aquatic environments. Plants began to move onto land about 460 million years ago. The evolutionary history of the plant kingdom involves various adaptations to changing terrestrial conditions.

■ **Structural and reproductive adaptations made colonization of land possible:** *an overview of plant evolution*

General Characteristics of Plants Plants are multicellular, photoautotrophic eukaryotes. Nearly all plants are terrestrial, although some have returned to water during their evolution. Adaptations to life on land include a waxy **cuticle** on the stems and leaves that prevents desiccation and **stomata** or pores to allow for the exchange of gases through the cuticle. Plant chloroplasts contain chlorophylls *a* and *b* and a variety of carotenoids. Plant cell walls consist of cellulose; starch is the food storage compound.

The Embryophyte Condition Nearly all plants reproduce sexually; most can also propagate asexually. Gametes are produced in **gametangia**, organs with a protective jacket of cells. The egg is fertilized and develops into an embryo within the female gametangium, which was an important adaptation for terrestrial life and the basis for the name **embryophytes**.

Alternation of Generations: A Review The life cycle of plants involves the alternation of generations between the haploid gametophyte and the diploid sporophyte. In all extant plants, these generations are heteromorphic. In all but the mosses, the sporophyte is the more conspicuous stage.

Some Highlights of Plant Phylogeny Four major periods of plant evolution are linked to the evolution of structures that opened new adaptive zones on land. The first period involved the origin of plants from green algae, during the late Ordovician period of the Paleozoic era about 460 million years ago. Terrestrial adaptations included a cuticle, jacketed reproductive organs, and **vascular tissue**—specialized cells joined into tubes for transport throughout the plant. Molecular evidence suggests an early divergence between the nonvascular mosses and the lineages leading to vascular plants.

The second period of plant evolution took place 400 million years ago during the early Devonian period with the adaptive radiation of seedless vascular plants.

The third major period of plant evolution, near the end of the Devonian period (about 360 million years ago), was marked by the origin of vascular plants with **seeds**, which are embryos that are enclosed with a store of food inside a protective coat. The early seed plants gave rise to **gymnosperms** (naked seed plants), such as the conifers. The gymnosperms and seedless ferns dominated the landscape for over 200 million years.

The emergence of flowering plants about 130 million years ago, during the early Cretaceous period in the Mesozoic era, started the fourth major evolutionary episode. The majority of contemporary plants are **angiosperms**, or flowering plants that bear seeds in protective chambers called ovaries.

Classification of Plants Plant biologists use the term **division** in place of phylum for the major taxonomic groups within the plant kingdom. This textbook recognizes 12 divisions within the kingdom Plantae.

■ **INTERACTIVE QUESTION 27.1**

List the major characteristics of plants.

a.

b.

c.

d.

e.

f.

g.

h.

i.

j.

■ **Plants probably evolved from green algae called charophytes**

Chlorophyll *a*, chlorophyll *b*, and beta-carotene are found in green algae and plants, and both have thylakoid membranes stacked as grana. Some systematists suggest including green algae in kingdom Plantae on the basis of their evolutionary relationship.

Evidence that plants evolved from the green algae called **charophytes** includes the following homologies: (1) closely matched chloroplast DNA; (2) biochemical similarity in composition of cellulose cell walls and peroxisomes; (3) similarity in mechanisms of mitosis and involvement of microtubules, actin microfilaments, and vesicles in cell plate formation; (4) sperm ultrastructure similarities; and (5) genetic relationship as shown by molecular comparisons of nuclear genes and ribosomal RNA.

The Origin of Alternation of Generations in Plants Whereas alternation of generations seems to have evolved independently in some brown, red, and green algae, charophytes do not exhibit this trait. Thus, alternation of generations probably had a separate origin in green plants.

In the sexual reproduction of certain modern charophytes, the eggs and zygotes remain attached to the haploid thallus. Nonreproductive cells grow around each zygote, which then enlarges and undergoes meiosis, releasing haploid swimming spores.

Adaptations to Shallow Water as Preadaptations for Living on Land Ancient charophytes living along the edges of bodies of water may have been the first green plants to colonize the land. Fluctuations in water levels during the late Ordovician period may have resulted in selection for species that could survive exposed periods. The development of waxy cuticles and jacketed reproductive organs opened the adaptive zone of land.

■ **INTERACTIVE QUESTION 27.2**

How might alternation of generations have originated in the ancestor of plants?

■ **Bryophytes are embryophytes that generally lack vascular tissue and require environmental water to reproduce**

The mosses, liverworts and hornworts, formerly grouped into a single division, have been separated to reflect their lack of close relationship. They share many characteristics, however, and are still commonly referred to as **bryophytes**. Bryophytes have cuticles and protective gametangia. Sperm are produced in

antheridia; one egg is produced and fertilized within the **archegonium**, where the zygote develops into an embryo. They are restricted to damp, shady places due to both their flagellated sperm and, with a few exceptions, their lack of vascular tissue. The bryophytes date back at least 400 million years and have remained adapted to moist habitats.

Mosses (Division Bryophyta)   **Mosses** grow as water-absorbing mats with many plants in a tight pack. Each gametophyte plant attaches to the soil with elongated cells called rhizoids. Most photosynthesis occurs in small leaflike appendages. The embryo remains attached within the archegonium, and the retained sporophyte depends on the gametophyte for water and nutrients. At the tip of the sporophyte stalk, a **sporangium** produces haploid spores by meiosis. Germinating spores grow into the more dominant gametophyte plant.

Liverworts (Division Hepatophyta)   Some **liverworts** have bodies that are divided into lobes. Their life cycle is similar to that of mosses, although they also reproduce asexually by the formation of gemmae that grow in cups on the surface of the gametophyte.

Hornworts (Division Anthocerophyta)   Hornworts resemble liverworts, but their sporophytes are elongated capsules that grow like horns from the matlike gametophyte.

■ **INTERACTIVE QUESTION   27.3**

Review the life cycle of a typical moss plant by filling in the following blanks.

The dominant generation is the **a.** _____.
Female gametophytes produce eggs in **b.** _____ .
Male gametophytes produce sperm in **c.** _____ .
Sperm **d.** _____ through the damp environment to fertilize the egg. The zygote remains in the archegonium and grows into the **e.** _____ still attached to the female gametophyte. Spores are formed by the process of **f.** _____ in the **g.** _____ When shed, spores develop into the **h.** _____ .

■ **The origin of vascular tissue was an evolutionary breakthrough in the colonization of land**

During the evolution of vascular plants, there was increasing differentiation of an underground root system to absorb water and minerals and an aerial shoot system of stems and leaves to receive light for photosynthesis. **Lignin**, a hard material embedded in the cellulose of plant cell walls, provides support for the aerial parts of the plant. The vascular system of **xylem** and **phloem** provides the conducting vessels to connect roots and leaves.

■ **INTERACTIVE QUESTION   27.4**

a. What is the structure and function of xylem?

b. What is the structure and function of phloem?

The Earliest Vascular Plants   Fossils of vascular plants are present in the sedimentary rocks of the late Silurian and early Devonian periods. *Cooksonia*, a simple, dichotomously branching plant, is the oldest that has been discovered.

■ **Ferns and other seedless plants dominated the Carboniferous "coal forests"**

Division Psilophyta   *Psilotum*, commonly called whiskfern, is one of the two genera of relatively simple, primitive vascular plants in this division. Rhizomes bear tiny rhizoids. The scales emerging from the dichotomously branching sporophyte stems lack vascular tissue and thus are not true leaves. Sporangia along the stems release haploid spores. The nonphotosynthetic gametophyte depends on symbiotic fungi.

Division Lycophyta (Lycopods)   Lycopods were a major part of the landscape during the Carboniferous period (340 to 280 million years ago). One evolutionary line, the giant lycopods, became extinct when the Carboniferous swamps dried up. The other line of small lycopods are represented today by the genera *Lycopodium* and *Selaginella*, commonly called club mosses or ground pines. Many tropical species grow on trees as **epiphytes**—plants that anchor to other organisms but are not parasites.

In the club moss, the sporangia are borne on **sporophylls**—leaves specialized for reproduction. Spores germinate and grow into inconspicuous, subterranean gametophytes, which depend on symbiotic fungi for nutrition. *Lycopodium* is said to be **homosporous** because it produces only one kind of spore, which develops into bisexual gametophytes with both archegonia and antheridia. *Selaginella* is **heterosporous**; it produces **megaspores** that develop into female gametophytes and **microspores** that develop into male gametophytes.

Division Sphenophyta (Horsetails) This division originated in the Devonian radiation and produced tall plants during the Carboniferous period. Today only the genus *Equisetum* is found. Commonly called horsetails, these homosporous plants grow in damp locations. The gametophyte is minute but free-living and photosynthetic.

Division Pterophyta (Ferns) Ferns were also found in the great forests of the Carboniferous period and are the most numerous of seedless plants in the modern flora. Fern leaves, or fronds, are much larger than those of lycopods. The leaves of lycopods, called microphylls, probably evolved as emergences from the stem containing a single strand of vascular tissue.

Fern leaves, called megaphylls, have branching systems of veins, perhaps having evolved by the formation of webbing between closely growing separate branches.

Some fern leaves are specialized sporophylls with sporangia, arranged into clusters called sori, on their undersides. Most ferns are homosporous, although the archegonia and antheridia on the small, photosynthetic gametophyte mature at different times. Sperm swim to fertilize the egg in neighboring archegonia, and the young sporophyte grows out from the archegonium.

The Coal Forests The seedless plants of the Carboniferous forests left behind extensive beds of coal. Dead plants did not completely decay in the stagnant swamp waters, and great accumulations of peat developed. When the sea later covered the swamps and marine sediments piled on top, heat and pressure converted the peat to coal.

■ INTERACTIVE QUESTION 27.5

In the following diagram of a fern life cycle, label the processes (in the boxes) and structures (on the lines). Indicate which portion of the life cycle is haploid and which is diploid.

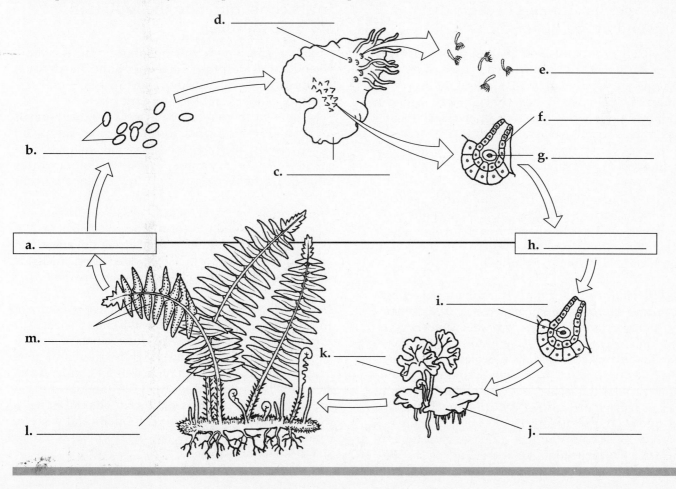

■ Reproductive adaptations catalyzed the success of the seed plants

Three life cycle modifications contributed to the success of seed plants on land: (1) the further reduced gametophyte generation was retained and protected within the sporophyte plant; (2) pollination replaced the swimming of sperm to egg; and (3) the seed, with embryo surrounded by a food supply within a seed coat, maintained dormancy through harsh conditions and functioned in dispersal.

■ **INTERACTIVE QUESTION   27.6**

a. What advantage would come from having the diploid sporophyte as the dominant generation?

b. What advantage would come from retaining the gametophyte generation in the life cycle?

■ Gymnosperms began to dominate landscapes as climates became drier at the end of the Paleozoic era

Of the two groups of seed plants, gymnosperms appear earlier in the fossil record. Three of its four divisions are relatively small: Cycadophyta, which includes the palmlike cycads; Ginkgophyta, with the deciduous, ornamental ginkgo tree; and Gnetophyta, which includes three genera: the bizarre *Welwitschia*, the tropical *Gnetum*, and the shrublike *Ephedra*. The largest division is Coniferophyta.

Division Coniferophyta   The name **conifer** refers to the reproductive structure, the cone. Most conifers are evergreens, with needle-shaped leaves covered with a thick cuticle. Coniferous trees are among the tallest, largest, and oldest living organisms.

The Life History of a Pine   The pine tree is a heterosporous sporophyte. Pollen cones consist of many tiny sporophylls that bear sporangia. Meiosis gives rise to microspores that develop into pollen grains—the immature male gametophyte. Scales of the ovulate cone hold ovules, each of which contains a sporangium, called a nucellus, within a protective integument. A megaspore mother cell undergoes meiosis, and one of the resulting megaspores undergoes repeated divisions to produce a female gametophyte in which a few archegonia develop.

Pollination occurs when a pollen grain is drawn through the micropyle, an opening in the integument around the nucellus. A pollen tube grows and digests its way through the nucellus. Fertilization occurs when a sperm nucleus joins with the egg nucleus. The zygote develops into a sporophyte embryo, which is nourished by the remaining female gametophyte tissue and is enclosed in a seed coat. Seed production takes 3 years.

The History of Gymnosperms   Gymnosperms probably developed from a group called progymnosperms and had evolved seeds by the end of the Devonian period. The divisions of gymnosperms arose during the Carboniferous and early Permian periods. Warmer and drier conditions gave selective advantage to the conifers and cycads by the end of the Permian at the close of the Paleozoic era. Conifers and the great palmlike cycads supported the giant dinosaurs of the Mesozoic. When the climate became cooler at the end of the Mesozoic, the dinosaurs and many plants became extinct, but some gymnosperms, particularly conifers, persisted.

■ The evolution of flowers and fruits contributed to the radiation of angiosperms

Angiosperms, or flowering plants, are the most diverse and widespread of modern flora. The division Anthophyta is divided into two classes: Monocotyledones and Dicotyledones.

Most angiosperms rely on insects or other animals for transferring pollen to female sex organs, an advance over the more random wind pollination of gymnosperms. Gymnosperms transport water through **tracheids**, relatively primitive, tapered cells that also function in mechanical support. In angiosperms, shorter, wider **vessel elements** are arranged end to end to form more specialized tubes for water transport. **Fibers**, which have thick lignified walls that add support, also evolved from tracheids and are found in both conifers and angiosperms.

The Flower   The **flower**, the reproductive structure of an angiosperm, is a compressed shoot with four whorls of modified leaves. The outer **sepals** are usually green, whereas **petals** are brightly colored in most flowers that are pollinated by insects and birds. A **stamen** consists of a stalk, called a **filament**, and an **anther**, in which pollen is produced. The **carpel** has a sticky **stigma**, which receives pollen, and a **style**, which leads to the **ovary**. The ovary contains ovules that develop into seeds.

■ **INTERACTIVE QUESTION 27.7**

In the diagram of the life cycle of a pine, label the indicated structures and show where meiosis, pollination, and fertilization take place. Which structures represent the gametophyte generation?

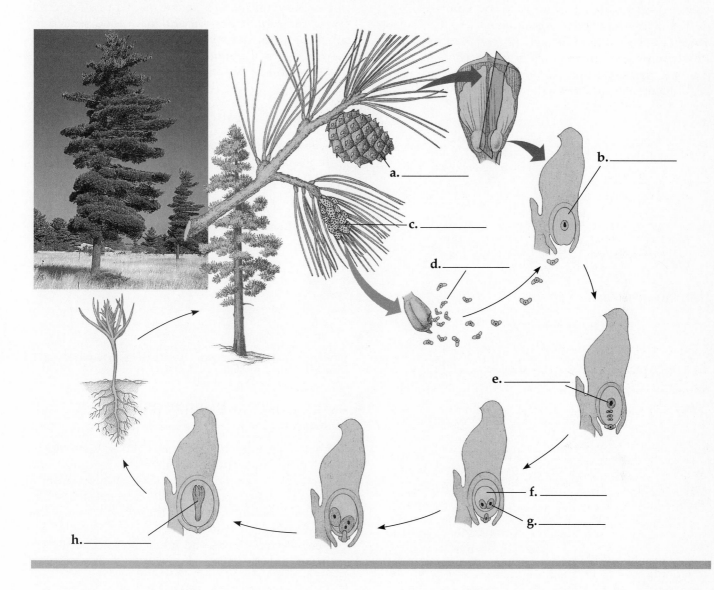

Four evolutionary trends in angiosperm flower structure include: (1) reduction of the number of floral parts; (2) fusion of floral parts (compound carpels are common—the term pistil refers to single or fused carpels); (3) change from radial to bilateral symmetry; and (4) lowering the position of the ovary to below the petals and sepals for better protection.

The Fruit A **fruit** is a mature ovary that functions in the protection and dispersal of seeds. Fruits may be modified in various ways to disperse seeds.

Life Cycle of an Angiosperm **Pollen grains**, consisting of two haploid cells, are immature male gametophytes that develop from microspores within the anthers. **Ovules** contain the female gametophyte, which consists of an **embryo sac** with seven haploid cells (one of which contains two nuclei).

Most flowers have some mechanism to ensure **cross-pollination**. A pollen grain germinates on the stigma and extends a pollen tube down the style to the ovule, where it releases two sperm cells into the embryo sac. In a process called **double fertilization**, one sperm unites with the egg to form the zygote, and the other sperm nucleus fuses with the large cell containing two nuclei in the center of the embryo sac. The **endosperm**, which develops from this triploid (3*n*) nucleus, serves as a food reserve for the embryo.

The embryo consists of a rudimentary root and one (in monocots) or two (in dicots) seed leaves, called **cotyledons**. The seed is a mature ovule containing an embryo, an endosperm, and a seed coat derived from the integuments of the ovule.

■ **INTERACTIVE QUESTION   27.8**

a. Why is cross-pollination considered advantageous?

b. What is a possible function of double fertilization?

The Rise of Angiosperms   Angiosperms appear rather suddenly in the fossil record about 120 million years ago in the early Cretaceous period. By the end of this period, 65 million years ago, angiosperms had radiated and become the dominant plants, as they are today. The abruptness of their appearance may be due to an imperfect fossil record or to the punctuated equilibrial nature of their evolution from a gymnosperm ancestor.

The Cretaceous was a period of climatic change and extinctions, marking the boundary between the Mesozoic and Cenozoic eras. Dinosaurs and many cycads and conifers disappeared and were replaced by mammals and flowering plants.

Relationships between Angiosperms and Animals   Animals influenced the evolution of plants and vice versa. Animal predation on plants may have provided selective pressure for plants to keep spores and vulnerable gametophytes on the plant. As flowers and fruits evolved, some predators became beneficial as pollinators and seed dispersers. The **coevolution**, or mutual evolutionary influence, of angiosperms and their pollinators is seen in the diversity of flowers, whose color, fragrance, and shape are often matched to the sense of sight and smell or to the particular morphology of a group of pollinators. As seeds

mature, fruits soften and increase in sugar content, attracting bird and mammal seed dispersers.

Angiosperms and Agriculture   All of our fruit and vegetable crops are angiosperms. Grains are grass fruits; their endosperm is the main food source for most people and their domesticated animals. Agriculture is a unique relationship between plants and the humans who breed and cultivate them.

■ **INTERACTIVE QUESTION   27.9**

a. What constitutes the gametophyte generation of an angiosperm?

b. What does a seed consist of?

c. What is a fruit?

■ **Plant diversity is a nonrenewable resource**

The growing human population and its demand for space, food, and natural resources are leading to the extinction of hundreds of species each year. Tropical rain forests, where plant diversity is greatest, could be eliminated within 25 years if the pace of destruction continues. Humans are destroying their potential supply of new food crops and medicines. Economic and political solutions are required to preserve plant diversity.

STRUCTURE YOUR KNOWLEDGE

1. Fill in the evolutionary events and modern plant groups (include some common name examples) on this phylogenetic tree that shows the major lines of plant evolution on a geological time frame. Indicate the periods of dominance for the major groups. What was the impact of the major climatic changes that occurred between the Paleozoic, Mesozoic, and Cenozoic eras?

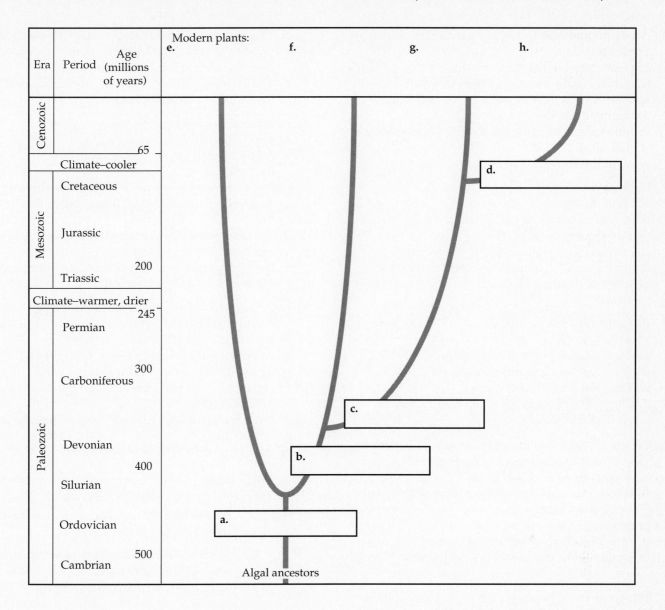

Era	Period	Age (millions of years)
Cenozoic		65
	Climate–cooler	
Mesozoic	Cretaceous	
	Jurassic	200
	Triassic	
	Climate–warmer, drier	245
Paleozoic	Permian	
		300
	Carboniferous	
	Devonian	400
	Silurian	
	Ordovician	
	Cambrian	500

Modern plants: e. f. g. h.

a. b. c. d.

Algal ancestors

2. The evolution of plants shows a trend of increasing adaptation to a terrestrial habitat. List the characteristics that were novel adaptations for the following major plant groups.

Plant Groups	Novel Adaptations for Life on Land
Bryophytes	a.
Ferns	b.
Gymnosperms	c.
Angiosperms	d.

3. Now indicate on the larger-scale time line in Chapter 24, page 166, the point at which plants moved onto land.

TEST YOUR KNOWLEDGE

MULTIPLE CHOICE: *Choose the one best answer.*

1. Adaptations for terrestrial life seen in all plants are
   a. chlorophylls *a* and *b*.
   b. cell walls of cellulose and lignin.
   c. cuticle and gametangia.
   d. vascular tissue and stomata.
   e. alternation of generations.

2. Plants are thought to have originated from charophytes based on
   a. their homologous chloroplasts.
   b. biochemical similarity of their cell walls.
   c. their mechanism of mitosis and cell plate formation.
   d. molecular similarities in their DNA and ribosomal RNA.
   e. all of the above.

3. Megaphylls
   a. are large leaves.
   b. are gametophyte plants that develop from megaspores.
   c. are leaves specialized for reproduction.
   d. are leaves with branching vascular systems.
   e. were a dominant group of the great coal forests.

4. Bryophytes differ from the other plant groups because
   a. Their gametophyte generation is dominant.
   b. They are lacking cuticle and lignin.
   c. They have flagellated sperm.
   d. They have rhizoids for water absorption.
   e. of all of the above.

5. Which of the following plant groups is *incorrectly* paired with its gametophyte generation?
   a. angiosperm—a pollen grain
   b. whiskfern (*Psilotum*)—nonphotosynthetic subterranean structure
   c. moss—green matlike plant
   d. fern—frond growing from rhizome
   e. conifer—multicellular haploid tissue in nucellus

6. Which of the following *incorrectly* pairs a sporophyte embryo with its food source?
   a. pine embryo—endosperm in nucellus
   b. corn embryo—$3n$ endosperm tissue in seed
   c. moss embryo—archegonium and gametophyte plant
   d. fern embryo—female gametophyte surrounding archegonium
   e. lycopod embryo—subterranean gametophyte

7. A difference between seedless vascular plants and plants with seeds is that
   a. The gametophyte generation is dominant in the seedless plants, whereas the sporophyte is dominant in the seeded plants.
   b. The spore is the agent of dispersal in the first, whereas the seed functions in dispersal in the second.
   c. The gametophyte is photoautotrophic in all seedless plants but dependent on the sporophyte generation in the other group.

d. The embryo is unprotected in the seedless plants but retained within the female reproductive structure in the other group.
   e. The vascular tissues are not strengthened by lignin in the seedless plants.

8. An example of coevolution is
   a. wind pollination in conifers.
   b. a flower with nectar guides that direct bees to its nectaries.
   c. the synchronization of nutrient development and fertilization resulting from double fertilization.
   d. the retention of the gametophyte generation to weed out harmful mutations.
   e. the clumping of moss plants to support each other and create an absorbent mat.

9. If you could take a time machine back to the Carboniferous period, which of the following scenarios would you most likely confront?
   a. creeping mats of low-growing bryophytes
   b. fields of tall grasses swaying in the wind
   c. swamps dominated by large lycopods, horsetails, and ferns
   d. huge forests of naked-seed trees filling the air with pollen
   e. the dominance of flowering plants

TRUE OR FALSE: *Indicate T or F, and then correct the false statements.*

_____ 1. All photoautotrophic, multicellular eukaryotes are plants.

_____ 2. Heteromorphic plants produce male and female gametophytes.

_____ 3. Bryophytes are terrestrial nonvascular plants.

_____ 4. The club mosses and horsetails are naked seed plants.

_____ 5. A sporangium produces spores, no matter what group it is found in.

_____ 6. A fruit consists of an embryo, nutritive material, and a protective coat.

_____ 7. Tracheids are wide, specialized cells arranged end to end for water transport and are found in angiosperms.

_____ 8. A stamen consists of a filament and anther in which pollen is produced.

_____ 9. The female gametophyte in angiosperms consists of haploid cells in which a few archegonia develop.

_____ 10. The male gametophyte in angiosperms consists of a pollen grain.

CHAPTER 28

FUNGI

FRAMEWORK

This chapter describes the morphology, life cycles, and ecological and economic importance of the kingdom Fungi. The divisions of fungi are established on the basis of variations in sexual reproduction. Lichens are symbiotic complexes of fungi and algae. Fungi play an important ecological role, both as decomposers and by their mycorrhizal association with plant roots. A flagellated protistan may have been the common ancestor to fungi and animals.

CHAPTER REVIEW

■ Structural and life history adaptations equip fungi for an absorptive mode of nutrition

Fungi were once classified with plants, but these eukaryotic, mostly multicellular organisms are now placed in their own kingdom.

Nutrition and Habitats Fungi are heterotrophs that obtain their nutrients by **absorption**; they secrete digestive enzymes into the surrounding food media and absorb the resulting small organic molecules. Fungi may be saprophytes that decompose nonliving organic material, parasites that absorb nutrients from living hosts, or mutualistic symbionts that feed on, but also benefit, their hosts. They are found in terrestrial and aquatic environments.

Structure Most fungi are composed of filaments (**hyphae**) that form a network called a **mycelium** that provides an extensive surface area for absorption. The hyphae of most fungi are divided into cells by crosswalls called **septa**, which usually have pores through which nutrients and cell organelles can pass. Cell

walls are composed of **chitin**. Some hyphae are aseptate, and these **coenocytic** fungi consist of a continuous cytoplasmic mass with many nuclei. Parasitic fungi may penetrate host cells with modified hyphae called **haustoria**.

Growth and Reproduction Although fungi are nonmotile, they rapidly enter into new food territory by extension of their hyphae.

Mitosis in fungi is unique. The spindle forms within the nucleus, which later divides after the chromosomes separate. Fungi produce huge quantities of asexual spores when conditions are favorable; sexual reproduction may occur when environmental conditions change.

Nuclei of the mycelia are haploid. Genetic heterogeneity may exist when hyphae with different nuclei join. Their haploid nuclei may stay in separate regions of the mycelia, or they may engage in a process similar to crossing over.

In the sexual cycle, syngamy occurs as two separate events. **Plasmogamy**, cytoplasmic fusion, leads to a **dikaryon** stage in which the two nuclei pair up but do not fuse. They may continue to divide in tandem, forming dikaryotic cells. After a period of time, the nuclei fuse (**karyogamy**), and the zygote undergoes immediate meiosis. (See Interactive Question 28.1, p. 194.)

■ The three major divisions of fungi differ in details of reproduction

Division Zygomycota (Zygote Fungi) Zygomycetes are mostly terrestrial fungi living in soil or on decaying matter. One group forms important mutualistic associations called **mycorrhizae** with plant roots. Hyphae are coenocytic. Resistant zygosporangia are formed in sexual reproduction.

Rhizopus stolonifer, the black bread mold, is a common zygomycete that spreads horizontal hyphae

■ INTERACTIVE QUESTION   28.1

Briefly define the following terms that relate to the structure and reproduction of fungi:

a. mycelium

b. septa

c. coenocytic

d. dikaryotic

e. plasmogamy

f. karyogamy

across its food substrate and erects hyphae with bulbous black sporangia containing hundreds of haploid spores. Sexual reproduction produces a dikaryotic zygosporangium with a tough protective coat that remains dormant until conditions are favorable. Karyogamy occurs between paired nuclei, followed immediately by meiosis to produce haploid spores.

Division Ascomycota (Sac Fungi)   The ascomycetes range in complexity from unicellular yeasts to elaborate cup fungi. They are found in a wide variety of habitats and symbiotic associations. Many species are in mutualistic relationships with algae to form lichens or with plants to form mycorrhizae. Others are the major nonbacterial saprobes in saltwater habitats.

Asexual spores, called **conidia**, are produced in chains or clusters at the ends of hyphae. The dikaryotic stage is associated with the formation of fruiting bodies called **ascocarps**. Karyogamy occurs in terminal cells, the saclike **asci**, and meiosis yields haploid spores called ascospores.

Division Basidiomycota (Club Fungi)   The basidiomycetes, which include the mushrooms, shelf fungi, puffballs, and rusts, produce a club-shaped **basidium** during a short diploid stage in the life cycle. The club fungi include important saprobes, mycorrhizae-forming mutualists, and plant parasites.

In response to environmental stimuli, an elaborate fruiting body called a **basidiocarp** is formed from the dikaryotic hyphae. Karyogamy and meiosis occur in cells at the tip of hyphae, forming basidia that produce haploid spores. Asexual reproduction of spores as conidia is less common in basidiomycetes than in ascomycetes.

■ INTERACTIVE QUESTION   28.2

Indicate whether the following diagrams are from a zygomycete, ascomycete, or basidiomycete life cycle. Identify the labeled structures.

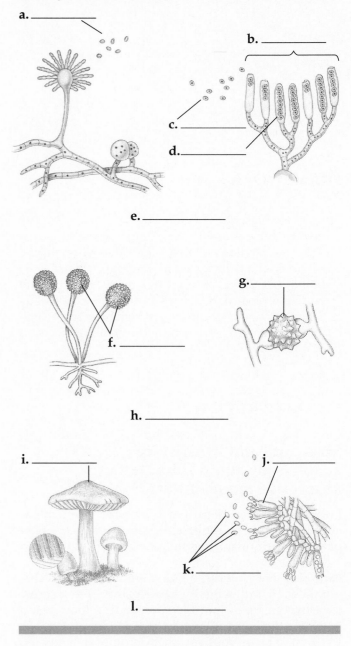

a. _____

b. _____

c. _____

d. _____

e. _____

f. _____

g. _____

h. _____

i. _____

j. _____

k. _____

l. _____

■ Molds, yeasts, lichens, and mycorrhizae represent unique lifestyles that evolved independently in all fungal divisions

Molds   **Molds** are rapidly growing, asexually reproducing fungi that are saprobes or parasites on a variety of substrates. In later life stages, sexual reproduction may occur in zygosporangia, ascocarps, or basidiocarps. If no sexual stage is known, the mold is classified as Deuteromycota, sometimes called **imperfect fungi** in reference to their lack of a sexual stage. These fungi reproduce by asexual conidia. Some molds are sources of antibiotics, such as penicillin.

Yeasts **Yeasts** are unicellular organisms that grow in liquid or moist habitats. Reproduction is commonly asexual by cell division or budding, but some yeasts produce asci or basidia and are classified accordingly. Yeasts placed in the division Deuteromycota have no known sexual stages. *Saccharomyces cerevisiae* is used in baking, brewing, and molecular genetic research.

Lichens **Lichens** are symbiotic associations of millions of algal cells in a lattice of fungal hyphae. The fungus is usually an ascomycete, and the alga is usually unicellular or filamentous green algae or cyanobacteria. The alga provides the fungus with food and, in the case of lichens containing cyanobacteria, nitrogen. The fungus creates the thallus that provides for mineral absorption, water retention, and gas exchange. Fungal compounds, including pigments, toxins, and acids that bind mineral elements, contribute to the well-being of the association.

Lichens reproduce asexually, either as fragments or by tiny clumps called **soredia**. In addition, it is common for the fungal component to reproduce sexually and for the algal component independently to reproduce asexually.

Lichens are important colonizers of bare rock and soil and can withstand desiccation and great cold. Lichens cannot, however, tolerate air pollution.

Mycorrhizae **Mycorrhizae** are very common and important mutualistic associations of plant roots and fungi in which the fungal hyphae provide the plant with minerals absorbed from the soil and the plant provides organic nutrients to the fungus. The fungi periodically form fruiting bodies for sexual reproduction.

▦ INTERACTIVE QUESTION 28.3

What is the key characteristic of the fungi placed in the division Deuteromycota?

■ Fungi have a tremendous ecological impact

Fungi as Decomposers Fungi and bacteria are the principal decomposers of organic matter, making possible the essential recycling of chemical elements between living organisms and their abiotic surroundings. Invasive hyphae, enzymes that work on cellulose and lignin, and prolific production of colonizing spores make fungi excellent decomposers of plant material.

Fungi as Spoilers As well as decomposing the organic litter in our ecosystem, fungi work on our food, clothing, and the wood used in buildings and boats. Fungi destroy a large proportion of the world's fruit harvest each year.

Pathogenic Fungi Pathogenic fungi cause athlete's foot, ringworm, vaginal yeast infections, and lung infections. Fungal diseases of plants are common. An ascomycete forms ergots on rye that can cause serious symptoms when accidentally milled into flour. Lysergic acid, the raw material of LSD, is one of the toxins in the ergots.

Edible Fungi Commercially cultivated species of *Agaricus* are eaten, as are wild mushrooms gathered by knowledgeable collectors. Truffles, the ascocarps of certain mycorrhizal ascomycetes, develop underground and release strong odors that signal their location.

■ Fungi and animals probably evolved from a common protistan ancestor

Recent molecular evidence places the funguslike **chytrids** in the division Chytridiomycota, a primitive group of fungi that diverged early and retained a flagellated stage perhaps typical of the protist ancestor of fungi.

Comparisons of several proteins and ribosomal RNA indicate that animals and fungi diverged from a common flagellated protistan ancestor, perhaps a choanoflagellate.

The oldest undisputed fossils of fungi date back 440 million years. The fungal divisions Zygomycota, Ascomycota, and Basidiomycota apparently lost their flagellated stages as they developed reproductive and dispersal adaptations for life on land. The first vascular plant fossils have petrified mycorrhizae, indicating that plants and fungi moved onto land together. Terrestrial communities appear to have always been dependent on fungi as decomposers and partners in symbiotic associations.

■ INTERACTIVE QUESTION   28.4

Why have chytrids been moved back and forth between kingdoms Protista and Fungi? Why are they now classified with the fungi?

STRUCTURE YOUR KNOWLEDGE

1. Fill in the following table that summarizes the characteristics of these major fungal types.

2. The kingdom Fungi contains members with saprobic, parasitic, and mutualistic modes of nutrition. How do these types of nutrition relate to the ecological and economic importance of this group?

3. What is the basis for saying that fungi and animals evolved from a common protistan ancestor?

4. Indicate on the time line in Chapter 24, page 166, the time when the first fossils of fungi appear.

TEST YOUR KNOWLEDGE

FILL IN THE BLANKS

_____ 1. division between cells in fungal hyphae

_____ 2. hyphal cells with two nuclei

_____ 3. component of cell walls in most fungi

_____ 4. asexual spores produced in chains at ends of hyphae

_____ 5. club-shaped reproductive structure found in mushrooms

_____ 6. sacs that contain sexual spores in cup fungi

_____ 7. mutualistic associations between plant roots and fungi

_____ 8. tough, protective zygote produced by zygomycetes

_____ 9. hyphae with many nuclei

_____ 10. most primitive fungal group

Fungal Type	Examples	Morphology	Asexual Reproduction	Sexual Reproduction
Zygomycetes	a.	b.	c.	d.
Ascomycetes	e.	f.	g.	h.
Basidiomycetes	i.	j.	k.	l.
Deuteromycetes	m.	n.	o.	p.
Lichens	q.	r.	s.	t.

MULTIPLE CHOICE: *Choose the one best answer.*

1. The major difference between fungi and plants is that fungi
   a. have an absorptive form of nutrition.
   b. do not have a cell wall.
   c. are not eukaryotic.
   d. are multinucleate but not multicellular.
   e. reproduce by spores.

2. Fungal mitosis
   a. produces spores in dikaryotic cells.
   b. does not involve the formation of a spindle.
   c. takes place within the nucleus.
   d. results in plasmogamy.
   e. involves all of the above.

3. A fungus that is both a parasite and a saprobe is one that
   a. digests only the nonliving portions of its host's body.
   b. lives off the sap within its host's body.
   c. first lives as a parasite but then consumes the host after it dies.
   d. lives as a mutualistic symbiont on its host.
   e. causes athlete's foot and vaginal infections.

4. The fact that karyogamy does not immediately follow plasmogamy
   a. is necessary to create coenocytic hyphae.
   b. allows for the development of more genetic variation.
   c. allows fungi to reproduce asexually most of the time.
   d. creates dikaryotic cells that may benefit from the presence of duplicate copies of alleles.
   e. is characteristic of deuteromycetes.

5. Deuteromycota
   a. represents the most ancient lineage of fungi.
   b. includes the fungal components of lichens.
   c. includes the imperfect fungi that have abnormal forms of sexual reproduction.
   d. is home to molds, yeasts, and lichens.
   e. includes molds and other types of fungi whose sexual stage is lacking or unknown.

6. In the Ascomycota,
   a. Sexual reproduction occurs by conjugation.
   b. Spores often line up in a sac in the order they were formed by meiosis.
   c. Asexual spores form in sporangia on erect hyphae.
   d. Most hyphae are dikaryotic.
   e. Sexual spores are produced in conidia.

7. Lichens are symbiotic associations that
   a. usually involve an ascomycete and a green alga or cyanobacterium.
   b. can reproduce sexually by forming soredia.
   c. require moist environments to grow.
   d. fix nitrogen for absorption by plant roots.
   e. are unusually resistant to air pollution.

8. The name given to each of the three divisions of fungi is based on
   a. the structure in which karyogamy occurs during sexual reproduction.
   b. the location of plasmogamy during sexual reproduction.
   c. the location of the dikaryotic stage in the life cycle.
   d. the structure that produces asexual spores.
   e. their ancestral origin.

9. Fungi and animals appear to have evolved from a common ancestor
   a. because neither of them are photosynthetic.
   b. based on similarities in cell structure.
   c. that was a flagellated protist.
   d. about the time that fungi and plants moved onto land.
   e. based on homologous ultrastructure of their flagella.

CHAPTER 29

INVERTEBRATES AND THE
ORIGIN OF ANIMAL DIVERSITY

FRAMEWORK

Animals are multicellular eukaryotic heterotrophs whose mode of nutrition is ingestion. Characteristics of animals include sexual reproduction; dominant diploid stage; specialized cells, tissues, and organs; embryonic development with a blastula stage; and the presence of muscles and nerves.

This chapter surveys the characteristics and representatives of the major animal phyla. Fossil evidence for kingdom Animalia's origin and phylogeny is lacking, so comparative anatomy, embryology, and molecular systematics are used to construct phylogenetic trees. The figure on p. 200 is one version of that phylogeny.

CHAPTER REVIEW

The evolution of multicellular organisms that eat other organisms opened a new adaptive zone. Animal life began in the Precambrian seas and then radiated, populating first the seas and eventually the land. Animals are grouped into about 35 phyla based primarily on anatomical and embryological criteria. All subphyla but one are **invertebrates**, animals that lack backbones, which include over 95 percent of animal species.

■ What is an animal?

Animals are multicellular heterotrophic eukaryotes, most of which use **ingestion** as their mode of nutrition. They store carbohydrates as glycogen. Animal cells lack walls and may have desmosomes, gap junctions, and tight junctions connecting adjacent cells. Muscle and nervous tissues are unique to animals.

The diploid stage is usually dominant and reproduction is primarily sexual, with a flagellated sperm fertilizing a larger, nonmotile egg. The zygote undergoes a series of mitotic divisions, called **cleavage**, usually passing through a **blastula** stage during embryonic development, followed by **gastrulation**, which produces embryonic tissues for all of the adult body parts. The life cycle of many animals includes a **larva**—a free-living, sexually immature form. **Metamorphosis** transforms a larva into a sexually mature adult.

The greatest numbers of animal phyla are found in the oceans. Terrestrial habitats have been extensively exploited by only a few animal phyla, notably the vertebrates and arthropods.

■ Comparative anatomy and embryology provide clues to animal phylogeny: *an overview of animal diversity*

Because of the inadequate fossil record of the rapid diversification of animal phyla during the Precambrian and early Cambrian periods, animal phylogeny depends on comparative anatomy, embryology, and molecular systematics. The animal kingdom seems to have evolved in the late Precambrian from a single protistan ancestor. All animal phyla had evolved by the early Cambrian period. The following are four key evolutionary branch points:

1. The Parazoa–Eumetazoa Split   The sponges (phylum Porifera) are called **parazoa** and are separated from other animals on the basis of anatomical simplicity. All other animal phyla are grouped into the **eumetazoa**.

2. The Radiata–Bilateria Split

The eumetazoa are divided into two groups, partly on the basis of body symmetry: The **radiata**, containing the hydras and jellyfishes, have **radial symmetry**. The **bilateria** include animals with **bilateral symmetry**, having distinct head (**anterior**) and tail (**posterior**) ends and left and right sides. Bilateral animals also have top (**dorsal**) and bottom (**ventral**) sides.

Bilateral symmetry is associated with **cephalization**, the concentration of sensory organs in the head end, which is an adaptation for unidirectional movement.

During gastrulation, a eumetazoan embryo develops layers of cells called the **germ layers**: Ectoderm develops into the outer body covering and, in some phyla, into the central nervous system; **endoderm** lines the rudimentary gut, or **archenteron**, and gives rise to the lining of the digestive tract and associated organs. The radiata (cnidarians and ctenophores) are **diploblastic**, forming only these two layers. The bilateria are **triploblastic**, producing a middle layer, the **mesoderm**, from which arise muscles and most other organs.

3. The Acoelomate–Coelomate Split

Triploblastic animals that have solid bodies are called **acoelomates**. These include the flatworms (Platyhelminthes) and a few other phyla. A tube-within-a-tube body plan with a fluid-filled cavity separating the digestive tract from the outer body wall is found in the other triploblastic animals. Some sort of blood vascular system is found in animals with body cavities and is not found in acoelomates.

In **pseudocoelomates**, including the rotifers (Rotifera) and roundworms (Nematoda), the cavity is not completely lined by mesoderm and is called a **pseudocoelom**. **Coelomates** have a true **coelom**, a body cavity completely lined by mesoderm. Mesenteries connect the inner and outer lining of the coelom and suspend the internal organs. A fluid-filled body cavity cushions internal organs, allows organs to grow and move independently of the outer body wall, and also functions as a hydrostatic skeleton in soft-bodied animals. Coeloms evolved independently in the protostomes and deuterostomes.

4. The Protostome–Deuterostome Split

The coelomates divide into two evolutionary lines—the **proto-stomes** (annelids, mollusks, and arthropods) and the **deuterostomes** (echinoderms and chordates).

Most protostomes have **spiral cleavage**, in which the planes of cell division are such that newly formed cells fit in the grooves between cells of adjacent tiers. The **determinate cleavage** of some protostomes sets the developmental fate of each embryonic cell very early. Many deuterostomes exhibit **radial cleavage**, in which parallel and perpendicular cleavage planes result in aligned tiers of cells in the dividing zygote. **Indeterminate cleavage** in most deuterostomes means that early embryonic cells retain the capacity to develop into a complete embryo.

The **blastopore** is the opening of the developing archenteron. In typical protostomes ("first mouth"), the blastopore develops into the mouth, and a second opening forms at the end of the archenteron to produce an anus. In deuterostomes, the blastopore becomes the anus, and the second opening develops into the mouth.

In protostomes, the coelom forms from splits within solid masses of mesoderm, called **schizocoelous** development. In deuterostomes, the mesoderm begins as lateral outpocketings from the archenteron, called **enterocoelous** development. (See Interactive Question 29.1, p. 200.)

■ Sponges are sessile animals lacking true tissues

Sponges are sessile, mostly marine animals. Water is drawn through pores in the body wall of this saclike animal into a central cavity, the **spongocoel**, and flows out through the **osculum**. Sponges are suspension-feeders, collecting food particles by the action of collared, flagellated **choanocytes** lining the inside of the body. Sponges may have evolved from colonial choanoflagellates.

In the **mesohyl**, or gelatinous matrix between the two body-wall layers, are **amoebocytes**. These cells take up food from the choanocytes, digest it, and carry nutrients to other cells. Amoebocytes also form skeletal fibers, which may be sharp spicules or flexible fibers.

Most sponges are **hermaphrodites**, producing both eggs and sperm. Sperm, released into the spongocoel and carried out through the osculum, cross-fertilize eggs retained in the mesohyl of neighboring sponges. Flagellated larvae disperse through the osculum. Sponges are capable of extensive regeneration, replacing damaged body parts and reproducing asexually from fragments.

■ INTERACTIVE QUESTION 29.1

In this proposed animal phylogenetic tree, indicate the characteristics that determine the major evolutionary branch points and the general names associated with each branch.

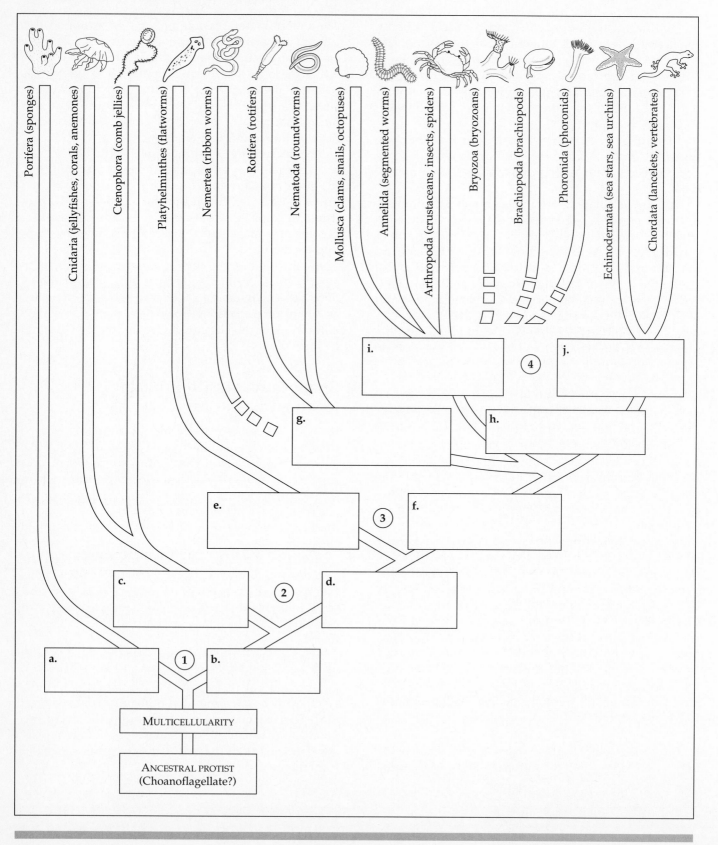

■ **INTERACTIVE QUESTION 29.2**

Give the locations and functions of the following:

a. choanocytes

b. amoebocytes

■ **INTERACTIVE QUESTION 29.3**

Name these two cnidarian body plans and identify the indicated structures.

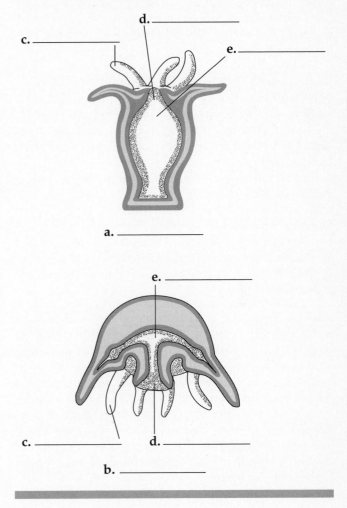

■ Cnidarians and ctenophores are radiate, diploblastic animals with gastrovascular cavities

Phylum Cnidaria The phylum Cnidaria includes hydras, jellyfishes, sea anemones, and coral animals. These diploblastic, radially symmetric animals had an early origin in eumetazoan history. Their simple anatomy consists of a sac with a central **gastrovascular cavity** and a single opening serving as both mouth and anus. This body plan has two forms: **polyps**, which are sessile, cylindrical forms with mouth and tentacles extending upward, and **medusae**, which are flattened, mouth-down polyps that move by passive drifting and weak body contractions. Both these body forms occur in the life histories of some cnidarians.

Cnidarians use their ring of tentacles, armed with **cnidocytes** containing stinging capsules called **nematocysts**, to capture prey. Cells of the epidermis and gastrodermis have bundles of microfilaments arranged into contractile fibers acting as simple muscles. A nerve net is associated with simple sensory receptors and coordinates the contraction of cells against the hydrostatic skeleton of the gastrovascular cavity, resulting in movement.

The class Hydrozoa includes animals that alternate between polyp and medusa forms, although the polyp stage is usually more conspicuous. The common, freshwater *Hydra* exists only in polyp form.

In the class Scyphozoa, the medusa stage is more prevalent. The sessile polyp stage often does not occur in the jellyfishes of the open ocean. The class Anthozoa includes sea anemones and coral animals. They occur only as polyps. Coral polyps secrete calcified external skeletons, and the accumulation of such skeletons produces coral.

Phylum Ctenophora Comb jellies of the phylum Ctenophora are the largest animals that use cilia for locomotion. Nerves running from a sensory organ to the combs of cilia coordinate movement of these small, transparent marine animals.

■ Flatworms and other acoelomates are bilateral, triploblastic animals lacking body cavities

Phylum Platyhelminthes These flattened, bilateral animals have moderate cephalization, and their mesoderm gives rise to organs and muscles. Like cnidarians, however, typical flatworms have a gastrovascular cavity with only one opening. Their lack of a body cavity indicates an early divergence in the bilateria line.

The class Turbellaria includes freshwater **planarians** and other, mostly marine, free-living flatworms. Their branching gastrovascular cavity functions for both digestion and circulation. Gas exchange and diffusion of nitrogenous wastes occur across the body wall. Ciliated flame cells function in osmoregulation.

Planarians move using cilia to glide on secreted mucus or body undulations to swim.

Eyespots on the head detect light, and lateral head flaps function for smell. Their more complex and centralized nervous system enables planarians to modify their behavior. Planarians reproduce asexually by regeneration or sexually by copulation between hermaphroditic worms.

Flukes are placed in the classes Trematoda and Monogenea. Adapted to live as parasites in or on other animals, flukes have a tough outer covering, suckers, and extensive reproductive organs. The life cycles of flukes are complex, usually including asexual and sexual stages and intermediate hosts in which larvae develop.

Tapeworms of the class Cestoda are parasites, mostly of vertebrates. Tapeworms consist of a scolex, with suckers and hooks for attaching to the host's intestinal lining, and a ribbon of proglottids packed with reproductive organs. Absorption of predigested food from the host eliminates the need for a digestive system. The life cycles of tapeworms may include intermediate hosts.

Phylum Nemertea   Most ribbon or proboscis worms in the phylum Nemertea are marine. Although their body is acoelomate, a fluid-filled sac allows these worms to hydraulically operate an extensible proboscis to capture prey. These worms probably evolved from flatworms, based on similarities of their excretory, sensory, and nervous systems, although their phylogenetic position is currently uncertain. They show, however, two new anatomical features: a **complete digestive tract** with separate mouth and anus, and a blood vascular system.

■ **INTERACTIVE QUESTION   29.4**

a. Describe the digestive system of a planarian.

b. Why do tapeworms, which are also platyhelminthes, lack a digestive system?

■ **Rotifers, nematodes, and other pseudocoelomates have complete digestive tracts and blood vascular systems**

The evolutionary relationships of the pseudocoelomate phyla are unclear, although they appear to be more closely related to protostomes than to deuterostomes.

Phylum Rotifera   Rotifers are smaller than many protists but have a complete digestive tract, other organ systems, and a pseudocoelom that functions as a hydrostatic skeleton and a blood vascular system. A crown of cilia draws microscopic food organisms into the mouth. Some species reproduce by **parthenogenesis**, in which female offspring develop from unfertilized eggs. In other species, two types of egg develop parthenogenically: one type forming females and the other developing into degenerate males that produce sperm. The resulting resistant zygotes survive harsh conditions in a dormant state.

Phylum Nematoda   Roundworms are among the most abundant of all animals, found inhabiting water, soil, and the bodies of plants and animals. These cylindrical worms are covered with tough cuticles and have complete digestive tracts. Their thrashing movement is produced by contraction of longitudinal muscles. The pseudocoelom serves as a blood vascular system. Reproduction is usually sexual, fertilization is internal, and most zygotes form resistant cells.

Numerous nematode species are ecologically important decomposers. Other nematodes are serious agricultural pests and animal parasites.

■ **INTERACTIVE QUESTION   29.5**

a. What is parthenogenesis?

b. In which phylum is parthenogenesis common?

c. Name some of the nematodes that are parasitic in humans.

■ **Mollusks and annelids are among the major variations on the protostome body plan**

Phylum Mollusca   Mollusks are mostly soft-bodied marine animals, many of which are protected by a shell. The molluscan body plan has three main parts: a muscular **foot** used for movement, a **visceral mass** containing the internal organs, and a **mantle** that covers the visceral mass and may secrete a shell. The **mantle cavity**, the water-filled chamber formed by the mantle, encloses the gills, anus, and excretory pores. A rasping **radula** is used for feeding by many mollusks. Most mollusks have separate sexes.

Some marine mollusks have a life cycle that includes a ciliated larva called the **trochophore**. Mollusks may have branched early on the protostome line. Four of the eight molluscan classes are discussed in the text.

Chitons, class Polyplacophora, are oval marine animals with shells that are divided into eight dorsal plates. Chitons cling tightly to rocks in the intertidal zone, where they feed on algae.

Gastropoda is the largest molluscan class. Most members are marine, although there are many freshwater species, and some snails and slugs are terrestrial. A distinctive feature of this class is **torsion**, the embryonic asymmetric growth of the visceral mass that results in the anus and mantle cavity being above the head. Most gastropods have single spiraled shells, although slugs and nudibranchs have no shells. Many gastropods have distinct heads with eyes at the tips of tentacles. Moving by the rippling of the elongated foot, most gastropods graze on plant material. Land snails lack gills; their vascularized mantle cavity functions as a lung.

The clams, oysters, mussels, and scallops of the class Bivalvia have the two halves of their shell hinged at the mid-dorsal line. Most bivalves are suspension-feeders; water flows into and out of the mantle cavity through siphons, and food particles are trapped in the mucus that coats the gills and then swept to the mouth by cilia.

The active squids and octopuses of the class Cephalopoda are rapidly moving carnivores. The mouth has beaklike jaws to crush prey and is surrounded by tentacles. The shell is reduced and internal in squids, absent in octopuses, and external only in the chambered nautilus. The foot has been modified to form the muscular siphon used to jet-propel the squid when water from the mantle cavity is expelled.

Cephalopods are the only mollusks with **closed circulatory systems**. They have well-developed nervous systems, sense organs, and complex brains—important features for active predators. The ancestors of octopuses and squids were probably shelled, predaceous mollusks. Shelled **ammonites** were the dominant invertebrate predators until their extinction at the end of the Cretaceous period.

The Lophophorate Animals   The three phyla of **lophophorate animals** (Phoronida, Bryozoa, and Brachiopoda) all have a **lophophore**, a horseshoe-shaped or circular fold bearing ciliated tentacles, which surrounds the mouth and functions in suspension-feeding. This complex structure suggests that all

■ **INTERACTIVE QUESTION   29.6**

a. Describe the three main parts of the molluscan body plan.

   1.

   2.

   3.

b. Compare the feeding behavior and activity level of snails, clams, and squid.

   1. Snails:

   2. Clams:

   3. Squid:

three phyla are related. They share other characteristics that may have evolved separately as adaptations to a sessile life. The lophophorate animals exhibit both deuterostome and protostome traits in their embryonic development, although molecular systematics suggests that these phyla are more closely related to protostomes.

Phoronids are marine, tube-dwelling worms that often live buried in sand with their lophophore extended. **Bryozoans** are tiny, mostly marine animals living in colonies that are often encased in a hard exoskeleton, with pores through which their lophophores extend. **Brachiopods**, or lamp shells, attach to the substrate by a stalk and open their shell to allow water to flow through the lophophore.

Phylum Annelida   Annelids are segmented worms found in most marine, freshwater, and soil habitats. The earthworm has a complete digestive system with specialized regions and a closed circulatory system. Respiration occurs across the moist, highly vascularized skin. Septa partition the coelom into segments, in each of which is found a pair of excretory **metanephridia**, which filter metabolic wastes from the blood and coelomic fluid.

The nervous system consists of a pair of cerebral ganglia, a subpharyngeal ganglion, and segmental ganglia along the fused ventral nerve cords. Earthworms are hermaphrodites; sperm are exchanged between worms during copulation. A mucous cocoon, secreted by the clitellum, slides off the worm after picking up its eggs and stored sperm.

Noncompressible coelomic fluid is enclosed by septa within the body segments, serving as a hydrostatic skeleton; circular and longitudinal muscles contract alternately to extend the body and pull it forward. Segmentation may first have evolved as an adaptation for this type of burrowing or creeping movement.

The class Oligochaeta includes the earthworms and aquatic species. Earthworm castings improve soil texture.

The mostly marine worms in the class Polychaeta have parapodia on each segment that function in gas exchange and locomotion. Polychaetes may be planktonic, bottom burrowers, or tube-dwellers.

Most of the leeches in the class Hirudinea inhabit fresh water. Many feed on small invertebrates, whereas others are parasites that temporarily attach to animals, slit or digest a hole through the skin, and suck the blood of their host.

Annelids exhibit two evolutionary innovations. A coelom provides a hydrostatic skeleton and many other advantages. Segmentation allows for regional specialization.

■ **INTERACTIVE QUESTION 29.7**

Identify the structures shown in this body segment of an earthworm.

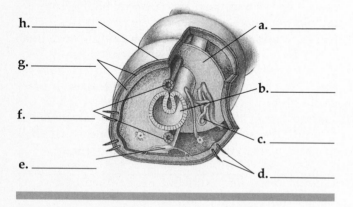

h._____
g._____
f._____
e._____
a._____
b._____
c._____
d._____

■ **The protostome phylum Arthropoda is the most successful group of animals ever to live**

General Characteristics of Arthropods Characteristics of arthropods include segmentation, which allows for regional specialization, a hard exoskeleton, and jointed appendages. Appendages are modified for walking, feeding, sensing, mating, and defense. A **cuticle** of chitin and protein completely covers the

body as an **exoskeleton**, providing protection and points of attachment for the muscles that move the appendages. To grow, an arthropod must **molt**, shedding its old exoskeleton and secreting a larger one. Arthropods have extensive cephalization and well-developed sensory organs.

A heart pumps hemolymph through an open circulatory system consisting of short arteries and a network of sinuses known as the hemocoel. Respiratory gas exchange in most aquatic species occurs through gills, whereas terrestrial arthropods have tracheal systems of branching internal ducts.

Arthropod Phylogeny and Classification Arthropods may have evolved from annelids or from a common ancestor of both phyla. Some molecular systematists believe that these two phyla are not so closely related and segmentation evolved independently in both. Arthropods diverged into four subphyla: Trilobitomorpha, Cheliceriformes, Uniramia, and Crustacea.

Trilobites **Trilobites**, with pronounced segmentation and uniform appendages, were common early arthropods throughout the Paleozoic era but were lost in the Permian extinctions.

Spiders and Other Chelicerates Existing during and beyond this same period were large **eurypterids**, or sea scorpions, marine predators that were **chelicerates**. The chelicerate body is divided into a cephalothorax and abdomen, with the most anterior appendages modified as pincers or fangs, called **chelicerae**. Most marine chelicerates are now extinct.

Most modern chelicerate species are in the terrestrial **class Arachnida**, which includes spiders, scorpions, ticks, and mites. The cephalothorax has six pairs of appendages: four pairs of walking legs, the chelicerae, and sensing or feeding appendages called pedipalps. **Book lungs**, consisting of stacked internal plates, function in gas exchange. Many spiders spin characteristic webs of silk from special abdominal glands.

■ **INTERACTIVE QUESTION 29.8**

a. What are chelicerae?

b. How do spiders trap, kill, and eat their prey?

Comparison of Chelicerates to Uniramians and Crustaceans Another evolutionary line produced uniramians and crustaceans with jawlike **mandibles**, one or two pairs of sensory **antennae**, and usually a pair of **compound eyes**. **Uniramians** (insects, millipedes, and centipedes) have one pair of antennae and unbranched (uniramous) appendages and are believed to have evolved on land. The aquatic **crustaceans**, with two pairs of antennae and branched appendages, are believed to have evolved in the ocean.

The exoskeleton of early marine arthropods, which probably functioned in protection and anchorage for muscles, helped preadapt arthropods for terrestrial life by providing the support and protection from desiccation needed on land. During the early Devonian period, both chelicerates and uniramians spread onto land. The oldest evidence of terrestrial animals is burrows of millipedelike arthropods about 450 million years old.

Insects and Other Uniramians   The millipedes of the **class Diplopoda** are wormlike, segmented vegetarians with two pairs of walking legs per segment. The centipedes of the **class Chilopoda** are terrestrial carnivores with appendages modified as jawlike mandibles and poison claws. Each segment of the trunk has one pair of legs.

The **class Insecta** is divided into about 26 orders and has more species than all other forms of life combined. The study of insects, called **entomology**, is a large field with many subspecialties. The oldest insect fossils are from the Devonian period, but a major diversification occurred in the Carboniferous and Permian periods with the evolution of flight and modification of mouth parts for specialized feeding on plants. The major diversification of insects preceded and probably influenced the radiation of flowering plants.

Flight is a major key to the success of insects. Many insects have one or two pairs of wings that are extensions of the cuticle of the dorsal thorax. Dragonflies were among the first winged insects.

■ **INTERACTIVE QUESTION   29.9**

Describe the three regions of a generalized insect body.

a.

b.

c.

The digestive tract of an insect has several specialized regions. The circulatory system is open. **Malpighian tubules** are outpocketings of the gut that function as excretory organs, removing metabolic wastes from the hemolymph. Gas exchange is accomplished by a **tracheal system** of chitin-lined tubes that ramify throughout the body and open to the outside through spiracles.

The nervous system consists of a pair of ventral nerve cords and a dorsal brain. The complex behavior of insects appears to be largely innate.

In **incomplete metamorphosis,** the young are smaller versions of the adult and pass through several molts. In **complete metamorphosis,** the larvae look entirely different from the adult. The larvae eat and grow; adults mate and reproduce. Reproduction is usually sexual; fertilization is usually internal.

Insects affect humans as pollinators of crops, vectors of disease, and competitors for food.

Crustaceans   Animals in the subphylum Crustacea have remained mostly in aquatic habitats. Crustaceans have two pairs of antennae, three or more pairs of mouthpart appendages, including mandibles, walking legs on the thorax, and appendages on the abdomen. Larger crustaceans have gills. A heart pumps hemolymph through arteries into sinuses that bathe the organs. Nitrogenous wastes pass by diffusion through thin areas of the cuticle, and a pair of glands regulates salt balance. Sexes usually are separate. One or more swimming larval stages occur in most aquatic crustaceans.

Lobsters, crayfish, crabs, and shrimp are relatively large crustaceans called decapods. Their cuticle is

■ **INTERACTIVE QUESTION   29.10**

Compare and contrast the following features for insects and crustaceans.

Feature	Insects	Crustaceans
Habitat	**a.**	**b.**
Locomotion	**c.**	**d.**
Respiration	**e.**	**f.**
Excretion	**g.**	**h.**
Antennae #	**i.**	**j.**
Appendages	**k.**	**l.**

hardened by calcium carbonate, and the dorsal side of their cephalothorax is covered by a carapace. Isopods are mostly small marine crustaceans but include terrestrial sow bugs and pill bugs. The small, very numerous copepods are important members of plankton communities that form the foundation of marine and freshwater food chains. Barnacles are sessile crustaceans that strain food from the water with their appendages.

The deuterostome lineage includes echinoderms and chordates

The echinoderms and chordates are grouped together based on their common embryological traits of radial cleavage, coelom formation from the archenteron, and origin of the mouth opposite the blastopore.

Phylum Echinodermata Most **echinoderms** are sessile or slow-moving marine animals with radial symmetry. They have a thin skin covering an endoskeleton of hard calcareous plates. A **water vascular system** with a network of hydraulic canals controls extensions called **tube feet** that function in locomotion, feeding, and gas exchange. Sexual reproduction usually involves separate sexes and external fertilization. Bilateral larvae metamorphosize into radial adults.

The sea stars, **class Asteroidea**, have five arms radiating from a central disc. Sea stars use tube feet lining the undersurfaces of their arms to creep slowly and to open bivalves for food. They evert their stomach through their mouth and slip it into a slightly opened shell.

Brittle stars, **class Ophiuroidea**, have distinct central discs and move by lashing their long, flexible arms.

Sea urchins and sand dollars, **class Echinoidea**, have no arms but are able to move slowly using their five rows of tube feet. Long spines also aid in movement. A sea urchin's mouth is ringed by complex jaw-like structures used to eat seaweeds and other foods.

Sea lilies, **class Crinoidea**, may live attached to the substrate by stalks. Their long, flexible arms extend upward from around the mouth and are used in suspension-feeding. Their form has changed little in 500 million years of evolution.

Sea cucumbers, **class Holothuroidea**, are elongated animals that bear little resemblance to other echinoderms other than having five rows of tube feet.

The recently discovered sea daisies of the **class Concentricycloidea** are small animals that live on waterlogged wood in the deep sea.

Phylum Chordata This phylum contains two subphyla of invertebrates plus the subphylum Vertebrata. Echinoderms and chordates share deuterostome

developmental characteristics but have existed as separate phyla for at least 500 million years.

The Cambrian explosion produced all the major animal body plans

Animals probably originated from a flagellated colonial protist related to choanoflagellates, which perhaps first became a hollow sphere of cells, then developed cell specialization, and later invaginated to form a gastrula-like "protoanimal." During the Cambrian explosion, occurring in only 5 to 10 million years at the beginning of this period (544 million years ago), all major body plans of animals evolved.

A less diverse group of animals dates back 700 million years to the **Ediacaran period** of the Precambrian era. Most of these fossils appear to be cnidarians, although fossilized burrows indicate that bilateral worms may have evolved in this period. The phylogenetic connection between the Ediacaran fauna and those of the Cambrian explosion is unclear.

Fossil sites in Greenland and the Yunnan region of China predate the famous Burgess Shale fossil bed and point to a remarkable animal diversification within 5 million years of the beginning of the Cambrian. The bizarre-looking fossils of the Burgess Shale may represent extinct "experiments" or may simply be ancient forms of modern phyla.

Explanations for the Cambrian explosion include adaptive radiation into previously unfilled ecological niches resulting from the evolution of the first animals, the emergence of predator–prey relationships that triggered various adaptations, the accumulation of sufficient atmospheric oxygen to support active metabolisms, or changes within animals (such as development of mesoderm or developmental genes) that facilitated the evolution of complex body forms. Such developmental patterns, while allowing rapid diversification, may then have constrained morphological evolution such that no new phyla evolved after the Cambrian explosion.

STRUCTURE YOUR KNOWLEDGE

1. Take some time to organize the kingdom Animalia for yourself. The diverse phyla are structured into broad groups based on a series of evolutionary characteristics (symmetry, coelom, and embryology). Create a concept map or diagram that shows these divisions, the criteria used to determine them, and the major phyla included in each group. Include common examples for each taxon.

2. **a.** What was the Cambrian explosion?

   b. List some of the proposed hypotheses to explain this explosion.

3. Indicate on the time line in Chapter 24, page 166, the date of the oldest known fossils of animals.

TEST YOUR KNOWLEDGE

MATCHING: *Match the following organisms with their classes and phyla. Answers may be used more than once or not at all.*

Phylum	Class	Organism
_____	_____	1. jellyfish
_____	_____	2. crayfish
_____	_____	3. snail
_____	_____	4. leech
_____	_____	5. tapeworm
_____	_____	6. cricket
_____	_____	7. scallop
_____	_____	8. tick
_____	_____	9. sea urchin
_____	_____	10. hydra
_____	_____	11. planaria
_____	_____	12. chambered nautilus

Phyla	Classes
A. Annelida	a. Arachnida
B. Arthropoda	b. Bivalvia
C. Cnidaria	c. Cephalopoda
D. Echinodermata	d. Cestoda
E. Mollusca	e. Crustacea
F. Nematoda	f. Echinoidea
G. Nemertea	g. Gastropoda
H. Platyhelminthes	h. Hirudinea
I. Porifera	i. Hydrozoa
J. Rotifera	j. Insecta
	k. Oligochaeta
	l. Scyphozoa
	m. Turbellaria

MULTIPLE CHOICE: *Choose the one best answer.*

1. Invertebrates include
   a. all animals except for the phylum Vertebrata.
   b. all animals without backbones.
   c. only animals that use hydrostatic skeletons.
   d. members of the parazoa, radiata, and protostomes, but not of the deuterostomes.
   e. all the animals that evolved in the Cambrian explosion.

2. Sponges differ from the rest of the Animalia because
   a. They are completely sessile.
   b. They have radial symmetry and are suspension-feeders.
   c. Their simple body structure has no tissues or organs.
   d. They have an unusual feeding structure called a lophophore and their embryology combines both protostome and deuterostome characteristics.
   e. They have no flagellated cells.

3. An insect larva
   a. is a miniature version of the adult.
   b. is transformed into an adult by molting.
   c. ensures more genetic variation in the insect life cycle.
   d. is a sexually immature organism specialized for eating and growth.
   e. is all of the above.

4. Cephalization
   a. is the development of bilateral symmetry.
   b. is the formation of a coelom by the outpocketing of the archenteron.
   c. is a diagnostic characteristic of deuterostomes.
   d. is common in radially symmetrical animals.
   e. is associated with motile animals that concentrate sensory organs in a head region.

5. A true coelom
   a. is found in deuterostomes.
   b. is found in protostomes.
   c. is a fluid-filled cavity completely lined by mesoderm.

d. may be used as a hydrostatic skeleton by soft-bodied coelomates.

e. is all of the above.

6. Which of the following is descriptive of proto-stomes?

   a. radial and determinate cleavage, blastopore becomes mouth

   b. spiral and indeterminate cleavage, coelom forms as split in solid mass of mesoderm

   c. spiral and determinate cleavage, blastopore becomes mouth, schizocoelous development

   d. spiral and indeterminate cleavage, blastopore becomes mouth, enterocoelous development

   e. radial and determinate cleavage, enterocoelous development, blastopore becomes anus

7. Hermaphrodites

   a. contain male and female sex organs but usually cross fertilize.

   b. include sponges, earthworms, and most insects.

   c. are characteristically found in parthenogenic rotifers.

   d. are both a and b.

   e. are a, b, and c.

8. A gastrovascular cavity

   a. functions in both digestion and circulation and has a single opening.

   b. has a large incurrent siphon and smaller excurrent pores.

   c. is found in the phyla Cnidaria, Platy-helminthes, and Rotifera.

   d. develops from the hollow blastula stage.

   e. is all of the above.

9. Which of the following is *not* true of cnidarians?

   a. An alternation of medusa and polyp stage is common in the class Hydrozoa.

   b. They use a ring of tentacles armed with stinging cells to capture prey.

   c. They include hydras, jellyfishes, sea anemones, and barnacles.

   d. They have a nerve net that coordinates contraction of microfilaments for movement.

   e. They have a gastrovascular cavity.

10. Which of the following combinations of phylum and characteristics is incorrect?

    a. Nemertea—proboscis worm, first complete digestive tract

    b. Rotifera—parthenogenesis, crown of cilia, microscopic animals

    c. Nematoda—gastrovascular cavity, tough cuticle, ubiquitous

    d. Annelida—segmentation, closed circulation, hydrostatic skeleton

    e. Echinodermata—radial symmetry, endoskeleton, water vascular system

11. Which of the following is either not an excretory structure or is incorrectly matched with its class?

    a. metanephridia—Oligochaeta

    b. Malpighian tubules—Echinoidea

    c. flame cells—Turbellaria

    d. thin region of cuticle—Crustacea

    e. diffusion across cell membranes—Hydrozoa

12. Torsion

    a. is embryonic asymmetric growth that results in a U-shaped digestive tract in gastropods.

    b. is characteristic of mollusks.

    c. is responsible for the spiral growth of bivalve shells.

    d. describes the thrashing movement of nematodes.

    e. is responsible for the metamorphosis of insects.

13. Bivalves differ from other mollusks in that they

    a. are predaceous.

    b. have no heads and are suspension-feeders.

    c. have shells.

    d. have open circulatory systems.

    e. use a radula to feed as they burrow through sand.

14. The exoskeleton of arthropods

    a. functions in protection and anchorage for muscles.

    b. is composed of chitin and cellulose.

    c. is absent in millipedes and centipedes.

    d. expands at the joints when the arthropod grows.

    e. functions in respiration and movement.

15. Which of the following is true of the subphylum Uniramia?

    a. The horseshoe crab is the one surviving marine member of this group.

    b. It contains the primarily aquatic crustaceans.

    c. It contains insects, centipedes, and millipedes, characterized by their unbranched appendages.

    d. It is characterized by jawlike mandibles, antennae, compound eyes, and includes all arthropods except the chelicerates.

    e. It includes the extinct trilobites and eurypterids.

16. The oldest known animals

    a. were colonies of choanoflagellates.

    b. were acoelomate worms.

    c. were sponges of the Radiata.

    d. were soft-bodied creatures from the Ediacaran period, from 700 million years ago.

    e. were from the Cambrian explosion at the beginning of the Paleozoic era.

CHAPTER 30

THE VERTEBRATE GENEALOGY

FRAMEWORK

This chapter focuses on the origin and characteristics of the subphylum Vertebrata and its classes. One version of chordate evolution is illustrated on p. 211. Fossil evidence and speculation about human ancestry are described.

CHAPTER REVIEW

■ Vertebrates belong to the phylum Chordata

There are three subphyla of **chordates**: urochordates and cephalochordates, which are without a backbone, and **vertebrates**, which have a cranium and backbone.

Chordate Characteristics Chordates have four distinguishing anatomical features, which may appear only during the embryonic stage: (1) a flexible rod called a **notochord** running the length of the animal as a simple skeleton; (2) a dorsal, hollow nerve cord that develops into the brain and spinal cord of the central nervous system; (3) pharyngeal slits that open from the anterior region of the digestive tube and function in suspension feeding or are modified for gas exchange and other functions; and (4) a muscular postanal tail containing skeletal elements and muscles.

■ Invertebrate chordates provide clues to the origin of vertebrates

Subphylum Urochordata **Tunicates**, or sea squirts, are the common names for these mostly sessile marine **urochordates**. Water moves through the saclike, tunic-covered animal from the incurrent siphon, through

pharyngeal slits where food is filtered by a mucous net, to the excurrent siphon. The larval traits of notochord, nerve cord, and tail are lost in the adult animal.

Subphylum Cephalochordata **Cephalochordates**, known as lancelets, are tiny marine animals that retain all four chordate characteristics into the adult stage. They filter food particles through a mucous net secreted across the pharyngeal slits. The lancelet swims by coordinated contractions of serial muscle segments that flex the notochord from side to side. Blocks of mesoderm, called **somites**, develop into these muscle segments and indicate that chordates are segmented animals.

■ INTERACTIVE QUESTION 30.1

What is the common name for the animal shown in this diagram, and to which subphylum does it belong? Identify the labeled structures in this diagram, and indicate which are the four chordate characteristics.

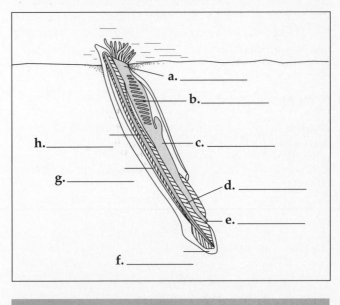

a. _____
b. _____
c. _____
d. _____
e. _____
f. _____
g. _____
h. _____

Relationship of Invertebrate Chordates to the Vertebrates   Fossils resembling cephalochordates appear in the Burgess Shale of the Cambrian period and have been dated at 550 million years old; vertebrates appear in the fossil record 50 million years later. Molecular systematics indicate that vertebrates are most closely related to cephalochordates. Both subphyla may have evolved from a common sessile ancestor that underwent **paedogenesis**, the development of sexual maturity while still a larva.

■ The evolution of vertebrate characteristics is associated with increased size and activity

Derived shared characters (synapomorphies) of vertebrates include cephalization, a skeleton that includes a cranium and vertebral column, and anatomical adaptations to support an active metabolism.

The axial skeleton includes the cranium, which encloses and protects the brain, and the vertebral column, which provides a jointed attachment for segmental swimming muscles. Many vertebrates also have ribs and an appendicular skeleton supporting two pairs of limbs. The vertebrate skeleton is made of bone and/or cartilage, which is secreted and maintained by living cells and can grow with the animal. A group of embryonic cells called the **neural crest** is found along the margin of the embryonic folds that meet to form the dorsal, hollow nerve cord. These cells migrate in the embryo and contribute to various skeletal components and other vertebrate structures.

Increased size and activity level require a high level of cellular respiration, supported by adaptations of the respiratory and circulatory systems. Vertebrates have a closed circulatory system with a ventral, chambered heart pumping blood to capillaries that ramify through body tissues. An active lifestyle is also supported by adaptations for feeding, digestion, and nutrient absorption.

■ **INTERACTIVE QUESTION   30.2**

List the three shared derived characteristics of vertebrates that facilitate increased size and activity.

a.

b.

c.

■ Vertebrate diversity and phylogeny: *an overview*

Four classes of the subphylum Vertebrata—Amphibia, Reptilia, Aves, and Mammalia—are called **tetrapods** because they have two pairs of limbs. The other three extant classes—Agnatha, Chondrichthyes, and Osteichthyes—are commonly called fishes.

Reptiles, birds, and mammals are called **amniotes** in reference to the **amniotic egg**, a shelled egg that provides a fluid-filled sac around the embryo and enables these vertebrates to reproduce on land. (See Interactive Question 30.3, p. 211.)

■ Agnathans are jawless vertebrates

The oldest vertebrate fossils are those of jawless creatures, including the fishlike **ostracoderms** with an armor of bony plates, which are classified with some extant vertebrates in the **class Agnatha**. Agnathans first appear in the Cambrian strata, but most are found from the Ordovician and Silurian periods. Early agnathans were generally small, lacked paired fins, and probably filtered food by taking in mud or water through their jawless mouths and passing it out through gill slits. The gills probably also functioned in gas exchange.

Ostracoderms and most agnathan groups disappeared during the Devonian period. Jawless fishes are represented today by lampreys, which may be suspension-feeders as larvae and blood-sucking parasites as adults, and marine hagfishes, which are scavengers.

■ Placoderms were armored fishes with jaws and paired fins

Armored fishes called **placoderms** replaced most agnathans during the late Silurian and early Devonian periods. These larger fish had paired fins that enhanced swimming and jaws that evolved from the skeletal supports of the anterior pharyngeal slits. Both these traits facilitated a predatory lifestyle. The remaining gill slits maintained their function in gas exchange.

The Devonian period is known as the age of fishes. Placoderms and acanthodians, another class of jawed fishes, radiated and then disappeared almost completely by the Carboniferous period. Bony fishes and sharks diverged separately from the ancestral line. (See Interactive Question 30.4, p. 212.)

■ INTERACTIVE QUESTION 30.3

This phylogenetic tree shows one set of hypotheses of chordate relationships. As you work through this chapter, fill in the large groupings at the top, the boxed shared derived features for each branch point, and the class that makes up each branch. You should be able to fill in letters **a–j** now.

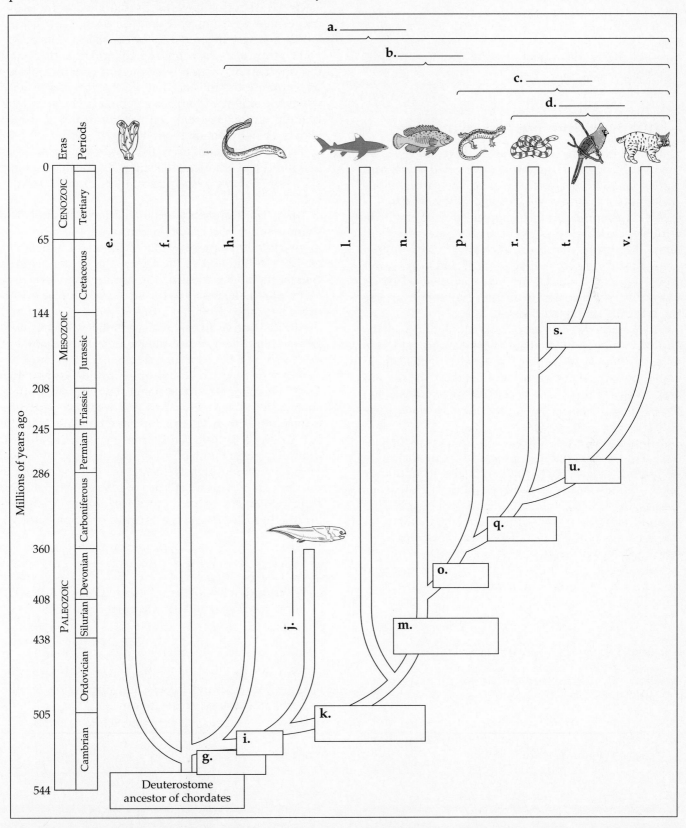

■ **INTERACTIVE QUESTION 30.4**

What two modifications found in placoderms enabled them to be active predators?

Sharks and their relatives have adaptations for powerful swimming

The vertebrates of **class Chondrichthyes** have skeletons made of cartilage, well-developed jaws, and paired fins with skeletal elements. The cartilaginous skeleton of sharks evolved secondarily, perhaps as a result of a modification of normal development in which a cartilaginous skeleton becomes ossified. Sharks swim to maintain buoyancy and to move water past their gills. Although most sharks are carnivores, the largest sharks and rays suspension-feed on plankton. Shark teeth probably evolved from the jagged scales that cover the skin.

Sharks have keen senses: sharp vision, nostrils, auditory organs that can sense percussion, and skin regions in the head that can detect electric fields generated by muscle contractions of nearby fish. The **lateral line system** is a row of microscopic organs along the sides of the body that are sensitive to water pressure changes.

Following internal fertilization, **oviparous** species of sharks lay eggs that hatch outside the mother; **ovoviviparous** species retain the fertilized eggs until they hatch within the uterus; and, in the few **viviparous** species, developing young are nourished by a placenta until born. The **cloaca** is the common chamber for the openings of the reproductive tract, excretory system, and digestive tract.

Rays, which propel themselves with their enlarged pectoral fins, are primarily bottom-dwellers that feed on mollusks and crustaceans.

■ **INTERACTIVE QUESTION 30.5**

a. What is the function of the spiral valve in a shark's intestine?

b. All vertebrates except placental mammals have cloacas. What is a cloaca?

Bony fishes are the most abundant and diverse vertebrates

Bony fishes, of the **class Osteichthyes**, have skeletons reinforced with a hard matrix of calcium phosphate. Their skin is often covered by flattened bony scales and coated with mucus to reduce drag. They have a lateral line system similar to that of sharks. The movement of muscles surrounding the gill chambers and the **operculum**, or flap over the gill chamber, draws water across the gills so that a bony fish can breathe while not moving. Most bony fishes also have a **swim bladder**, an air sac that controls buoyancy and also allows a fish to remain motionless.

The flexible fins of bony fish are better for steering and propulsion than are the stiffer fins of sharks. Most species of fish are oviparous and fertilization is external.

Bony fishes probably originated in freshwater. The swim bladder was modified from lungs used to augment gills in stagnant water. Two extant subclasses evolved by the end of the Devonian: the **ray-finned fishes**, including most of the familiar modern fish, many of which spread to the seas during their evolution and some of which returned to fresh water; and the **fleshy-finned fishes**, which remained in freshwater and continued to use their lungs. Three orders of fleshy-finned fishes evolved: coelocanths, rhipidistians (both referred to as **lobe-finned fishes)**, and **lungfishes**. Lobe-fins and some lungfishes had muscular pectoral and pelvic fins that were supported by extensions of the skeleton and may have enabled the fish to "walk" along the bottom or, occasionally, on land. Three present-day genera of lungfishes are found in the Southern Hemisphere. The coelocanth is the only known extant species of lobe-finned fish. The **rhipidistians** are extinct, but a Devonian rhipidistian is thought to have been the ancestor of amphibians.

■ **INTERACTIVE QUESTION 30.6**

a. How and why are the body shapes of sharks, bony fishes, and aquatic mammals similar?

b. Go back to the evolutionary tree in Interactive Question 30.3 and fill in the information for the vertebrate classes up to letter **n**.

Amphibians are the oldest class of tetrapods

Members of the **class Amphibia** were the first vertebrates on land.

Early Amphibians   Lobe-finned fishes with lungs, living in freshwater habitats subject to drought, may have used their paired appendages to move on land. The oldest amphibian fossils date back to the late Devonian, about 365 million years ago. Amphibians have always been predators, eating insects and other invertebrates that preceded them onto land. The Carboniferous period, during which a diversity of new forms and sizes evolved, is known as the Age of Amphibians. Amphibians declined during the late Carboniferous; by the Mesozoic era, the survivors resembled modern amphibians.

Modern Amphibians   There are three extant orders of amphibians. **Urodeles** (salamanders), which may be aquatic or terrestrial, swim or walk with a lateral bending of the body. **Anurans** (frogs and toads) are more specialized for moving on land, using their powerful hind legs for hopping. Protective adaptations of frogs include camouflage coloring and distasteful or poisonous skin mucus that may be advertised by bright coloration. Tropical **apodans** (caecilians) are burrowing, legless, nearly blind, wormlike animals.

Many frogs undergo a metamorphosis from a tadpole, the aquatic, herbivorous larval form with gills and a tail, to the carnivorous, terrestrial adult form with legs and lungs. Many amphibians do not have an aquatic tadpole stage, and some species are exclusively aquatic or terrestrial. Amphibians are most abundant in damp habitats, since much of their gas exchange occurs across their moist skin.

Fertilization is generally external. Oviparous species lay their eggs in aquatic or moist environments. Ovoviviparous and viviparous species are also known. Some species show varying amounts of parental care. During breeding season, some frogs communicate with vocalizations. Some species migrate to specific breeding sites, using various forms of communication and navigation.

■ INTERACTIVE QUESTION   30.7

a. From what group are amphibians thought to have evolved?

b. When was the "Age of Amphibians"?

The evolution of the amniotic egg expanded the success of vertebrates on land

The amniotic egg, enclosed in a shell, permitted vertebrates to complete their life cycles on land. **Extraembryonic membranes** within the egg are involved in gas exchange, waste storage, and nutrient transfer from the egg to the embryo. The amnion encases the embryo in amniotic fluid. Birds and mammals evolved from reptiles. Birds have shelled, amniotic eggs, and placental mammals retain the extraembryonic membranes during embryonic development within the uterus.

■ INTERACTIVE QUESTION   30.8

Identify the four extraembryonic membranes in this sketch of an amniotic egg. What are the functions of these membranes?

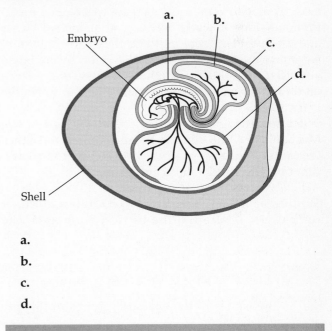

a.

b.

c.

d.

A reptilian heritage is evident in all amniotes

Reptilian Characteristics   The skin of lizards, snakes, turtles, and crocodilians of **class Reptilia** is covered with keratinized, waterproof scales. Lungs are the main vehicle for gas exchange. Most species lay shelled amniotic eggs, and fertilization is internal. Some species are viviparous, in which case the extraembryonic membranes form a placenta.

Reptiles are **ectotherms**. They absorb external heat rather than generating their own and thus have approximately one-tenth the caloric needs as do mammals of similar sizes. Reptiles regulate their body temperature through behavioral adaptations.

The Age of Reptiles   The oldest reptilian fossils date from the upper Carboniferous period, 300 million years ago. During the Permian period, the first reptilian radiation gave rise to two evolutionary branches: the **synapsids**, which included mammal-like reptiles called therapsids, and the **sauropsids** that gave rise to all modern amniotes except the mammals. The sauropsids split into the **anapsids**, of which turtles are the only survivors, and the **diapsids**, including the lizards, snakes, crocodilians, and the extinct dinosaurs. The birds are descendants of a dinosaur ancestor.

The second great radiation occurred a little over 200 million years ago during the late Triassic period, when the terrestrial dinosaurs and flying pterosaurs originated and diversified. The highly diverse dinosaurs included the largest animals ever to live on land. The pterosaurs had wings formed from a membrane of skin stretched between an elongated finger and the body. Evidence indicates that many dinosaurs were active, agile, and, in some species, social. Some scientists postulate that dinosaurs were **endothermic**—able to use metabolic heat to warm the body.

During the Cretaceous period at the end of the Mesozoic era, the climate became cooler and more variable. Probably within 5 to 10 million years, dinosaurs declined and became extinct.

Reptiles of Today   The three largest orders of extant reptiles are **Chelonia**, the turtles; **Squamata**, the lizards and snakes; and **Crocodilia**, the alligators and crocodiles. Turtles evolved among the anapsids during the Mesozoic era and have changed little since then. They are protected by their hard shell.

Lizards, the most numerous and diverse reptiles, are relatively small. Snakes, apparently descended from lizards that adopted a burrowing lifestyle, are limbless, although primitive species have vestigial pelvic and limb bones. Snakes are carnivorous, with adaptations for locating (heat-detecting and olfactory organs), killing (sharp teeth, often with toxins), and swallowing (loosely articulated jaws) their prey.

Crocodiles and alligators, among the largest living reptiles, spend most of their time in water. Crocodilians are the reptiles most closely related to the dinosaurs.

■ Birds began as flying reptiles

Class Aves evolved during the reptilian radiation of the Mesozoic era. The amniote egg and scales on the legs are two of the reptilian traits that birds display.

Characteristics of Birds   Bird anatomy has evolved to enhance flight. Weight is reduced by honeycombed bones, air sacs off the lungs, and lack of teeth (food is ground in the gizzard). The keratinized beak has assumed a great variety of shapes associated with different diets. Birds are endothermic; feathers and a fat layer help retain metabolic heat. The four-chambered heart segregates oxygenated and oxygen-poor blood, which facilitates a high metabolic rate.

Birds have excellent vision. Their relatively large brains permit detailed visual processing, motor coor-

---

▓ INTERACTIVE QUESTION   30.9

Fill in the names of the ancestral groups in this phylogeny of the amniotes. Why are reptiles not considered a monophyletic taxon?

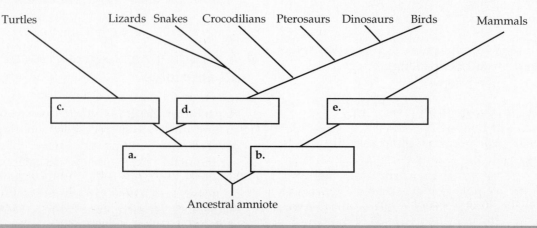

dination, and complex behavior, particularly in courtship rituals and parenting. Eggs are brooded during development by the female and/or male bird.

Aerodynamic wings are flapped by large pectoral muscles attached to a keel on the sternum. Feathers made of keratin, a diagnostic characteristic of birds, are extremely light and strong, and shape the wing into an airfoil.

The evolution of strong flying ability enhanced hunting, foraging, escape from predators, and migration to favorable habitats.

The Origin of Birds   Many dinosaurs, including some theropods, built nests, cared for their young, and were most likely endothermic. According to a strictly cladistic taxonomy, birds and reptiles are placed in the same vertebrate class.

Fossils of *Archaeopteryx lithographica*, an ancient feathered creature with clawed forelimbs, teeth, and long tail, have been found from 150 million years ago. This weak flyer was probably a tree-dwelling glider, perhaps the earliest mode of flight. Much about the origin of birds remains unknown.

Modern Birds   Flightless birds called **ratites**, such as the ostrich, kiwi, and emu, lack a keeled breastbone. **Carinates** have a sternal keel to which attach large breast muscles. **Passeriforms**, or perching birds, are an order of carinate birds that includes 60 percent of living bird species.

■ **INTERACTIVE QUESTION   30.10**

List several adaptations for flight found in birds.

■ **Mammals diversified extensively in the wake of the Cretaceous extinctions**

Mammalian Characteristics   Vertebrates of **class Mammalia** have hair made of keratin, which helps to insulate these endothermic animals. Active metabolism is provided by an efficient respiratory system that uses a diaphragm to help ventilate the lungs and a four-chambered heart that separates oxygenated and oxygen-poor blood.

Milk, produced in mammary glands, is used to nourish mammalian young. Fertilization is internal, and the embryo develops in the uterus. In placental mammals, a **placenta**, formed from the uterine lining and extraembryonic membranes of the embryo, nourishes the developing embryo.

Mammals have large brains and seem the most capable of learning. Extended parental care of the young provides time to learn complex skills.

Mammalian teeth come in a diverse assortment of shapes and sizes that are specialized for eating a variety of foods. Part of the evolutionary remodeling of the mammalian jaw produced a stronger, single-bone lower jaw and the incorporation of two jaw bones into the middle ear.

Evolution of Mammals   The oldest fossils of mammals date back to the Triassic period, 220 million years ago. Reptilian **therapsids**, part of the anapsid branch of reptiles, were mammalian ancestors. Small Mesozoic mammals, which were probably nocturnal and insectivorous, coexisted with dinosaurs. Mammals underwent an extensive radiation, filling the adaptive zones created by the extinction of dinosaurs and the rise of flowering plants in the late Cretaceous period. There are three major groups of mammals today.

Monotremes   The platypus and echidnas (spiny anteaters) are the only extant **monotremes**, egg-laying mammals. The egg is reptilian in structure, but monotremes have hair and produce milk for their young. Monotremes seem to have descended from a very early mammalian branch.

Marsupials   **Marsupials**, including opossums, kangaroos, and the koala, complete their embryonic development in a maternal pouch called a marsupium. Born very early in development, the neonate crawls into the marsupium, fixes its mouth to a teat and completes its development while nursing.

Australian marsupials have radiated and filled the niches occupied by placental mammals in other parts of the world. Marsupials apparently originated in what is now North America, spreading southward. After the breakup of Pangaea, South America and Australia became island continents where marsupials diversified in isolation from placental mammals. Placental mammals reached South America in the Cenozoic era, whereas Australia has remained isolated.

Placental Mammals   **Placental mammals** complete development attached to a placenta within the maternal uterus. Placentals and marsupials may have diverged from a common ancestor 80 to 100 million years ago. The major orders of placental mammals radiated during the late Cretaceous and early Tertiary, about 70 to 45 million years ago.

There appear to be at least four main evolutionary lines of placental mammals. The orders Chiroptera and Insectivora, containing bats and shrews, respectively, resemble early mammals. A lineage of medium-sized herbivores gave rise to the lagomorphs (rabbits), perissodactyls (odd-toed ungulates such as horses and rhinos), artiodactyls (even-toed ungulates such as deer and swine), sirenians (sea cows), proboscideans (elephants), and cetaceans (porpoises and whales).

The order Carnivora, the third evolutionary line, includes cats, dogs, raccoons, skunks, and the pinnipeds (seals, sea lions, and walruses), and first appeared during the early Cenozoic era. The fourth radiation resulted in the order Rodentia (rats, squirrels, and beavers) and the order Primates (monkeys, apes, and humans).

■ INTERACTIVE QUESTION   30.11

a. List the main characteristics of mammals.

b. Now go back to the evolutionary tree in Interactive Question 30.3 and fill in the information for the rest of the classes and derived characters.

■ Primate evolution provides a context for understanding human origins

Evolutionary Trends in Primates   The first primates probably were small, arboreal animals, descended from insectivores late in the Cretaceous. Characteristics that were shaped by this arboreal ancestry include limber shoulder joints that make it possible to brachiate (swing from one hold to another), dexterous hands with sensitive fingers, eye–hand coordination, forward-facing and closely-set eyes to provide depth perception, and parental care.

Modern Primates   There are two suborders of Primates. **Prosimians**, such as lemurs, lorises, and tarsiers, probably resemble early arboreal primates. The **anthropoids** include monkeys, apes, and humans.

Prosimian fossils have been divided into two groups that split at least 50 million years ago.

Recently discovered fossils in Asia and Africa may represent another lineage that may be more similar to the anthropoids than to the other two groups of prosimian fossils.

Fossils of monkeylike primates have been found from about 40 million years ago in Africa or Asia. Somehow ancestral monkeys reached South America. These strictly arboreal New World monkeys have been evolving separately from Old World monkeys for millions of years. Most monkeys in both groups are diurnal and live in social bands.

The anthropoid suborder also includes four genera of apes: gibbons, orangutans, gorillas, and chimpanzees, all of which live in the Old World tropics. Modern apes generally are larger than monkeys, with relatively long legs, short arms, and no tails. Ape behavior is more adaptable than that of monkeys, and gorillas and chimpanzees are highly social.

■ Humanity is one very young twig on the vertebrate tree

Paleoanthropology is the study of human origins and evolution.

Some Common Misconceptions   Humans are not descended from chimpanzees; the two represent divergent branches from a common, less-specialized anthropoid ancestor. Human evolution has not occurred as phyletic change within an unbranched hominid line; there have been times when several different human species coexisted. Different human features have evolved at different rates, known as **mosaic evolution**. Thus, for example, bipedalism or erect posture evolved before an enlarged brain developed.

Early Anthropoids   The oldest known ape fossils are of a cat-sized tree-dweller that lived about 35 million years ago. About 25 million years ago, ape descendants diversified and spread into Eurasia. About 20 million years ago, the Himalayan range arose, the climate became drier, and the forests contracted, separating these two regions of anthropoid evolution. Fossil evidence and DNA comparisons indicate that humans and chimpanzees diverged from a common African ancestor only about 6 to 8 million years ago.

Australopithecines: The First Humans   In 1924, a British anthropologist found a skull of an early human that he called *Australopithecus africanus*. It appears that *Australopithecus* was an upright hominid, with humanlike hands and teeth but a small brain. Various species of *Australopithecus* existed for over 3 million years, beginning about 4.4 million years ago.

In 1974, a petite *Australopithecus* skeleton was discovered in the Afar region of Ethiopia. Lucy, as this

3.18 million year old fossil was named, and similar fossils have been designated as a separate species, *Australopithecus afarensis*, which appears to have existed as a species for about 1 million years. Fossils have recently been discovered from a bipedal species, named *Australopithecus ramidus*, that lived about 4.4 million years ago. It is not known how *A. ramidus* and *A. afarensis* are related to each other or to *A. africanus*. An adaptive radiation that began about 3 million years ago produced *A. africanus* and several heavier-boned *Australopithecus* species, as well as *Homo habilis*, which appeared in the fossil record about 2.5 million years ago. Hominids walked erect for 2 million years with no enlargement of the brain.

Homo habilis Larger-brained fossils date back about 2.5 million years and sometimes have been found with simple stone tools. Most paleoanthropologists place these fossils in the genus *Homo*, calling them *Homo habilis*, or "handy man."

Homo habilis and *Australopithecus*, including the more stocky *A. robustus*, coexisted for a million years. One hypothesis is that these were distinct lines of hominids, one of which was an evolutionary dead-end, while the other, *Homo habilis*, led to modern humans.

■ **INTERACTIVE QUESTION 30.12**

a. How long ago did humans and chimpanzees diverge from a common ape ancestor?

b. Name and give the age of the oldest discovered *Australopithecus* species.

***Homo erectus* and Descendants** *Homo erectus*, the first hominid to migrate out of Africa into Asia and Europe, is represented by fossils known as Java Man and Beijing Man, which range in age from 1.8 million years to 300,000 years old. The spread of *H. erectus* from Africa may be related to a change to a carnivorous diet that required more hunting territory. *H. erectus* had greater height and brain capacity than *H. habilis*. Their greater intelligence allowed these humans to survive in the colder climates of the north, by living in shelters, building fires, wearing animal skins, and using more elaborate stone tools.

The Neanderthals were among the several descendants of *H. erectus*. They lived in Europe, the Middle East, and parts of Asia from about 130,000 to 35,000 years ago. They had heavier brow ridges, less pronounced chins and larger brains than modern humans. They were skilled tool-makers and left evidence of burials and other rituals. Many paleoanthropologists group post-*H. erectus* fossils dating back 300,000 years in Africa with Neanderthals and other Asian and Australasian fossils as the earliest forms of our species and call them "archaic *Homo sapiens*."

The Emergence of* Homo sapiens: *Out of Africa . . . but When? Science as a Process According to the *multiregional model*, these archaic *H. sapiens* evolved into modern humans in parallel in different parts of the world. Geographical diversification began between 1 and 2 million years ago, although gene flow maintained genetic similarity among the populations.

The oldest, fully modern *Homo sapiens* fossils have been found in Africa, dating back 100,000 years. Two groups of fossils found in Israeli caves indicate that Neanderthals may have overlapped with modern humans for about 40,000 years. According to Stringer, this overlap means that Neanderthals and the modern forms did not interbreed, nor were the Neanderthals ancestors to those modern humans. Stringer postulates that Neanderthals and other archaic *H. sapiens* died out and that modern humans evolved from *H. erectus* in Africa, then migrated to other areas. According to this "out of Africa" or *monogenesis model*, modern humans dispersed from Africa relatively recently, with geographic diversification of modern humans occurring in the past 100,000 years.

A group of geneticists recently claimed that mitochondrial DNA comparisons indicated that humans could be traced back to Africa, with the divergence of mtDNA beginning just 200,000 years ago. Some paleoanthropologists continue to support the multiregional evolution of modern humans as the best interpretation of the fossil evidence. Additional comparisons of nuclear DNA and methods for using mtDNA data, however, have strengthened the monogenesis model. Perhaps DNA samples recovered from fossils will allow molecular systematists and paleontologists to test the predictions made by the two competing models of the origin of modern humans.

Cultural Evolution: A New Force in the History of Life Evolution of an erect stance required remodeling the foot, pelvis, and vertebral column. Enlargement of the human brain was made possible by a prolonged period of growth after birth. An extended period of parental care helps to make possible the basis of culture—the transmission of knowledge accumulated over generations. Language, spoken and written, is the major means for this transmission.

Cultural evolution has included several stages: the communal efforts of nomads who hunted and gathered food on African grasslands 2 million years ago, the development of agriculture about 10,000 to 15,000 years ago, and the Industrial Revolution, which began in the eighteenth century. Since then, new technology has been growing exponentially.

The biological evolution of life has been linked with environmental changes. Cultural evolution has made *Homo sapiens* a new force in the history of life—a species that no longer needs to wait to adapt to its environment, but simply changes the environment to meet its needs. As agents of rapid environmental change, we are altering the planet faster than many species can adapt. The next and perhaps most devastating crisis in the history of life may be the habitat destruction, pollution, resource decline, and climatic changes resulting from the actions of the huge population of *H. sapiens*.

STRUCTURE YOUR KNOWLEDGE

1. List the seven extant classes of vertebrates. Put check marks by the tetrapod classes, and star the amniotes.

2. Describe several examples from vertebrate evolution that illustrate the common evolutionary theme that new adaptations usually evolve from preexisting structures.

3. How did various vertebrate groups meet the challenges of a terrestrial habitat?

4. Contrast the multiregional and monogenesis models of the evolution of modern humans. What evidence is used to support each of these models?

TEST YOUR KNOWLEDGE

FILL IN THE BLANKS

_____ 1. chordate subphylum of sessile, suspension-feeding marine animals

_____ 2. chordate subphylum that includes lancelets

_____ 3. blocks of mesoderm along notochord that develop into muscles

_____ 4. extinct jawless fish with armor of bony plates

_____ 5. flap over the gills of bony fishes

_____ 6. adaptation that allows reptiles to reproduce on land

_____ 7. reptilian ancestor of lizards, snakes, crocodilians, and dinosaurs

_____ 8. group from which birds descended

_____ 9. egg-laying mammals

_____ 10. structure that helps mammals ventilate their lungs

_____ 11. suborder that includes monkeys, apes, and humans

_____ 12. genus in which the fossil Lucy is placed

MULTIPLE CHOICE: *Choose the one best answer.*

1. Pharyngeal slits appear to have functioned first as
   a. suspension-feeding devices.
   b. gill slits for respiration.
   c. components of the jaw.
   d. portions of the inner ear.
   e. mouth openings.

2. Which of the following characteristics differentiates the vertebrates from the invertebrate chordates?
   a. notochord
   b. extensive cephalization
   c. postanal tail
   d. dorsal hollow nerve cord
   e. motility

3. Modern agnathans are represented by
   a. caecilians.
   b. sharks and rays.
   c. the lobe-finned coelocanth.
   d. lungfishes.
   e. hagfishes and lampreys.

4. Jaws first occurred in the class
   a. Agnatha.
   b. Chondrichthyes.
   c. Osteichthyes.
   d. Placodermi.
   e. Cephalochordata.

5. Which of the following is incorrectly paired with its gas exchange mechanism?
   a. amphibians—skin and lungs
   b. lobe-finned fishes—gills and lungs
   c. reptiles—lungs

    d. bony fishes—swim bladder

    e. mammals—lungs with diaphragm

6. Urodeles include
   a. limbless, burrowing amphibians.
   b. frogs that have a tadpole stage.
   c. frogs and toads that may or may not have a tadpole stage.
   d. salamanders.
   e. turtles.

7. Reptiles have lower caloric needs than do mammals of comparable size because they
   a. are ectotherms.
   b. have waterproof scales.
   c. have an amniotic egg.
   d. move by bending their vertebral column back and forth.
   e. have a more efficient respiratory system.

8. Which of the following best describes the earliest mammals?
   a. large, herbivorous
   b. large, carnivorous
   c. small, insectivorous
   d. small, herbivorous
   e. small, carnivorous

9. Oviparity is a reproductive strategy that
   a. allows mammals to bear well-developed young.
   b. is used by both birds and reptiles.
   c. is necessary for vertebrates to reproduce on land.
   d. is a necessity for all flying vertebrates.
   e. protects the embryo inside the mother and uses the food resources of the egg.

10. In Australia, marsupials fill the niches that placental mammals fill in other parts of the world because
    a. They are better adapted and have outcompeted placental mammals.
    b. Their offspring complete their development attached to a teat in a marsupium.
    c. They originated in Australia.
    d. They evolved from monotremes that migrated to Australia about 12 million years ago.

    e. After Pangaea broke up, they diversified in isolation from placental mammals.

11. *Homo habilis* is believed to have coexisted with
    a. *Australopithecus ramidus.*
    b. *A. afarensis.*
    c. *A. robustus.*
    d. *Homo sapiens neanderthalensis.*
    e. *H. sapiens sapiens.*

12. In accordance with the model of mosaic evolution,
    a. Biogeography reflects the adaptive radiation of many groups.
    b. Erect posture preceded the enlargement of the brain in human evolution.
    c. Modern humans evolved in parallel in different parts of the world, laying the groundwork for geographic differences more than a million years ago.
    d. Modern humans first evolved in Africa and later dispersed to other regions.
    e. The rapid divergence of hominids about 2.5 million years ago was associated with climatic changes.

13. Which species is believed to have persisted for the longest period of time?
    a. *Australopithecus afarensis*
    b. *A. robustus*
    c. *Homo habilis*
    d. *H. erectus*
    e. *H. sapiens*

14. According to the monogenesis model, modern humans
    a. have been diverging from *H. erectus* for over 1 million years.
    b. evolved from many regional descendants of *H. erectus.*
    c. evolved from an "Eve" who has been dated from Africa 90,000 years ago.
    d. evolved from African descendants of *H. erectus* and began spreading out 100,000 years ago.
    e. descended from the following sequence: *A. ramidus, A. afarensis, A. africanus, H. habilis, H. erectus.*

PLANTS: FORM AND FUNCTION

CHAPTER 31

PLANT STRUCTURE AND GROWTH

FRAMEWORK

A rather large new vocabulary is needed to name the specialized cells and structures in a study of plant morphology and anatomy. Focus your attention on how the roots, stems, and leaves of a plant are specialized to function in absorption, support, transport, protection, and photosynthesis. Plants exhibit indeterminate growth. Apical meristems at the tips of roots and shoots create primary growth; the primary meristems produce dermal, ground, and vascular tissues. The lateral meristems, vascular cambium and cork cambium, create secondary growth that adds girth to stems and roots.

CHAPTER REVIEW

■ Plant biology reflects the major themes in the study of life

Modern plant biology is using new methods and new research organisms, such as *Arabidopsis thaliana*, to study the genetic control of plant development. Plants show adaptation to their environments on two time scales: the evolutionary interactions that have shaped plants as terrestrial organisms and the physiological and structural responses of individual plants to their environments. As always, the correlation of structure and function will be obvious in our study of plants.

■ A plant's root and shoot systems are evolutionary adaptations to living on land

Plant morphology considers the external structure of plants, whereas **plant anatomy** is the study of internal structure. The most diverse and widespread group of plants is the angiosperms, characterized by flowers and fruits, which function in reproduction and seed dispersal. This group is split into two classes: **monocots**, which have one cotyledon in each seed, and **dicots**, which have two of these "seed leaves."

To be able to acquire dispersed resources in a terrestrial environment, plants have evolved an underground **root system** for obtaining water and minerals from the soil and an aerial **shoot system** of stems, leaves, and flowers for absorbing light and carbon dioxide for photosynthesis. Vascular tissue transports materials between the interdependent roots and shoots. **Xylem** carries water and dissolved minerals from roots to shoots, whereas **phloem** conducts food made in the leaves to roots and nonphotosynthetic plant parts.

The Root System The structure of roots is adapted for anchorage, absorption, conduction, and storage of food. A **taproot** system is found commonly in dicots, whereas monocots generally have **fibrous root** systems. Most absorption of water and minerals occurs through the tiny **root hairs** that are clustered near the root tips. Mycorrhizae also contribute to water and mineral absorption. **Adventitious** roots arise from stems or leaves and may serve as props to help support stems.

The Shoot System   The shoot system includes vegetative shoots, which bear leaves, and floral shoots. A stem consists of alternating **nodes**, the points at which leaves are attached, and segments between nodes, called **internodes**. An **axillary bud** is found in the angle between a leaf and the stem. A **terminal bud**, consisting of developing leaves and compacted nodes and internodes, is found at the tip or apex of a shoot. The terminal bud may exhibit **apical dominance**, inhibiting the growth of axillary buds and producing a taller plant. Axillary buds may develop into reproductive shoots with flowers or vegetative branches with their own terminal buds, leaves, and axillary buds.

Modifications of stems include stolons, horizontal stems that grow along the surface of the ground; rhizomes, horizontal stems growing underground that may end in enlarged tubers; and bulbs, vertical underground shoots functioning in food storage.

Leaves, the main photosynthetic organs of most plants, usually consist of a flattened **blade** and a **petiole**, or stalk. Many monocots lack petioles. Monocot leaves usually have parallel major veins, whereas dicot leaves have networks of branched veins.

■ INTERACTIVE QUESTION   31.1

a. What is a node?

b. Where are axillary buds found?

c. List some of the differences between monocots and dicots. Add to this list as you go through the chapter.

Monocots	Dicots

■ The many types of plant cells are organized into three major tissue systems

Types of Plant Cells   Specialized cells of a multicellular plant have particular adaptations of the cell wall and/or of the **protoplast**, the cell contents excluding the cell wall.

Parenchyma cells, relatively unspecialized cells that usually lack secondary walls and have large central vacuoles, carry on most of the metabolism of the plant, functioning in photosynthesis and food storage. Mature parenchyma cells usually do not divide but usually retain the ability to divide and differentiate into other types of plant cells.

Collenchyma cells lack secondary walls but have thickened primary walls. Strands or cylinders of collenchyma cells function in support for young parts of the plant and elongate along with the plant.

Sclerenchyma cells have thick secondary walls strengthened with lignin. These specialized supporting elements often lose their protoplasts at maturity. **Fibers** are long, tapered schlerenchyma cells that usually occur in bundles. **Sclereids** are shorter and irregular in shape.

After the cells of xylem form secondary walls and mature, the protoplast disintegrates, leaving behind a conduit through which water flows. In growing parts of a plant, the secondary walls may be deposited in spiral or ring patterns; in areas that are no longer elongating, the secondary walls may have scattered **pits**, which are thinner regions consisting only of primary walls. **Tracheids** are long, thin, tapered xylem cells with lignin-strengthened walls. Water passes through pits from cell to cell. **Vessel elements** are wider, shorter, and thinner walled, with perforations in their abutting ends.

Phloem consists of chains of cells called **sieve-tube members**, which remain alive at functional maturity but lack nuclei, ribosomes, and vacuoles. In angiosperms, sucrose, other organic compounds, and some mineral ions flow through pores in the **sieve plates** in the end walls between cells. The nucleus and ribosomes of an adjacent **companion cell**, which is connected to a sieve-tube member by numerous plasmodesmata, may serve both cells.

■ INTERACTIVE QUESTION   31.2

a. Which types of plant cells are dead at functional maturity?

b. Which types of plant cells lack nuclei at functional maturity?

The Three Tissue Systems of a Plant The **dermal tissue system**, or **epidermis**, is a single layer of cells that covers and protects the young parts of the plant. Root hairs are extensions of epidermal cells that increase the surface area for absorption. The epidermis of leaves and most stems is covered with a **cuticle**, a waxy coating that prevents excess water loss. The **vascular tissue system** consists of xylem and phloem and functions in transport and support. The **ground tissue system**, making up the bulk of a young plant, functions in photosynthesis, support, and storage; it consists predominantly of parenchyma cells.

Meristems generate cells for new organs throughout the lifetime of a plant: *an overview of plant growth*

Most plants exhibit indeterminate growth, continuing to grow as long as they live. Animals usually have determinate growth and stop growing after reaching a certain size. A plant's lifespan may be genetically programmed or environmentally determined. Plants may be **annuals**, which complete their life cycle in one growing season or year; **biennials**, which have a life cycle spanning two years; or **perennials**, which live many years.

Plants have tissues called **meristems** that remain embryonic and divide to form new cells. Cells that remain to divide in the meristem are called initials, whereas cells that will specialize into developing tissue are called derivatives.

Apical meristems, located at the tips of roots and in the buds of shoots, produce **primary growth**, resulting in the elongation of roots and shoots. **Secondary growth** is an increase in diameter as new cells are produced by **lateral meristems**, creating a thicker secondary dermal tissue and new layers of vascular tissue.

■ INTERACTIVE QUESTION 31.3

What types of growth occur in woody plants?

Apical meristems extend roots and shoots: *a closer look at primary growth*

The **primary plant body** is produced by primary growth in herbaceous plants and the youngest parts of a woody plant and consists of dermal, vascular, and ground tissues.

Primary Growth of Roots The meristem of the root tip is protected by a **root cap** which secretes a polysaccharide slime to lubricate the growth route. The **zone of cell division** includes the apical meristem and the primary meristems. At the center of the apical meristem is the **quiescent center**, a slowly dividing, damage-resistant reserve of meristematic tissue. Derived from the apical meristem are three concentric cylinders of tissue: the **protoderm**, **procambium**, and **ground meristem**, the primary meristems that continue to divide and give rise to the cells of the three tissue systems of the root.

In the **zone of elongation**, cells lengthen to more than ten times their original size, which pushes the root tip through the soil. Cells begin to specialize in structure and function as the zone of elongation grades into the **zone of maturation**.

The primary meristems produce the tissue systems of the root. The protoderm gives rise to the single cell layer of the epidermis. The procambium forms a central vascular cylinder or **stele**. In most dicots, xylem cells radiate from the center of the stele in spokes, with phloem in between. The stele of a monocot may have **pith**, a central core of parenchyma cells, inside the xylem and phloem.

The ground meristem produces the ground tissue system, filling the **cortex**, or region of the root between the stele and epidermis. These parenchyma cells store food and participate in the uptake of minerals from the soil solution entering the root. The innermost layer of cortex is the one-cell-thick **endodermis**, which regulates the passage of materials into the stele. **Lateral roots** may develop from the **pericycle**, a layer of cells just inside the endodermis with meristematic potential. A lateral root pushes through the cortex and retains its vascular connection with the central stele.

Primary Growth of Shoots The dome-shaped mass of apical meristem in the terminal bud also gives rise to the primary meristems: protoderm, procambium, and ground meristem. Leaf primordia form on the sides of the apical dome, and axillary buds develop from clumps of meristematic cells left at the bases of the leaf primordia. Elongation of the shoot occurs by cell division and cell elongation within young internodes. In the grasses and some other plants, internodes have intercalary meristems that enable shoots to continue elongation. Axillary buds, with their connection to the vascular tissue located near the stem's surface, may form branches later in development.

Unlike the single central stele of a root, vascular tissue runs through the stem in several **vascular bundles**, each surrounded by ground tissue. In dicots, the vascular bundles may be arranged in a ring, with the

■ INTERACTIVE QUESTION 31.4

Label the tissues in this cross section of a monocot root and list their functions.

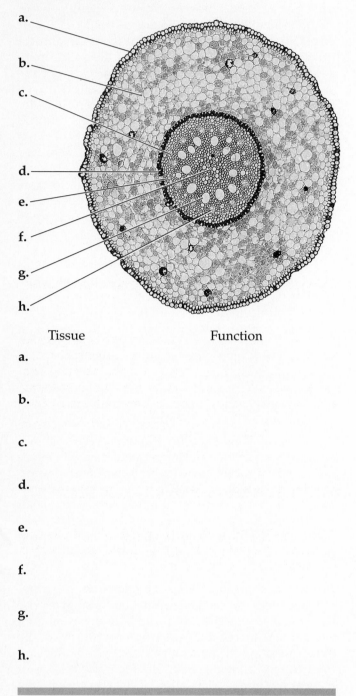

Tissue	Function
a.	
b.	
c.	
d.	
e.	
f.	
g.	
h.	

pith inside the ring connected by thin rays of ground tissue to the cortex outside the ring. Xylem is located internal to the phloem in the vascular bundles. In most monocot stems, the vascular bundles are scattered throughout the ground tissue. A layer of collenchyma just beneath the epidermis may strengthen the stem.

The leaf is covered by the wax-coated, tightly-interlocking cells of the epidermis. **Stomata**, tiny pores

flanked by **guard cells**, permit both gas exchange and **transpiration**, the evaporation of water from the leaf.

Mesophyll consists of parenchyma ground tissue cells containing chloroplasts. In many dicot leaves, columnar palisade parenchyma is located above spongy mesophyll, which has loosely packed, irregularly shaped cells surrounding many air spaces.

A leaf trace, which is a branch of a vascular bundle in the stem, continues into the petiole and divides repeatedly within the blade of the leaf, providing support and vascular tissue to the photosynthetic mesophyll.

■ INTERACTIVE QUESTION 31.5

Identify the indicated structures in this diagram of a leaf.

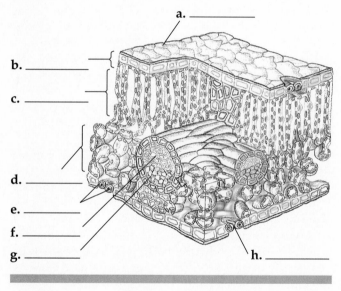

A shoot has a modular construction of serial segments of stem with leaves and axillary buds, resulting from the development of successive nodes and internodes from the apical meristem. The developmental phase of the apical meristem can change from a juvenile vegetative state to a mature vegetative state. The leaves and axillary buds that develop from juvenile nodes retain their juvenile characteristics. The shoot apex may also make a transition to a reproductive (flower-producing) state, which terminates its primary growth.

■ INTERACTIVE QUESTION 31.6

Return to the table in Interactive Question 31.1 to add anatomical differences between monocot and dicot roots and stems.

■ Lateral meristems add girth to stems and roots: *a closer look at secondary growth*

The **secondary plant body** consists of the tissues produced by lateral meristems in secondary growth. **Vascular cambium** produces secondary xylem and phloem, and **cork cambium** produces a tough covering for stems and roots. Secondary growth occurs in all gymnosperms and most dicots.

Secondary Growth of Stems   Vascular cambium forms a continuous cylinder from a band of parenchyma cells that become meristematic, located between the xylem and phloem of each vascular bundle (fascicular cambium), and in the rays of ground tissue between bundles (interfascicular cambium). **Ray initials**, the meristematic cells of the interfascicular cambium, produce **xylem rays** and **phloem rays**, which function in storage and lateral transport of water and nutrients. **Fusiform initials** are the cells of the fascicular cambium that produce secondary xylem to the inside and secondary phloem to the outside. Wood is the accumulation of secondary xylem cells with thick, lignified walls. Annual growth rings result from the seasonal cycle of xylem production in temperate regions. Older secondary phloem forms part of the bark and splits and sloughs off as a tree increases in girth.

The epidermis splits off during secondary growth and is replaced by new protective tissues produced by the **cork cambium**. Initials of the cork cambium produce parenchyma cells called **phelloderm** to the inside and **cork cells**, with waxy walls impregnated with suberin, to the outside. The protective coat formed by the layers of cork, cork cambium, and phelloderm is called the **periderm**. **Bark** refers to phloem and periderm. As secondary growth continually splits the periderm, a new cork cambium develops, eventually forming from parenchyma cells in the secondary phloem. **Lenticels** are spongy regions in the bark through which gas exchange occurs.

In older trees, a central supportive column of heartwood consists of older xylem with resin-filled cell cavities; the sapwood consists of secondary xylem that still functions in transport.

■ INTERACTIVE QUESTION   31.7

Starting from the outside, place the letters of the tissues in the order in which they are located in a woody tree trunk.

___ ___ ___ ___ ___ ___ ___ ___ ___

A. primary phloem	**F.** cork cambium
B. secondary phloem	**G.** vascular cambium
C. primary xylem	**H.** cork cells
D. secondary xylem	**I.** phelloderm
E. pith	

Secondary Growth of Roots   The two lateral meristems also function in the secondary growth of roots. Vascular cambium produces xylem internal and phloem external to itself, and a cork cambium forms from the pericycle and produces the periderm. Older roots resemble old stems, with annual rings and a thick, tough bark.

STRUCTURE YOUR KNOWLEDGE

1. Fill in the following diagram that shows the origins of the tissues involved in the primary and secondary growth of a stem.

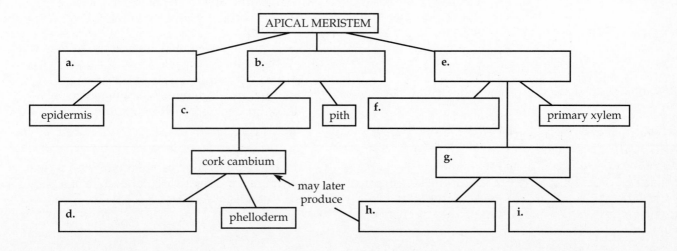

2. In the following cross section of a stem, label the indicated structures. Is this a stem of a monocot or dicot? How can you tell?

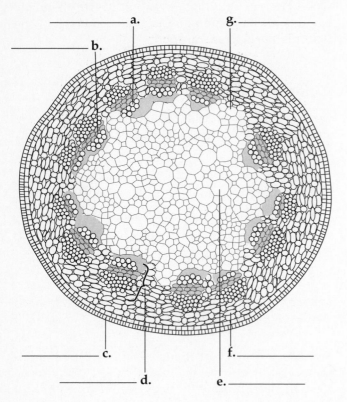

_____ a.

_____ b.

g._____

_____ c.

_____ d.

_____ e.

f._____

TEST YOUR KNOWLEDGE

MATCHING: *Match the plant tissue with its description.*

_____ 1. sclerenchyma

_____ 2. collenchyma

_____ 3. tracheid

_____ 4. fibers

_____ 5. initials

_____ 6. pericycle

_____ 7. mesophyll

A. layer from which lateral roots originate

B. tapered xylem cells with lignin in cell walls

C. parenchyma cells with chloroplasts in leaves

D. protective coat made of cork, cork cambium, and phelloderm

E. bundles of long sclerenchyma cells

F. supporting cells with thickened primary walls

_____ 8. periderm

_____ 9. endodermis

_____ 10. pith

G. parenchyma cells inside vascular ring in dicot stem

H. supporting cells with thick secondary walls

I. newly formed cells that then remain to divide in meristem

J. cell layer in root regulating movement into central vascular cylinder

MULTIPLE CHOICE: *Choose the one best answer.*

1. Which of the following is *incorrect*? Monocots have
   a. a taproot rather than a fibrous root system.
   b. leaves with parallel veins rather than branching venation.
   c. no secondary growth.
   d. scattered vascular bundles in the stem rather than bundles in a ring.
   e. pith in the center of the vascular stele in the root.

2. Which of the following is *incorrectly* paired with its function?
   a. fusiform initials—form secondary xylem and phloem
   b. lenticel—spongy area for gas exchange in woody stem
   c. root hairs—absorption of water and dissolved minerals
   d. root cap—protects root as it pushes through soil
   e. procambium—meristematic tissue that forms protective layer of cork

3. Axillary buds
   a. may exhibit apical dominance over the terminal bud.
   b. form at nodes in the angle where leaves join the stem.
   c. grow out from the pericycle layer.
   d. are formed from intercalary meristems.
   e. only develop into vegetative shoots.

4. A leaf trace is
   a. a petiole.
   b. the outline of the vascular bundles in a leaf.
   c. a branch from a vascular bundle at a node that extends into a leaf.

d. a tiny bulge on the flank of the apical dome that grows into a leaf.

e. a system of plant identification based on leaf morphology.

5. Ground meristem
   a. produces the root system.
   b. produces the ground tissue system.
   c. produces secondary growth.
   d. is meristematic tissue found at the nodes in monocots.
   e. develops from protoderm.

6. The zone of cell elongation
   a. is responsible for pushing a root through the soil.
   b. has a quiescent center that can undergo mitosis should the apical meristem be destroyed.
   c. produces the protoderm, procambium, and ground meristem tissues.
   d. is further from the root tip than the zone of maturation.
   e. is or does all of the above.

7. Which of the following is *incorrectly* paired with its type of life cycle?
   a. pine tree—perennial
   b. rose bush—perennial
   c. carrot—biennial
   d. marigold—annual
   e. wheat—biennial

8. Secondary xylem and phloem is produced in a root by the
   a. pericycle.
   b. endodermis.
   c. vascular cambium.
   d. apical meristem.
   e. ray initials.

9. Bark consists of
   a. secondary phloem.
   b. phelloderm.
   c. cork cells.
   d. cork cambium.
   e. all of the above.

10. Sieve-tube members
    a. are responsible for lateral transport through a woody stem.
    b. control the activities of phloem cells that have no nuclei or ribosomes.
    c. have spiral thickenings that allow the cell to grow along with a young shoot.
    d. are vascular cells with sieve plates in the end walls between cells.
    e. are tapered transport cells with pits.

CHAPTER 32

TRANSPORT IN PLANTS

FRAMEWORK

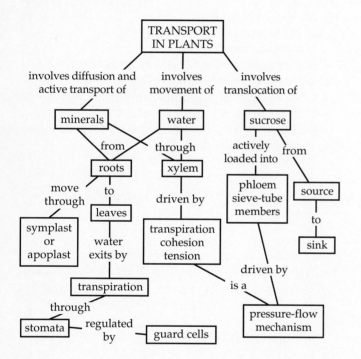

CHAPTER REVIEW

The move onto land involved the evolutionary specialization of roots for the uptake of water and minerals, shoots for the uptake of CO_2 and exposure to light, and vascular tissue for the transport of materials between roots and shoots.

■ The traffic of water and solutes occurs on cellular, organ, and whole-plant levels: *an overview of transport in plants*

Transport in plants involves the absorption of water and solutes by individual cells, the movement of sub-stances from cell to cell, and the long-distance transport of sap in xylem and phloem.

Transport at the Cellular Level   Solutes may move across the selectively permeable plasma membrane by passive transport when they diffuse down their concentration gradients. **Transport proteins** speed passive transport and may be either specific **carrier proteins**, which selectively bind and transport a solute, or **selective channels**, which allow specific solutes to cross the membrane. Some channels are gated: They open and close in response to environmental stimuli.

Active transport requires the cell to expend energy to move a solute against its gradient. A plant **proton pump** uses ATP to pump hydrogen ions out of the cell, generating an energy-storing proton gradient and a membrane potential due to the separation of charges. The proton gradient and membrane potential are used by plant cells to drive the transport of many different solutes. The membrane potential helps drive positively charged ions down their electrochemical gradient and into the negatively charged cell. In **cotransport**, a solute can be pumped against its concentration gradient when a transport protein links its passage to the "downhill" movement of H^+. **Chemiosmosis** links energy-releasing and energy-consuming processes using a transmembrane proton gradient like that created by the proton pump.

The passive transport of water across a membrane is called **osmosis**. In animal cells, water moves by osmosis in the hypotonic → hypertonic direction. **Water potential**, designated by Ψ (*psi*), is a useful measurement for predicting the direction that water will move when a plant cell is surrounded by a solution. It takes into account both solute concentration and the physical pressure exerted by the plant cell wall. Water will flow from a region of higher water potential to one of lower water potential. Water potential is measured in **megapascals** (MPa); 1 MPa is equal to about 10 atmospheres of pressure.

The water potential of pure water in an open container is defined as 0 MPa. Adding solutes lowers water potential; therefore, a solution at atmospheric pressure has a negative water potential. The addition of pressure increases water potential because it increases the potential for water to move. A **tension**, or negative pressure, will lower Ψ. **Bulk flow** is the movement of water due to a pressure difference.

The combined effect of pressure and solute concentration is shown by the equation for water potential: $\Psi = \Psi_P + \Psi_S$ (Ψ_P = pressure potential, Ψ_S = solute potential or osmotic potential). Ψ_P can be a positive or negative value, whereas Ψ_S is always a negative number.

A plant cell bathed in a more concentrated solution (hypertonic) will lose water by osmosis because the solution has a lower Ψ. When bathed in pure water, the cell has the lower Ψ. Water will enter the cell until enough **turgor pressure** builds up so that $\Psi_P = \Psi_S$, and $\Psi = 0$ both inside and outside the cell. Net movement of water will then stop.

■ INTERACTIVE QUESTION 32.1

a. A flaccid plant cell has a water potential of -0.6 MPa. Fill in the water potential equation for this cell.

$$\Psi_P =$$
$$+ \Psi_S = \underline{\hspace{2cm}}$$
$$\Psi =$$

b. The cell is then placed in a beaker of distilled water ($\Psi = 0$). What happens to the water potential of this cell? Fill in the equation for the cell after it is bathed in pure water.

$$\Psi_P =$$
$$+ \Psi_S = \underline{\hspace{2cm}}$$
$$\Psi =$$

c. What would happen to the same cell if it is placed in a solution that has a water potential of -0.8 MPa? Fill in the equation for the cell after it is moved to this hypertonic solution and reaches equilibrium.

$$\Psi_P =$$
$$+ \Psi_S = \underline{\hspace{2cm}}$$
$$\Psi =$$

The **tonoplast**, the membrane of the central vacuole, also influences transport into the cell. Its proton pumps move H^+ from the cytoplasm into the vacuole, helping to maintain a low cytosolic H^+ concentration and generating a membrane potential that drives transport of certain solutes into the vacuole.

Short-Distance (Lateral) Transport at the Level of Tissues and Organs Short-distance or lateral transport of water and solutes within plant tissues can occur by three routes: cell to cell by crossing plasma membranes and cell walls; the **symplast**, the continuum of cytoplasm between cells connected through **plasmodesmata**; and the **apoplast**, the extracellular pathway along cell walls. Solutes and water may change routes during transit. The direction of short-distance transport is largely determined by diffusion due to concentration gradients and bulk flow due to pressure differences.

Long-Distance Transport at the Whole-Plant Level Long-distance transport throughout plants occurs by bulk flow. Water and minerals move through xylem vessels as the result of tension created by transpiration. Sap is forced through phloem sieve tubes by hydrostatic pressure.

■ Roots absorb water and minerals from soil

Most absorption of water and minerals occurs through root hairs, extensions of epidermal cells located along young root tips. The soil solution soaks into the hydrophilic walls of epidermal cells and along the apoplast into the root cortex, exposing a large surface area of plasma membrane for the uptake of water and minerals. Mycorrhizae also increase absorptive surface area. Selective absorption of minerals, using transport proteins in the plasma membrane and tonoplast, allows cells to accumulate essential minerals.

The **endodermis**, the innermost layer of cells surrounding the stele, selectively screens all minerals entering the vascular tissue. A ring of suberin around each endodermal cell, called the **Casparian strip**, prevents water from the apoplast from entering the stele without passing through the selective plasma membrane of an endodermal cell. Water and minerals that had already entered the symplast through a cortex cell pass through plasmodesmata of endodermal cells into the stele.

■ **INTERACTIVE QUESTION   32.2**

Label the diagram of the cell layers and routes of transport of water and minerals from the soil through the root. (**a** and **b** refer to routes of transport; letters **c–i** refer to cell layers or structures.)

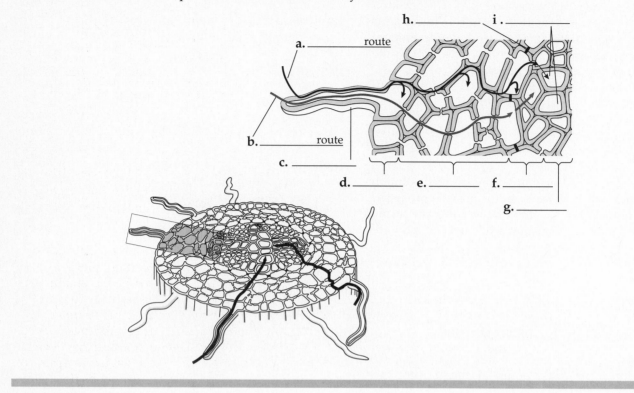

By a combination of diffusion and active transport, minerals leave the symplast, enter the nonliving xylem tracheids and vessel elements along with water, and move again through the apoplast system.

The ascent of xylem sap depends mainly on transpiration and the physical properties of water

Through **transpiration**, plants lose a tremendous amount of water that must be replaced by water transported up from the roots.

Pushing Xylem Sap: Root Pressure   As minerals are actively pumped into the stele and prevented from leaking out by the endodermis, water potential in the stele is lowered. Water flows into the stele by osmosis, resulting in **root pressure**, which forces fluid up the xylem. Root pressure may cause **guttation**, the exudation of water droplets through specialized structures in the leaves when more water is forced up the xylem than is transpired by the plant.

Pulling Xylem Sap: The Transpiration-Cohesion-Tension Mechanism   Water vapor from saturated

air spaces within a leaf exits to drier air outside the leaf by way of stomata. The thin layer of water that coats the mesophyll cells lining the air spaces begins to evaporate. The adhesion of the remaining water to the hydrophilic walls and the cohesion between water molecules causes menisci to form that increase the surface tension of this water layer. The negative pressure pulls water from the xylem and through the apoplast and symplast of the mesophyll to the cells and surface film lining the air spaces. Water moves along a gradient of decreasing water potential from xylem to neighboring cells to air spaces to the drier air outside the leaf.

The transpirational pull on xylem sap is transmitted from the leaves to the root tips by the cohesiveness of water that results from hydrogen bonding between molecules. Adhesion of water molecules to the hydrophilic walls of the narrow xylem elements and tracheids also helps the upward movement against gravity.

The upward transpirational pull on the cohesive sap creates tension within the xylem, further decreasing water potential so that water flows passively from the soil, across the cortex, and into the stele.

A break in the chain of water molecules by the formation of a water vapor pocket in a xylem vessel, called cavitation, breaks the transpirational pull, and the vessel cannot function in transport.

■ **INTERACTIVE QUESTION 32.3**

Explain the contribution of each of the following to the long-distance transport of water.

a. Transpiration:

b. Cohesion:

c. Adhesion:

d. Tension:

Review: Long-Distance Transport of Water in Plants by Solar-Powered Bulk Flow The transpiration-cohesion-tension mechanism results in the bulk flow of water from roots to leaves. Water potential differences caused by solute concentration and pressure contribute to the movement of water from cell to cell, but solar-powered tension caused by transpiration is responsible for long-distance transport.

■ Guard cells mediate the transpiration-photosynthesis compromise

The Photosynthesis-Transpiration Compromise A plant's tremendous requirement for water is partly a consequence of making food by photosynthesis. To obtain sufficient CO_2 for photosynthesis, leaves must exchange gases through the stomata, thus exposing a large internal surface area from which water may evaporate.

A common transpiration-to-photosynthesis ratio is 600 grams of water transpired for each gram of CO_2 incorporated into carbohydrate (600:1). Plants such as corn that use the more efficient C_4 pathway for photosynthesis have ratios of 300:1 or less.

Transpiration assists in transferring minerals and other substances to the leaves and results in evaporative cooling. When transpiration exceeds the water available, leaves wilt as cells lose turgor pressure.

How Stomata Open and Close Guard cells regulate the size of stomatal openings and control the rate of transpiration. When the kidney-shaped guard cells of dicots become turgid and swell, their radially oriented microfibrils cause them to buckle outward and increase the size of the gap between them. When the guard cells become flaccid, they sag and close the space. The dumbbell-shaped monocot guard cells operate by the same basic mechanism.

Guard cells can lower their water potential by actively accumulating potassium ions (K^+), which leads to osmosis and the resulting increase of turgor pressure. The movement of K^+ across the guard cell membrane and into the vacuole is probably coupled with the generation of membrane potentials by proton pumps that transport H^+ out of the cell. A technique called patch clamping allows plant physiologists to study proton pumps and K^+ channels in guard cells; a small patch of membrane is held across the opening of a micropipette while an electrode records ion fluxes across the membrane.

The opening of stomata at dawn is related to at least three factors: Light stimulates guard cells to accumulate K^+, perhaps triggered by the illumination of blue-light receptors which activate the proton pumps of the plasma membrane. Light also drives photosynthesis in the guard cell chloroplasts, making ATP available for proton pumping. Second, stomata are stimulated to open when CO_2 within air spaces of the leaf is depleted as photosynthesis begins in the mesophyll. The third factor is a daily rhythm of opening and closing that is endogenous to guard cells. Cycles that have intervals of approximately 24 hours are called **circadian rhythms**.

Environmental stress can cause stomata to close during the day. Guard cells lose turgor when water is in short supply. A hormone called abscisic acid, produced by mesophyll cells in response to a lack of water, signals guard cells to close stomata. High temperatures induce closing, probably because increasing cellular respiration raises CO_2 concentrations in the air spaces in the leaf.

Evolutionary Adaptations that Reduce Transpiration Many xerophytes, plants adapted to arid climates, have leaves that are small and thick, limiting water loss by reducing their surface-to-volume ratio. Leaf cuticles may be thick, and the stomata may be sheltered in depressions. Some desert plants lose their leaves in the driest months.

Succulent plants of the family Crassulaceae and some members of other plant families assimilate CO_2 into organic acids during the night by a pathway known as CAM (crassulacean acid metabolism) and then release it for photosynthesis during the day. Thus, the stomata are open at night and are closed during the day when water loss would be greatest.

■ **INTERACTIVE QUESTION   32.4**

What three factors contribute to the opening of stomata at dawn?

a.

b.

c.

A bulk-flow mechanism translocates phloem sap from sugar sources to sugar sinks

Translocation, the transport of photosynthetic products throughout the plant, occurs in the end-to-end sieve-tube members as sap flows through porous sieve plates. Phloem sap may have a sucrose concentration as high as 30 percent and may also contain minerals, amino acids, and hormones.

Source-to-Sink Transport   Phloem sap flows from a **sugar source**, where it is produced by photosynthesis or the breakdown of starch, to a **sugar sink**, an organ that consumes or stores sugar. The direction of transport in any one sieve tube depends on the location of the source and sink connected by that tube, and the direction may change with the season or needs of the plant.

■ **INTERACTIVE QUESTION   32.5**

Explain how a root or tuber can serve as both a sugar source and a sugar sink.

Phloem Loading and Unloading   In some species, sugar in the leaf moves through the symplast of the mesophyll cells to sieve-tube members. In other species, sugar first moves through the symplast and then into the apoplast in the vicinity of sieve-tube members and companion cells, which actively accumulate sugar. Some companion cells, specialized as **transfer cells**, have ingrowths of their wall that increase surface area for movement of solutes from apoplast to symplast.

Phloem loading requires active transport in plants that accumulate high sugar concentration in the sieve tubes. Proton pumps and the cotransport of sucrose through membrane proteins along with the returning protons is the mechanism used for active transport. Sucrose may move by active transport or diffusion out of sieve tubes at the sink end.

Pressure Flow (Bulk Flow) of Phloem Sap   The rapid movement of phloem sap from source to sink is due to a pressure-flow mechanism. High solute concentration at the source lowers water potential, and the resulting movement of water into the sieve tube produces hydrostatic pressure. At the sink end, the osmotic loss of water following the exodus of sucrose into the surrounding tissue results in a lower hydrostatic pressure. The difference in these pressures causes water to flow from source to sink, transporting sugar. Innovative tests of this model support it as the explanation for the flow of sap in the phloem of angiosperms.

STRUCTURE YOUR KNOWLEDGE

1. Describe the ways in which solutes may move across the plasma membrane in plants.
2. Both xylem sap and phloem sap move by bulk flow in an angiosperm. Compare and contrast the mechanisms for their movement.

TEST YOUR KNOWLEDGE

MULTIPLE CHOICE: *Choose the one best answer.*

1. Which of the following is *not* a component of the symplast?
   a. sieve-tube members
   b. xylem tracheids
   c. endodermal cells
   d. cortex cells
   e. companion cells

2. Proton pumps in the plasma membranes of plant cells may
   a. generate a membrane potential that helps drive cations into the cell through their specific carriers.
   b. be coupled to the movement of K^+ into guard cells.
   c. drive the accumulation of sucrose in sieve-tube members.
   d. contribute to the movement of anions through a cotransport mechanism.
   e. be involved in all of the above.

3. The Casparian strip prevents water and minerals from entering the stele through the
   a. plasmodesmata.
   b. endodermal cells.
   c. symplast.
   d. apoplast.
   e. xylem vessels.

4. The water potential of a plant cell
   a. is equal to 0 when the cell is in pure water and is turgid.
   b. is greater than that of air.
   c. is equal to −0.23 MPa.
   d. becomes greater when K^+ ions are actively moved into the cell.
   e. becomes 0 due to loss of turgor pressure in a hypertonic solution.

5. Guttation results from
   a. the pressure flow of sap through phloem.
   b. a water vapor break in the column of xylem sap.
   c. root pressure causing water to flow up through xylem faster than it can be lost by transpiration.
   d. a higher water potential of the leaves than that of the roots.
   e. specialized structures in transport cells that accumulate sucrose.

6. Which of these is *not* a major factor in the movement of xylem sap up a tall tree?
   a. transpiration
   b. root pressure
   c. adhesion
   d. cohesion
   e. tension

7. Adhesion is a result of
   a. hydrogen bonding between water molecules.
   b. the pull on the water column as water evaporates from the surfaces of mesophyll cells.
   c. tension within the xylem caused by a lowered water potential.
   d. attraction of water molecules to hydrophilic walls of narrow xylem tubes.
   e. the high surface tension of water.

8. A plant with a low transpiration-to-photosynthesis ratio
   a. would lose more water through transpiration for each gram of CO_2 fixed.

   b. could be a C_4 plant.
   c. could be a CAM plant.
   d. b and c could be correct.
   e. All of the above could be correct.

9. Which of these does not stimulate the opening of stomata?
   a. release of abscisic acid by mesophyll cells
   b. depletion of CO_2 in the air spaces of the leaf
   c. stimulation of proton pumps that results in the movement of K^+ into the guard cells
   d. the circadian rhythm of the guard cell opening
   e. an increase in the turgor of guard cells

10. Your favorite spider plant is wilting. What is the most likely cause and remedy for its declining condition?
    a. Water potential is too low; apply sugar water.
    b. The stomata won't open; no remedy available.
    c. Plasmolysis of its cells; water the plant.
    d. Cavitation; perform a xylem vessel bypass.
    e. Circadian rhythm has stomata closed; place it in bright light.

11. A plant cell placed in a solution in an open beaker becomes flaccid.
    a. The water potential of the cell was higher than that of the solution.
    b. The water potential of the cell was lower than that of the solution.
    c. The pressure (Ψ_P) of the cell was initially lower than that of the solution.
    d. The cell was hypertonic to the solution.
    e. Turgor pressure disappears since the cell no longer needs support.

12. What facilitates the movement of K^+ into epidermal cells of the root?
    a. cotransport through a membrane protein
    b. bulk flow of water into the root
    c. passage through selective channels, aided by the membrane potential created by proton pumps
    d. active transport through a potassium pump
    e. simple diffusion across the cell membrane down its concentration gradient

CHAPTER 33

PLANT NUTRITION

FRAMEWORK

The nutritional requirements of plants include essential macronutrients and micronutrients. Carbon dioxide enters the plant through the leaves, but water and minerals must be absorbed through the roots. Soil fertility is influenced by its texture and composition. Nitrogen assimilation by plants is made possible by the decomposition of humus by microbes and nitrogen fixation by bacteria. Mycorrhizae are important associations between fungi and plant roots that increase mineral and water absorption.

CHAPTER REVIEW

Plants, as photoautotrophs, make their own organic compounds and are critical to the energy flow and chemical cycling of an ecosystem. Roots, shoots, and leaves are structurally adapted to obtain water, minerals, and carbon dioxide from the soil and air.

■ Plants require at least seventeen essential nutrients

Chemical Composition of Plants Early scientists speculated on whether soil, water, or the air provides the substance for plant growth. Minerals, although essential, make only a small contribution to the mass of a plant. Water supplies most of the hydrogen and some of the oxygen incorporated into organic compounds. More than 90 percent of the water absorbed is lost by transpiration, however, and most of the water retained is used for growth by cell elongation or support through turgor pressure. By weight, CO_2 from the air is the source of most of the organic material of a plant.

Organic substances, most of which are carbohydrates (especially cellulose), make up 95 percent of the dry weight of plants. Thus, carbon, oxygen, and hydrogen are the most abundant elements. Nitrogen, sulfur, and phosphorus—also ingredients of organic compounds—are relatively abundant.

Essential Nutrients **Essential nutrients** are those required for a plant to complete its life cycle from a seed to an adult that produces more seeds. Hydroponic culture has been used to determine which of the mineral elements found in plants are essential nutrients. Seventeen elements have been identified as essential in all plants.

Nine **macronutrients** are required by plants in relatively large amounts and include the six major elements of organic compounds as well as calcium, potassium, and magnesium.

Eight **micronutrients** have been identified as needed by plants in very small amounts, functioning mainly as cofactors of enzymatic reactions.

■ INTERACTIVE QUESTION 33.1

a. What is a function of the macronutrient magnesium in plants?

b. What is a function of the macronutrient phosphorus?

c. What is a function of the micronutrient iron?

■ The symptoms of a mineral deficiency depend on the function and mobility of the element

A mobile nutrient will move to young, growing tissues, so that a deficiency will show up first in older parts of the plant. A deficiency in magnesium, a component of chlorophyll, causes chlorosis (yellowing of the leaves).

Symptoms of a mineral deficiency may be distinctive enough for the cause to be diagnosed by a plant physiologist or farmer. Soil and plant analysis can confirm a specific deficiency. Nitrogen, potassium, and phosphorus deficiencies are most common.

▓ INTERACTIVE QUESTION   33.2

Where would you expect a deficiency of a relatively immobile element to be seen first?

Availability of Soil Water and Minerals   Water containing dissolved minerals binds to hydrophilic soil particles and is held there in small spaces, available for uptake by plant roots. Positively charged minerals, such as K^+, Ca^{2+}, and Mg^{2+}, adhere to the negatively charged surfaces of finely divided clay particles. Negatively charged minerals, such as nitrate (NO_3^-), phosphate ($H_2PO_4^-$), and sulfate (SO_4^{2-}), tend to leach away more quickly. Roots release acids that facilitate **cation exchange**, in which hydrogen ions displace positively charged mineral ions from the clay particles, making the ions available for absorption.

▓ INTERACTIVE QUESTION   33.3

Roots deplete the supply of water and minerals in their vicinity. How does a plant maintain a supply of nutrients?

■ Soil characteristics are key environmental factors in terrestrial ecosystems

Plants that grow well in a particular region are adapted to the texture and mineral content of that soil.

Texture and Composition of Soils   The formation of soil begins with the weathering of rock and accelerates with the secretion of acids by lichens, fungi, bacteria, and plant roots. **Topsoil** is a mixture of decomposed rock, living organisms, and humus (decomposing organic matter). Several other distinct soil layers, or **horizons**, are found under the topsoil layer.

The size of soil particles varies from coarse sand to fine clay. **Loams**, made up of a mixture of sand, silt, and clay, are often the most fertile soils, having enough fine particles to provide a large surface area for retaining water and minerals but enough coarse particles to provide air spaces with oxygen for respiring roots.

The activities of the numerous soil inhabitants, such as bacteria, fungi, algae, other protists, insects, worms, nematodes, and plant roots, affect the physical and chemical properties of soil.

Humus builds a crumbly soil that retains water, provides good aeration of roots, and supplies mineral nutrients.

■ Soil conservation is one step toward sustainable agriculture

Without good soil conservation, agriculture can quickly destroy the fertility of a soil that has built up over centuries. Agriculture diverts essential elements from the chemical cycles when crops are harvested, thus depleting the mineral content of the soil.

Fertilizers   Historically, farmers used manure to fertilize their crops. Today in developed nations, commercially produced fertilizers, usually containing nitrogen, phosphorus, and potassium, are used. Manure, fishmeal, and compost are called organic fertilizers because they contain organic material that is in the process of decomposing. These fertilizers slowly decompose into inorganic nutrients, so that they are taken up by the plant in the same form supplied by commercial fertilizers. A disadvantage of commercial fertilizers is that they may be rapidly leached from the soil, polluting streams and lakes.

The acidity of the soil affects cation exchange and can alter the chemical form of minerals and thus their ability to be absorbed by the plant. Managing the pH of soil is an important aspect of maintaining fertility.

Irrigation   Irrigation can make farming possible in arid regions, but it places a huge drain on water resources and raises soil salinity. New methods of irrigation and new varieties of plants that can tolerate less water or more salinity may reduce some of these problems.

Erosion   In the U.S., topsoil from thousands of acres of farmland is lost to water and wind erosion each year. Agricultural use of cover crops, windbreaks, and terracing can minimize erosion.

■ INTERACTIVE QUESTION   33.4

What is sustainable agriculture?

■ The metabolism of soil bacteria makes nitrogen available to plants

Nitrogen is the mineral that usually limits plant growth and crop yields. To be absorbed by plants, nitrogen must be converted to nitrate (NO_3^-) or ammonia (NH_3) by the action of microbes, such as ammonifying bacteria, that decompose humus. Some nitrate is lost to the atmosphere by the action of denitrifying bacteria. **Nitrogen-fixing bacteria** convert atmospheric nitrogen into ammonia through the process of **nitrogen fixation**.

Nitrogen Fixation   Bacteria capable of nitrogen fixation contain **nitrogenase**, an enzyme complex that reduces N_2 by adding H^+ and electrons to form ammonia (NH_3). This process is energetically expensive, and nitrogen-fixing bacteria rely on organic material in soils for their cellular respiration. In the soil solution, ammonia forms ammonium (NH_4^+), which plants can absorb. Nitrifying bacteria oxidize ammonium, producing nitrate, the form of nitrogen most readily absorbed by roots. Most plants incorporate nitrogen into amino acids or other organic compounds in the roots and then transport it through the xylem in organic form.

Symbiotic Nitrogen Fixation   Plants of the legume family have root swellings, called **nodules**, composed of plant cells with nitrogen-fixing bacteria in a form called **bacteroids** contained in vesicles. The coevolution of the symbiotic relationship between these bacteria of the genus *Rhizobium* and legumes is illustrated by their cooperative synthesis of leghemoglobin. This oxygen-binding protein releases oxygen for the respiration needed to supply energy for nitrogen fixation and keeps the free oxygen concentration in the nodules low to prevent inhibition of the enzyme nitrogenase.

The root nodules use most of the symbiotically fixed nitrogen to make amino acids, which are then transported throughout the plant. Excess ammonium may be secreted into the soil, increasing the fertility of the soil for nonlegumes, which are grown in rotation with the legumes. Rice farmers culture a water fern that has symbiotic cyanobacteria that fix nitrogen, improving the fertility of rice paddies.

■ INTERACTIVE QUESTION   33.5

Fill in the types of bacteria (**a–d**) that participate in the nitrogen nutrition of plants. Indicate the form (**e**) in which nitrogen is transported to the shoot system.

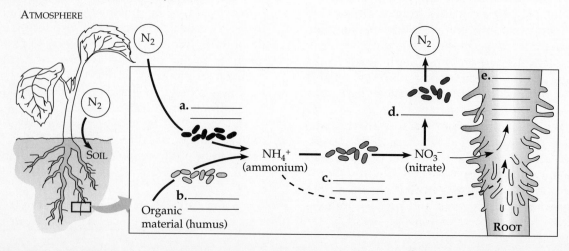

■ Improving the protein yield of crops is a major goal of agricultural research

Protein deficiency is the most common form of human malnutrition. Agricultural research attempts to improve the quality and quantity of proteins in crops. Varieties of corn, wheat, and rice have been developed that are enriched in protein, but they require the addition of large quantities of expensive nitrogen fertilizer. Other strategies for increasing protein yields include improving the productivity of symbiotic nitrogen fixation so that more of the photosynthetic energy goes into legume production, and incorporating into root nodules mutant strains of *Rhizobium* that do not shut off production of nitrogenase when fixed nitrogen accumulates.

The tools of genetic engineering are being used to create varieties of *Rhizobium* that can infect non-legumes and to transplant the genes for nitrogen fixation into other bacteria and, possibly, directly into plant genomes.

■ Predation and symbiosis are evolutionary adaptations that enhance plant nutrition

Parasitic Plants   Parasitic plants, such as the mistletoe, produce haustoria that may invade a host plant and siphon xylem sap from its vascular tissue. Epiphytes are plants that grow on the surface of another plant but do not take nourishment from it.

Carnivorous Plants   Living in acid bogs or other nutrient-poor soils, carnivorous plants obtain nitrogen and minerals by killing and digesting insects that are caught in traps formed from modified leaves.

Mycorrhizae   Most plants have modified roots called **mycorrhizae**, mutualistic associations between the roots and fungi. The fungus secretes growth hormones that stimulate root growth and branching. The fungus extends hyphae among or into cortex cells, while mycelial projections of the fungus provide the root with a large surface area for the absorption of minerals and water. The plant provides food to the fungus, and the fungus supplies absorbed minerals to the plant and may help to protect the plant from some soil pathogens.

Most plants form mycorrhizae when they grow in their natural habitat. The fossil record shows that the earliest land plants had mycorrhizae. Plant nutrition reinforces the concept that organisms have adapted through evolution to their physical and biological environment.

STRUCTURE YOUR KNOWLEDGE

1. Develop a concept map that organizes your understanding of the basic nutritional requirements of plants.

2. List the key properties of a fertile soil.

3. What are the differences between root nodules and mycorrhizae? How are each beneficial to plants?

TEST YOUR KNOWLEDGE

MULTIPLE CHOICE: *Choose the one best answer.*

1. The inorganic compound that contributes most of the mass to a plant's organic matter is
   a. H_2O.
   b. CO_2.
   c. NO_3^-.
   d. O_2.
   e. $C_6H_{12}O_6$.

2. The effects of mineral deficiencies involving fairly mobile nutrients will first be observed in
   a. older portions of the plant.
   b. new leaves and shoots.
   c. the root system.
   d. the color of the leaves.
   e. the flowers.

3. Most macronutrients are
   a. cofactors in enzymes.
   b. readily available from air and water.
   c. components of organic compounds.
   d. identified by hydroponic culture.
   e. components of cytochromes.

4. The most fertile type of soil is usually
   a. sand because its large particles allow room for air spaces.
   b. loam, which has a mixture of fine and coarse particles.
   c. clay, because the fine particles provide much surface area to which minerals and water adhere.
   d. humus, which is decomposing organic material.
   e. wet and alkaline.

5. Chlorosis is
   a. a symptom of a mineral deficiency indicated by yellowing leaves due to decreased chlorophyll production.
   b. the uptake of the micronutrient chlorine by a plant, which is facilitated by symbiotic bacteria.
   c. the production of chlorophyll within the thylakoid membranes of a plant.
   d. a contamination of glassware in hydroponic culture.
   e. a mold of roots caused by wet soil conditions.

6. Negatively charged minerals
   a. are released from clay particles by cation exchange.
   b. are reduced by cation exchange before they can be absorbed.
   c. are converted into amino acids before they are transported through the plant.
   d. are bound when roots release acids into the soil.
   e. are leached away by the action of rainwater more easily than positively charged minerals.

7. Nitrogenase is
   a. an enzyme that reduces atmospheric nitrogen to ammonia.
   b. found in *Rhizobium* and other nitrogen-fixing bacteria.
   c. inhibited by high concentrations of oxygen.

   d. providing a model for chemical engineers to design catalysts to make nitrogen fertilizers.
   e. all of the above.

8. Epiphytes
   a. have haustoria for anchoring to their host plants and obtaining xylem sap.
   b. are symbiotic relationships between roots and fungi.
   c. live in poor soil and digest insects to obtain nitrogen.
   d. grow on other plants but do not obtain nutrients from their hosts.
   e. are able to fix their own nitrogen.

9. The nitrogen content of some agricultural soils may be improved by
   a. the synthesis of leghemoglobin by ammonifying bacteria.
   b. mycorrhizae on legumes.
   c. water ferns with symbiotic cyanobacteria.
   d. cation exchange.
   e. the action of denitrifying bacteria.

10. An advantage of organic fertilizers over chemical fertilizers is that they
    a. are more natural.
    b. release their nutrients over a longer period of time and are less likely to be lost to runoff.
    c. provide nutrients in the forms most readily absorbed by plants.
    d. are easier to mass produce and transport.
    e. are all of the above.

CHAPTER 34

PLANT REPRODUCTION AND DEVELOPMENT

FRAMEWORK

This chapter describes the sexual and asexual reproduction of flowering plants. The flower, with leaves modified for reproduction, produces the haploid gametophyte stages of the life cycle: microspores in the anther develop into pollen grains, and a megaspore in the ovule produces an embryo sac. Pollination and the double fertilization of egg and polar nuclei are followed by the development of a seed with a quiescent embryo and endosperm, protected in a seed coat and housed within a fruit. Seed dormancy is broken following proper environmental cues and the imbibition of water.

Vegetative propagation allows successful plants to clone themselves. Agriculture makes extensive use of this type of plant reproduction by using cuttings, grafts, and test-tube cloning.

Development is the result of the overlapping processes of growth, morphogenesis, and cellular differentiation. Pattern formation, the development of tissues and structures in specific locations, is linked to positional information.

CHAPTER REVIEW

◼ Sporophyte and gametophyte generations alternate in the life cycles of plants: *an overview*

Plants exhibit an **alternation of generations** between haploid (*n*) and diploid (2*n*) generations. The diploid plant, the **sporophyte**, produces haploid spores by meiosis. Spores develop into multicellular haploid male or female **gametophytes** which produce gametes by mitosis. Fertilization yields diploid zygotes which grow into new sporophyte plants. In angiosperms, the gametophytes have become reduced to tiny structures that remain dependent on the sporophyte plant.

The flower, the reproductive structure of the sporophyte, evolved from a compressed shoot with four whorls of modified leaves: **sepals**, **petals**, **stamens**, and **carpels**. Sepals—the outermost, usually green whorl—enclose and protect a floral bud before it opens. Petals are generally brightly colored and may attract pollinators. Stamens and carpels contain sporangia in which male and female gametophytes, respectively, develop. Male gametophytes are sperm-containing pollen grains, which develop in an anther at the tip of a stamen. A carpel consists of a sticky stigma at the top of a slender style, which leads to an ovary. The ovary encloses one or more **ovules** in which an embryo sac, the female gametophyte, develops.

Pollination is the arrival of pollen onto the stigma. The pollen grain germinates and grows a tube down the neck of the carpel, releasing its sperm within the embryo sac. Following fertilization, the zygote develops into an embryo as the surrounding ovule develops into a seed. The entire ovary forms a fruit, which aids in seed dispersal.

More About Flowers   A **complete flower** has sepals, petals, stamens, and carpels. **Incomplete flowers** have eliminated one or more of these parts. A **perfect flower** has both stamens and carpels; an **imperfect flower** is missing one of these two. Imperfect flowers may be either staminate or carpellate. In **monoecious** plant species, both staminate and carpellate flowers are on the same plant; in **dioecious** species, these flowers are on separate plants. Floral variations include fused carpels, inflorescences, and composite flowers, as well as diverse shapes, colors, and odors adapted to attract different pollinators.

■ **INTERACTIVE QUESTION   34.1**

Identify the flower parts in the following diagram. Indicate where pollen is produced and where pollination and fertilization occur.

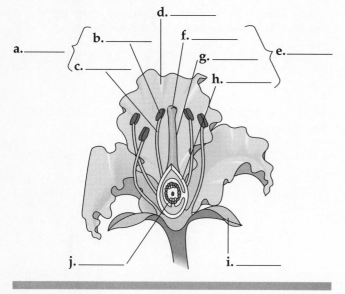

■ **Male and female gametophytes develop within anthers and ovaries, respectively**

Pollen Development Diploid cells called microsporocytes undergo meiosis to form four haploid **microspores**. The nucleus of a microspore divides once by mitosis to produce a generative cell and a tube cell. The wall surrounding the two cells thickens into the durable coat of the pollen grain. A pollen grain is an immature male gametophyte.

Ovule Development Ovules form within the ovary. The megasporocyte in the sporangium of each ovule grows and undergoes meiosis to form four haploid **megaspores**, only one of which usually survives. This megaspore divides by mitosis three times, forming the female gametophyte, called the **embryo sac**, which typically consists of eight nuclei contained in seven cells. At one end of the embryo sac, an egg cell is lodged between two cells called synergids; three antipodal cells are at the other end; and two nuclei, called polar nuclei, are in a large central cell. The ovule consists of the embryo sac and its surrounding protective sporophyte layers called integuments.

■ **INTERACTIVE QUESTION   34.2**

a. Describe the male gametophyte.

b. Describe the female gametophyte.

■ **Pollination brings female and male gametophytes together**

Pollination **Pollination** is the arrival of pollen onto the stigma, carried there by wind or animals. Self-pollination is usually prevented by temporal or structural mechanisms. In flowers that are **self-incompatible**, a biochemical block prevents the development of pollen that does land on a stigma from the same plant.

The Molecular Basis of Self-Incompatibility The S-locus, which codes for several proteins, forms the basis for self-incompatibility. If a pollen grain lands on a stigma with a matching S-locus allele, the pollen does not adhere strongly or does not develop a pollen tube. The products of the multiple transcription units of the S-locus include a ribonuclease, a receptor protein, and a regulatory protein called a kinase. According to a signal pathway hypothesis, molecules secreted by pollen of matching S-type bind with the receptor protein that is secreted into the cell wall of a stigma cell, and the regulatory kinase, probably built into the plasma membrane of the stigma cell, relays the message to the cytoplasm. The stigma cell may then release ribonuclease that destroys RNA in the pollen, preventing the formation of a pollen tube. Further research on the molecular basis of self-incompatibility may allow plant breeders to manipulate crop species to assure hybridization.

Double Fertilization The pollen grain grows a tube through the style, and the generative cell divides to form two sperm, the male gametes. The pollen tube probes through the micropyle, an opening through the integuments of the ovule, and releases its two sperm within the embryo sac. By **double fertilization**, one sperm fertilizes the egg to form the zygote, and the other combines with the polar nuclei to form a triploid nucleus, which will develop into a food-storing tissue called the **endosperm**. Each ovule develops into a seed, and the ovary develops into a fruit.

■ **INTERACTIVE QUESTION   34.3**

What function may double fertilization serve?

■ **The ovule develops into a seed containing a sporophyte embryo and a supply of nutrients**

Endosperm Development The triploid nucleus divides to form the endosperm, a multicellular mass rich in nutrients, which are provided to the developing embryo and stored for later use when the seed germinates. In many dicots, endosperm is transferred to the cotyledons before the seed matures.

Embryo Development (Embryogenesis) In the zygote, the first mitotic division creates a basal cell and a terminal cell. The basal cell divides to produce a thread of cells, called the suspensor, that anchors the embryo and transfers nutrients to it. The terminal cell divides to form a spherical proembryo, on which the cotyledons begin to form as bumps. The embryo elongates and apical meristems develop at the apexes of the embryonic shoot and root. Embryogenesis also establishes the radial arrangement of the primary meristems—protoderm, ground meristem, and procambium.

Structure of the Mature Seed As it matures, the seed dehydrates and the embryo becomes dormant. The embryo and its food supply, the endosperm and/or enlarged cotyledons, are enclosed in a **seed coat** formed from the ovule integuments.

In a dicot seed, such as a bean, the embryo is an elongated embryonic axis attached to fleshy cotyledons. The axis below the cotyledonary attachment is called the **hypocotyl**; it terminates in the **radicle**, or embryonic root. The upper axis is the **epicotyl**; it terminates as a plumule, a shoot tip with a pair of leaves.

■ **INTERACTIVE QUESTION 34.4**

Label the parts in these diagrams of a bean and corn seed.

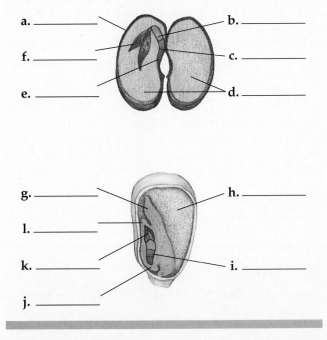

a. _____
b. _____
f. _____
c. _____
e. _____
d. _____

g. _____
h. _____
l. _____
k. _____
i. _____
j. _____

In some dicots, the cotyledons remain thin and absorb nutrients from the endosperm when the seed germinates.

A monocot seed, such as a corn kernel, has a single thin cotyledon, called a **scutellum**, which absorbs nutrients from the endosperm during germination. A sheath called a **coleorhiza** covers the root, and a **coleoptile** encloses the embryonic shoot.

■ **The ovary develops into a fruit adapted for seed dispersal**

The ovary of the flower ripens into a **fruit**, which both protects and helps to disperse the seeds. Other floral parts may contribute to what we commonly call a fruit. Hormonal changes following pollination cause the ovary to enlarge, its wall becoming the **pericarp**, or thickened wall of the fruit. Fruit usually does not set if a flower has not been pollinated.

A **simple fruit** develops from a single ovary. An **aggregate fruit** results from a single flower that has several separate carpels. A **multiple fruit**, such as a pineapple, develops from an inflorescence, a group of tightly clustered flowers. Fruits usually ripen as the seeds are completing their development. Fruits are adapted to disperse seeds, enlisting the aid of wind or animals.

Humans have selectively bred edible fruits. Cereal grains—the wind-dispersed fruits of grasses—are staple foods for humans.

■ **INTERACTIVE QUESTION 34.5**

What changes usually occur when a fleshy fruit ripens?

■ **Evolutionary adaptations in the process of germination increase the probability that seedlings will survive**

At germination, the plant resumes the growth and development that was suspended when the seed matured.

Seed Dormancy Dormancy is an adaptation to terrestrial habitats that increases the chances that the seed will germinate when and where the embryo has a good chance of surviving. The specific cues for breaking dormancy vary with the environment and may include heavy rain, intense heat from fires, cold, light, or chemical breakdown of the seed coat. The viability of a dormant seed may vary from a few days to decades or longer.

From Seed to Seedling   **Imbibition**, the absorption of water by the dry seed, causes the seed to expand, rupture its coat, and begin a series of metabolic changes. Stored compounds are digested by enzymes, and nutrients are sent to growing regions. Soon after hydration in cereal seeds, the thin outer layer of the endosperm, called the aleurone, is cued by a hormone produced by the embryo to begin making α-amylase and other enzymes that digest starch stored in the endosperm.

The radicle emerges from the seed first, followed by the shoot tip. In many dicots, a hook that forms in the hypocotyl is pushed up through the ground, pulling the delicate shoot and cotyledons behind it. Light stimulates the straightening of the hook, and the first foliage leaves begin photosynthesis. In peas, a hook forms in the epicotyl, lifting the shoot tip out of the soil while the pea cotyledons remain in the ground. In monocots, the coleoptile pushes through the soil, and the shoot tip is protected as it grows up through the tubular sheath.

■ **INTERACTIVE QUESTION   34.6**

What happens to a bean seedling grown in the dark? Why?

■ Many plants can clone themselves by asexual reproduction

Many plant species can clone themselves through asexual or **vegetative reproduction**—an extension of the indeterminate growth of plants in which meristematic tissues can grow indefinitely and parenchyma cells can divide and differentiate into specialized cells. A common type of vegetative reproduction is **fragmentation**, the separation of a plant into parts that then form whole plants. In some dicot species, the root system gives rise to many adventitious shoots that develop into a clone with separate shoot systems.

Some plants, such as dandelions, can produce seeds asexually, a process called **apomixis**.

■ **INTERACTIVE QUESTION   34.7**

What would be an advantage of apomixis?

■ Vegetative reproduction of plants is common in agriculture

Clones from Cuttings   New plants develop from stem cuttings when a **callus**, or mass of dividing cells, forms at the cut end of the shoot and adventitious roots develop from the callus.

Twigs or buds of one plant can be grafted onto a plant of a different variety or closely related species. The plant that provides the root system is called the **stock**, and the twig is called the **scion**. Grafting can combine the best qualities of different plants.

Test-Tube Cloning and Related Techniques   In test-tube cloning, whole plants can develop from pieces of tissue, called explants, or even from single parenchyma cells. A single plant can be cloned into thousands of plants by subdividing the undifferentiated calluses as they grow in tissue culture. Stimulated by proper hormone balances, calluses sprout shoots and roots and develop into plantlets, which can be transferred to soil to develop. Some cultured explants develop into "somatic embryos," which can be packaged along with nutrients into gels to create artificial seeds. With the use of a gene gun or other techniques, foreign DNA is inserted into individual plant cells, which then grow into plants by test-tube culture.

A technique called **protoplast fusion** is being coupled with tissue culture to create new plant varieties. Protoplasts are screened for agriculturally beneficial mutations and then cultured. Protoplasts from different species can be fused and cultured to form hybrid plantlets.

Benefits and Risks of Monoculture   Much of the genetic variability in agricultural crops has been eliminated by plant breeders. **Monoculture**, the cultivation of a single plant variety on large areas of land, produces a fragile ecosystem in which there is little genetic variability and little adaptability to resist disease or parasites. "Gene banks" have been created in which plant breeders maintain seeds of many different plant varieties, so that new varieties can be developed if current ones fail.

What are some advantages of monoculture?

■ Sexual and asexual reproduction are complementary in the life histories of many plants: *a review*

Sexual reproduction in plants generates variation in a population, an advantage when the environment changes. Sexual reproduction also produces seeds, a means of dispersal to new locations and dormancy during harsh conditions. By asexual reproduction, plants well suited to a certain environment can clone exact copies. The progeny of vegetative propagation are usually not as frail as seedlings. Both modes of reproduction have been useful in the evolutionary adaptation of plant populations to their environments.

■ Growth, morphogenesis, and differentiation produce the plant body: *an overview of developmental mechanisms in plants*

Development includes the overlapping processes of growth, morphogenesis, and cellular differentiation that lead to a differentiated body. **Growth** results from cell division and enlargement. **Morphogenesis**, the development of form, accounts for the formation of roots, shoots, and leaves. **Cellular differentiation** creates a cell's specific structural and functional features.

It is said that animals move through their environments, whereas plants _____ through their environments. What makes this possible?

■ The cytoskeleton guides the geometry of cell division and expansion: *a closer look*

Unlike animal morphogenesis in which cells migrate, plant morphogenesis depends on the oriented division and expansion of cells that are immobilized by their cell walls.

Orienting the Plane of Cell Division　The plane of cell division is determined by the **preprophase band**, a ring of cytoskeletal microtubules that forms during late interphase. The microtubules disperse before metaphase, but leave behind an ordered array of actin microfilaments that first orient the nucleus and later direct the movement of the vesicles that form the cell plate.

Orienting the Direction of Cell Expansion　About 90 percent of plant cell growth is due to the uptake of water. Acids secreted by the cell break cross-links between cellulose microfibrils in the cell wall, weakening the wall so that water enters the hypertonic cell. The plant cell stops expanding when the wall again becomes rigid enough to offset the osmotic potential of the cell. This economical means of cell elongation produces the rapid growth of shoots and roots.

Cells expand in a direction perpendicular to the orientation of cellulose microfibrils in the inner layers of the cell wall. The orientation of cellulose microfibrils parallels the orientation of microtubules on the other side of the plasma membrane. These microtubules may control the movement of cellulose-producing enzymes and thus determine the alignment of microfibrils in the wall.

Review the role of microtubules in the orientation of plant cell division and expansion.

■ Cellular differentiation depends on the control of gene expression: *a closer look*

Differentiation arises from the synthesis of different proteins in different types of cells. The control of gene expression, and thus the regulation of transcription and translation of specific proteins, leads to the development of different structures and functions of cells, all of which share a common genome.

Mechanisms of pattern formation determine location and tissue organization of plant organs: *a closer look*

Pattern formation is the characteristic development of structures and organs in specific locations.

Positional Information   Pattern formation depends on **positional information**, signals that indicate a cell's location within an embryonic structure or developing organ and thus direct its division, expansion, and differentiation. An embryonic cell may detect its location by gradients of molecules, probably proteins, that diffuse from specific locations in a developing structure.

Clonal Analysis of the Shoot Apex   Using clonal analysis, researchers are able to identify cells in the apical meristem and follow their lineages to pinpoint when the developmental fate of a cell becomes determined. Apparently the cells of the meristem are not committed to the formation of specific organs and tissues. A cell's final position in a developing organ determines what type of cell it will become.

The Genetic Basis of Pattern Formation in Flower Development   In the transition from a vegetative shoot tip to a floral meristem, growth changes from indeterminate to determinate. Some of the genes that respond to positional information and determine what floral organ will develop have been identified. Mutations of these **organ-identity genes** result in the placement of one type of floral organ where another type would normally develop.

Organ-identity genes have been identified and cloned from *Arabidopsis thaliana*, a tiny wild mustard plant with a rapid life cycle and small genome. Similar genes have been isolated in snapdragon, suggesting that these development-controlling genes have been conserved in evolution. These genes code for transcription factors, which are regulatory proteins that bind to specific DNA sites and affect RNA synthesis. Positional information may determine which organ-identity genes are expressed. The resulting transcription factors then influence the transcription of those genes that control development of a specific organ in its proper location.

■ **INTERACTIVE QUESTION   34.11**

Consider the hypothesis for floral-part development due to expression of organ-identity genes illustrated in textbook Figure 34.18.

a. Explain why the mutant flower lacking gene activity C consists of only sepals and petals.

b. What would a double mutant flower that had no gene activity for B and C look like?

STRUCTURE YOUR KNOWLEDGE

1. Draw yourself a diagram of the major events in the life cycle of angiosperms.

2. List the advantages and disadvantages of sexual and asexual reproduction in plants.

TEST YOUR KNOWLEDGE

FILL IN THE BLANKS

_____   1. ring of microtubules that determine future plane of cell division

_____   2. generation that produces spores by meiosis

_____   3. species with male and female flowers on the same plant

_____   4. female gametophyte of angiosperms

_____   5. embryonic root

_____   6. embryonic axis above attachment of cotyledon

_____   7. protects dicot shoot as it breaks through the soil

_____   8. twig or stem portion of a graft

_____   9. fruit formed from a flower with several separate carpels

_____   10. mass of dividing cells at cut end of a shoot

MULTIPLE CHOICE: *Choose the one best answer.*

1. A flower on a dioecious plant would be
   a. complete.
   b. perfect.
   c. imperfect.
   d. asexual.
   e. carpellate.

2. Which of the following structures is haploid?
   a. embryo sac
   b. anther
   c. endosperm
   d. microsporocyte
   e. both a and b

3. The terminal cell in a plant zygote
   a. develops into the shoot apex of the embryo.
   b. forms the suspensor that anchors the embryo and transfers nutrients.
   c. develops into the endosperm when fertilized by a sperm nucleus.
   d. divides to form the proembryo.
   e. develops into the cotyledons.

4. In angiosperms, sperm are formed by
   a. meiosis in the anther.
   b. meiosis in the pollen grain.
   c. mitosis in the anther.
   d. mitosis in the pollen tube.
   e. double fertilization in the embryo sac.

5. The endosperm
   a. may be absorbed by the cotyledons in the seeds of dicots.
   b. is usually a triploid tissue.
   c. is digested by enzymes in monocot seeds following hydration.
   d. develops in concert with the embryo as a result of double fertilization.
   e. is all of the above.

6. A seed consists of
   a. an embryo, a seed coat, and a nutrient supply.
   b. an embryo sac.
   c. a gametophyte and a nutrient supply.
   d. an enlarged ovary.
   e. an ovule.

7. Which structure protects a monocot shoot as it breaks through the soil?
   a. hypocotyl hook
   b. epicotyl hook
   c. coleoptile
   d. coleorhiza
   e. aleurone

8. Which of the following is a form of vegetative reproduction?
   a. apomixis
   b. grafting
   c. test-tube cloning
   d. fragmentation
   e. all of the above

9. A disadvantage of monoculture is that
   a. The whole crop ripens at one time.
   b. The whole crop could be annihilated by a change in conditions, a new pest, or disease.
   c. It predominantly uses vegetative reproduction.
   d. Cross-pollination is eliminated.
   e. It is not organic.

10. Protoplast fusion
    a. is used to develop gene banks to preserve genetic variability.
    b. is the method used for test-tube cloning.
    c. can be used to form new plant species.
    d. occurs within a callus.
    e. is done with a gene gun.

11. In the plant embryo, the suspensor
    a. determines the orientation of mitotic divisions in the early cells.
    b. connects the early root and shoot apexes.
    c. develops into the endosperm.
    d. is analogous to the umbilical cord in mammals.
    e. is the point of attachment of the cotyledons.

12. Flower parts have evolved from modified
    a. leaves.
    b. branches.
    c. sporangia.
    d. sporophytes.
    e. apical meristems.

13. Self-incompatibility
    a. prevents cross-pollination.
    b. involves the S-locus that consists of multiple alleles for a regulatory kinase.
    c. occurs when the pollen secretes ribonuclease after binding to a matching S-type receptor.
    d. occurs when pollen and stigma have different S-locus alleles.
    e. insures genetic variability of offspring.

14. Morphogenesis in plants is a direct result of
    a. position effects.
    b. the plane of cell divisions and the direction of cell expansion.
    c. chemicals that create gradients in developing organs.
    d. organ-identity genes.
    e. migration of cells from the three primary meristems.

15. Clonal analysis of cells of the shoot apex indicates that
   a. Organ-identity gene mutations substitute one body part for another.
   b. A cell's developmental fate is more influenced by position effects than by its meristematic lineage.
   c. Cellular differentiation results from regulation of gene expression resulting in production of different proteins in different cells.
   d. All cells were derived from the same parent cell.
   e. The apical meristem of the root tip produces the protoderm, ground meristem, and procambium.

CHAPTER 35

CONTROL SYSTEMS IN PLANTS

FRAMEWORK

Plant hormones—auxin, cytokinins, gibberellins, abscisic acid, and ethylene—control growth, development, movement, flowering, and senescence, as plants respond and adapt to their environments. Plant movements in response to environmental stimuli include phototropism, gravitropism, thigmotropism, and turgor movements. The biological clock of plants controls circadian rhythms, such as stomatal opening and sleep movements, and may use the phytochrome system to time night length in the photoperiodic control of flowering.

CHAPTER REVIEW

Plants have various mechanisms that enable them to sense and adaptively respond to their environments, generally by altering their patterns of growth and development. These intricate control systems are the product of the evolutionary history of plants interacting with their environments.

■ Research on how plants grow toward light led to the discovery of plant hormones: *science as a process*

Hormones, chemical signals that coordinate the parts of an organism, are translocated through the body, where minute concentrations are able to trigger responses in target cells and tissues.

The growth of a shoot toward light is called positive **phototropism**. A coleoptile, enclosing the shoot of a grass seedling, bends toward the light when illuminated from one side because of the greater elongation of cells on the darker side.

Darwin and his son observed that a grass seedling would not bend toward light if its tip were removed or covered by an opaque cap. They postulated that some signal must be transmitted from the tip down to the elongating region of the coleoptile. Boysen-Jensen demonstrated that the signal was a mobile substance, capable of being transmitted through a block of gelatin separating the tip from the rest of the coleoptile.

■ INTERACTIVE QUESTION 35.1

A block of agar is placed beneath the tip of a coleoptile that is illuminated from the right side. The block is then divided in two and each half is placed on a decapitated coleoptile as illustrated below. What growth response would you predict for these coleoptiles grown in the dark? Explain.

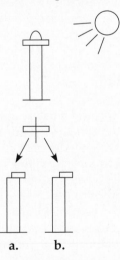

In 1926, Went placed coleoptile tips on blocks of agar to extract the chemical messenger. From his studies he concluded that the chemical produced in the tip, which he called auxin, promoted growth and that it was in higher concentration on the side away from the light.

■ Plant hormones help coordinate growth, development, and responses to environmental stimuli

Five classes of plant hormones have been identified: auxin, cytokinins, gibberellins, abscisic acid, and ethylene. These hormones affect cell division, elongation, and differentiation. Depending on the site of action, the developmental stage of the plant, and relative hormone concentrations, the effect of a hormone will vary. Hormones are effective in very small concentrations, indicating that their signal must be amplified in the cell in some way. They may act by affecting the expression of genes, the activity of enzymes, or the properties of membranes.

Auxin   **Auxins** include any substance, including synthetic compounds, that stimulate elongation of coleoptiles. The natural auxin extracted from plants is indoleacetic acid (IAA).

A major site of auxin synthesis is in the apical meristem of a shoot. Within a certain concentration range, auxin stimulates cell elongation. Auxin is transported through parenchyma tissue in only one direction, from the shoot tip down the shoot. This polar transport involves a chemiosmotic mechanism of ATP-driven proton pumps that generate a membrane potential favoring the exit of auxin anions through specific carriers located only at the basal ends of cells. In the more acidic environment outside the cell, also created by proton pumps, auxin picks up a hydrogen ion and the now neutral auxin molecule can move across the plasma membrane into the next parenchyma cell.

According to the *acid growth hypothesis*, auxin initiates shoot growth in the region of elongation by stimulating proton pumps. The proton pumps lower pH in the cell wall, breaking cross-links between cellulose microfibrils. The turgor pressure of the cell then exceeds the restraining pressure of the wall, and the cell elongates. For continued growth after this relatively fast elongation, the cell must produce more cytoplasm and wall material, processes also stimulated by auxin.

Auxin stimulates cell division in the vascular cambium, differentiation of secondary xylem, and formation of adventitious roots at the cut base of stems. Auxin produced by developing seeds promotes fruit growth; synthetic auxins induce seedless fruit development. The herbicide 2,4-D is a synthetic auxin used to disrupt the normal balance of plant growth in dicot weeds.

Cytokinins   In tissue culture, coconut milk and degraded DNA were found to induce plant cell growth; later, **cytokinins** were identified as the active ingredients. Cytokinins are modified forms of adenine, named because they stimulate cytokinesis. Cytokinins are produced in actively growing roots, embryos, and fruits. Acting with auxin, they stimulate cell division and affect differentiation.

When cytokinin and auxin are added to parenchyma tissue grown in tissue culture, the cells divide. The ratio of the two hormones controls differentiation: equal concentrations produce undifferentiated cells, more cytokinin results in shoot buds, whereas more auxin leads to root formation. Cytokinins stimulate RNA and protein synthesis, and the resulting proteins may be involved in cell division.

The control of apical dominance involves the interaction between auxin, transported down from the terminal bud, which restrains axillary bud development, and cytokinins, transported up from the roots, which stimulate bud growth. The interaction between auxin and cytokinins works the opposite way in the development of lateral roots. Together, these hormones may coordinate the growth of shoot and root systems.

Cytokinins can retard aging of some plant organs, perhaps because they stimulate RNA and protein synthesis, mobilize nutrients, and inhibit protein breakdown.

Gibberellins   In the 1930s, Japanese scientists determined that the fungus *Gibberella* secreted a chemical that caused the hyperelongation of rice stems or "foolish seedling disease." Over 70 different gibberellins have now been identified, many occurring naturally in plants.

Gibberellins, which are produced by roots and young leaves, stimulate growth in both leaves and stem, affecting cell elongation and cell division in stems. The application of different concentrations of gibberellin to dwarf plants has demonstrated a positive correlation between growth and the concentration of hormone added. Experimental results such as these can be used in a bioassay to determine hormone concentration in a sample of unknown concentration.

Bolting, the growth of an elongated floral stalk, is caused by a surge of gibberellins. In some plants, both auxin and gibberellins contribute to fruit set.

In many plants, the release of gibberellins from the embryo signals the seed to break dormancy. The germination of grain is triggered when gibberellins stimulate the synthesis of messenger RNA that codes for the digestive enzymes that break down stored nutrients. Gibberellins are also involved in the breaking of dormancy of apical buds in spring.

Abscisic Acid   The hormone **abscisic acid (ABA)**, which is produced in the terminal bud, slows growth, inhibits cell division in the vascular cambium, and induces leaf primordia to develop into scales that protect the dormant bud during winter.

Abscisic acid may act as a growth inhibitor of the embryo when the seed becomes dormant. For dormancy to be broken in some seeds, ABA must be removed or inactivated, or the ratio of gibberellins to ABA must increase. Abscisic acid also acts to help the plant cope with adverse conditions. In a wilting plant, ABA causes stomata in the leaves to close.

Ethylene   Plants produce the gas **ethylene**, which acts as a hormone promoting fruit ripening and inhibiting growth in roots and axillary buds. Ethylene production is induced by a high concentration of auxin. Ethylene contributes to the aging, or **senescence**, of parts of the plant. It initiates or hastens the degradation of cell walls and the decrease in chlorophyll content that is associated with fruit ripening. Many commercial fruits are ripened in huge containers perfused with ethylene gas.

Deciduous leaf loss protects against winter dehydration. Before leaves abscise in the autumn, many of their compounds are stored in the stem awaiting recycling to new leaves. The leaf stops making chlorophyll; fall colors result from a combination of pigments that had been concealed by chlorophyll and new pigments that are made during autumn.

A change in the balance of auxin and ethylene initiates changes in the abscission layer located near the base of the petiole, including the production of enzymes that hydrolyze polysaccharides in cell walls. A layer of cork forms a protective covering on the twig side of the abscission layer. Shortening days and cooler temperatures are the stimuli for leaf abscission.

■ INTERACTIVE QUESTION   35.2

Match the functions and site of production to the plant hormone.

Hormone	Function	Location
Auxin	_____	_____
Cytokinins	_____	_____
Gibberellins	_____	_____
Abscisic acid	_____	_____
Ethylene	_____	_____

Function	Location
A. promotes fruit ripening, senescence, leaf drop	**a.** apical buds, roots, young leaves, embryo
B. stimulate growth, germination, delay senescence	**b.** apical buds, young leaves, embryo
C. inhibit growth, maintain dormancy, close stomata	**c.** ripening fruits, senescent leaves
D. stimulates stem and root growth, fruit development	**d.** leaves, stems, green fruit
E. promote seed and bud germination, stem elongation, leaf growth, flowers and fruit	**e.** roots

■ Tropisms orient the growth of plant organs toward or away from stimuli

Tropisms are growth responses in which a plant organ curves toward or away from a stimulus as a result of differential cell elongation on opposite sides of the organ.

Phototropism   In coleoptiles, cells on the darker side of a stem elongate faster in response to a larger concentration of auxin moving down from the shoot tip. In dicots, the concentration of growth inhibitors on the light side may produce the phototropic response. The photoreceptor is believed to be a blue-light-sensitive pigment.

Gravitropism   Roots exhibit positive **gravitropism**, whereas shoots show negative gravitropism. According to one hypothesis, the settling of **statoliths**, plastids containing dense starch grains, in cells of the root cap triggers movement of calcium and the lateral transport of auxin. Both of these accumulate on the lower side of the growing root and inhibit cell elongation, causing the root to curve downward. An alternative hypothesis is that the settling of the protoplast signals gravitational direction.

Thigmotropism   Most climbing plants have tendrils that exhibit **thigmotropism**, directional growth in response to touch. The stunting of growth in height and increase in girth of plants that are exposed to wind or mechanical stimulation is called **thigmomorphogenesis**.

a. Name the type of tropisms that a stem may exhibit.

b. What growth mechanism produces the coiling of a tendril?

What are free-running periods? How are they determined?

■ Turgor movements are relatively rapid, reversible plant responses

Rapid Leaf Movements   The sensitive plant *Mimosa* folds its leaves after being touched due to the rapid loss of turgor by cells in specialized motor organs called pulvini, located at the joints of the leaf. These cells lose potassium when stimulated, resulting in osmotic water loss. The message travels through the plant from the point of stimulation, perhaps as the result of chemical messengers and electrical impulses, called **action potentials**. These electrical messages may be used in plants as a form of internal communication.

Sleep Movements   Many members of the legume family exhibit **sleep movements**, the daily raising and lowering of leaves caused by changes in the turgor pressure of motor cells in pulvini. Massive migration of potassium ions from one side of the pulvinus to the other causes these reversible osmotic changes in the motor cells.

■ Biological clocks control circadian rhythms in plants and other eukaryotes

Many physiological processes in animals, fungi, protists, and plants fluctuate with the time of day and are controlled by biological clocks—internal oscillators that keep accurate time. A **circadian rhythm** is a physiological cycle with about a 24-hour frequency. These rhythms persist, even when the organism is sheltered from environmental cues. Research indicates that the oscillator for circadian rhythms is endogenous, although the clock is set (entrained) to a 24-hour period by daily environmental signals.

The nature of the oscillator of the biological clock is still unknown, but it is most likely regulated at the cellular level, perhaps in membranes or protein-synthesizing machinery.

■ Photoperiodism synchronizes many plant responses to changes of season

Seasonal events in the life cycle of plants usually are cued by photoperiod, the relative length of night and day. A physiological response to day length is called **photoperiodism**.

Photoperiodic Control of Flowering   Garner and Allard discovered that a variety of tobacco plant flowered only when the day length was 14 hours or shorter. They termed it a **short-day plant**, because it seemed to need a light period shorter than a critical length to flower. **Long-day plants** flower when days are longer than a certain number of hours, and the flowering of **day-neutral plants** is unaffected by photoperiod.

Researchers have found that it is night length, not day length, that controls flowering and other photoperiod responses. If the dark period is interrupted by even a few minutes of light, a short-day (long-night) plant such as the cocklebur will not flower. Photoperiodic responses thus depend on a critical night length. Short-day plants require a minimum number of hours of uninterrupted darkness, and long-day plants will flower only if they receive less than a maximum number of hours of darkness.

Some plants bloom after a single exposure to the required photoperiod. Others respond to photoperiod only after exposure to another environmental stimulus. The need for pretreatment with cold before flowering is called vernalization.

Leaves detect the photoperiod. The signal for flowering is believed to be a hormone that travels from leaves to buds, but it has yet to be identified and may be a mixture of several substances.

■ Phytochrome functions as a photoreceptor in many plant responses to light and photoperiod

Red light is the most effective in interrupting a plant's perception of night length. A brief exposure to red light breaks a dark period of sufficient length and prevents short-day plants from flowering, whereas a flash of red light during a dark period longer than the critical length will induce flowering in a long-day plant. A subsequent flash of light from the far-red part of the spectrum negates the effect of the red light.

The photoreceptor pigment **phytochrome** alternates between two forms, one of which absorbs red light and the other, far-red light. These two variations of phytochrome are photoreversible; the P_r to P_{fr} interconversion acts as a switch, controlling various events in the life of the plant.

Ecological Significance of Phytochrome as a Photoreceptor Phytochrome tells the plant that light is present by the conversion of P_r, which is the form the plant synthesizes, to P_{fr} in the presence of sunlight. P_{fr} triggers many plant responses to light, such as breaking seed dormancy. The amount and type of available light, specifically the relative amounts of red and far-red light, is communicated to a plant by the ratio of the two forms of phytochrome. Evidence

shows that the widespread results of the photoconversion of the relatively little phytochrome in a plant is caused by two mechanisms: changes in membrane permeability and changes in gene expression.

Interaction of Phytochrome and the Biological Clock in Photoperiodism In darkness, the phytochrome ratio shifts toward P_r, in part because P_{fr} is converted to P_r in some plants, and also because P_{fr} is degraded and new pigment is synthesized as P_r. The conversion to P_r is complete within a few hours of darkness; therefore, the plant cannot use the phytochrome ratio to measure night length. The role of phytochrome may be to synchronize the biological clock by signaling when the sun sets and rises.

■ Control systems enable plants to cope with environmental stress

Stress is defined as an environmental fluctuation severe enough to threaten a plant's growth, reproduction, and survival.

■ INTERACTIVE QUESTION   35.5

Indicate whether a short-day plant (**a–f**) and a long-day plant (**g–l**) would flower or not flower under the indicated light conditions.

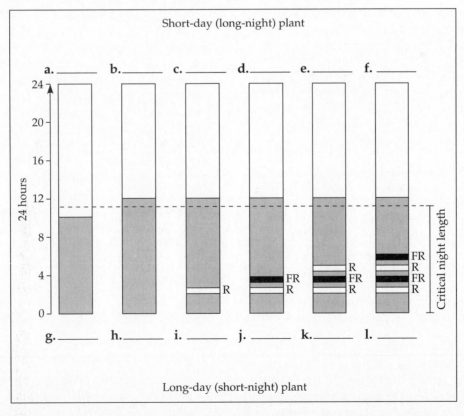

Responses to Water Deficit   Mechanisms that reduce transpiration help a plant respond to water deficit. Guard cells lose turgor and stomata close. Mesophyll cells produce abscisic acid, which acts on guard cell membranes to keep stomata closed. Expansion of young leaves is inhibited and wilted leaves may roll up, further reducing transpiration. During a drought, root growth in shallow, dry soil decreases while deeper roots continue to grow.

Responses to Oxygen Deprivation   Plants adapted to wet habitats may provide oxygen to their roots from aerial roots. When roots of other plants are submerged, oxygen deprivation may stimulate ethylene production, causing some root cortex cells to die and open up air tubes within the roots.

Responses to Salt Stress   Excess salts in the soil may lower the water potential of the soil solution below that of roots, causing the roots to lose water. The plasma membranes of root cells can reduce the uptake of sodium and some other ions that are toxic to plants in high concentrations. Plants may respond to moderate soil salinity by producing compatible solutes that lower the water potential of root cells.

Responses to Heat Stress   Transpiration creates evaporative cooling for a plant but may be reduced on hot, dry days when stomata close to reduce water loss. Above critical temperatures, plant cells may produce **heat-shock proteins** that may provide temporary support to reduce protein denaturation.

Responses to Cold Stress   Plants respond to cold stress by increasing the proportion of saturated fatty acids in membrane lipids. Some woody plants are able to change solute composition to prevent the formation of ice crystals in the cytosol.

Responses to Herbivores   Physical and chemical defenses may help plants cope with herbivory. Some plants produce canavanine, which resembles arginine and may be incorporated into an insect's proteins, altering protein conformation and causing death.

Defense Against Pathogens: Systemic Acquired Resistance   Once a pathogen gets past the first line of defense of the epidermis or periderm, an infected plant produces **phytoalexins**, antibiotics that inhibit or kill microorganisms. Pathogen molecules and compounds released from injured plant tissue induce rapid responses that contain the infection. The hormone **salicylic acid**, released from cells in the infected tissue, triggers phytoalexin production throughout the plant, creating a **systemic acquired resistance (SAR)** that helps protect uninfected tissue.

■ **INTERACTIVE QUESTION   35.6**

a. A plant is exposed to a spell of very hot and dry weather. How might it survive this stress?

b. How might a plant adapt to an unusually cold and very wet fall?

c. What is systemic acquired resistance?

■ **Signal-transduction pathways mediate the responses of plant cells to environmental and hormonal stimuli**

A **signal-transduction pathway**, which links stimulus to cell response, consists of three main steps: reception, transduction, and induction. Reception may be the absorption of a specific wavelength of light by a pigment molecule or the binding of a hormone to a specific receptor. **Target cells** have receptors for a particular hormone and are thus the only cells able to respond to that hormone message.

The transduction step amplifies the stimulus. A **second messenger** is a substance that serves to amplify the environmental message. Calcium ions (Ca^{2+}) appear to be common second messengers that increase in concentration in response to a signal reception and bind to the protein **calmodulin**; this complex then activates other cellular proteins and molecules. The specificity of hormone function is due to the types of proteins present in a cell that are activated by second messengers.

The amplified signal now induces a specific cell response. Relatively rapid responses include stomatal closing and cell elongation resulting from cell wall acidification. Slower responses to a signal-transduction pathway involve changes in gene expression, as in the increased transcription of the genes activated in response to mechanical touch.

STRUCTURE YOUR KNOWLEDGE

1. List some agricultural uses of plant hormones.
2. Develop a concept map to illustrate your understanding of the photoperiodic control of flowering. Do not forget to include the role of the biological clock.

3. Explain the three steps in the signal-transduction pathway that link stimulus to response in plant cells.

TEST YOUR KNOWLEDGE

TRUE or FALSE: *Indicate T or F and then correct the false statements.*

_____ 1. Abscisic acid is necessary for a seed to break dormancy.

_____ 2. Roots exhibit negative phototropism and positive gravitropism.

_____ 3. Gibberellins, synthesized in the root, counteract apical dominance.

_____ 4. Bolting occurs when ethylene stimulates the degradation of cell walls, decreases chlorophyll content, and causes fruit to drop.

_____ 5. The application of gibberellins or auxin may induce the development of seedless fruit.

_____ 6. Abscisic acid induces the synthesis of enzymes that hydrolyze polysaccharides in cell walls of the abscission layer.

_____ 7. Action potentials are electrical impulses that may be used as internal communication in plants.

_____ 8. Thigmotropism is involved in the rapid movements of *Mimosa* leaves.

_____ 9. A physiological response to day or night length is called a circadian rhythm.

_____ 10. Vernalization is the need for pretreatment with cold before flowering.

MULTIPLE CHOICE: *Choose the one best answer.*

1. The body form within a species of plants may vary more than that within a species of animals because
   a. Growth in animals is predominantly indeterminate.
   b. Plants respond adaptively to their environments by altering their patterns of growth and development.
   c. Plant growth and development are governed by many hormones.

d. Plants are able to respond to environmental stress.
   e. All of the above are true.

2. Polar transport of auxin involves
   a. the accumulation of higher concentrations on the side of a shoot away from light.
   b. the reverse movement of auxin from roots to shoot.
   c. movement of auxin ions through carrier proteins located at the basal end of cells, facilitated by the membrane potential.
   d. the unidirectional active transport of auxin into and out of parenchyma cells.
   e. the settling of statoliths.

3. According to the acid growth hypothesis,
   a. Auxin stimulates membrane proton pumps.
   b. A lowered pH breaks cross-links between cellulose microfibrils.
   c. The turgor pressure of the cell exceeds the lowered restraining wall pressure.
   d. Cells elongate when they take up water by osmosis.
   e. All of the above are involved in cell elongation.

4. Which of the following situations would stimulate the development of axillary buds?
   a. a large quantity of auxin traveling down from the shoot and a small amount of cytokinin produced by the roots
   b. a small amount of auxin traveling down from the shoot and a large amount of cytokinin traveling up from the roots
   c. an equal ratio of gibberellins to auxin
   d. the absence of cytokinins caused by the removal of the terminal bud
   e. an infusion of ABA from a neighboring plant

5. The growth inhibitor in seeds is usually
   a. abscisic acid.
   b. ethylene.
   c. gibberellin.
   d. a small amount of ABA combined with a larger ratio of gibberellins.
   e. a high cytokinin-to-auxin ratio.

6. A circadian rhythm
   a. is controlled by an external oscillator.
   b. is a physiological cycle of approximately a 24-hour frequency.
   c. involves a biological clock that is not influenced by daily environmental signals.
   d. provides the signal for seasonal flowering.
   e. involves all of the above.

7. A bioassay
   a. uses chemical means to measure the activity of biological compounds.
   b. determines agricultural applications of plant hormones.
   c. determines the components present in a mixture of biologically active compounds.
   d. measures the rate of growth of dwarf plants.
   e. measures the response of living systems to determine the concentration of a biologically active chemical compound.

8. In order to flower, a short-day plant needs
   a. a burst of red light in the middle of the night.
   b. a night that does not exceed a certain length.
   c. a night that does exceed a minimal length.
   d. a higher ratio of P_r :P_{fr}.
   e. a lower ratio of P_r :P_{fr}.

9. A flash of far-red light during a critical-length dark period
   a. will induce flowering in a long-day plant.
   b. will induce flowering in a short-day plant.
   c. will not influence flowering.
   d. will increase the P_{fr} level suddenly.
   e. will be negated by a flash of red light.

10. The conversion of P_r to P_{fr}
    a. occurs slowly at night.
    b. may be part of the way in which plants sense light.
    c. is the timekeeper or oscillator for the circadian rhythms in plants.
    d. occurs when P_r absorbs far-red light.
    e. occurs within flower buds and induces flowering.

11. The function of calmodulin in a signal-transduction pathway is
    a. to activate other second messengers in the transduction step.
    b. as a hormone that induces the selective activation of genes.
    c. as a membrane-bound receptor that causes an influx of Ca^{2+}.
    d. to form a complex with Ca^{2+} and activate specific proteins and cellular molecules.
    e. to stimulate the release of phytoalexin throughout the plant.

12. A plant may withstand salt stress by
    a. releasing abscisic acid that closes stomata to salt accumulation.
    b. the production of compatible solutes that lower the water potential of root cells.
    c. wilting that reduces water and salt uptake by reducing transpiration.
    d. producing canavanine that reduces the toxic effect of sodium ions.
    e. developing systemic acquired resistance to salt ions.

ANIMALS: FORM AND FUNCTION

CHAPTER 36

AN INTRODUCTION TO ANIMAL STRUCTURE AND FUNCTION

FRAMEWORK

The body structures of an animal include organs that are composed of specialized cells grouped into the four basic tissues: epithelial, connective, muscle, and nervous. Organs function together in organ systems. Structure correlates with function in these hierarchical levels of organization, and the functions of an animal are powered by chemical energy derived from food. Metabolic rate determines the amount of energy needed and is higher for endothermic animals and inversely related to body size. Body proportions and posture are related to size.

All cells must be bathed in an aqueous solution. Compact animal bodies have highly folded exchange surfaces and a circulatory system that distributes materials throughout the body. The internal environment is carefully regulated by the process of homeostasis.

CHAPTER REVIEW

The comparative study of animals illustrates three general themes: the correlation of structure and function, the capacity of organisms to adapt to their environments, both by short-term physiological adjustments and long-term evolutionary changes, and bioenergetics as the basis for understanding animal physiology.

■ **The functions of animal tissues and organs are correlated with their structures**

Hierarchical levels of organization characterize life. Life emerges at the level of the cell. Multicellular organisms have specialized cells grouped into tissues, which may be combined into organs. Various organs may function together in organ systems.

Animal Tissues   **Tissues** are collections of cells with a common structure and function, held together by a sticky extracellular matrix or fibers. Histologists—biologists who study tissues—classify tissues into four categories.

Epithelial tissue lines the outer and inner surfaces of the body in protective sheets of tightly packed cells. Cells at the base of an epithelium are attached to a **basement membrane**, a dense layer of extracellular matrix. A **simple epithelium** has one layer, whereas a **stratified epithelium** has multiple layers of cells. The shape of cells at the free surface may be **squamous** (flat), **cuboidal** (boxlike), or **columnar** (pillarlike).

Some epithelia are specialized for absorption or secretion. **Mucous membranes** lining the gut and the air passages secrete mucus. The small intestine lining also releases digestive enzymes and absorbs nutrients.

255

Cilia on the epithelium lining the trachea sweep particles trapped in mucus away from the lungs.

Name the two types of epithelium illustrated below. One of these epithelial types forms the outer skin and the other lines the digestive tract. Explain why each would be found in its location.

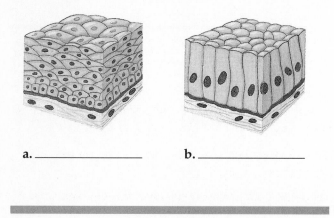

a. _____   b. _____

Connective tissue connects and supports other tissues and is characterized by having relatively few cells suspended in an extracellular matrix of fibers, which may be embedded in a liquid, jellylike or solid ground substance.

Loose connective tissue attaches epithelia to underlying tissues and holds organs in place. Its loosely woven fibers are of three types: **Collagenous fibers** are made of collagen and have great tensile strength that resists stretching. **Elastic fibers**, made of the protein elastin, can stretch and provide resilience. Branched and thin **reticular fibers** are composed of collagen and form a tightly woven connection with adjacent tissues. The most common types of cells enmeshed in loose connective tissue are **fibroblasts**, which secrete the protein of the extracellular fibers, and **macrophages**, amoeboid cells that engulf bacteria and cellular debris by phagocytosis.

Adipose tissue is a special form of loose connective tissue that pads and insulates the body and stores fuel reserves. Adipose cells each contain a large fat droplet.

Fibrous connective tissue, with its dense arrangement of parallel collagenous fibers, is found in **tendons**, which attach muscles to bones, and in **ligaments**, which join bones together at joints.

Cartilage is composed of collagenous fibers embedded in a rubbery ground substance called chondroitin sulfate, both secreted by **chondrocytes** found in scattered lacunae in the ground substance.

Cartilage is a strong but somewhat flexible support material, making up the skeleton of sharks and vertebrate embryos.

Bone is a mineralized connective tissue formed by **osteocytes** that deposit a matrix of collagen and calcium phosphate, which hardens into hydroxyapatite. **Haversian systems** consist of concentric layers of matrix deposited around a central canal containing blood vessels and nerves. Osteocytes are located in lacunae within the matrix and are connected to one another by thin cellular extensions. In long bones, the hard outer region is compact bone, whereas the interior is a spongy bone tissue filled with bone marrow. Red bone marrow, near the ends of long bones, manufactures blood cells.

Blood is a connective tissue that has a liquid extracellular matrix called plasma, containing water, salts, and dissolved proteins. Erythrocytes (red blood cells) carry oxygen, leukocytes (white blood cells) function in defense, and cell fragments called platelets are involved in the clotting of blood.

Identify the types of connective tissue and their components in the following three diagrams.

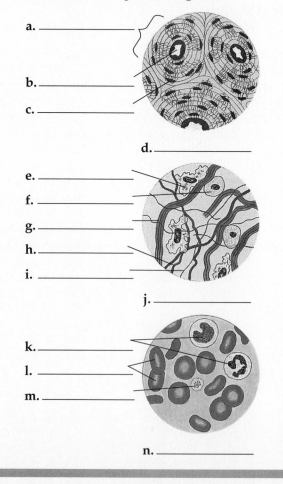

a. _____

b. _____

c. _____

d. _____

e. _____

f. _____

g. _____

h. _____

i. _____

j. _____

k. _____

l. _____

m. _____

n. _____

Muscle tissue consists of long, contractile cells that are packed with microfilaments of actin and myosin. **Skeletal muscle**—also called **striated muscle** because it looks striped due to the arrangement of overlapping filaments—is responsible for voluntary body movements. **Cardiac muscle**, forming the wall of the heart, is also striated, but its cells are branched, joined at their ends by intercalated discs that relay signals to synchronize the heartbeat. **Smooth muscle** is composed of spindle-shaped cells lacking striations. It is found in the walls of the digestive tract, arteries, and other internal organs. Unlike voluntary skeletal muscle, smooth muscle is often called involuntary because it is not generally under conscious control.

■ **INTERACTIVE QUESTION   36.3**

Identify these types of vertebrate muscle. What are the dark bands between fibers in figure **a**, and what is their function?

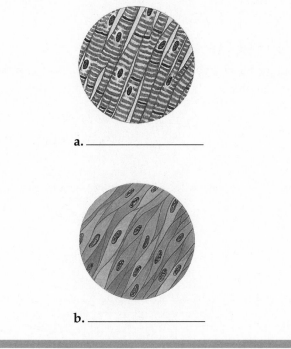

a. _____

b. _____

Nervous tissue senses stimuli and transmits electrical signals. The **neuron**, or nerve cell, consists of a cell body and two or more processes that conduct impulses toward (dendrites) and away from (axons) the cell body.

Organs and Organ Systems   In all animals but sponges and cnidarians, tissues are organized into specialized units of function called **organs**. Organs often consist of a layered arrangement of tissues. Many vertebrate organs are suspended by **mesenteries** in fluid-filled body cavities. Mammals have a **thoracic cavity** separated by a muscular diaphragm from an **abdominal cavity**.

Groups of organs are integrated into **organ systems**, which perform the major functions required for life. The organ systems are coordinated to create the functional integration needed by an organism.

■ **INTERACTIVE QUESTION   36.4**

There are 11 organ systems in mammals. How many of them can you name?

■ **Bioenergetics is fundamental to all animal functions**

Animals exchange energy with the environment by taking in food, from which they harvest ATP for cellular work and chemical energy, and carbon skeletons for biosynthesis, and returning heat to the environment.

Metabolic Rate   The total energy an animal uses in a unit of time is its **metabolic rate**. Energy is measured in **calories** (cal) or **kilocalories** (kcal). Metabolic rate can be measured by placing an animal in a calorimeter and measuring heat production. The rate of oxygen consumption, also a measure of metabolic rate, can be determined with a respirometer.

Metabolic rates are influenced by age, sex, size, activity level, time of day, and other variables. Endothermic animals (birds and mammals) require more energy to sustain minimal life functions than do ectotherms that do not use metabolic heat to maintain a constant body temperature.

The **basal metabolic rate (BMR)** for an endotherm is described as the number of kcal needed per hour when totally at rest, fasting, and nonstressed. The **standard metabolic rate (SMR)** is the metabolic rate of a resting, fasting ectotherm determined at a specific temperature. Metabolic rates during intense physical exercise may be five to ten times the BMR or SMR.

Body Size and Metabolic Rate   The energy required to maintain each gram of body weight is inversely related to body size. Smaller animals have higher metabolic rates, breathing rates, blood volume, and heart rates. With a greater surface-to-volume ratio, small animals may have a higher energy cost to maintain a stable body temperature. The fact that ectotherms also show this inverse relationship indicates that other factors must contribute to its cause.

■ INTERACTIVE QUESTION   36.5

a. Which animal would have a higher BMR, a rabbit or a bear?

b. Which animal would have a higher SMR, a frog or an alligator?

c. Which animal of the four above would consume the most cal/g of body weight?

d. Which animal would consume the most total calories?

■ An animal's size and shape affect its interactions with the external environment

Body Size, Proportions, and Posture   Body proportions of large and small animals vary, partly as a result of the physical relationship between the diameter of a support and the strain produced by increasing size and weight. Posture appears to be the most important factor in supporting body weight. Legs of large animals are held more upright and are more extended when running, thereby reducing strain.

Body Plans and Exchange with the Environment Every cell must be in an aqueous medium to maintain the integrity of the plasma membrane and allow for exchange across it. Cell size is limited by the need for a sufficient surface-to-volume ratio. Single-celled organisms or animals with two-layered saclike bodies or thin flat bodies can maintain sufficient cellular contact with the aqueous environment.

Animals with compact bodies, however, must provide extensive moist internal membranes for exchanging materials with the environment. The circulatory system connects these exchange surfaces with the aqueous environment bathing the body's cells.

■ Homeostatic mechanisms regulate an animal's internal environment

The internal environment of vertebrates is the **interstitial fluid** surrounding the cells, through which oxygen, nutrients, and wastes are exchanged with blood in capillaries. The "steady state" or constancy of this environment is called **homeostasis**. The mechanisms by which animals maintain homeostasis involve a *receptor* that detects a change in the internal environment and a *control center* that processes information and directs an *effector* to respond. Most homeostatic control mechanisms operate by **negative feedback**. When some variable moves above or below a **set point**, a control mechanism is turned on or off to return the condition to normal.

Positive feedback in a physiological function is a mechanism in which a change in a variable serves to amplify rather than reverse the activity. The stimulation of uterine contractions during childbirth is an example.

■ INTERACTIVE QUESTION   36.6

One of the negative feedback mechanisms controlling body temperature in humans involves the **(a)**_____ _____, which are signaled to increase their activity when the **(b)**_____ senses a rise in body temperature. When body temperature falls below the **(c)**_____ _____, the **(d)**_____ stops sending "sweat" signals.

STRUCTURE YOUR KNOWLEDGE

1. Fill in the table below on the structure and function of the four types of animal tissues.

Tissue	Structural Characteristics	General Functions	Specific Examples

2. Organs are composed of layers of several different tissues. Which of the four major animal tissues do you think would be included in all organs? In what types of organs would you predict the remaining tissues would be included? Give an example of an organ composed of several layers of tissues.

TEST YOUR KNOWLEDGE

MULTIPLE CHOICE: *Choose the one best answer.*

1. Which of the following is *not* an organ system?
   a. skeletal
   b. connective
   c. digestive
   d. excretory
   e. immune and lymphatic

2. A stratified squamous epithelium would be composed of
   a. several layers of flat cells attached to a basement membrane.
   b. a layer of ciliated, mucus-secreting, flattened cells.
   c. a hierarchical arrangement of boxlike cells.
   d. an irregularly arranged layer of pillarlike cells.
   e. several layers of flat cells underneath columnar cells.

3. Which of the following is *not* true of connective tissue?
   a. It consists of few cells surrounded by fibers and a ground substance.
   b. It includes such diverse tissues as bone, cartilage, tendons, adipose, and loose connective tissue.
   c. It connects and supports other tissues.
   d. It forms the internal and external lining of many organs.
   e. It can have a matrix that is a liquid, gel, or solid.

4. Which of the following are incorrectly paired?
   a. blood—erythrocytes, leukocytes, and platelets in plasma
   b. bone—osteocytes embedded in hydroxyapatite in Haversian systems
   c. loose connective tissue—collagenous, elastic, reticular fibers
   d. adipose tissue—loose connective tissue with fat-storing cells
   e. fibrous connective tissue—chondrocytes embedded in chondroitin sulfate

5. Which is the best description of smooth muscle?
   a. striated, branching cells; involuntary control
   b. spindle-shaped cells; involuntary control
   c. spindle-shaped cells connected by intercalated discs
   d. striated cells containing overlapping filaments; involuntary control
   e. spindle-shaped striated cells; voluntary control

6. The diaphragm
   a. is a mesentery.
   b. increases the surface area of the lungs.
   c. is part of the mammalian reproductive system.
   d. separates the thoracic and abdominal cavities in mammals.
   e. separates the lungs from the heart cavity.

7. The interstitial fluid of animals
   a. is the internal environment within cells.
   b. bathes cells and provides for the exchange of nutrients and wastes.
   c. makes up the plasma of blood.
   d. surrounds unicelluar and flat, thin animals.
   e. is less in ectotherms than in endotherms.

8. Negative feedback circuits are
   a. mechanisms that most commonly maintain homeostasis.
   b. activated only when a physiological variable rises above a set point.
   c. analogous to a radiator that heats a room.
   d. involved in maintaining contractions during childbirth.
   e. all of the above.

9. The basal metabolic rate
   a. is constant for each species.
   b. may vary depending on the sex or size of an organism.
   c. is highest when an animal is actively exercising.
   d. is lower than the standard metabolic rate for ectotherms.
   e. may be measured from the quantity of food an animal eats.

10. Which of the following would most likely have the highest metabolic rate?
    a. whale
    b. dog
    c. snake
    d. tuna
    e. bat

CHAPTER 37

ANIMAL NUTRITION

FRAMEWORK

Animals eat other organisms to obtain fuel for respiration, organic raw materials for biosynthesis, and essential nutrients. Digestion is the enzymatic hydrolysis of macromolecules into monomers that can be absorbed across cell membranes.

Gastrovascular cavities are digestive sacs in which some extracellular digestion takes place before food particles are phagocytized by cells lining the cavity. Alimentary canals are one-way tracts with specialized regions for mechanical breakdown of food, storage, digestion, absorption of nutrients, and elimination of wastes.

This chapter also details the structures, functions, enzymes, and hormones of the human digestive tract. A nutritionally adequate diet provides sufficient calories, essential amino acids, vitamins, and minerals.

CHAPTER REVIEW

Animals, as heterotrophs, obtain energy and raw materials from the organic compounds in their food.

■ Diets and feeding mechanisms vary extensively among animals

Herbivores eat autotrophs; **carnivores** eat animals; and **omnivores** consume both autotrophs and animals.

Many aquatic animals are **suspension-feeders,** sifting small food particles from the water. **Substrate-feeders** live in or on their food, eating their way through it. **Deposit-feeders** consume decaying organic matter. **Fluid-feeders** suck fluids from a living plant or animal host. Most animals are **bulk-feeders,** eating relatively large pieces of food.

■ Ingestion, digestion, absorption, and elimination are the four main stages of food processing

Ingestion, or eating, is the first stage of food processing.

Digestion splits macromolecules into monomers by **enzymatic hydrolysis**, the addition of a water molecule when the bond between monomers is broken. Hydrolysis occurs in a specialized compartment to protect an animal's cells from its own hydrolytic enzymes.

In the last two stages of food processing, monomers and small molecules are **absorbed** into the cells of the animal, and the undigested remainder of the food is **eliminated**.

■ INTERACTIVE QUESTION   37.1

Why must macromolecules be digested into monomers?

■ Digestion occurs in food vacuoles, gastrovascular cavities, and alimentary canals

Intracellular Digestion in Food Vacuoles **Intracellular digestion** of food molecules in food vacuoles is typical of protozoa and sponges. Most animals, however, break down their food, at least initially, by **extracellular digestion** within a separate compartment of the body that connects to the external environment.

Digestion in Gastrovascular Cavities Single-opening **gastrovascular cavities** function in both digestion and transport of nutrients throughout the body. In the small cnidarian *Hydra*, digestive enzymes secreted into the gastrovascular cavity initiate food break-down. Gastrodermal cells take in food particles by phagocytosis, and hydrolysis of macromolecules occurs by intracellular digestion within food vacuoles. Undigested materials are expelled through the mouth/anus. Flatworms, such as planaria, also digest their food in gastrovascular cavities.

Digestion in Alimentary Canals More complex animals have **complete digestive tracts** or **alimentary canals**, with two openings, a mouth and an anus, and specialized regions allowing for the sequential digestion and absorption of nutrients. Food, ingested through the mouth and pharynx, passes through an esophagus that leads to either a crop, stomach, or gizzard—organs specialized for storing or grinding food. In the intestine, digestive enzymes hydrolyze macromolecules, and nutrients are absorbed across the tube lining. Undigested material exits through the **anus**.

■ **INTERACTIVE QUESTION 37.2**

Label and list the functions for the five organs of an earthworm's digestive tract. What is the function of the typhlosole?

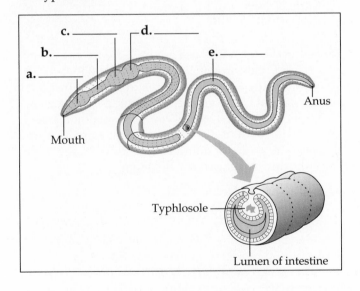

c. _____ d. _____

b. _____

a. _____

e. _____

Anus

Mouth

Typhlosole

Lumen of intestine

■ **A tour of the mammalian digestive system**

The mammalian digestive tract has a four-layered wall: an inner mucosa, a connective-tissue layer, smooth muscle, and connective tissue attached to the body cavity membrane. Rhythmic waves of muscular contraction called **peristalsis** push food through the tract. Ringlike valves called **sphincters** regulate the passage of material between some segments.

Accessory glands, the **salivary glands**, **pancreas**, and **liver** with its **gall bladder**, deliver digestive enzymes to the alimentary canal through ducts. The following paragraphs describe the human digestive system.

The Oral Cavity Physical and chemical digestion begins in the mouth, where teeth grind food to expose a greater surface area to enzyme action. The presence of food in the **oral cavity** triggers the release of saliva. Saliva has several components: mucin, a glycoprotein that protects the mouth lining from abrasion and lubricates food for swallowing; buffers to neutralize acidity; antibacterial agents; and **salivary amylase**, which begins the hydrolysis of starch and glycogen into smaller polysaccharides and maltose. The tongue is used to taste, to manipulate food, and to push the food ball, or **bolus**, into the pharynx for swallowing.

The Pharynx The **pharynx** is the intersection leading to both the esophagus and trachea. During swallowing, the top of the windpipe moves so that its opening is blocked by the cartilaginous **epiglottis**.

The Esophagus Food moves down through the narrow, flexible **esophagus** to the stomach, squeezed along by a wave of smooth muscle contraction called peristalsis.

The Stomach The epithelium lining the stomach secretes **gastric juice**, a digestive fluid containing hydrochloric acid that breaks down food tissues, kills bacteria, and denatures proteins, and **pepsin**, an enzyme that hydrolyzes specific peptide bonds in proteins. Pepsin is synthesized and secreted in an inactive form called **pepsinogen**. It is activated by hydrochloric acid and by pepsin itself—an example of positive feedback. Protein-digesting enzymes are often secreted in inactive forms called **zymogens**.

The mucous coating secreted by the epithelium protects the stomach lining from digestion. Gastric ulcers, mainly caused by bacteria, may worsen when the lining is eroded faster than it can be regenerated.

The sight, smell, or taste of food sends a nervous message from the brain to the stomach that initiates secretion of gastric juice. Food then stimulates the stomach wall to release the hormone **gastrin** into the circulatory system. This hormone stimulates further secretion of gastric juice. If the pH of the stomach contents becomes too low, the release of gastrin is inhibited.

Smooth muscles mix the contents of the stomach. **Acid chyme** is the nutrient broth produced by the action of the stomach and its secretions on ingested food. The stomach is usually closed off by two sphincters: one at the cardiac orifice prevents backflow into the esophagus, and a **pyloric sphincter** regulates passage of acid chyme into the intestine.

■ INTERACTIVE QUESTION 37.3

List the two types of macromolecules that have been partially digested by the time acid chyme moves into the intestine. Where did this digestion take place and what enzymes were involved?

a.

b.

The Small Intestine   Most enzymatic hydrolysis of macromolecules and nutrient absorption into the blood takes place in the **small intestine**, the longest section of the alimentary canal.

The pancreas produces hydrolytic enzymes and a bicarbonate-rich alkaline solution that offsets the acidity of the chyme. The liver produces **bile**, which is stored in the gallbladder until needed. Bile aids in the digestion of fats and contains pigments that are byproducts of the breakdown of red blood cells in the liver.

Bile and digestive juices (from the pancreas and gland cells of the intestinal wall) are mixed with the chyme in the **duodenum**, the first section of the small intestine. Regulatory hormones coordinate the release of digestive secretions: **Secretin** is released by cells in the intestinal wall in response to the acidic pH of the chyme and stimulates the pancreas to release bicarbonate. **Cholecystokinin (CCK)** stimulates gallbladder contraction and the release of pancreatic enzymes. A fat-rich chyme causes the duodenum to release **enterogastrone**, a hormone that inhibits peristalsis in the stomach, slowing the entry of chyme into the duodenum.

The digestion of starch and glycogen into disaccharides is continued by pancreatic amylases. Disaccharidases, enzymes specific for hydrolysis of different disaccharides, are built into the membranes and extracellular matrix of epithelial cells, facilitating sugar absorption through the intestinal wall.

Protein digestion is completed in the small intestine by **trypsin** and **chymotrypsin**, enzymes specific for peptide bonds adjacent to certain amino acids; **carboxypeptidase**, which splits amino acids off the free carboxyl end; and **aminopeptidase**, which works from the amino end. **Dipeptidases** are attached to the intestinal lining and split small peptides. The pancreatic protein-digesting enzymes are secreted as zymogens and activated by the intestinal enzyme **enteropeptidase**.

Nucleases are a group of enzymes that hydrolyze DNA and RNA into their nucleotide monomers. Other enzymes dismantle nucleotides.

The digestion of fats is aided by bile salts, which coat or **emulsify** tiny fat droplets so they do not coalesce, leaving a greater surface area for **lipase** to hydrolyze the fat molecules.

Most digestion is completed while the chyme is still in the duodenum. The **jejunum** and **ileum** are regions of the small intestine specialized for nutrient absorption. Circular folds of the small intestine lining are covered with fingerlike projections called **villi**, on which the epithelial cells have microscopic extensions called **microvilli**. This **brush border** of microvilli creates a huge surface area adapted for absorption.

The core of each villus has a net of capillaries and a lymph vessel called a **lacteal**. Nutrients are absorbed across the epithelium of the villus and then across the single-celled walls of the capillaries or lacteal. Transport may be passive or active. Active transport of sodium into the intestinal lumen and its passive re-entry into epithelial cells leads to the cotransport of certain nutrients.

Amino acids and sugars enter capillaries and are carried to the liver by the bloodstream. Glycerol and fatty acids are absorbed by epithelial cells where they recombine to form fats and are mixed with cholesterol and coated with proteins to make tiny globules called **chylomicrons**. These packages are transported by exocytosis out of the epithelial cells and into a lacteal.

The nutrient-laden blood from the small intestine is carried directly to the liver by the large **hepatic portal vessel**. The liver interconverts molecules and regulates the nutrient content of the blood.

■ INTERACTIVE QUESTION 37.4

List the enzymes that hydrolyze the following macromolecules in the lumen of the intestine. Also indicate the enzymes that are attached to the brush border of the intestine.

a. polysaccharides

b. polypeptides

c. DNA, RNA

d. fats

The Large Intestine   The small intestine leads into the **large intestine**, or **colon**, at a junction with a sphincter. A pouch called the **cecum**, with a fingerlike extension, the **appendix**, attaches at this juncture. The colon finishes the reabsorption of the large quantity of water secreted with digestive enzymes into the digestive tract.

Escherichia coli and other mostly harmless bacteria live on organic material in the **feces**. Some of these bacteria produce vitamin K that is absorbed by the host. The feces contain cellulose, other undigested ingredients of food, salts excreted by the colon, and a large proportion of intestinal bacteria. Feces are stored in the **rectum**. A voluntary and an involuntary sphincter between the rectum and anus control the elimination of feces, which is initiated by strong contractions of the colon.

■ Vertebrate digestive systems exhibit many evolutionary adaptations associated with diet

Dentition, the type and arrangement of teeth, correlates with diet.

Herbivores have longer alimentary canals because plant material is more difficult to digest than meat and contains less concentrated nutrients. Specialized structures, such as the spiral valve in a shark's intestine, functionally increase intestinal length by providing additional surface area.

Many herbivorous mammals have special fermentation chambers filled with symbiotic bacteria and protozoa. These microorganisms, often housed in the cecum, digest cellulose into simple sugars and produce a variety of essential nutrients for the animal.

■ INTERACTIVE QUESTION   37.5

Label the indicated structures in this diagram of the human digestive system.

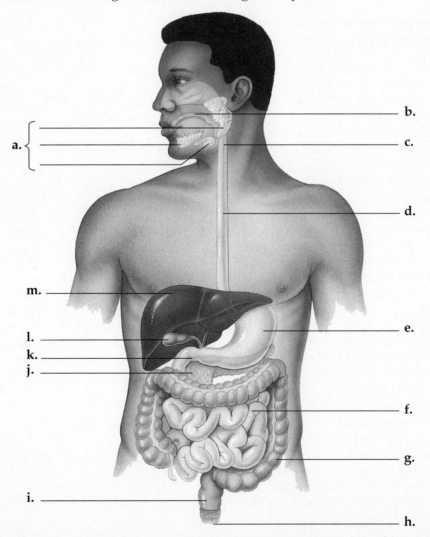

Rabbits and rodents ingest some of their feces so that the nutrients produced by the symbiotic bacteria in the large intestine may be absorbed.

Ruminants have the most elaborate adaptations for an herbivorous diet. The stomach is divided into four chambers, two of which contain symbiotic bacteria that digest cellulose. The cud is regurgitated, rechewed, and swallowed into the other chambers, where both the microbially digested cellulose and the microorganisms themselves are digested and their nutrients absorbed.

■ INTERACTIVE QUESTION 37.6

Compare and contrast the dentition and alimentary canals of carnivores and herbivores.

Eight of the 20 amino acids required to make proteins are **essential amino acids** in the adult human diet. Protein deficiency develops from a diet that lacks one or more essential amino acids. In Africa, the resulting syndrome of retarded physical and mental development in children is called **kwashiorkor.**

Meat, eggs, and cheese contain complete proteins with all essential amino acids in proportions that meet human requirements. Most plant proteins are incomplete, and diets built on a single staple, such as corn, beans, or rice, can result in protein deficiency.

■ INTERACTIVE QUESTION 37.7

How can vegetarians avoid protein deficiencies? Explain why such dietary consideration is necessary.

■ An adequate diet provides fuel, carbon skeletons for biosynthesis, and essential nutrients

Food as Fuel   The monomers of carbohydrates, fats, and proteins can be used as fuel for cellular respiration, although the first two are used preferentially.

When an animal consumes more calories than are needed to meet its energy requirements, excess calories are stored in the liver and muscles as glycogen, or in adipose tissue as fat when the glycogen stores are full.

An **undernourished** person or other animal has a diet deficient in calories. With severe deficiency, the body breaks down its own proteins for energy, eventually causing irreversible damage.

Overnourishment, or obesity, increases the risk of heart attack, diabetes, and other disorders. Successful weight control requires balancing caloric intake with caloric demand.

Food for Biosynthesis   Animals can fabricate most of the organic molecules they need by using enzymes to rearrange the carbon skeletons and organic nitrogen acquired from food. In vertebrates, the interconversion of organic molecules occurs mainly in the liver.

Essential Nutrients   Molecules that an animal requires but cannot make are called **essential nutrients.** These requirements vary from species to species. When the diet is missing one or more essential nutrients, the animal is said to be **malnourished.**

Animals are able to make most of the fatty acids they need. Linoleic acid, an unsaturated fatty acid used to make some membrane phospholipids, is required in the human diet. Deficiencies of **essential fatty acids** are rare.

Vitamins are essential organic molecules required in small amounts in the diet. The first vitamin isolated was thiamine, a deficiency of which results in the disease beriberi.

Thirteen vitamins essential to humans have been identified. Water-soluble vitamins include the B-complex, most of which function as coenzymes in key metabolic processes, and vitamin C, required for production of connective tissue. The fat-soluble vitamins are A, incorporated into visual pigments; D, aiding in calcium absorption and bone formation; E, seeming to protect phospholipids in membranes from oxidation; and K, required for blood clotting.

Minerals are inorganic nutrients, usually needed in very small amounts. Vertebrates require relatively large quantities of calcium and phosphorus for bone construction. Calcium is also needed for normal nerve and muscle function, and phosphorus is needed for ATP and nucleic acids. Iron is a component of the cytochromes and hemoglobin. Other minerals function as cofactors of enzymes. Iodine is needed by vertebrates to make the metabolism-regulating hormone, thyroxin. Sodium, potassium, and chlorine are important in nerve function and osmotic balance.

STRUCTURE YOUR KNOWLEDGE

1. Food provides fuel, organic raw materials, and essential nutrients. Complete the following concept map that summarizes the nutritional needs of animals.

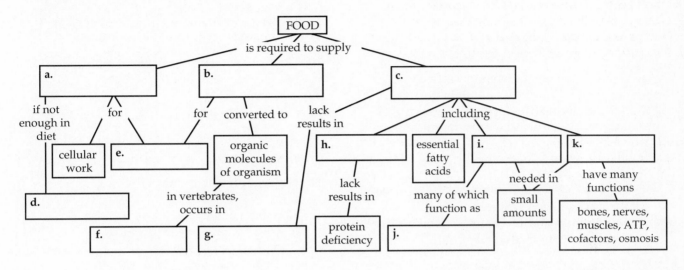

TEST YOUR KNOWLEDGE

MATCHING: *Match the description with the correct enzyme or hormone.*

_____ 1. enzyme that hydrolyzes peptide bonds, works in the stomach

_____ 2. hormone that stimulates secretion of gastric juice

_____ 3. hormone that stimulates release of bile and pancreatic enzymes

_____ 4. enzyme that begins digestion of starch in the mouth

_____ 5. enzymes specific for hydrolyzing various disaccharides

_____ 6. hormone that inhibits peristalsis in the stomach

_____ 7. enzyme that hydrolyzes fats

A. aminopeptidase

B. amylase

C. bile salts

D. cholecystokinin

E. chymotrypsin

F. disaccharidases

G. enterogastrone

H. enteropeptidase

I. gastrin

J lipase

K. maltase

L. nuclease

M. pepsin

N. secretin

_____ 8. intestinal enzyme that activates zymogens

_____ 9. hormone that stimulates secretion of bicarbonate ions from the pancreas

_____ 10. enzyme specific for peptide bonds adjacent to specific amino acids, works in the duodenum

MULTIPLE CHOICE: *Choose the one best answer.*

1. Deposit-feeders
   a. feed mostly on mineral substrates.
   b. filter small organisms from water.
   c. eat autotrophs.
   d. feed on detritus.
   e. live in or on their food, eating their way through.

2. The energy content of fats
   a. is released by bile salts.
   b. may be lost unless an herbivore eats some of its feces.
   c. is approximately two times that of carbohydrate or protein.
   d. can reverse the effects of malnutrition.
   e. Both c and d are correct.

3. Zymogens are
   a. hormones that stimulate release of pancreatic enzymes.
   b. enzymes attached to the brush border that hydrolyze nucleosides.
   c. hydrolytic enzymes manufactured in inactive forms to protect the cells that produce them.
   d. protein-digesting enzymes that are activated by hydrochloric acid.
   e. mucus-secreting cells of the stomach mucosa.

4. Which of the following statements is *false*?
   a. The average human has enough stored glycogen to supply calories for several weeks.
   b. Eating less and/or exercising more can result in weight loss.
   c. Conversion of glucose and glycogen takes place in the liver.
   d. Excessive calories are stored as fat, regardless of their original food source.
   e. Carbohydrates and fats are preferentially used as fuel before proteins are used.

5. The purpose of antidiarrhea medicine would most likely be to
   a. speed up peristalsis in the small intestine.
   b. speed up peristalsis in the large intestine.
   c. kill *E. coli* in the intestine.
   d. increase water reabsorption in the large intestine.
   e. increase salt secretion into the feces.

6. Incomplete proteins are
   a. lacking in essential vitamins.
   b. a cause of undernourishment.
   c. found in meat, eggs, and cheese.
   d. lacking in one or more essential amino acids.
   e. a result of overcooking vegetables.

7. Vitamins
   a. may be produced by intestinal microorganisms.
   b. are consistent from one species to the next.
   c. are stored in the liver.
   d. are inorganic nutrients, needed in small amounts, that usually function as cofactors.
   e. are all water soluble and must be replaced every day.

8. Which of the following is more likely to be found in animals with gastrovascular cavities than in those with alimentary canals?
   a. a large surface area for absorption of predigested nutrients
   b. extracellular digestion
   c. intracellular digestion
   d. suspension feeding
   e. a sac for storing food called a crop

9. Ruminants
   a. have teeth adapted for an omnivorous diet.
   b. use microorganisms to digest cellulose.
   c. eat their feces to obtain nutrients digested from cellulose by microorganisms.
   d. house symbiotic bacteria and microorganisms in a cecum.
   e. get all of their nutrition from digested plant material.

10. The brush border of the small intestine is formed by
    a. circular folds.
    b. cilia.
    c. cecae.
    d. villi.
    e. microvilli.

11. The acid pH of the stomach
    a. hydrolyzes proteins.
    b. is regulated by the release of gastrin.
    c. is neutralized by gastric juice.
    d. is produced by pepsin.
    e. triggers the release of enterogastrone.

12. After a meal of greasy french fries, which hormones and enzymes would you expect to be most active?
    a. salivary and pancreatic amylase, disaccharidases
    b. bile salts, lipase, enterogastrone
    c. pepsin, trypsin, chymotrypsin, amino- and carboxypeptidases, dipeptidases, CCK
    d. both a and b
    e. a, b, and c

13. Chylomicrons are
    a. lipoproteins transported by the circulatory system.
    b. small branches of the lymphatic system.
    c. protein-coated fat globules excreted out of epithelial cells into a lacteal.
    d. small peptides acted upon by chymotrypsin.
    e. fats emulsified by bile salts.

14. Which of the following is *not* a common component of feces?
    a. intestinal bacteria
    b. cellulose
    c. saturated fats
    d. bile pigments
    e. salts

15. The hepatic portal vessel
    a. supplies the capillaries of the intestines.
    b. carries absorbed nutrients to the liver for processing.
    c. carries blood from the liver to the heart.
    d. drains the lacteals of the villi.
    e. supplies oxygenated blood to the liver.

CHAPTER 38

CIRCULATION AND GAS EXCHANGE

FRAMEWORK

This chapter surveys the basic approaches to circulation and gas exchange found in the animal kingdom, with special attention to the human systems and the problems of cardiovascular disease. The following concept map organizes some of the chapter's key ideas.

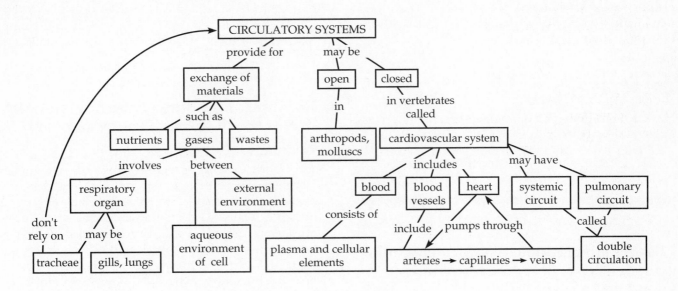

CHAPTER REVIEW

Every living cell must reside in an aqueous environment, which provides oxygen and nutrients and permits disposal of carbon dioxide and metabolic wastes.

■ Transport systems functionally connect body cells with the organs of exchange: *an overview*

The circulatory system provides an internal transport system to move substances across macroscopic distances in animals. Blood exchanges materials with the external environment across the thin and extensive epithelia of organs specialized for gas exchange, nutrient absorption, and waste removal and then carries materials throughout the body for delivery to the interstitial fluid that bathes the body cells.

Most invertebrates have a gastrovascular cavity or a circulatory system for internal transport

Gastrovascular Cavities   The central gastrovascular cavity inside the two-cell-thick body wall of cnidarians serves for both digestion and transport of materials. The internal fluid exchanges directly with the aqueous environment through the single opening. Flatworms also have gastrovascular cavities that branch throughout the body.

Open and Closed Circulatory Systems   In the **open circulatory system** found in insects, other arthropods, and in most mollusks, **hemolymph** in **sinuses**, or spaces between organs, bathes the internal tissues.

Annelids, some mollusks, and vertebrates have **closed circulatory systems**, in which the blood remains in vessels and exchanges materials with the interstitial fluid bathing the cells.

■ INTERACTIVE QUESTION   38.1

How is the hemolymph of an open circulatory system moved throughout the body?

Diverse adaptations of a cardiovascular system have evolved in vertebrates

The closed circulatory system of vertebrates, also called the **cardiovascular system**, consists of the heart, blood vessels, and blood. The heart has one or more **atria**, which receive blood, and one or more **ventricles**, which pump blood out of the heart. **Arteries**, carrying blood away from the heart, branch into tiny **arterioles** within organs, which then divide into the microscopic **capillaries**. Exchange of substances between blood and interstitial fluid occurs within **capillary beds**. Capillaries converge to form **venules**, which meet to form the **veins** that return blood to the heart.

The ventricle of a fish's two-chambered heart pumps blood first to the capillary beds of the gills, from which the oxygen-rich blood flows through a vessel to the systemic capillary beds in the other organs. Veins return the oxygen-poor blood to the atrium. Passage through two capillary beds slows the flow of blood, but body movements help to maintain circulation.

In the three-chambered heart of most amphibians, the single ventricle pumps blood through a forked artery into the **pulmonary circuit**, which leads to lungs and skin and then back to the left atrium, and the **systemic circuit**, which carries blood to the rest of the body and back to the right atrium. This **double circulation** repumps blood after it returns from the capillary beds of the lungs, ensuring a strong flow of oxygen-rich blood to the brain, muscles, and body organs. A ridge in the ventricle diverts most of the oxygen-rich blood from the left atrium into the systemic circuit and most of the oxygen-poor blood from the right atrium into the pulmonary circuit. The three-chambered reptilian heart has a septum that partially divides the single ventricle. In crocodiles, the septum completely divides the ventricle into two chambers.

Delivery of oxygen for cellular respiration is most efficient in birds and mammals, which, as endotherms, have high oxygen demands. The left side of the four-chambered heart handles only oxygen-rich blood, whereas the right side receives and pumps oxygen-poor blood. (See Interactive Question 38.2, p. 269.)

Rhythmic pumping of the human heart drives blood through pulmonary and systemic circuits

The Heart: General Form and Function   The human heart is enclosed in a two-layered, fluid-filled sac just beneath the sternum. The atria have relatively thin muscular walls, whereas the ventricles have thicker muscular walls. The **cardiac cycle** consists of the **systole**, during which the cardiac muscle contracts and the chambers pump blood, and the **diastole**, when the heart chambers are relaxed and filling with blood.

Atrioventricular valves between each atrium and ventricle are snapped shut when blood is forced back against them as the ventricles contract. The **semilunar valves** at the exit of the aorta and pulmonary artery are forced open by ventricular contraction and close when the ventricles relax and blood is forced back against them by the recoil of the elastic walls of the arteries. The heart-beat sounds are caused by the closing of these valves. A heart murmur is the detectable hissing sound of blood leaking back through a defective valve.

Heart rate can be measured by taking the **pulse**. Studies indicate that human heart rate normally fluctuates a great deal. The inverse relationship between

■ **INTERACTIVE QUESTION   38.2**

The diagram below will help you review the flow of blood through a mammalian circulatory system. Label the indicated parts, color the vessels that carry oxygen-rich blood red, and then trace the flow of blood by numbering the circles from 1–11. Start with number 1 in the right ventricle.

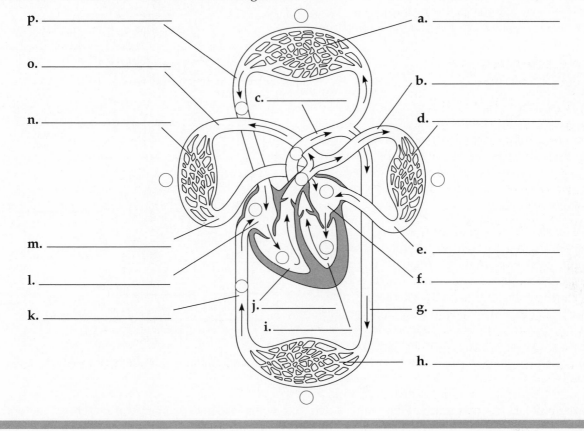

p. _____

o. _____

n. _____

m. _____

l. _____

k. _____

c. _____

j. _____

i. _____

a. _____

b. _____

d. _____

e. _____

f. _____

g. _____

h. _____

size and heart rate is attributable to the higher metabolic rate per gram of tissue of smaller animals.

The **cardiac output**, or volume of blood pumped per minute into the systemic circuit, depends on heart rate and **stroke volume**, or quantity of blood pumped by each contraction of the left ventricle.

Control of the Heart   Cells of cardiac muscle are self-excitable or myogenic; they have an intrinsic ability to contract. The rhythm of contractions is coordinated by the **sinoatrial (SA) node**, or **pacemaker**, a region of specialized muscle tissue located in the wall of the right atrium near the superior vena cava. The SA node initiates an excitation impulse that spreads via the intercalated discs of the cardiac muscle cells, and the two atria contract. The **atrioventricular (AV) node**, located between the right atrium and ventricle, relays the impulse after a 0.1-second delay to the ventricles. The electrical currents produced during the heart

cycle can be detected by electrodes placed on the skin and recorded in an **electrocardiogram (EKG or ECG)**. The SA node is controlled by two sets of nerves with antagonistic signals and is influenced by hormones, temperature, and exercise.

■ **INTERACTIVE QUESTION   38.3**

a. The valve between an atrium and ventricle is

_____.

b. The valve between a ventricle and the aorta or pulmonary artery is_____.

c. The node that directly controls contraction of the atria is_____.

d. The node that directly controls contraction of the ventricles is _____.

Blood Vessel Structure   The wall of an artery or vein consists of three layers: an outer connective tissue zone with elastic fibers; a middle layer of smooth muscle and more elastic fibers; and an inner lining of **endothelium**, a simple squamous epithelium. The outer two layers are thicker in arteries, which must be stronger and more elastic. Capillaries have only the endothelial layer.

Velocity of Blood Flow   The flow of blood decelerates with the increasing cross-sectional area of the branching vessel system. The enormous number of capillaries creates a large total diameter, and friction is greater than in larger vessels; as a result, blood flow is very slow in capillaries, improving opportunity for exchange with interstitial fluid. Blood flow speeds up within venules and veins, due to the decrease in total cross-sectional area.

Blood Pressure   **Blood pressure**, the hydrostatic force exerted against the wall of a blood vessel, is much greater in arteries than in veins and is greatest during systole. This hydrostatic pressure drives blood from the heart to the capillary beds. **Peripheral resistance**, caused by the narrow openings of the arterioles impeding the exit of blood from arteries, causes the swelling of the arteries during systole. The snapping back of the elastic arteries during diastole maintains a continuous blood flow into arterioles and capillaries. Arterial blood pressure may be measured with a sphygmomanometer. The first number is the pressure during systole (when blood first spurts through the artery that was closed off by the cuff), and the second is the pressure during diastole (when blood flows smoothly through the artery).

Together, cardiac output and peripheral resistance determine blood pressure. Contraction of smooth muscles in arteriole walls increases resistance and thus increases blood pressure, whereas dilation of arterioles as smooth muscles relax lowers blood pressure. Nervous and hormonal signals control these muscles.

Blood pressure drops to almost zero beyond the capillary beds. The one-way valves in veins and the contraction of skeletal muscles between which veins are embedded force blood to flow back to the heart. Pressure changes during breathing also draw blood into the large veins in the thoracic cavity.

Blood Flow Through Capillary Beds   Only about 5% to 10% of the body's capillaries have blood flowing through them at any one time. Capillaries branch off thoroughfare channels that are direct connections between arterioles and venules. The distribution of blood to capillary beds varies with need and is regulated by contraction of smooth muscle in arteriole

■ **INTERACTIVE QUESTION   38.4**

Complete the following concept map to help you organize your understanding of blood pressure.

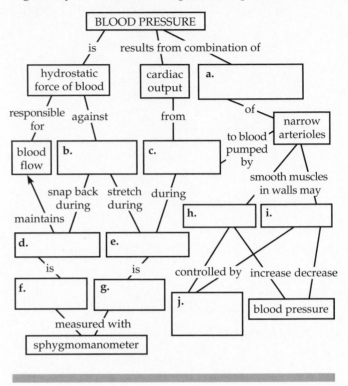

walls and precapillary sphincters at the entrance to capillary beds.

Capillary Exchange   The exchange of substances between blood and interstitial fluid may involve transport by endocytosis and exocytosis by the endothelial cells, passive diffusion of small molecules, and bulk flow through the clefts between cells due to hydrostatic pressure. Water, sugars, salts, oxygen, and urea move through capillary clefts. Blood cells and proteins are too large to pass easily through the endothelium. Blood pressure at the upstream end of a capillary forces fluid out, whereas osmotic pressure at the downstream end tends to draw about 85% of that fluid back into the capillary.

■ The lymphatic system returns fluid to the blood and aids in body defense

The fluid that does not return to the capillary and any proteins that may have leaked out are returned to the blood through the **lymphatic system**. Fluid diffuses into lymph capillaries intermingled in the blood capillary net. The fluid, called **lymph**, moves through lymph vessels with one-way valves, mainly as a result of the movement of skeletal muscles, and is returned to the circulatory system at two locations near the shoulders. In **lymph nodes**, the lymph is filtered, and white blood cells attack viruses and bacteria. Lymph capillaries called lacteals transport fats from the small intestine to the circulatory system.

▓ INTERACTIVE QUESTION   38.5

a. When the lymphatic system does not return sufficient interstitial fluid to the circulatory system, fluid accumulates in tissues, creating a condition known as _____.

b. What is a common cause of this condition?

■ Blood is a connective tissue with cells suspended in plasma

Blood is a type of connective tissue with cells in a liquid matrix called **plasma**.

Plasma   The plasma consists of a large variety of solutes dissolved in water. The collective and individual concentrations of electrolytes, or inorganic salts in the form of dissolved ions, are important to osmotic balance, pH levels, and the functioning of muscles and nerves. The kidney is responsible for the homeostatic regulation of ion concentrations.

Plasma proteins function in the osmotic balance of the blood and as buffers, antibodies, escorts for lipids, and clotting factors or fibrinogens. Blood plasma from which clotting factors have been removed is called serum. Nutrients, metabolic wastes, gases, and hormones are transported in the plasma. Except for its higher protein concentration, plasma is very similar in composition to interstitial fluid.

Cellular Elements   The **red blood cells** of mammals lack nuclei, and all **erythrocytes** lack mitochondria and generate their ATP by anaerobic metabolism. Their small size and biconcave shape create a large surface area of plasma membrane across which oxygen can diffuse. Erythrocytes are packed with **hemoglobin**, an iron-containing protein that binds oxygen.

Erythrocytes are formed in the red marrow of bones, and their production is controlled by a negative feedback mechanism involving the hormone **erythropoietin**, which is secreted by the kidney in response to low oxygen supply in tissues.

There are five major types of **leukocytes**, or **white blood cells**, all of which fight infections. Some white cells are phagocytes that engulf bacteria and cellular debris; some are lymphocytes that give rise to cells that produce the immune response. The majority of leukocytes are located in interstitial fluid and lymph nodes.

Platelets, pinched-off fragments of large cells in the bone marrow, are involved in the blood-clotting mechanism.

The Formation of Blood Cells and Platelets   Erythrocytes circulate for about 3 to 4 months before they are phagocytized by cells in the liver and spleen and their components recycled. Red and white blood cells and platelets are produced from **multipotent stem cells** in red bone marrow.

Blood Clotting   The clotting process usually begins when platelets, clumped together along a damaged endothelium, release clotting factors. By a series of steps, the plasma protein **fibrinogen** is converted to its active form, **fibrin**. The threads of fibrin form a patch. An inherited defect in any step of the complex clotting process causes **hemophilia**, a disease characterized by prolonged bleeding from even minor injuries. Anticlotting factors normally prevent clotting of blood in the absence of injury. A **thrombus** is a clot that occurs within a blood vessel and blocks the flow of blood. (See Interactive Question 38.6, p. 272.)

■ Cardiovascular diseases are the leading cause of death in the United States and many other developed nations

More than half of all deaths in the United States are caused by **cardiovascular disease**. Blockage of coronary arteries leads to the death of cardiac muscle in a **heart attack**; blockage of arteries in the head leads to a **stroke**, the death of nervous tissue in the brain. A heart attack or stroke may result from blockage by a thrombus or an embolus, which is a clot formed elsewhere that moves through the circulatory system.

Most heart-attack and stroke victims had been suffering from a chronic cardiovascular disease known as **atherosclerosis**, in which growths called plaques develop within arteries and narrow the vessels. Plaques that become hardened by calcium deposits result in **arteriosclerosis**, or hardening of the arteries. An embolus is more likely to be trapped in narrowed vessels, and plaques are common sites of thrombus formation. Occasional chest pains, known as angina pectoris, may warn that a coronary artery is partially blocked.

■ **INTERACTIVE QUESTION   38.6**

Complete this concept map of the components of vertebrate blood and their functions.

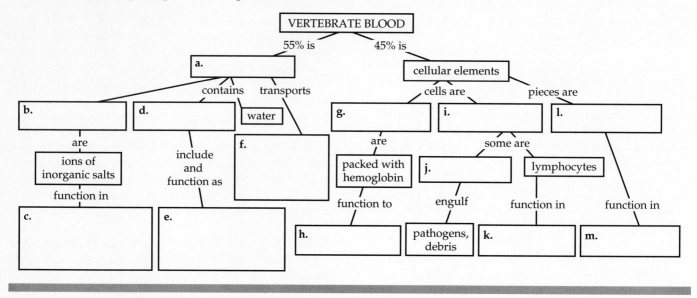

Hypertension, or high blood pressure, is thought to damage the endothelium and initiate plaque formation, promoting atherosclerosis and increasing the risk of heart attack and stroke. This condition can be easily diagnosed and controlled by drugs, diet, and exercise. Hypertension and atherosclerosis tend to be inherited.

The cholesterol level in blood plasma is an indicator of potential atherosclerosis. Most cholesterol is carried as **low-density lipoproteins (LDLs)**. **High-density lipoproteins (HDLs)** are another form of cholesterol carriers that appear to reduce cholesterol deposition in plaques. A high ratio of LDL to HDL is an indication of potential cardiovascular disease.

The death rate from cardiovascular disease has been significantly declining in the United States. Diagnosis and treatment of hypertension, improved intensive care for cardiovascular patients, and healthier lifestyles may be contributing to this trend.

■ **INTERACTIVE QUESTION   38.7**

What lifestyle choices have been correlated with increased risk of cardiovascular disease?

■ Gas exchange supplies oxygen for cellular respiration and disposes of carbon dioxide

Gas Exchange and the Respiratory Medium   The **respiratory medium** is air for a terrestrial animal and water for an aquatic one. The **respiratory surface**, where gas exchange with the respiratory medium occurs, must be moist and large enough to supply the whole body.

Respiratory Surfaces: Form and Function   A localized region of the body surface is usually specialized as a respiratory surface with a thin, moist epithelium separating a rich blood supply and the respiratory medium. When the entire outer skin serves as a respiratory organ, as in an earthworm, the animal must live in damp places and have a relatively small, long or flattened body to provide a high ratio of surface area to volume. Most animals use an extensively branched or folded localized region of the body surface for gas exchange, which is usually external in aquatic animals and internal in terrestrial animals.

Gills: Respiratory Adaptations of Aquatic Animals   **Gills** are outfoldings of the body surface, ranging from simple bumps on echinoderms to the more complex and often localized gills of annelids, mollusks, crustaceans, fishes, and tadpoles. Due to the low oxygen concentration in a water environment, gills usually require **ventilation**, or movement of the respiratory medium across the respiratory surface—often an energy-intensive process.

In the gills of a fish, blood flows through capillaries in a direction opposite to the flow of water. This arrangement sets up a **countercurrent exchange**, in which the diffusion gradient favors the movement of oxygen into the blood throughout the length of the capillary.

Tracheae: Respiratory Adaptations of Insects The advantages of air as a respiratory medium include its higher concentration of oxygen, faster diffusion rate of O_2 and CO_2, and lower density. To prevent the disadvantageous water loss from large, moist surfaces, the respiratory surfaces of most terrestrial organisms are invaginated.

The **tracheae** of insects are tiny air tubes that branch throughout the body to come into contact with nearly every cell. Thus, the open circulatory system does not function in O_2 and CO_2 transport. Rhythmic body movements of large insects and the contraction and relaxation of flight muscles serve to ventilate tracheal systems.

■ Lungs are the respiratory adaptation of most terrestrial vertebrates

Lungs are invaginated respiratory surfaces restricted to one location from which oxygen is transported to the rest of the body by the circulatory system. Mammalian lungs have a spongy texture with a honeycombed moist epithelium that provides a large surface area for gas exchange.

Form and Function of Mammalian Respiratory Systems The lungs of mammals are located in the thoracic cavity and enclosed in a double-walled sac. Surface tension in a thin layer of fluid holds the two layers together and they adhere to the lungs and the chest-cavity wall.

Air, entering through the nostrils, is filtered, warmed, humidified, and smelled in the nasal cavity. Air passes through the pharynx and enters the windpipe through the glottis. This opening is pushed against the epiglottis when food is swallowed. The **larynx** functions as a voicebox in humans and many other mammals; exhaled air vibrates a pair of **vocal cords**. The **trachea**, or windpipe, branches into two **bronchi**, which then branch repeatedly into **bronchioles** within the lungs. Ciliated, mucus-coated epithelium lines much of the respiratory tree. Multilobed air sacs encased in a web of capillaries are at the tips of the tiniest bronchioles. Gas exchange takes place across the thin, moist epithelium of these **alveoli**.

Ventilating the Lungs Vertebrate lungs are ventilated by **breathing**, the alternate inhalation and exhalation of air. A frog uses **positive pressure breathing** by expanding its mouth cavity, which draws air into the mouth, and then raising the jaw, which pushes air into the lungs.

Mammals ventilate their lungs by **negative pressure breathing**. The volume of the lungs is increased by expansion of the rib cage. Air pressure is reduced within this increased volume, and air is drawn through the nostrils down to the lungs. Relaxation of the rib muscles compresses the lungs, increases the pressure, and forces air out. During shallow breathing, contraction and relaxation of the **diaphragm** is responsible for most of the decrease and increase in pressure that ventilate the lungs.

Tidal volume is the volume of air inhaled and exhaled by an animal during normal breathing. The maximum volume that can be inhaled and exhaled by forced breathing is called **vital capacity**. The **residual volume** is the air that remains in the alveoli and lungs after forceful exhaling.

Birds have air sacs that penetrate their abdomen, neck, and wings. The air sacs and lungs are ventilated through a circuit that includes a one-way passage through tiny channels, called **parabronchi**, in the lungs. Air sacs act as bellows to maintain air flow through the lungs and also lower the density of the bird.

■ INTERACTIVE QUESTION 38.8

For the following animals, indicate the respiratory medium, respiratory surface, and means of ventilation.

Animal	Respiratory Medium	Respiratory Surface	Ventilation
Fish	a.	b.	c.
Grasshopper	d.	e.	f.
Frog	g.	h.	i.
Human	j.	k.	l.

The Control of Breathing   Breathing is under automatic regulation by the **breathing control centers** in the medulla oblongata and the pons. Nerve impulses from the medulla instruct the rib muscles and diaphragm to contract. In a negative feedback loop, stretch sensors in the lungs respond to the expansion of the lungs and send nervous impulses that inhibit the breathing control center in the medulla. A drop in the pH of the blood and cerebrospinal fluid when CO_2 concentration increases is sensed by the medulla's control center, which increases the depth and rate of breathing. Although breathing centers respond primarily to CO_2 levels, they are alerted by oxygen sensors in key arteries that react to severe deficiencies of O_2.

Loading and Unloading of Oxygen and Carbon Dioxide   The concentration of gases in air or dissolved in water is measured as **partial pressure**. At sea level, the partial pressure of oxygen, which makes up 21% of the atmosphere, is 160 mm Hg (0.21×760 mm—atmospheric pressure at sea level). The partial pressure of CO_2 is 0.23 mm Hg. A gas will diffuse from a region of higher partial pressure to lower partial pressure. Blood entering the lungs has a lower P_{O_2} and a higher P_{CO_2} than does the air in the alveoli, so oxygen diffuses into the capillaries and carbon dioxide diffuses out. In the systemic capillaries, pressure differences favor the diffusion of O_2 out of the blood into the interstitial fluid and CO_2 into the blood.

Respiratory Pigments and Oxygen Transport   In most animals, O_2 is carried by **respiratory pigments** in the blood. Copper is the oxygen-binding component in the blue respiratory protein **hemocyanin**, common in arthropods and many mollusks.

Hemoglobin, usually located in red blood cells, is the respiratory pigment of almost all vertebrates. Hemoglobin is composed of four subunits, each of which has a cofactor called a heme group with iron at its center. The binding of O_2 to the iron atom of one subunit induces a change in shape of the other subunits, and their affinity for oxygen increases. Likewise, the unloading of the first O_2 effects a conformational change that lowers the other subunits' affinity for oxygen.

The **dissociation curve** for hemoglobin shows the relative amounts of oxygen bound to hemoglobin under varying oxygen partial pressures. In the steep part of this S-shaped curve, a slight change in partial pressure will cause hemoglobin to load or unload a substantial amount of oxygen.

Carbon Dioxide Transport   Over 20% of CO_2 is transported in blood bound to amino groups of hemoglobin. Seventy percent is transported as bicarbonate

The following graph shows dissociation curves for hemoglobin at two different pH values. Explain the significance of these two curves.

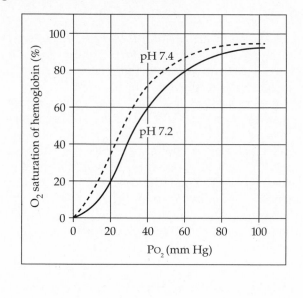

ions. Carbon dioxide enters into the red blood cells where it first reacts to form carbonic acid and then dissociates into H^+ and a bicarbonate ion. Bicarbonate moves into the plasma for transport. The hydrogen ions are bound to hemoglobin and other proteins, and the pH value of the blood is not greatly lowered during the transport of carbon dioxide. In the lungs, the diffusion of CO_2 out of the blood shifts the equilibrium in favor of the conversion of bicarbonate back to CO_2, which is unloaded from the blood.

Adaptation of Diving Mammals   Special physiological adaptations have enabled some air-breathing mammals to make sustained underwater dives. Weddell seals store twice the amount of O_2/kg body weight as do humans, mostly by having a larger volume of blood, a huge spleen that stores blood, and a higher concentration of **myoglobin**, an oxygen-storing muscle protein. A diving reflex slows the pulse of diving mammals, reroutes blood to the brain and essential organs, and restricts blood supply to the muscles, which use anaerobic respiration during long dives.

STRUCTURE YOUR KNOWLEDGE

1. Identify the labeled structures in this diagram of a human heart. Draw arrows to trace the flow of blood.

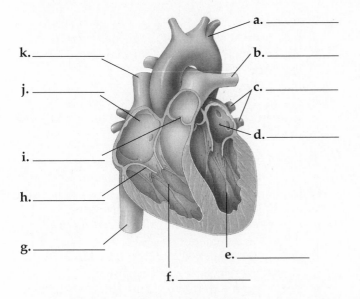

2. Trace the path of a molecule of oxygen being inhaled through the nasal cavity to delivery to the kidney.

TEST YOUR KNOWLEDGE

MULTIPLE CHOICE: *Choose the one best answer.*

1. A gastrovascular cavity
   a. is found in cnidarians and annelids.
   b. functions to pump fluid in short vessels from which it spreads throughout the body.
   c. functions in both digestion and distribution of nutrients.
   d. has a single opening for ingestion and elimination and a separate opening for gas exchange in larger animals.
   e. involves all of the above.

2. Which of the following is *not a similarity* between open and closed circulatory systems?
   a. Some sort of pumping device helps to move blood through the body.
   b. Some of the circulation of blood is a result of movements of the body.
   c. The blood and interstitial fluid are distinguishable from each other.
   d. All tissues come into close contact with the circulating body fluid so that the exchange of nutrients and wastes can take place.

   e. All of these apply to both open and closed circulatory systems.

3. In a system with double circulation,
   a. Blood is pumped at two separate locations as it circulates through the body.
   b. There is a countercurrent exchange within the gills.
   c. Hemolymph circulates both through a pumping blood vessel and through body sinuses.
   d. Blood is pumped to the gas-exchange organ and returning blood is then pumped through systemic vessels.
   e. There are always two atria and two ventricles.

4. During diastole,
   a. The atria fill with blood.
   b. Blood flows passively into the ventricles.
   c. The elastic recoil of the arteries maintains hydrostatic pressure on the blood.
   d. The semilunar valves are closed but the atrioventricular valves are open.
   e. All of the above are occurring.

5. An atrioventricular valve prevents the backflow or leakage of blood from
   a. the right ventricle into the right atrium.
   b. the left atrium into the left ventricle.
   c. the aorta into the left ventricle.
   d. the pulmonary vein into the right atrium.
   e. the right atrium into the vena cava.

6. Breathing rate will increase when ____ CO_2 in your blood causes a ____ in pH.
   a. increased/rise
   b. increased/drop
   c. decreased/rise
   d. decreased/drop
   e. It is the level of O_2 in the blood that normally signals changes in breathing rate.

7. Heart beat is initiated by the
   a. SA node.
   b. AV node.
   c. right atrium.
   d. superior and inferior vena cavae.
   e. left ventricle.

8. Blood flows more slowly in the arterioles than in the arteries because the arterioles
   a. have thoroughfare channels to venules that are often closed off.
   b. collectively have a larger cross-sectional area than do the arteries.
   c. must provide opportunity for exchange with the interstitial fluid.
   d. have sphincters that restrict flow to capillary beds.
   e. do or have all of the above.

9. If all the body's capillaries were open at the same time,
   a. Blood pressure would fall dramatically.
   b. Peripheral resistance would increase.
   c. Blood would move too rapidly through the capillary beds.
   d. The amount of blood returning to the heart would increase.
   e. The increased gas exchange would allow for strenuous exercise.

10. Which of the following is *not* a factor in the exchange of substances in capillary beds?
    a. endocytosis and exocytosis
    b. passive diffusion through clefts between endothelial cells
    c. hydrostatic pressure
    d. bulk flow through the apoplast
    e. osmotic pressure

11. Edema is a result of
    a. swollen lymph glands.
    b. a loss of interstitial fluid.
    c. not enough lymph being returned to the circulatory system.
    d. too high a concentration of blood proteins.
    e. swollen feet.

12. The functions of the kidney include
    a. control of the lymphatic system.
    b. maintenance of electrolyte balance.
    c. destruction and recycling of red blood cells.
    d. secretion of erythropoietin to increase production of red blood cells.
    e. both b and d.

13. Fibrinogen is
    a. a blood protein that escorts lipids through the circulatory system.
    b. a cell fragment involved in the blood-clotting mechanism.
    c. a blood protein that is converted to fibrin to form a blood clot.
    d. a leukocyte involved in trapping viruses and bacteria.
    e. a lymph protein that regulates osmotic balance in the tissues.

14. In countercurrent exchange,
    a. The flow of fluids in opposite directions maintains a favorable diffusion gradient along the length of an exchange surface.
    b. Oxygen is exchanged for carbon dioxide.
    c. Double circulation keeps oxygenated and deoxygenated blood separate.
    d. Oxygen moves from a region of high partial pressure to one of low partial pressure, but carbon dioxide moves in the opposite direction.
    e. The capillaries of the lung pick up a higher concentration of oxygen than in concurrent exchange.

15. A thrombus
    a. may form at a region of plaque in an artery.
    b. is a traveling embolism that may cause a heart attack or stroke.
    c. may cause hypertension.
    d. may calcify and result in arteriosclerosis.
    e. is often associated with high levels of HDLs in the blood.

16. The tracheae of insects
    a. are stiffened with chitinous rings and lead into the lungs.
    b. are filled by positive pressure breathing.
    c. ramify along capillaries of the circulatory system for gas exchange.
    d. are highly branched, coming into contact with almost every cell for gas exchange.
    e. provide an extensive, fluid-filled system of tubes for gas exchange.

17. Which of the following is *not* involved in speeding up breathing?
    a. a drop in the pH of the blood
    b. stretch receptors in the lungs
    c. impulses from the breathing centers in the medulla
    d. severe deficiencies of oxygen
    e. a drop in the pH of cerebrospinal fluid

18. The binding of an O_2 atom to the first iron atom of a hemoglobin molecule
    a. occurs at a very low partial pressure of oxygen.
    b. produces a conformational change that lowers the other subunits' affinity for oxygen.
    c. is an example of cooperativity as O_2 readily binds to the other subunits.
    d. occurs more readily at a lower pH value.
    e. is the signal for hemoglobin to release H^+ so that bicarbonate converts to CO_2.

19. Fluid is moved back into the capillaries at the downstream end of a capillary bed by
    a. active transport.
    b. hydrostatic pressure.
    c. osmosis.
    d. bulk transport.
    e. facilitated diffusion.

20. The dissociation curve for hemoglobin indicates that
  a. The percent saturation of hemoglobin decreases with the amount of oxygen to which it is bound.
  b. There is a range in which a slight change in P_{O_2} will cause hemoglobin to load or unload a large amount of oxygen.
  c. The P_{O_2} in the tissues is lower than that in the alveoli.
  d. A drop in pH value raises the affinity of hemoglobin for O_2.
  e. The fetal dissociation curve is shifted to the right of the maternal curve.

21. In which animal does blood flow through vessels from the respiratory organs to the rest of the body without returning to the heart?
  a. fish
  b. insect
  c. frog
  d. mammal
  e. both a and b

22. As a general rule, blood leaving the right ventricle of a mammal's heart will pass through how many capillary beds before it returns to the right ventricle?
  a. one
  b. two
  c. three
  d. one or two, depending on the circuit it takes
  e. at least two, but most often three

CHAPTER 39

THE BODY'S DEFENSES

FRAMEWORK

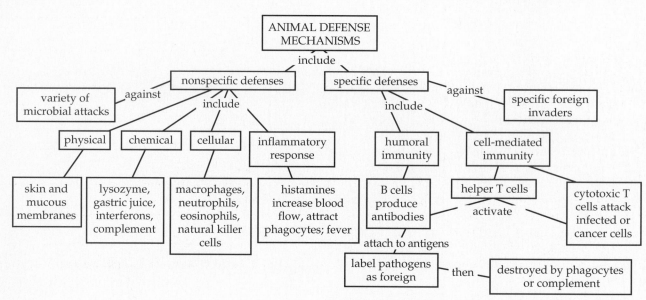

CHAPTER REVIEW

The defense system protects an animal from bacteria, viruses, and early-stage cancer cells. Nonspecific defense mechanisms include the physical barriers of the skin and mucous membranes, antimicrobial proteins, phagocytic cells, and the inflammatory response. The immune system consists of descendants of lymphocytes that react specifically to threat with the production of antibodies and various cell-mediated responses.

■ Nonspecific mechanisms provide general barriers to infection

The Skin and Mucous Membranes   Skin is a physical barrier to microbes, reinforced by oil and sweat secre-

tions with a low pH. **Lysozyme**, an enzyme that attacks bacterial cell walls, is present in tears, saliva, and mucus. The ciliated, mucus-coated epithelial lining of the respiratory tract traps and removes particles. The acidity of gastric juice kills most bacteria that reach the stomach.

Phagocytic White Cells and Natural Killer Cells
The major internal nonspecific defense is **phagocytosis** by white blood cells. Short-lived **neutrophils** leave the blood in response to chemical signals and phagocytize microbes in infected tissue. **Monocytes** migrate into tissues and develop into **macrophages**, large, long-lived amoeboid cells that engulf microbes and destroy them with digestive enzymes and reactive forms of oxygen. Some microbes have capsules that macrophages cannot grasp or are resistant to their lytic enzymes. Macrophages may migrate through the

body or become permanently attached in organs such as the lungs, liver, lymph nodes, and spleen.

Eosinophils are white blood cells that attack larger parasitic invaders with destructive enzymes. **Natural killer cells** destroy the body's infected or aberrant cells by attacking their membranes.

Antimicrobial Proteins   The **complement system** is a group of at least 20 proteins that cooperate with other nonspecific and specific defense mechanisms, resulting in lysis of microbes or attraction of phagocytes. **Interferons** are proteins produced by virus-infected cells that diffuse to neighboring cells, stimulating production of proteins that inhibit viral replication in those cells. One of the three known types of interferons, interferon-gamma, also activates phagocytes. Interferons produced by recombinant DNA technology are being tested for their effectiveness in treating viral infections and cancer.

The Inflammatory Response   Physical injury or microorganisms can trigger an **inflammatory response**, characterized by redness, swelling, and heat. Injured cells release chemical signals: **basophils** in the blood and **mast cells** in connective tissue release **histamine**, which triggers vasodilation and leakiness of blood vessels; white blood cells and damaged tissue cells release prostaglandins and other chemicals, which promote blood flow to the damaged area. Blood-clotting elements delivered to the area begin the repair process. Vasodilation and chemotaxis result in the congregation of phagocytic white blood cells in the interstitial fluid near a wound. Accumulated pus is a combination of dead cells and fluid from the capillaries.

A systemic inflammatory response to an infection may include an increase in the number of circulating white blood cells and a fever. Fever may be triggered by toxins produced by pathogens or by pyrogens released by certain white blood cells. Fevers stimulate phagocytosis and inhibit growth of microorganisms.

■ The immune system defends the body against specific invaders: *an overview*

Key Features of the Immune System   The **immune system** differs from nonspecific defenses on the basis of its specificity, diversity, self/nonself recognition, and memory.

The immune system mounts specific attacks against invaders when special lymphocytes recognize **antigens**, components of foreign molecules, and produce specific proteins called **antibodies**. The diversity of the immune system results from the huge variety of lymphocytes, each of which has receptors for a specif-

■ **INTERACTIVE QUESTION   39.1**

Complete the following table that summarizes the functions of the cells and compounds of the nonspecific defense system.

Cells or Compounds	Functions
Neutrophils	a.
Monocytes	b.
Macrophages	c.
Eosinophils	d.
Natural killer cells	e.
Basophils and mast cells	f.
Histamine	g.
Interferons	h.
Complement system	i.
Lysozyme	j.
Prostaglandins	k.
Pyrogens	l.

ic antigenic marker to which it responds. The immune system distinguishes self from nonself by recognizing foreign molecules and the body's own molecules. The immune system "remembers" antigens and can react against them more promptly upon reexposure, a property known as **acquired immunity**.

Active versus Passive Acquired Immunity   **Active immunity** can be acquired when the body produces antibodies and develops immunological memory from either exposure and recovery from an infectious disease, or from vaccination with an inactivated pathogen. In **passive immunity**, antibodies are supplied through the placenta to a fetus, through milk to a nursing infant, or by an antibody injection providing temporary immunity.

Humoral Immunity and Cell-Mediated Immunity   **Humoral immunity** involves the production of antibodies that circulate in the blood and lymph and

defend against free bacteria and viruses. **Cell-mediated immunity** involves specialized lymphocytes that react against body cells infected by bacteria or viruses and against fungi, protozoa, and worms. The cell-mediated branch also responds to tissue transplants and cancerous cells.

Cells of the Immune System   **B cells** (or B lymphocytes) produce antibodies; **T cells** (or T lymphocytes) function in cell-mediated immunity. Blood cells, including lymphocytes, develop from multipotent stem cells located in the bone marrow. Lymphocytes either migrate to the thymus and differentiate into T cells or continue to develop in the bone marrow as B cells.

B cells and T cells have specific **antigen receptors** on their surfaces that allow them to react to particular antigens. **T cell receptors** recognize antigens as specifically as do the membrane-bound antibodies that serve as receptors on B cells. When an antigen binds to a complementary antigenic receptor, the lymphocyte is activated to proliferate a clone of **effector cells**. Activated B cells give rise to antibody-producing **plasma cells**. T effector cells include **cytotoxic T cells** (T_c), which kill infected cells and cancer cells, and **helper T cells** (T_H), which stimulate humoral and cell-mediated immunity by secreting **cytokines**, molecules that regulate neighboring cells.

■ **INTERACTIVE QUESTION   39.2**

The immune system includes both humoral and cell-mediated immunity. Describe these specific defenses.

a. Humoral

b. Cell-mediated

■ **Clonal selection of lymphocytes is the cellular basis for immunological specificity and diversity**

The ability of the immune system to respond specifically to the enormous variety of different antigens depends on the presence of a great diversity of lymphocytes that are each destined during embryonic development to respond to a particular antigen. If such an antigen binds to its antigen receptor, the selected cell then gives rise to a clone of millions of effector cells. Through this **clonal selection**, the body mounts an immune response against a specific antigen.

■ **Memory cells function in secondary immune responses**

Five to 10 days after the first encounter with an antigen, enough B and T effector cells have proliferated to produce a **primary immune response**. If the body reencounters the same antigen, the **secondary immune response** is more rapid, effective, and longer in duration. This immunological memory is due to the production of **memory cells** along with effector cells. These long-lived cells are able to produce effector cells and more memory cells rapidly when they reencounter their antigen.

■ **Molecular markers on cell surfaces function in self/nonself recognition**

During embryonic development, lymphocytes carrying antigen receptors for molecules present in the body are destroyed, thus ensuring **self-tolerance**. The **major histocompatibility complex (MHC,** or **HLA** in humans) is a set of genes that code for glycoproteins embedded in the plasma membrane. This biochemical fingerprint, coded for by at least 20 MHC genes with as many as 100 alleles for each gene, is unique to each individual (except for identical twins). **Class I MHC** molecules are found on every nucleated body cell, whereas **class II MHC** markers are found on macrophages, B cells, and activated T cells, allowing these cells to interact with each other in immune responses.

■ **INTERACTIVE QUESTION   39.3**

What would happen if a lymphocyte carrying antigen receptors for its own class II MHC markers was not destroyed during embryonic development?

■ **In the humoral response, B cells defend against pathogens in body fluids by generating specific antibodies**

The Activation of B Cells   The humoral immune response involves the production of antibodies by plasma cells. Selective activation of a B cell occurs when an antigen binds to a membrane-bound antibody. This process may be assisted by macrophages called **antigen-presenting cells (APCs)** that engulf

pathogens and then display antigen fragments in a complex with their class II MHC molecules. A helper T cell's receptor recognizes and binds to a specific MHC–antigen complex, which activates the T cell to form a clone of helper T cells. B cells that have also encountered the antigen display antigen fragments in combination with a class II MHC molecule. Activated helper T cells bind with a B cell's MHC–antigen complex and secrete cytokines, which stimulate the B cell to proliferate into plasma cells and memory cells.

T-Dependent and T-Independent Antigens Antigens that require the cooperation of macrophages, helper T cells and B cells for antibody production are called **T-dependent antigens**.

T-independent antigens trigger antibody production by B cells without the help of macrophages or T cells. Usually found in bacterial capsules and flagella, these antigenic molecules are usually long chains of repeating units that apparently bind to sufficiently numerous antigen receptors on a B cell to activate the cell.

The Molecular Basis of Antigen–Antibody Specificity Most antigens are proteins or polysaccharides that are part of the capsule, cell wall, or coat of viruses or bacteria. Antigens also may be part of the surface membrane of transplanted cells. Antibodies recognize a localized region, called an antigenic determinant, or **epitope**, of an antigen. Antigens may have many different epitopes that stimulate different B cells to produce antibodies.

Antibodies are a class of proteins called **immunoglobulins (Igs)**. The Y-shaped antibody molecule consists of two pairs of polypeptide chains: two identical short light chains and two identical longer heavy chains, linked together by disulfide bridges. Both heavy and light chains have constant regions located in the tail of the Y, and variable regions at the ends of the two arms of the Y, which form two identical antigen-binding sites. The amino acid composition of the variable region creates the unique contours and specific antibody–antigen binding potential of these sites.

In mammals, there are five types of constant regions, creating five classes of antibodies: IgM, IgG, IgA, IgD, and IgE. The constant region of each class determines its function.

How Antibodies Work Antibodies label foreign molecules and cells for destruction by one of several effector mechanisms. Antibodies may cover the binding sites of a virus or neutralize a bacterial toxin by coating it. Because each antibody molecule has at least two antigen-binding sites, the formation of antigen–antibody complexes may cause bacteria to

Label the components in this diagram of the immune response to a T-dependent antigen.

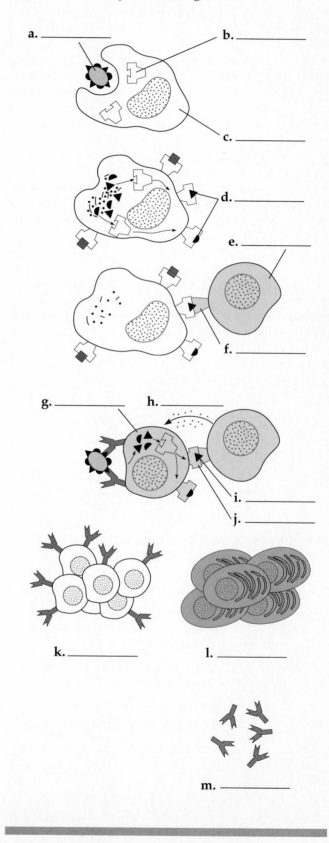

a. _____

b. _____

c. _____

d. _____

e. _____

f. _____

g. _____

h. _____

i. _____

j. _____

k. _____

l. _____

m. _____

agglutinate or soluble antigen molecules to precipitate. The resulting clumps are then engulfed by phagocytes. Antigen–antibody complexes may activate the complement system.

Monoclonal Antibody Technology   Antibodies obtained from the blood of an immunized animal are *polyclonal* because they were formed by several different B cell clones, each specific for a different epitope. A technique for making **monoclonal antibodies,** developed in 1975, can supply quantities of identical antibodies for biological research, clinical testing, and medical applications.

Monoclonal antibodies are produced from a culture of hybrid cells called **hybridomas,** made by fusing tumor cells, which can be cultured indefinitely, with antibody-producing plasma cells obtained from the spleen of animals that have been exposed to the antigen of interest.

■ **INTERACTIVE QUESTION   39.5**

List four ways in which antibodies inactivate antigens. Which of these enhance phagocytosis by macrophages?

a.

b.

c.

d.

■ In the cell-mediated response, T cells defend against intracellular pathogens

Whereas the humoral immune response reacts to extracellular pathogens, cell-mediated immunity acts against pathogens that have already entered cells of the body.

The Activation of T Cells   Helper T cells and cytotoxic T cells have membrane-bound T-cell receptors that recognize a self/nonself complex of a specific antigenic epitope and MHC markers. T_H-cell receptors recognize an antigen fragment presented by a class II MHC on the surface of a macrophage or B cell.

T_C-cell receptors recognize an antigen fragment and a class I MHC molecule. T cell surface molecules enhance the interaction with MHC–antigen complexes. CD4, found on T_H cells and CD8, found on T_C, have affinities for class II MHC and class I MHC molecules, respectively. T cells that recognize a specific MHC–antigen complex displayed by infected cells respond by proliferating clones of activated helper and cytotoxic T cells to fight that specific pathogen.

Helper T cells activate both humoral and cell-mediated responses by releasing cytokines. An antigen-presenting macrophage releases **interleukin-1,** which signals the T_H cell to release **interleukin-2,** a cytokine that stimulates helper T cells to grow more rapidly. Interleukin-2 and other cytokines also help to activate B cells and T_C cells.

How Cytotoxic T Cells Work   Cytotoxic T cells attach to infected body cells displaying a complex of a specific antigen and class I MHC markers, and they release *perforin*, a protein that forms a hole in the target cell's membrane and causes the cell to lyse. Pathogens are thus released and attacked by circulating antibodies. Cytotoxic T cells also attack cancer cells, which carry markers that identify them as nonself to the immune system.

Another type of lymphocyte, the **suppressor T cell (T_S),** probably suppresses the immune response when an antigen is no longer present.

■ **INTERACTIVE QUESTION   39.6**

a. What surface molecule of a helper T cell facilitates the interaction with a class II MHC of an APC and the helper T cell? _____

b. What surface molecule on a cytotoxic T cell assists in the interaction with class I MHC proteins displayed on infected cells? _____

c. What does an activated helper T cell release? _____

d. What does a cytotoxic T cell attached to an infected body cell release? _____

■ Complement proteins participate in both nonspecific and specific defenses

In the *classical pathway*, when antibody molecules target an invader such as a bacterial cell by attaching to it, a complement protein forms a bridge between two antibody molecules. This molecular association acti-

vates complement proteins to form a membrane attack complex that produces a lesion in the cell membrane, causing the foreign cell to lyse.

The *alternative pathway* does not involve antibodies and is, therefore, nonspecific. Complement may be activated by substances found in some bacteria, yeasts, viruses, virus-infected cells, and protozoa to form membrane attack complexes without the aid of antibodies. Complement can amplify the inflammatory response by stimulating histamine release. Several complement proteins attract phagocytes to infection sites. Some complement proteins coat microorganisms in a process called opsonization, which stimulates phagocytosis. In immune adherence, microbes coated with antibodies and complement proteins adhere to the walls of blood vessels, facilitating their destruction by phagocytic cells.

The immune system's capacity to distinguish self from nonself is critical in blood transfusion and transplantation

Blood Groups The immune response to the chemical markers that determine **ABO blood groups** must be considered in blood transfusions. Antibodies to blood group antigens arise in response to normal bacterial flora and circulate in the blood plasma, where they will destroy transfused blood cells with matching antigens by complement-mediated lysis. These antibodies are generally in the IgM class and do not cross the placenta.

An Rh-negative mother may develop antibodies against the **Rh factor**, another red-blood-cell antigen, if fetal blood from an Rh-positive child leaks across the placenta. Should she carry a second Rh-positive fetus, her immunological memory may result in the production of antibodies that cross the placenta and destroy fetal red blood cells. Treatment of the mother with anti-Rh antibodies just after delivery destroys any fetal cells that may have leaked into her circulation and prevents the mother's immunological response to the antigen.

Tissue Grafts and Organ Transplants Transplanted tissues and organs are rejected because the foreign MHC molecules are antigenic and trigger cytotoxic T-cell responses. The use of closely related donors, as well as drugs, such as cyclosporine, that suppress cell-mediated immunity help to reduce the immune response after a transplant operation.

■ **INTERACTIVE QUESTION   39.7**

Label the components of the illustrated classical complement pathway. Explain each step.

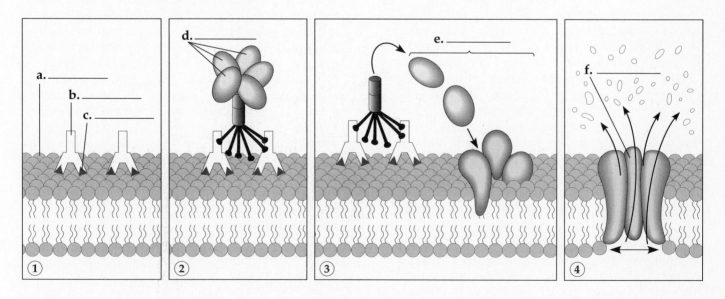

■ INTERACTIVE QUESTION   39.8

Fill in the following table to review your understanding of the antigens and antibodies of the ABO blood groups. Remember to compare the antigens on the donor cells with the antibodies in the recipient's plasma.

Blood Type	Antigens on RBCs	Antibodies in Plasma	Can Receive Blood From	Can Donate Blood To

■ Abnormal immune function leads to disease states

Autoimmune Diseases   Sometimes the immune system turns against self, leading to **autoimmune diseases**, such as lupus erythematosis, rheumatoid arthritis, rheumatic fever, insulin-dependent diabetes, and Grave's disease.

Allergy **Allergies** are hypersensitivities to certain environmental antigens, or allergens. IgE antibodies bound to mast cells (stationary cells in connective tissue) trigger allergic reactions. When antigens bind to these cell-surface antibodies, the mast cells degranulate and release histamines, which create an inflammatory response that may include sneezing, a runny nose, and difficulty in breathing due to smooth muscle contractions. Antihistamines are drugs that combat these symptoms by counteracting histamine. Anaphylactic shock is a severe allergic response in which the abrupt dilation of peripheral blood vessels caused by a rapid release of histamines leads to a life-threatening drop in blood pressure.

Immunodeficiency   A deficiency may occur in any of the components of the immune system. In the rare congenital disease known as severe combined immunodeficiency (SCID), both humoral and cell-mediated immune systems are nonfunctional. Gene therapy has been used to treat the T lymphocytes of individuals with a SCID caused by a deficiency of the enzyme adenosine deaminase. Certain cancers, such as Hodgkin's disease, and AIDS depress the immune system, making a person susceptible to infections.

There is growing evidence that general emotional health and immunity are related. Hormones secreted during stress affect the number of white blood cells; nerve fibers penetrate deep into lymphoid tissue, and receptors for chemical signals from nerve cells have been found on lymphocytes. Physiologists are now looking at the connections between general health, state of mind, and immunity.

Acquired Immunodeficiency Syndrome (AIDS) Investigation of the increasing incidence of Kaposi's sarcoma and *Pneumocystis* pneumonia in the early 1980s led to the recognition of **acquired immunodeficiency syndrome (AIDS)**. The infectious agent responsible for AIDS, known as **HIV (human immunodeficiency virus)**, was identified in 1983. This lethal pathogen probably arose by mutation in central Africa. The virus has been identified in preserved blood samples from as early as 1959.

Glycoproteins on the surface of HIV bind to CD4 molecules found on the surface of helper T cells. The virus enters the cell and replicates, and new viruses bud from the host cell. Destruction of T_H cells by the virus or by the immune system cripples both the humoral and cell-mediated branches of the immune system. Some macrophages and B cells also carry CD4 and may be infected.

HIV can remain latent as a provirus in the genomes of infected cells. HIV undergoes rapid antigenic changes during replication, complicating the ability of the immune system to respond to the infection. As the number of helper T cells decreases, the immune system becomes severely weakened. The collapse of cell-mediated immunity allows for the development of opportunistic infections. It takes an average of about 10 years for an HIV infection to progress to AIDS, characterized by a specified reduction of T cells and the appearance of characteristic secondary infections.

Most individuals exposed to HIV have circulating antibodies to the virus and will test as **HIV-positive**. This test for HIV antibodies is used to screen blood donations. HIV is transmitted by transfer of body fluids such as blood or semen. Sharing needles between intravenous drug users, unscreened blood supplies, and unprotected sex are all means by which AIDS is spread.

■ INTERACTIVE QUESTION   39.9

a. Why is AIDS such a deadly disease?

b. Why has it proved so difficult to prevent and cure this disease?

■ Invertebrates exhibit a rudimentary immune system

The ability to distinguish self from nonself is well developed in invertebrates. Many invertebrates have amoeboid cells called coelomocytes that identify and destroy foreign substances. Tissue graft experiments in earthworms have established that their defense systems reject foreign tissue and develop a memory response to these grafts.

STRUCTURE YOUR KNOWLEDGE

This chapter contains a wealth of information that is probably fairly new to you. If you take a little time and pull out the key players of the immune system and organize them first into very basic concept clusters and then develop more interrelated concept maps, you will find that this information is both understandable and fascinating.

1. Fill in the table below on some of the molecules involved in the immune system.

2. Create a concept map outlining specific defense mechanisms, showing the cells involved in humoral and cell-mediated immunities and their functions.

3. Describe the structure of an antibody molecule and how this structure relates to its function.

4. Briefly explain the clonal selection theory.

TEST YOUR KNOWLEDGE

MULTIPLE CHOICE: *Choose the one best answer.*

1. Which of the following is *incorrectly* paired with its effect?
   a. gastric juice—kills bacteria in the stomach
   b. fever—stimulates phagocytosis and inhibits microbial growth
   c. histamine—causes blood vessels to dilate
   d. vaccination—creates passive immunity
   e. lysozyme—attacks cell walls of bacteria

2. Monoclonal antibodies
   a. are produced by mast cells and basophils.
   b. explain the ability of the immune system to create a large number of identical T or B cells.
   c. are produced in tissue culture by hybridomas.
   d. initiate the secondary immune response.
   e. are used to treat autoimmune diseases.

3. Antibodies are
   a. proteins or polysaccharides usually found on the cell surface of invading bacteria or viruses.
   b. proteins that consist of two light and two heavy polypeptide chains.
   c. proteins circulating in the blood that tag foreign blood cells for complement-mediated lysis.
   d. proteins embedded in B-cell membranes.
   e. b, c, and d are correct.

Molecules	Where Produced or Found	Action
Complement	a.	b.
Antibodies	c.	d.
Interleukin-1	e.	f.
Interleukin-2	g.	h.
Perforin	i.	j.
Class I MHC	k.	l.
Class II MHC	m.	n.

4. A secondary immune response is more rapid and greater in effect than a primary immune response because
   a. Histamines and prostaglandins cause rapid vasodilation.
   b. The second response is an active immunity, whereas the primary one was a passive immunity.
   c. Helper T cells are available to activate other blood cells.
   d. Interleukins cause the rapid accumulation of phagocytic cells.
   e. Memory cells respond to the pathogen and rapidly clone more effector cells.

5. Lymphocytes capable of reacting against "self" molecules
   a. are usually not a problem until a woman's second pregnancy.
   b. are usually destroyed before birth.
   c. are usually kept separate from the immune system.
   d. contribute to immunodeficiency diseases.
   e. are characterized by class I MHC molecules.

6. The major histocompatibility complex
   a. is involved in the ability to distinguish self from nonself.
   b. is a set of several cell-surface antigens.
   c. may trigger T-cell responses after transplant operations.
   d. presents antigen fragments on infected cells.
   e. All of the above are correct.

7. In opsonization,
   a. Complement proteins coat microorganisms and help phagocytes bind to and engulf the invading cell.
   b. A set of complement proteins lyses a hole in the foreign cell's membrane.
   c. Antibodies cause cells to agglutinate, and the resulting clumps are engulfed by phagocytes.
   d. A flood of histamines is released that may result in anaphylactic shock.
   e. Microbes coated with antibodies and complement proteins adhere to vessel walls.

8. Severe combined immunodeficiency
   a. is an autoimmune disease.
   b. is a form of cancer in which the membrane surface of the cell has changed.
   c. is a disease in which both T and B cells are absent or inactive.
   d. is an immune disorder in which the number of helper T cells is greatly reduced.
   e. results from a few types of cancers, such as Hodgkin's disease.

9. A transfusion of type B blood given to a person who has type A blood would result in
   a. the recipient's anti-B antibodies reacting with the donated red blood cells.
   b. the recipient's B antigens reacting with the donated anti-B antibodies.
   c. the recipient forming both anti-A and anti-B antibodies.
   d. no reaction, because B is a universal donor type of blood.
   e. the introduced blood cells being destroyed by nonspecific defense mechanisms.

10. Which of the following are *incorrectly* paired?
   a. variable region—determines antibody specificity for an epitope
   b. immunoglobulins—glycoproteins forming antibodies
   c. constant region—determines class and function of antibody
   d. IgG—most abundant circulating antibodies, confer passive immunity to fetus
   e. IgE—receptor molecules attached to mast cells and basophils

11. Which of the following does *not* destroy a target cell by creating a hole in the membrane that causes the cell to lyse?
   a. neutrophil
   b. cytotoxic T cell
   c. natural killer cell
   d. complement proteins
   e. perforin

12. A T-independent antigen
   a. does not need the aid of T_H cells to bind to antibody molecules, whereas T-dependent antigens do.
   b. needs to bind to T_H cells to activate antibody production.
   c. is often a large molecule that simultaneously binds with several antigenic receptors on a B cell at one time.
   d. will result in the production of both plasma cells and memory cells.
   e. is responsible for the primary immune response.

13. What do antibodies, T-cell receptors, and MHC proteins have in common?
   a. They are found exclusively in cells of the immune system.
   b. They are all part of the complement system.
   c. They are antigen-presenting molecules.
   d. They are or can be membrane-bound proteins.
   e. They are involved in the cell-mediated portion of the immune system.

14. Place the following steps in the T_H-cell activation of cell-mediated and humoral immunity in the correct order:
   1. T_H cell secretes interleukin-2 and cytokines.
   2. Macrophage engulfs pathogen and presents antigen in class II MHC.
   3. Plasma cells secrete antibodies and T_C cells attack cells with class I MHC–antigen complex.
   4. T-cell receptor recognizes class II MHC–antigen complex.
   5. Macrophage secretes interleukin-1.
   6. Activated B cells form plasma and memory cells, activated T cells form T_H and T_C cells.
      a. 1, 3, 5, 6, 2, 4
      b. 5, 1, 2, 6, 4, 3
      c. 2, 4, 5, 1, 6, 3
      d. 5, 2, 4, 1, 3, 6
      e. 2, 4, 1, 5, 6, 3

15. Which of the following would be an effective defense against bacteria but not against viral particles?
   a. secretion of interferon by an infected cell
   b. neutralization by antibodies
   c. agglutination or precipitation by antibodies
   d. a secondary immune response
   e. membrane attack complex formed by complement

CHAPTER 40

CONTROLLING THE INTERNAL ENVIRONMENT

FRAMEWORK

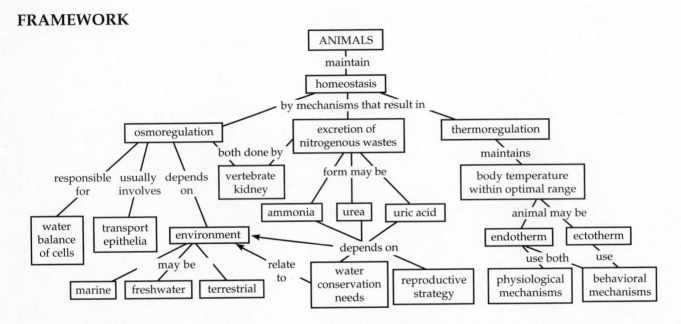

CHAPTER REVIEW

Animals are able to survive large fluctuations in their external environment by maintaining a relatively constant internal environment. The physiological adjustments that maintain homeostasis, such as **osmoregulation**, **excretion**, and **thermoregulation**, enable organisms to cope with short-term environmental changes. These mechanisms developed by natural selection as populations evolved in specific environments.

■ Homeostatic mechanisms protect an animal's internal environment from harmful fluctuations: *an overview*

Homeostatic mechanisms temper changes in the internal body fluid that bathes the cells, either hemolymph in animals with open circulatory systems or interstitial fluid serviced by blood in those with closed circulatory systems.

■ Cells require a balance between water uptake and loss

Whether an animal lives in salt water, fresh water, or on land, water gain must balance water loss in body cells. Interstitial fluid must be in osmotic balance with the cytosol.

Osmosis is the diffusion of water across a selectively permeable membrane that separates two solutions differing in **osmolarity** (moles of solute per liter). Osmolarity is expressed in units of milliosmoles per liter (10^{-3} moles/L). Isosmotic solutions are equal in osmolarity, and there is no net osmosis between them. There is a net flow of water across a membrane from a hypoosmotic (more dilute) to a hyperosmotic (more concentrated) solution.

Osmoconformers and Osmoregulators **Osmoconformers** are isosmotic with their aqueous surroundings and do not regulate their osmolarity. **Osmoregulators** must get rid of excess water if they live in a hypoosmotic medium or take in water to offset osmotic loss if they inhabit a hyperosmotic environment. Osmoregulation is energetically costly because animals must actively transport solutes in order to maintain osmotic gradients needed to gain or lose water.

Most animals are **stenohaline**, able to tolerate only small changes in external osmolarity. Animals that are **euryhaline** can survive in different osmotic environments by conforming or by maintaining a constant internal osmolarity.

Maintaining Water Balance in Different Environments Most marine invertebrates are osmoconformers, whereas most marine vertebrates are osmoregulators. Sharks have an internal salt concentration lower than that of seawater because their rectal glands pump salt out of the body. They maintain an osmolarity slightly higher than that of seawater, however, by retaining urea (a nitrogenous waste product) and trimethylamine oxide (which protects proteins from the damaging effects of urea) within their bodies. They dispose of excess water in **urine** produced in **kidneys**.

Many marine bony fishes, having evolved from freshwater ancestors, are hypoosmotic to seawater and must drink large quantities of seawater to replace the water they lose by osmosis. Excess salts are pumped out through salt glands and other ions are excreted in the scanty urine.

Freshwater animals constantly take in water by osmosis. Protozoa use contractile vacuoles to pump out excess water. Freshwater animals excrete large quantities of dilute urine. Salt supplies are replaced from their food or, in some fish, by active uptake of ions across the gills.

Some animals are capable of **anhydrobiosis**, or cryptobiosis—surviving dehydration in a dormant state. Anhydrobiotic roundworms produce large quantities of trehalose, a disaccharide that replaces water around membranes and proteins during dehydration.

Adaptations to prevent dehydration in terrestrial animals include water-impervious coverings, drinking and eating food with high water content, nervous and hormonal control of thirst, behavioral adaptations such as nocturnal lifestyles, and water-conserving excretory organs.

■ Osmoregulation depends on transport epithelia

Osmolar-ity of the internal environment is usually regulated by transporting solutes, followed by the osmotic movement of water, across a **transport epithelium**. Located at the tissue–environment boundary, this single sheet of cells is linked by impermeable tight junctions and is able to regulate the passage of solutes between the extracellular fluid and the environment. Variations in molecular composition of the plasma membrane of the transport epithelium determine its passive permeability to water and salts and its active transport of solutes by various membrane proteins.

The transport epithelia of salt glands function exclusively in osmoregulation, whereas the transport epithelia of excretory organs, often arranged in tubular networks, function additionally in metabolic waste excretion.

■ **INTERACTIVE QUESTION 40.1**

Indicate whether the following are osmoregulators or osmoconformers and whether they are isosmotic, hyperosmotic, or hypoosmotic to their environment.

Animal	Osmoregulator or Osmoconformer?	Osmotic Relation to Environment
Marine invertebrate	a.	b.
Shark	c.	d.
Marine fish	e.	f.
Freshwater fish	g.	h.
Freshwater protozoan	i.	j.
Terrestrial animal	k.	l.

■ Tubular systems function in osmoregulation and excretion in many invertebrates

Protonephridia: Flame-Bulb System of Flatworms The **protonephridia** of flatworms are branched systems of closed tubules embedded in the body tissues. Interstitial fluids passing into the flame bulbs at the ends of the smallest tubules are propelled by cilia of the flame bulbs along the tubules to excretory ducts that empty the dilute fluid into the external environment by way of nephridiopores. The flame-bulb system of freshwater flatworms is primarily osmoregulatory; metabolic wastes diffuse through the body surface or are excreted through the gastrovascular cavity.

In parasitic flatworms, however, protonephridia function mainly in excretion of nitrogenous wastes.

Metanephridia of Earthworms   Excretory tubules with internal openings that collect body fluids are found in most annelids. These **metanephridia** occur in pairs in each segment of the worm. An open ciliated funnel called a nephrostome collects coelomic fluid, which then moves through a folded tubule encased in capillaries. The transport epithelium of the tubule pumps salts out of the tubule, and the salts are reabsorbed into the blood. Hypoosmotic urine, carrying nitrogenous wastes and excess water absorbed by osmosis, exits by way of nephridiopores.

Malpighian Tubules of Insects   In insects and other terrestrial arthropods, **Malpighian tubules** remove nitrogenous wastes from the body fluid (hemolymph) and function in osmoregulation. Transport epithelia lining these blind sacs, which open into the digestive tract at the end of the midgut, pump salts and nitrogenous wastes from the hemolymph into the tubule. The fluid passes through the hindgut and into the rectum, where a transport epithelium pumps most of the salt back into the hemolymph, and water follows by osmosis. In this water-conserving system, nitrogenous wastes are eliminated as dry matter along with the feces.

■ **INTERACTIVE QUESTION   40.2**

Indicate whether these tubular systems function in osmoregulation, excretion, or both. If they function in osmoregulation, do they help conserve water or remove excess water that had entered by osmosis?

   a. Protonephridia of freshwater planaria

   b. Metanephridia of earthworms

   c. Malpighian tubules of insects

■ **The kidneys of most vertebrates are compact organs with many excretory tubules**

The excretory tubules of most vertebrates are enclosed in a network of capillaries and arranged into compact kidneys.

The Mammalian Excretory System   Blood enters the pair of bean-shaped kidneys through the **renal**
arteries and leaves by way of the **renal veins**. Urine exits through the **ureter** and is temporarily stored in the **urinary bladder**. The body periodically releases urine through the **urethra**.

The Nephron and Associated Structures   The outer **renal cortex** and inner **renal medulla** of the kidney are packed with excretory tubules (called nephrons), collecting ducts, and blood vessels. The **nephron** consists of a long tubule with a cuplike **Bowman's capsule** at the blind end that encloses a ball of capillaries called the **glomerulus**. Blood pressure forces water, urea, salts, and small molecules from the blood into Bowman's capsule, forming the **filtrate**. Filtrate passes through the **proximal tubule**, the **loop of Henle** with a descending limb and an ascending limb, and a **distal tubule**. **Collecting ducts** receive filtrate from many nephrons and pass the urine into the renal pelvis.

In the human kidney, 80 percent of the nephrons are **cortical nephrons** located entirely in the renal cortex. **Juxtamedullary nephrons**, found only in mammals and birds, have long loops of Henle that extend into the renal medulla and participate in the formation of hyperosmotic urine. A transport epithelium lining the nephron and collecting duct processes the filtrate to produce urine.

An **afferent arteriole** supplies each nephron, subdividing to form the glomerulus and converging to form an **efferent arteriole**. This arteriole then forms a second capillary network—the **peritubular capillaries**—which surrounds the proximal and distal tubules. The **vasa rectae** includes descending and ascending capillaries that parallel the loop of Henle. Exchange between the tubules and capillaries is via the interstitial fluid.

■ **The kidney's transport epithelia regulate the composition of blood**

Production of Urine from a Blood Filtrate   Filtration occurs when blood pressure forces water and small solutes out of the porous capillaries and between **podocytes**, epithelial cells that form part of a filtration membrane of Bowman's capsule, and into the nephron tubule.

Secretion of substances from the interstitial fluid across the tubule epithelium occurs as the filtrate moves through the proximal and distal tubules. This selective secretion involves both passive and active transport.

Most of the water, sugar, vitamins, and other organic nutrients in the filtrate are **reabsorbed** across the tubule epithelium into the interstitial fluid, from which they are returned to the blood.

■ **INTERACTIVE QUESTION   40.3**

a. What is found in the filtrate that moves from Bowman's capsule into the tubule?

b. What remains in the capillaries?

Transport Properties of the Nephron and Collecting Duct
(1) Reabsorption and secretion both take place in the proximal tubule. Its transport epithelium selectively secretes hydrogen ions, ammonia, and drugs and poisons processed by the liver into the filtrate, and it actively transports glucose and amino acids out of the filtrate. Potassium and the buffer bicarbonate are also reabsorbed. Salt diffuses from the filtrate into the cells of the proximal tubule, which then actively transport Na^+ across the membrane into the interstitial fluid; Cl^- is transported passively out of the cell to balance the positive charge; and water follows by osmosis. Salt and water diffuse from the interstitial fluid into the peritubular capillaries.

(2) The descending limb of the loop of Henle is freely permeable to water but not to salt or other small solutes. The interstitial fluid is increasingly hyperosmotic toward the inner medulla, so water exits by osmosis, leaving a filtrate with a high NaCl concentration.

(3) The ascending limb of the loop of Henle is not very permeable to water but is permeable to salt, which diffuses out of the thin lower segment of the loop, adding to the high osmolarity of the medulla. NaCl is actively transported out of the thick upper portion of the ascending limb. As salt leaves, the filtrate becomes less concentrated.

(4) The distal tubule is specialized for selective secretion and reabsorption, contributing to homeostasis of the body fluids. It regulates K^+ secretion and NaCl reabsorption, and it helps to regulate pH by secreting H^+ and reabsorbing bicarbonate.

(5) As the filtrate moves in the collecting duct back through the increasing osmotic gradient of the medulla, more and more water exits by osmosis. Urea diffuses out of the lower portion of the duct and helps to maintain the osmotic gradient. Salt excretion is controlled by the transport epithelium actively reabsorbing NaCl.

■ The water-conserving ability of the mammalian kidney is a key terrestrial adaptation

Conservation of Water by Two Solute Gradients
The osmolarity gradient of NaCl and urea in the inter-stitial fluid of the renal medulla is produced by the juxtamedullary nephrons and enables the kidney to produce urine up to four times as concentrated as blood and normal interstitial fluid. Filtrate leaving Bowman's capsule has an osmolarity of about 300 mosm/L, the same as blood. Both water and salt are reabsorbed in the proximal tubule. In the trip down the descending limb of the loop of Henle, water exits by osmosis, and the filtrate becomes more concentrated. Salt, which is now in high concentration in the filtrate, diffuses out as the filtrate moves up the salt-permeable but water-impermeable ascending limb—helping to create the osmolarity gradient.

The vasa recta also has a countercurrent flow through the loop of Henle. As the blood in these vessels moves down into the inner medulla, water leaves by osmosis and salt enters; as the blood moves back up toward the cortex, water moves back in and salt diffuses out. Thus the capillaries can carry nutrients and other supplies to the medulla without disrupting the osmolarity gradient.

The filtrate makes one final pass through the medulla, this time in the collecting duct, which is permeable to water and urea but not to salt. Water flows out by osmosis, and, as the filtrate becomes more concentrated, urea leaks into the interstitial fluid, adding to the osmolarity of the inner medulla. The resulting concentrated urine is isosmotic to the interstitial fluid of the inner medulla, which can be as high as 1200 mosm/L.

Regulation of Kidney Function by Feedback Circuits
Osmolarity of urine in humans can vary from 70 to 1200 mosm/L, depending on the body's hydration needs. Osmoregulation results from nervous and hormonal control of water and salt reabsorption in the kidneys.

Antidiuretic hormone (ADH) is produced in the hypothalamus and stored in the pituitary gland. When blood osmolarity rises above a set point of 300 mosm/L, osmoreceptor cells in the hypothalamus trigger the release of ADH. This hormone increases water permeability of the distal tubules and collecting ducts, increasing water reabsorption. After consuming water in food or drink, negative feedback decreases the release of ADH. When blood osmolarity is low, the absence of ADH results in the production of large volumes of dilute urine, called diuresis. Alcohol inhibits ADH release and can cause dehydration.

The **juxtaglomerular apparatus (JGA)**, located near the afferent arteriole leading to the glomerulus, responds to a drop in blood pressure or volume by releasing renin, an enzyme that converts the plasma protein angiotensinogen to **angiotensin II**. Angiotensin II functions as a hormone and constricts arterioles, stimulates the proximal tubules to reabsorb more NaCl and water, and stimulates the adrenal

■ **INTERACTIVE QUESTION   40.4**

In the diagram of a nephron and collecting tubule, label the parts on the indicated lines. Label the arrows to indicate the movement of salt, water, nutrients, K⁺, bicarbonate (HCO_3^-), H⁺, and urea into or out of the tubule.

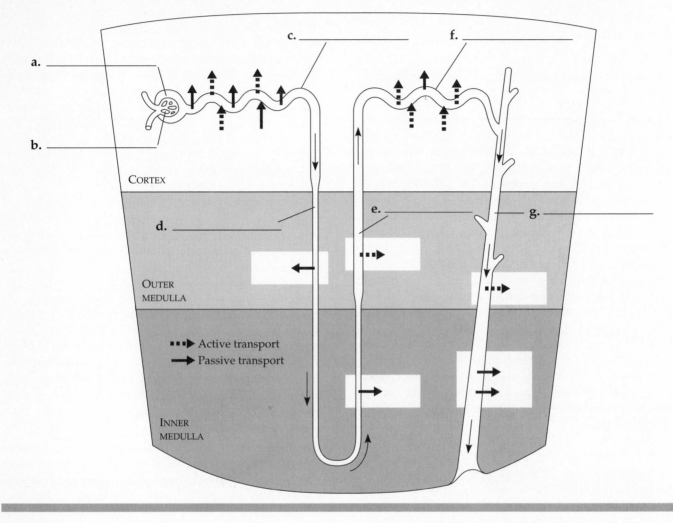

glands to release **aldosterone**. This hormone stimulates NaCl and water reabsorption in the distal tubules. The **renin-angiotensin-aldosterone system (RAAS)** is a homeostatic feedback circuit that maintains adequate blood pressure and volume.

Atrial natriuretic factor (ANF) counters RAAS. ANF, released by the atria of the heart in response to increased blood volume and pressure, inhibits the release of renin from the juxtaglomerular apparatus, inhibits NaCl reabsorption by the collecting ducts, and reduces the release of aldosterone from the adrenals.

■ Diverse adaptations of the vertebrate kidney have evolved in different habitats

Modifications of nephron structure and function correspond to osmoregulatory and excretory requirements in various habitats. Mammals in dry habitats,

needing to conserve water and excrete hyperosmotic urine, have long loops of Henle. Birds also have juxtamedullary nephrons, but their shorter loops of Henle prevent them from producing as hyperosmotic urine as can mammals. The kidneys of reptiles have only cortical nephrons, and their urine is isosmotic to body fluids; water is conserved, however, through reabsorption by the epithelium of the cloaca and the production of uric acid, an insoluble nitrogenous waste.

Freshwater fishes are hyperosmotic to their environment and must excrete large quantities of very dilute urine. Salts are conserved by the efficient reabsorption of ions from the filtrate. When in fresh water, frogs and other amphibians actively absorb certain salts across their skin and excrete a dilute urine. When on land, water is reabsorbed from the urinary bladder.

The kidneys of marine bony fishes excrete very little urine and function mainly to get rid of divalent ions taken in by drinking seawater. Monovalent ions

■ INTERACTIVE QUESTION   40.5

Complete this concept map to help organize your understanding of the nervous and hormonal control of water and salt reabsorption in the kidneys.

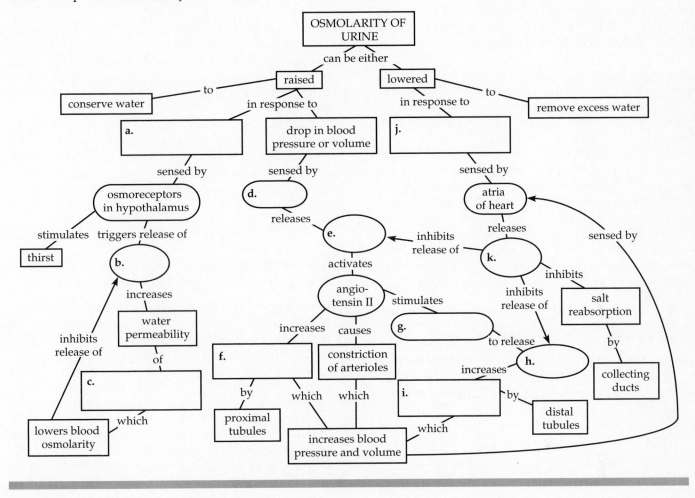

■ INTERACTIVE QUESTION   40.6

a. What type of nephron allows the production of hyperosmotic urine?

b. Which animal groups have this type of nephron?

and nitrogenous wastes in the form of ammonium are excreted by the epithelia of the gills.

The original function of the kidney was osmoregulation; excretion of nitrogenous wastes became a second function over the course of evolutionary time.

■ An animal's nitrogenous wastes are correlated with its phylogeny and habitat

Ammonia, a small and toxic molecule, is produced when proteins and nucleic acids are metabolized for energy or converted to carbohydrates or fats. Some animals excrete ammonia; others expend energy to convert it to less toxic wastes such as urea or uric acid.

Ammonia   Aquatic animals can excrete nitrogenous wastes as **ammonia** because it is very soluble and easily permeates membranes. Soft-bodied invertebrates lose ammonia across their whole body surface. In fishes, most of the ammonia is passed as NH_4^+ ions across the epithelia of the gills.

Urea Mammals, most adult amphibians, and many marine fishes and turtles produce **urea**, a much less toxic compound than ammonia. Urea can be tolerated in more concentrated form and excreted with less loss of water. Ammonia and carbon dioxide are combined in the liver to produce urea, which is then carried by the circulatory system to the kidneys.

Uric Acid Land snails, insects, birds, and many reptiles produce **uric acid**, a compound of low solubility in water that can be excreted as a precipitate after nearly all the water has been reabsorbed from the urine.

The mode of reproduction of terrestrial animals seems to have determined whether they excrete uric acid or urea as their nitrogenous waste product.

The form of nitrogenous wastes also relates to habitat. Terrestrial reptiles excrete uric acid, whereas crocodiles excrete ammonia and uric acid, and aquatic turtles excrete both urea and ammonia. Some animals actually shift their nitrogenous waste product depending on environmental conditions.

■ INTERACTIVE QUESTION   40.7

Vertebrates that produce shelled eggs excrete _____ . Mammals produce _____ . What adaptive advantage do these types of nitrogenous wastes provide?

■ Thermoregulation maintains body temperature within a range conducive to metabolism

Heat Transfer between Organisms and Their Surroundings Heat is exchanged between an organism and the external environment by the physical processes of **conduction**, the direct transfer of thermal motion between surfaces in contact; **convection**, the transfer of heat by the flow of air or water past a body; **radiation**, the emission of electromagnetic waves by all objects warmer than absolute zero; and **evaporation**, the loss of heat due to the conversion of the surface molecules of a liquid to a gas. Convection and evaporation are the most variable causes of heat loss; wind and the evaporation of sweat can greatly increase the loss of heat.

■ Ectotherms derive body heat mainly from their surroundings and endotherms derive it mainly from metabolism

Animals such as invertebrates, fishes, amphibians, and reptiles absorb their body heat from their surroundings and are called **ectotherms**. Mammals, birds, and some fishes and insects, which derive their body heat from their own metabolism, are called **endotherms**.

Endothermy allows terrestrial animals to maintain a constant internal body temperature regardless of external fluctuations. A warm body temperature both requires and facilitates active metabolism, which is needed for the energetically expensive movement of animals on land.

■ Thermoregulation involves physiological and behavioral adjustments

Four categories of adaptations enable ectothermic and endothermic animals to thermoregulate. (1) The rate of heat exchange with the environment can be adjusted by body insulation, **vasodilation**, **vasoconstriction**, and a **countercurrent heat exchanger**, which can minimize heat loss in body parts immersed in cold water. Dilation of superficial blood vessels increases heat loss, whereas constriction reduces blood flow and heat loss. (2) Cooling can occur by evaporative heat loss, which can be increased by panting or sweating. (3) Behavioral responses, such as basking or moving to a warmer or cooler environment, can regulate heat loss. (4) Many mammals and birds can change the rate of metabolic heat production.

■ Comparative physiology reveals diverse mechanisms of thermoregulation

Invertebrates Some invertebrates adjust their temperature by behavioral or physiological mechanisms. Large endothermic flying insects may "warm up" before taking off by contracting their flight muscles. Noctuid moths, bumblebees, and honeybees have a countercurrent heat exchanger that maintains a high thoracic temperature. The social organization of honeybees enables them to regulate body heat in cold temperatures by huddling together and to cool their hive by fanning their wings to promote evaporation and convection.

Amphibians and Reptiles Amphibians produce little heat and easily lose heat by evaporation from their moist skin. Behavioral adaptations usually enable them to maintain acceptable internal temperatures.

The metabolic rate of ectothermic reptiles is very low and contributes little to body temperature. Behavioral adaptations, such as orienting the body to the sun or finding suitable microclimates, allow these animals to regulate their temperature within an optimal range. In some reptiles, vasoconstriction of surface vessels sends more blood to the body core to conserve heat. A few reptiles use shivering to generate heat while incubating eggs.

Fishes The body temperature of fishes is determined mainly by the water temperature. Some large, active, endothermic fish are able to retain the heat produced by their swimming muscles with a countercurrent heat exchanger.

Mammals and Birds Mammals and birds, which maintain high body temperatures within a narrow range, are able to regulate the rate of metabolic heat production as well as control the rate of heat gain or loss. Heat production can be increased by muscle contraction (by moving or shivering) and by **nonshivering thermogenesis**, a rise in metabolic rate and the production of heat instead of ATP. **Brown fat** in some mammals is specialized for heat production.

Vasodilation and vasoconstriction regulate heat exchange. Feathers and fur may be raised to trap more air and increase their insulation against heat loss. Marine mammals maintain their high body temperature by efficient heat-conserving mechanisms, including a thick layer of insulating blubber and countercurrent heat exchange between arterial and venous blood in the extremities. When in warmer waters, these animals can dissipate metabolic heat by dilating superficial blood vessels. Terrestrial mammals and birds use evaporative cooling, enhanced by panting, sweating, and spreading saliva on body surfaces.

Thermoregulation in Humans Both warm and cold temperature-sensing nerve cells are located in the skin, hypothalamus, and some other areas of the nervous system. Control of body temperature involves feedback mechanisms, with the hypothalamus functioning as the thermostat.

Torpor: Conserving Energy during Environmental Extremes **Torpor** is a physiological state characterized by decreases in metabolic, heart, and respiratory rates. In **hibernation**, the body temperature is maintained at a lower level, allowing the animal to withstand long periods of cold temperature and decreased food supply. In **aestivation**, animals survive long stretches of elevated temperature and diminished water supply.

Some small mammals and birds that have very high metabolic rates enter a daily torpor controlled by the biological clock during the periods when they are not feeding. Hibernation and aestivation may be triggered by seasonal changes in the length of daylight.

Temperature Rate Adjustments Many animals are capable of **acclimatization**, a physiological adjustment to a different temperature, such as occurs in seasonal changes.

Cells can make rapid adjustments to temperature increases and other stresses by producing **stress-induced proteins**. These molecules protect proteins that might otherwise be denatured by the assault and may prevent cell death while the organism adjusts to environmental changes.

■ **INTERACTIVE QUESTION 40.8**

List some of the physiological changes that can occur during acclimatization to cooler temperatures.

■ **Regulatory systems interact in the maintenance of homeostasis**

A major challenge of animal physiology is determining how the various regulatory systems are controlled and how they interact to maintain internal homeostasis.

Roles of the Liver in Homeostasis The vertebrate liver interacts with most organ systems and is important to homeostasis. Its functions include synthesis of nitrogenous wastes, plasma proteins, growth factors, bile, lipids, cholesterol, and freeze-protectant chemicals; storage of fat-soluble vitamins, iron, and glycogen; interconversion of glucose and glycogen; recycling of red blood cells; and detoxification of drugs and toxins.

STRUCTURE YOUR KNOWLEDGE

1. Fill in the table below with a brief description of the general osmoregulatory, excretory, and thermoregulatory mechanisms for the animals listed.

Animal	Osmoregulation	Excretory System	Thermoregulation
a. Marine invertebrate			
b. Marine bony fish			
c. Freshwater fish			
d. Flatworm			
e. Earthworm			
f. Insect			
g. Reptile			
h. Bird			
i. Mammal			

2. List the substances that are filtered, secreted, and reabsorbed in the production of urine in mammals.
   a. Substances filtered:

   b. Substances secreted:

   c. Substances reabsorbed:

TEST YOUR KNOWLEDGE

MULTIPLE CHOICE: *Choose the one best answer.*

1. Contractile vacuoles most likely would be found in protists
   a. in a freshwater environment.
   b. in a marine environment.
   c. that are internal parasites.
   d. that are hypoosmotic to their environment.
   e. that are isosmotic to their environment.

2. Transport epithelia are responsible for
   a. pumping water across a membrane.

   b. forming an impermeable boundary at an interface with the environment.
   c. the exchange of solutes for osmoregulation or excretion.
   d. filtering blood to form the filtrate in Bowman's capsule.
   e. the passive transport of H^+ and HCO_3^- for the regulation of pH.

3. Which of the following is incorrectly paired with its excretory system?
   a. insect—Malpighian tubules
   b. flatworm—flame-bulb system
   c. earthworm—protonephridia
   d. amphibian—kidneys
   e. fish—kidneys

4. A freshwater fish would be expected to
   a. pump salt out through salt glands in the gills.
   b. produce copious quantities of dilute urine.
   c. diffuse urea out across the epithelium of the gills.
   d. have scales that reduce water loss to the environment.
   e. do all of the above.

5. Which of the following is *not* part of the filtrate entering Bowman's capsule?
   a. water, salt, and electrolytes
   b. glucose
   c. urea
   d. plasma proteins
   e. amino acids

6. Which is the correct pathway for the passage of urine in vertebrates?
   a. collecting tubule → ureter → bladder → urethra
   b. renal vein → renal ureter → bladder → urethra
   c. nephron → urethra → bladder → ureter
   d. cortex → medulla → bladder → ureter
   e. medulla → pelvis → bladder → urethra

7. Aldosterone
   a. is a hormone that stimulates thirst.
   b. is secreted by the adrenal glands in response to a high osmolarity of the blood.
   c. stimulates the active reabsorption of Na^+ in the distal tubules.
   d. causes diuresis.
   e. is converted from a blood protein by the action of renin.

8. Which of the following statements is *incorrect*?
   a. Long loops of Henle are associated with steep osmotic gradients and the production of hyperosmotic urine.
   b. Ammonia is a toxic nitrogenous waste molecule that passively diffuses out of the bodies of aquatic invertebrates.
   c. Uric acid is the form of nitrogenous waste that requires the least amount of water to excrete.
   d. In the mammalian kidney, urea diffuses out of the collecting duct and contributes to the osmotic gradient within the medulla.
   e. Ammonia produced by a mammalian fetus is removed through the placenta and converted to urea in the mother's liver.

9. The process of reabsorption in the formation of urine insures that
   a. Excess hydrogen ions are removed from the blood.
   b. Drugs and other poisons are removed from the blood.
   c. Urine is always hyperosmotic to interstitial fluid.
   d. Glucose, salts, and water are returned to the blood.
   e. pH is maintained with a balance of hydrogen ions and bicarbonate.

10. The peritubular capillaries
    a. form the ball of capillaries inside Bowman's capsule from which filtrate is forced by blood pressure into the renal tubule.
    b. intertwine with the proximal and distal tubules and exchange solutes with the interstitial fluid.
    c. form a countercurrent flow of blood through the medulla to supply nutrients without interfering with the osmolarity gradient.
    d. surround the collecting ducts and reabsorb water, helping to create a hyperosmotic urine.
    e. rejoin to form the efferent arteriole.

11. ADH and RAAS both increase water reabsorption, but they respond to different osmoregulatory problems. Which two of the following statements are true?
    1. ADH will be released in response to high alcohol consumption.
    2. ADH is released when osmoregulatory cells in the hypothalamus sense an increase in blood osmolarity.
    3. RAAS will increase the osmolarity of urine due to the cooperative action of renin, aldosterone, and ANP.
    4. RAAS is a response to a rise in blood pressure or volume.
    5. RAAS is most likely to respond following an accident or severe case of diarrhea.
    a. 1 and 4
    b. 1 and 5
    c. 2 and 3
    d. 2 and 4
    e. 2 and 5

12. Which of the following is used by terrestrial animals as a mechanism to dissipate heat?
    a. hibernation
    b. countercurrent exchange
    c. raising fur or feathers
    d. vasodilation
    e. vasoconstriction

13. Which of the following sections of the mammalian nephron is *incorrectly* paired with its function?
    a. Bowman's capsule and glomerulus—filtration of blood
    b. proximal tubule—secretion of ammonia and H^+ into filtrate and transport of glucose and amino acids out of tubule
    c. descending limb of loop of Henle—diffusion of urea out of filtrate
    d. ascending limb of loop of Henle—diffusion and pumping of NaCl out of filtrate
    e. distal tubule—regulation of pH and potassium level

14. The rate of metabolic heat production can be increased by
    a. nonshivering thermogenesis.
    b. storage of brown fat.
    c. shivering and vasoconstriction.
    d. thick layers of blubber and countercurrent heat exchangers.
    e. all of the above.

15. The function of stress-induced proteins is to
    a. provide proteins that function at a lower temperature.
    b. change the composition of plasma membranes to maintain fluidity.
    c. regulate blood osmolarity and volume following an injury.
    d. regulate daily torpor in small mammals with high metabolic rates.
    e. protect cellular proteins from denaturation during rapid temperature changes.

CHAPTER 41

CHEMICAL SIGNALS
IN ANIMALS

FRAMEWORK

This chapter introduces the intricate system of chemical control and communication within animals. The endocrine cells or neurosecretory cells produce hormones that regulate the activity of other organs. Steroid hormones bind with receptors within their target cells and move into the nucleus where they influence gene expression. The signal transduction pathway of most hormones derived from amino acids involves binding with cell surface receptors and producing second messengers that trigger a cascade of metabolic reactions in target cells.

The hypothalamus and pituitary gland play coordinating roles by integrating the nervous and endocrine systems and producing many tropic hormones that control the synthesis and secretion of hormones in other endocrine glands and organs.

CHAPTER REVIEW

Coordination and communication among the specialized parts of complex animals are achieved by the **nervous system** and the **endocrine system**. The nervous system conveys high-speed messages along neurons, whereas the endocrine system produces chemical messengers called hormones, which travel more slowly and influence bioenergetics, growth, development, metabolism, behavior, and homeostasis.

■ A variety of chemical signals coordinate body functions: *an overview*

Hormones An animal **hormone** is a chemical signal, secreted by endocrine cells or **neurosecretory cells** and usually carried through the circulatory system, which elicits a specific response from **target cells**. Endocrine cells are usually assembled into **endocrine glands**, which are ductless secretory organs that release hormones into body fluids.

Hormones can be grouped into two general classes based on chemical structure: **steroid hormones**, produced from cholesterol, and hormones derived from amino acids, which include amine, peptide, and protein hormones, and glycoproteins.

Hormonal response is triggered by the binding of a hormone to a complementary hormone receptor, which is either within the target cell or built into its plasma membrane. Most homeostatic functions regulated by hormones are controlled by a pair of hormones with antagonistic effects.

Pheromones **Pheromones** are chemical signals that communicate between animals of the same species. Serving as mate attractants, territorial markers, or alarm substances, these small and volatile molecules are active in minute amounts.

Local Regulators Local regulators are chemical signals that affect target cells near their points of secretion. The release of neurotransmitter into a synapse between a neuron and its target cell, called synaptic signaling, is a direct form of chemical communication. In paracrine signaling, a cell secretes local regulators, such as histamine and interleukins, into the interstitial fluid, affecting only nearby target cells.

Many cells produce nitric oxide (NO) or carbon monoxide (CO) as paracrine signals or neurotransmitters.

Growth factors are peptides and proteins that are required in the extracellular environment for many types of cells to grow and develop. The binding of growth factors to surface receptors on target cells initiates a signal transduction pathway that affects gene expression.

Prostaglandins (PGs) are modified fatty acids that are released from most cells and have a wide range of effects on nearby target cells. The balance of antagonistic prostaglandins is an important regulatory mechanism. In mammals, PGs help to induce labor. Aspirin inhibits the synthesis of prostaglandins and thus reduces their fever-inducing and pain-intensifying actions.

■ **INTERACTIVE QUESTION   41.1**

Briefly review the characteristics of the following:

a. hormone:

b. pheromone:

c. local regulator:

d. synaptic signaling:

e. paracrine signaling:

■ **Hormone binding to specific receptors triggers mechanisms at the cellular level**

Hormones act at very low concentrations and can have varying effects on different target cells and in different species. Two general mechanisms of hormone action have been identified; both involve the hormone binding to specific receptor proteins. Steroid hormones enter the target cell, whereas most nonsteroid hormones attach to the cell surface and stimulate release of intermediaries called second messengers in a **signal transduction pathway**.

Steroid Hormones and Gene Expression   Steroids appear to bind to receptor proteins in the cytoplasm, and the steroid–receptor complex then enters the nucleus. Attachment of this complex to regulatory sites along the chromosomes either induces or suppresses gene expression.

Two types of cells can respond differently to the same hormone, depending on the location within the genome of the regulatory site to which the hormone–receptor complex binds.

Peptide Hormones and Signal Transduction   Most peptide hormones and hormones derived from amino acids bind to specific receptor proteins embedded in the plasma membrane, initiating a signal transduction

pathway that converts an extracellular signal into intracellular signals that influence the reactions of the target cell.

■ **Many chemicals are relayed and amplified by second messengers and protein kinases**

Cyclic AMP   Sutherland found that the binding of epinephrine to a hormone receptor in the plasma membrane of a liver cell signaled the membrane enzyme **adenylyl cyclase** to convert ATP to cyclic adenosine monophosphate (**cyclic AMP or cAMP**). Cyclic AMP activates glycogen hydrolysis in the cell. The hormone is the first messenger, adenylyl cyclase is the effector, and cAMP is the second messenger, relaying the message to cytoplasmic enzymes. A decline in hormone levels and enzymatic degradation of cAMP terminates the hormonal response.

Binding of a hormone to its receptor first activates **G protein**, a membrane protein functioning as a relay, which hydrolyzes GTP to GDP, providing the energy to activate adenylyl cyclase. The same G protein stimulates cAMP production in response to several different hormone signals. Another type of G protein inhibits adenylyl cyclase in response to the binding of an inhibitory hormone. The balance of these antagonistic hormones allows the cell to precisely regulate its metabolism.

Signal Amplification and Specificity   The effect of a hormone is amplified when the second messenger sets off an **enzyme cascade**. A sequence of enzymes is activated in which the number of activated products greatly increases at each step. A **protein kinase** is an enzyme that catalyzes protein phosphorylation, which, depending on the protein, either increases or decreases its activity. In the liver cell's response to epinephrine, cAMP (produced by adenylyl cyclase) activates cAMP-dependent protein kinase, which activates phosphorylase kinase, an enzyme that adds a phosphate group to glycogen phosphorylase—the enzyme that hydrolyzes glycogen. (Or should we say that cAMP-dependent protein kinase phosphorylates phosphorylase kinase, which phosphorylates glycogen phosphorylase!) A few molecules of epinephrine can result in the release of millions of glucose molecules from glycogen.

This basic mechanism is used to mediate responses of many cell types to different hormones. The specificity of hormone action is maintained because different hormone receptors are found on specific target cells, cAMP-dependent protein kinases vary from tissue to tissue, and, in different cell types, cAMP-dependent protein kinases regulate different complements of proteins. Recent research has shown that cAMP also acts as a regulator of the signal transduction pathway of many growth hormones.

■ INTERACTIVE QUESTION 41.2

Identify the components in this diagram of a signal transduction pathway using cAMP as a second messenger.

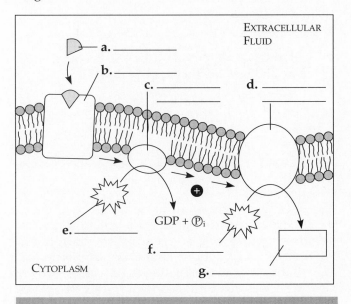

Cytoplasmic Ca²⁺ and Signal Transduction Inositol triphosphate (IP₃) is another second messenger that relays and amplifies a hormonal signal. The binding of a hormone to its receptor activates a relay G protein that stimulates phospholipase C, a membrane enzyme that cleaves a plasma membrane phospholipid into IP₃ and diacylglycerol, both of which may act as second messengers. Diacylglycerol activates protein kinase C, which phosphorylates specific proteins. IP₃ causes the release of Ca²⁺ from the endoplasmic reticulum or its uptake from extracellular fluid. The calcium either directly alters the activities of certain enzymes or binds to a protein called **calmodulin**, which then binds to and changes the activities of other enzymes and proteins. Neurotransmitters, growth factors, and some hormones use this mechanism of increasing the concentration of cytoplasmic Ca²⁺ to induce changes in their target cells.

Steroid hormones alter gene expression, generally resulting in the synthesis of proteins and creating a slower, but longer-lasting hormonal response. Hormones derived from amino acids use a signal transduction pathway and act through second messengers to alter the activity of enzymes and other proteins already in the cell, producing a more rapid hormonal response.

■ INTERACTIVE QUESTION 41.3

Identify the components in this diagram of a signal transduction pathway using IP₃ and diacylglycerol as second messengers. What constitutes a third messenger in this system?

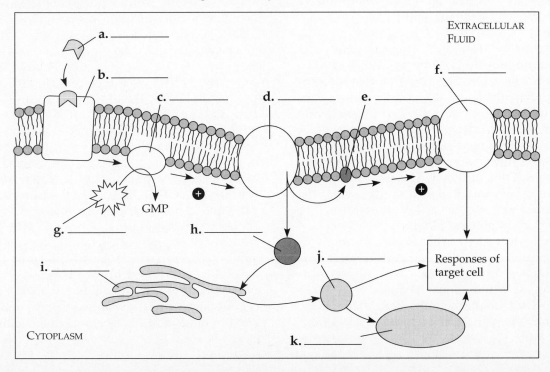

Invertebrate control systems often integrate endocrine and nervous systems

Arthropods have extensive endocrine systems. Insects have three hormones that interact to control molting and development of adult characteristics: The steroid hormone **ecdysone** stimulates the transcription of specific genes and functions to trigger molts and to promote development of adult characteristics. **Brain hormone** stimulates the prothoracic glands to secrete ecdysone. **Juvenile hormone (JH)** counters the action of ecdysone and promotes retention of larval characteristics during molting. When the JH level is high, ecdysone-induced molting produces additional larval stages; only after the JH level has decreased does molting result in a pupa.

The hypothalamus and pituitary integrate many functions of the vertebrate endocrine system

Vertebrate hormones may target a few or many tissues in the body. **Tropic hormones** target other endocrine glands and are particularly important to chemical coordination.

Structural and Functional Relationships of the Hypothalamus and the Pituitary Gland   The **hypothalamus**, situated in the lower brain, plays a key role in integrating the endocrine and nervous systems. Neurosecretory cells of the hypothalamus receive nerve signals from throughout the body and release hormones that are stored in or regulate the **pituitary gland** at the base of the hypothalamus.

Posterior Pituitary Hormone   The **posterior pituitary**, or **neurohypophysis**, stores and secretes two peptide hormones, oxytocin and ADH, that are produced by the hypothalamus; the **anterior pituitary**, or **adenohypophysis**, is composed of nonnervous endocrine tissue and produces its own hormones. A set of hypothalamic neurosecretory cells produces **releasing hormones** and **inhibiting hormones** that regulate the anterior pituitary. These hormones are released into capillaries in an area at the base of the hypothalamus and travel via a short portal vessel to capillary beds in the anterior pituitary.

Oxytocin induces uterine contractions during birth and milk ejection during nursing. **Antidiuretic hormone (ADH)** functions in osmoregulation. Osmoreceptor cells in the hypothalamus, responding to an increase in blood osmolarity, signal neurosecretory cells of the hypothalamus to release ADH from their tips, located in the posterior pituitary. Binding of ADH sets off a signal transduction pathway that results in increased permeability of the collecting ducts to water. The osmoreceptors also stimulate thirst. Drinking and the reabsorption of water from the urine lower blood osmolarity and cause the hypothalamus to stop the release of ADH, completing the negative feedback loop.

■ **INTERACTIVE QUESTION   41.4**

What hormones are produced by the hypothalamus? To where and how are they transported?

Anterior Pituitary Hormones   The anterior pituitary produces a number of hormones. **Growth hormone (GH)** affects a variety of target tissues, promoting growth directly and stimulating the release of growth factors, such as **insulinlike growth factors (IGFs)** that are produced by the liver and stimulate bone and cartilage growth. Gigantism and acromegaly are human growth disorders caused by excessive GH. Hypopituitary dwarfism can now be treated with genetically engineered GH.

The hormone **prolactin (PRL)** is a protein very similar to GH, with diverse effects in different vertebrate species, ranging from milk production and secretion in mammals, to delay of metamorphosis in amphibians, to osmoregulation in fish.

Three of the tropic hormones produced by the anterior pituitary are glycoproteins. **Thyroid-stimulating hormone (TSH)** regulates release of thyroid hormones. **Follicle-stimulating hormone (FSH)** and **luteinizing hormone (LH)**, also called **gonadotropins**, stimulate gonad activity.

Several hormones come from fragments of pro-opiomelanocortin, a large protein cleaved into short pieces inside the pituitary cells. **Adrenocorticotropic hormone (ACTH)** stimulates the adrenal cortex to produce and secrete its steroid hormones. **Melanocyte-stimulating hormone (MSH)** regulates the activity of pigment-containing cells in the skin of some vertebrates, as seen in the color changes of amphibians. **Endorphins** are hormones that inhibit pain perception. Endorphins are also produced by certain neurons in the brain. Morphine and other opiates bind to endorphin receptors in the brain.

■ INTERACTIVE QUESTION 41.5

Fill in the following table to review the hormones stored and released by the posterior pituitary (**a** and **b**) and secreted by the anterior pituitary (**c–i**).

Hormone	Main Actions
Oxytocin	a.
ADH	b.
GH	c.
Prolactin	d.
TSH	e.
FSH and LH	f.
ACTH	g.
MSH	h.
Endorphins	i.

■ The vertebrate endocrine system coordinates homeostasis and regulates growth, development, and reproduction

The Thyroid Gland The **thyroid gland** produces two hormones, triiodothyronine (T_3) and thyroxine (T_4). Thyroid hormones are critical to vertebrate development and maturation; an inherited deficiency in humans results in retarded skeletal and mental development, a condition known as cretinism.

The thyroid gland contributes to homeostasis in mammals, helping to maintain normal blood pressure, heart rate, muscle tone, digestion, and reproductive functions. T_3 and T_4 increase cellular metabolism. Excess thyroid hormone results in hyperthyroidism, with symptoms of weight loss, irritability, high body temperature, and high blood pressure. Hypothyroidism can cause cretinism or weight gain and lethargy in adults. A goiter is an enlarged thyroid gland and may be associated with a lack of iodine in the diet.

A negative feedback loop controls secretion of thyroid hormones. Secretion of TSH-releasing hormone, or TRH, by the hypothalamus stimulates the anterior pituitary to secrete thyroid-stimulating hormone. TSH binds to receptors on the thyroid gland, generating cAMP and triggering the synthesis and release of thyroid hormones. High levels of T_3, T_4, and TSH inhibit the secretion of TRH.

The mammalian thyroid gland also secretes **calcitonin**, a peptide hormone that lowers calcium levels in the blood.

The Parathyroid Glands The four **parathyroid glands** secrete **parathyroid hormone (PTH)**, which stimulates the uptake of Ca^{2+} in the small intestine, its reabsorption in the kidney, and its release from bone to raise blood calcium levels. A balance between the antagonistic PTH and calcitonin maintains calcium homeostasis. Vitamin D is needed for proper PTH function.

The Pancreas Scattered within the exocrine tissue of the **pancreas** are clusters of endocrine cells known as the **islets of Langerhans**. Within each islet are **alpha cells** that secrete the hormone **glucagon** and **beta cells** that secrete the hormone **insulin**. These antagonistic hormones regulate glucose concentration in the blood, and negative feedback controls their secretion.

Insulin lowers blood glucose levels by promoting the movement of glucose into body cells from the blood, by slowing the breakdown of glycogen in the liver, and by inhibiting the conversion of amino acids and fatty acids to sugar. Glucagon raises glucose concentrations by stimulating the liver to increase glycogen hydrolysis, convert amino acids and fatty acids to glucose, and release glucose to the blood.

In diabetes mellitus, the absence of insulin in the bloodstream or the loss of response to insulin in target tissues reduces glucose uptake by cells. Glucose accumulates in the blood and is excreted in the urine. The body must use fats for fuel, and acidic metabolites from fat breakdown may lower blood pH.

Type I diabetes mellitus, also known as insulin-dependent diabetes, is an autoimmune disorder in which pancreatic cells are destroyed. This type of diabetes is treated by regular injections of genetically engineered human insulin. More than 90 percent of diabetics have **type II diabetes mellitus**, or non-insulin-dependent diabetes, characterized either by insulin deficiency or reduced responsiveness of target cells. Exercise and dietary control are often sufficient to manage this disease.

The Adrenal Glands In mammals, the **adrenal glands** consist of two different glands: the outer

■ **INTERACTIVE QUESTION   41.6**

Complete this concept map on the regulation of blood glucose levels.

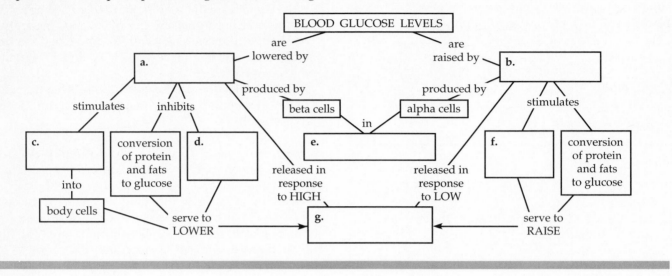

adrenal cortex and the central **adrenal medulla**. The adrenal medulla produces **epinephrine** (adrenalin) and **norepinephrine** (noradrenalin)—both of which are **catecholamines** synthesized from the amino acid tyrosine.

Epinephrine and norepinephrine, released in response to positive or negative stress, increase the availability of energy sources by stimulating glycogen hydrolysis in skeletal muscle and the liver, glucose release from the liver, and fatty acid release from fat cells. These hormones increase the rate and volume of the heart beat, dilate bronchioles in the lungs, and influence the contraction of smooth muscles to increase blood supply to the heart, brain, and skeletal muscles, while reducing the supply to the skin, digestive organs, and kidneys. The sympathetic division of the autonomic nervous system stimulates the release of epinephrine from the adrenal medulla. Norepinephrine also functions as a neurotransmitter in the nervous system.

The adrenal cortex responds to endocrine signals of stress. A releasing hormone from the hypothalamus causes the anterior pituitary to release ACTH. This tropic hormone stimulates the adrenal cortex to synthesize and secrete **corticosteroids**, a group of steroid hormones.

The two main human corticosteroids are the **glucocorticoids**, such as cortisol, and the **mineralocorticoids**, such as aldosterone. Glucocorticoids promote synthesis of glucose from noncarbohydrates and thus increase energy supplies during stress. Cortisone has been used to treat serious inflammatory conditions such as arthritis, but its immunosuppressive effects can be dangerous.

Mineralocorticoids affect salt and water balance. Aldosterone stimulates kidney cells to reabsorb sodium ions and water from the filtrate. Aldosterone secretion is regulated by the RAAS system in response to plasma ion concentrations, although ACTH secreted by the anterior pituitary in response to severe stress increases aldosterone secretion. Both glucocorticoids and mineralocorticoids appear to help maintain homeostasis during extended periods of stress.

A third group of corticosteroids produced by the adrenal cortex are sex hormones, in particular androgens, which appear to influence female sex drive.

■ **INTERACTIVE QUESTION   41.7**

Describe how the adrenal gland responds to short-term and long-term stress.

a. Short-term stress:

b. Long-term stress:

The Gonads   The testes of males and ovaries of females produce steroids that affect growth, development, and reproductive cycles and behaviors. The three major categories of gonadal steroids—androgens, estrogens, and progestins—are found in different proportions in males and females.

The testes primarily synthesize **androgens**, such as **testosterone**, which determine gender of the developing embryo and stimulate development of the male reproductive system and secondary sex characteristics. **Estrogens** regulate the development and maintenance of the female reproductive system and secondary sex characteristics. In mammals, **progestins** help prepare and maintain the uterus for the growth of an embryo.

A hypothalamic releasing hormone, GnRH, controls secretion of FSH and LH, gonadotropins from the anterior pituitary gland that control estrogen and androgen synthesis.

Other Endocrine Organs   Many organs whose main functions are nonendocrine, such as the digestive tract, kidney, and liver, secrete important hormones.

The **pineal gland**, located near the center of the mammalian brain, secretes **melatonin**, which regulates functions related to light and changes in day length. The secretion of melatonin at night may function with a biological clock for daily or seasonal activities such as reproduction.

The **thymus** plays a role in the immune system and is quite large during childhood. It secretes **thymosin** and other messengers that stimulate development and differentiation of T lymphocytes.

■ The endocrine system and the nervous system are structurally, chemically, and functionally related

The endocrine system and nervous system function together to control chemical communication and coordination in animals. They are structurally related in that many endocrine glands, such as the vertebrate hypothalamus, posterior pituitary, and the insect brain, are made of nervous tissue.

The endocrine and nervous systems are chemically related in that several vertebrate hormones are used by both systems. These systems are functionally related in that many physiological functions are coordinated by both nervous and hormonal communications, often in a type of neuroendocrine reflex. The nervous system often controls endocrine glands, and hormones affect the development and functioning of the nervous system.

STRUCTURE YOUR KNOWLEDGE

1. Create a concept map that illustrates your understanding of the mechanisms by which steroid and nonsteroid hormones affect their target cells.

2. Most homeostatic functions are maintained by a balance between hormones with antagonistic effects. Describe how the thyroid and parathyroid glands regulate calcium levels in the blood.

TEST YOUR KNOWLEDGE

MATCHING: *Match the hormone and gland or organ that produces it to the descriptions of hormone action. Choices may be used more than once, and not all choices are used.*

Hormones	Gland or Organ
A. ACTH	a. adrenal cortex
B. androgens	b. adrenal medulla
C. ADH	c. hypothalamus
D. calcitonin	d. pancreas
E. epinephrine	e. parathyroid
F. glucagon	f. pineal
G. glucocorticoids	g. pituitary
H. insulin	h. testis
I. melatonin	i. thymus
J. oxytocin	j. thyroid
K. thyroxine	

Hormone	Gland		Hormone Action
_____	_____	1.	involved in biological clock and seasonal activities
_____	_____	2.	break down muscle protein for conversion to glucose
_____	_____	3.	increase blood sugar, glycogen breakdown in liver
_____	_____	4.	stimulate development of male reproductive system
_____	_____	5.	stimulate adrenal cortex to synthesize corticosteroids
_____	_____	6.	increase available energy, heart rate, metabolism
_____	_____	7.	regulate metabolism, growth and development
_____	_____	8.	lower blood calcium levels
_____	_____	9.	increase reabsorption of water by kidney
_____	_____	10.	stimulate contraction of uterus, milk secretion

MULTIPLE CHOICE: *Choose the one best answer.*

1. Which of the following is *not* an accurate statement about hormones?
   a. Not all hormones are secreted by endocrine glands.
   b. Most hormones move through the circulatory system to their destination.
   c. Target cells have specific molecular receptors for hormones.
   d. All hormones are essential to homeostasis.
   e. Hormones produce their effects through signal transduction pathways involving second messengers.

2. The best description of the difference between pheromones and hormones is that
   a. Pheromones are small, volatile molecules, whereas hormones are steroids.
   b. Pheromones are involved in reproduction, whereas hormones are not.
   c. Pheromones are a form of olfactory communication; hormones are a form of chemical communication.
   d. Pheromones are signals that function between animals, whereas hormones communicate among the parts within an animal.
   e. Pheromones are local regulators, whereas hormones travel greater distances.

3. Which one of the following hormones is *incorrectly* paired with its origin?
   a. releasing hormones—hypothalamus
   b. growth hormone—anterior pituitary
   c. progesterone—ovary
   d. TSH—thyroid
   e. mineralocorticoids—adrenal cortex

4. In the second-messenger model of hormone action,
   a. A tropic hormone signals an endocrine gland to release a hormone.
   b. Another molecule relays the message from the membrane-bound hormone to the cytoplasmic enzymes.
   c. Negative feedback turns off the production of the hormone.
   d. Genes are turned on and new proteins are synthesized.
   e. IP_3 phosphorylates protein kinases.

5. Ecdysone
   a. is a steroid hormone produced in insects that promotes retention of larval characteristics.
   b. is responsible for the color changes in amphibians.
   c. is a peptide hormone secreted from "bag cells" that stimulates egg laying in the mollusk *Aplysia*.
   d. is secreted by prothoracic glands in insects and triggers molts and development of adult characteristics.
   e. is involved in metamorphosis in amphibians.

6. The role of cAMP in hormone action is to
   a. act as the second messenger and activate other enzymes.
   b. activate adenylyl cyclase.
   c. bind a specific hormone to the plasma membrane.
   d. release Ca^{2+} into the cell and initiate an enzyme cascade.
   e. serve as a relay between a hormone receptor complex and an effector.

7. Antidiuretic hormone (ADH)
   a. is produced by cells in the kidney and liver in response to low blood volume or pressure.
   b. stimulates the reabsorption of Na^+ from the urine.
   c. acts through a second-messenger system to increase the permeability of kidney collecting tubules to water.
   d. is a steroid hormone produced by the adrenal cortex.
   e. is part of the RAAS that regulates salt and water balance.

8. The anterior lobe of the pituitary
   a. stores oxytocin and ADH produced by the hypothalamus.
   b. is connected by portal vessels to capillary beds in the hypothalamus.
   c. produces several releasing hormones.
   d. produces several inhibitory hormones.
   e. does all of the above.

9. Acromegaly is caused by
   a. an autoimmune disorder.
   b. an excess of thyroxine.
   c. an excess of glucocorticoids.
   d. an abnormally high androgen-to-estrogen ratio.
   e. an excess of growth hormone.

10. The function of calmodulin is to
    a. begin an enzyme cascade by phosphorylating multiple proteins.
    b. bind with Ca^{2+} and regulate the activity of cellular proteins.
    c. lower blood calcium levels.
    d. serve as a second messenger in a signal transduction pathway.
    e. bind with vitamin D to increase Ca^{2+} uptake in the intestines.

11. Pro-opiomelanocortin
    a. is produced by the hypothalamus and is transported through a portal vessel.
    b. is cleaved into several hormones, including ACTH, MSH, and endorphins.
    c. produces several releasing hormones.
    d. mimics the effects of morphine and opiates.
    e. is a precursor to the hormones of the adrenal cortex.

12. Which of the following is *not* true of norepinephrine?
    a. It is secreted by the adrenal medulla.
    b. Its action is to increase the rate and volume of heart beat and constrict selected blood vessels.
    c. Its release is stimulated by ACTH.
    d. It serves as a neurotransmitter in the nervous system.
    e. It is part of the flight-or-fight response to stress.

Choose from the following hormones to answer questions 13–15.
    a. ACTH
    b. parathyroid hormone
    c. epinephrine
    d. glucagon
    e. insulin

13. Which of the above is a tropic hormone?

14. Which hormone lowers blood glucose levels?

15. Which hormone is released in response to a nervous impulse?

CHAPTER 42

ANIMAL REPRODUCTION

FRAMEWORK

This chapter covers the patterns and mechanisms of animal reproduction. In sexual reproduction, fertilization may occur externally or internally. Development of the zygote may take place internally within the female, externally in a moist environment, or in a protective, resistant egg. Placental mammals provide nourishment as well as shelter for the embryo, and they nurse their young.

The human reproductive system is described, including its organs, glands, hormones, gamete formation, and sexual response. The chapter also covers pregnancy and birth, contraception, and recently developed reproductive technologies.

CHAPTER REVIEW

■ Animals produce offspring by asexual or sexual reproduction

In **asexual reproduction**, a single individual produces offspring, relying totally on mitotic cell division. In **sexual reproduction**, two meiotically formed haploid **gametes** fuse to form the diploid **zygote** that develops into the offspring. Different individuals often provide the gametes.

Asexual Reproduction   Many invertebrates can reproduce asexually by **fission**, in which a parent is separated into two or more equal-sized individuals, by **budding**, in which a new individual grows out from the parent's body, or by **fragmentation**, in which the body is broken into several pieces, each of which develops into a complete animal. **Regeneration** is necessary for a fragment to develop into a new organism. Some invertebrates release specialized groups of cells that grow into new individuals, like the **gemmules** produced by sponges.

Sexual Reproduction   Gametes are usually a relatively large, nonmotile **ovum** and a small, flagellated **spermatozoon**. Sexual reproduction produces offspring with varying genotypes and phenotypes.

Reproductive Cycles and Patterns   Most animals show periodic cycles of reproductive activity. A combination of environmental and hormonal cues control the timing of these cycles, which may be linked to favorable conditions or energy supplies.

The freshwater crustacean *Daphnia*, like aphids and rotifers, produces eggs that develop by **parthenogenesis**, as well as eggs that are fertilized. In bees, wasps, and ants, males are produced parthenogenetically, whereas sterile worker females and reproductive females are produced from fertilized eggs. In a few parthenogenetic fishes, amphibians, and lizards, doubling of chromosomes creates diploid "zygotes."

In **hermaphroditism**, found in some sessile, burrowing, and parasitic animals, each individual has functioning male and female reproductive systems. Mating results in fertilization of both individuals. In some fishes and oysters, individuals reverse their sex during their lifetime, a pattern known as **sequential hermaphroditism**. Sex reversal may be related to age or to the relative advantage conferred by size. In a **protandrous** species, an organism is first a male; in a **protogynous** species, it is first a female.

■ **INTERACTIVE QUESTION   42.1**

a. What adaptive advantage would asexual reproduction provide?

b. What adaptive advantage would sexual reproduction provide?

In sexual reproduction, gametes unite in the external environment or within the female

In **external fertilization**, eggs and sperm are shed, and fertilization occurs in the environment. This type of reproduction occurs almost exclusively in moist habitats, where the gametes and developing zygote are not in danger of desiccation.

Internal fertilization involves the placement of sperm in or near the female reproductive tract so that the egg and sperm unite internally. Behavioral cooperation, as well as copulatory organs and sperm receptacles, are required for internal fertilization.

Protection of the Embryo Developing embryos may receive some type of protection. The amniote eggs of birds, reptiles, and monotremes protect the embryo in a terrestrial environment. The embryos of placental mammals develop within the uterus, nourished from the mother's blood supply through the placenta. Parental care of young is widespread among animals.

■ INTERACTIVE QUESTION 42.2

List three mechanisms that may help to ensure that gamete release is synchronized when fertilization is external.

a.

b.

c.

Diverse reproductive systems have evolved in the animal kingdom

Invertebrate Reproductive Systems Some invertebrates, such as polychaete annelids, have separate sexes but do not have distinct **gonads**; eggs and sperm develop from cells lining the coelom. Gametes are stored in the coelom and shed through excretory organs or released by the splitting of the parent.

Insects have separate sexes and complex reproductive systems. Sperm develop in the testes, are stored in the seminal vesicles, and are ejaculated into the female. Eggs, produced in the ovaries, pass through ducts to be fertilized in the vagina. Females may have a **spermatheca**, or sperm-storing sac.

Flatworms are hermaphroditic. The female reproductive system includes yolk and shell glands and a uterus, where eggs are fertilized and may begin development. The male system includes a copulatory apparatus. Mating involves mutual insemination.

Vertebrate Reproductive Systems With the exception of most mammals, vertebrates have a common opening, a **cloaca**, for the digestive, excretory, and reproductive systems. Nonmammalian vertebrates do not have well-developed penises.

■ INTERACTIVE QUESTION 42.3

List the openings to the digestive, excretory, and reproductive systems present

a. in most male mammals:

b. in most female mammals:

Human reproduction involves intricate anatomy and complex behavior

Reproductive Anatomy of the Human Male The external male **genitalia** include the scrotum and penis. Sperm are produced in the highly coiled **seminiferous tubules** of the **testes**. **Interstitial cells** produce testosterone and other androgens. Sperm production requires a cooler temperature than the internal body temperature of most mammals, so the **scrotum** suspends the testes below the abdominal cavity.

Sperm pass from a testis into the coiled **epididymis**, in which they mature, gain motility, and are stored. During **ejaculation**, sperm are propelled through the **vas deferens**, into a short **ejaculatory duct**, and out through the **urethra** which runs through the penis.

Three sets of accessory glands add secretions to the **semen**. The **seminal vesicles** contribute a fluid containing mucus, amino acids, fructose (as an energy source for the sperm), and prostaglandins (which stimulate uterine contractions). The **prostate gland** produces an alkaline secretion that balances the acidity of the urethra and the vagina. The **bulbourethral glands** produce a small amount of viscous fluid that may neutralize urine remaining in the urethra.

The **penis** is composed of spongy tissue that engorges with blood during sexual arousal, producing an erection that facilitates insertion of the penis into the vagina. Some mammals have a **baculum**, a bone that helps stiffen the penis. The head of the penis, called the **glans penis**, is covered by a fold of skin called the **prepuce**, or foreskin.

Reproductive Anatomy of the Human Female   The external genitalia include the clitoris and two sets of labia. The female gonads, the **ovaries**, contain many **follicles**, which are sacs of cells that nourish and protect the egg cell contained within each of them. The follicle produces estrogens, the primary female sex hormones. Following **ovulation**, the follicle forms a solid mass called the **corpus luteum**, which secretes progesterone and estrogen.

The egg cell is expelled into the abdominal cavity and swept by cilia into the **oviduct**, or fallopian tube, through which it is transported to the **uterus**. The **endometrium**, or lining of the uterus, is highly vascularized. The neck of the uterus, the **cervix**, opens into the **vagina**, the thin-walled birth canal and repository for sperm during copulation.

The vaginal orifice is initially covered by a vascularized membrane called the **hymen**. The separate openings of the vagina and urethra are in a region called the **vestibule**, enclosed by two pairs of skin folds, the inner **labia minora** and the outer **labia majora**. The **clitoris**, composed of erectile tissue, is at the top of the vestibule. **Bartholin's glands** secrete mucus into the vestibule during sexual arousal.

■ **INTERACTIVE QUESTION   42.4**

Label the indicated structures in these diagrams of the human male and female reproductive system.

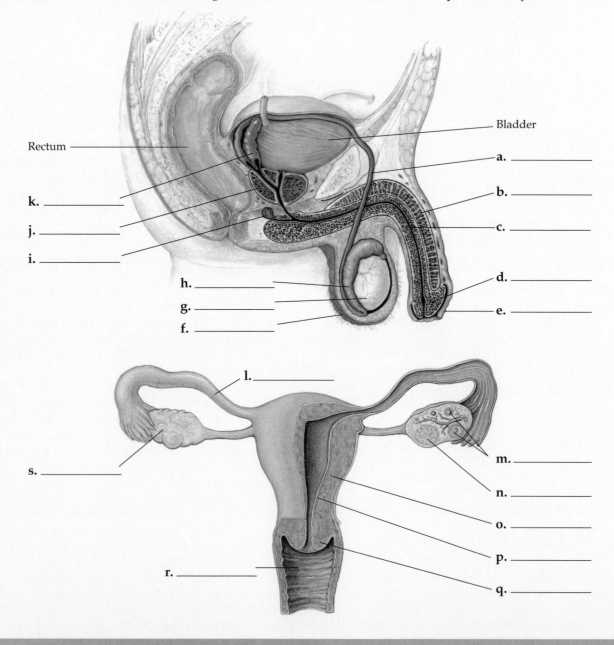

The **mammary gland**, or breast, is composed of fatty tissue and a series of small milk-secreting sacs that drain into ducts that join together and open at the nipple.

Human Sexual Response The human sexual response cycle includes two types of physiological reactions: **vasocongestion**, increased blood flow to a tissue, and **myotonia**, increased muscle tension. The excitement phase involves vasocongestion, myotonia, and vaginal lubrication, preparing the vagina and penis for **coitus**, or sexual intercourse. The plateau phase continues vasocongestion and myotonia, and breathing rate and heart rate increase. In both sexes, **orgasm** is characterized by rhythmic, involuntary contractions of reproductive structures. In males, emission deposits semen in the urethra, and ejaculation occurs when the urethra contracts and semen is expelled. In the resolution phase, vasocongested organs return to normal size, muscles relax, and nonreproductive reactions (breathing and heart rates) return to normal.

Spermatogenesis and oogenesis both involve meiosis but differ in three significant ways

Spermatogenesis, the production of mature sperm cells, occurs continuously in the seminiferous tubules of the testes. The haploid nucleus of a sperm is contained in a thick head, tipped with an **acrosome**, which contains enzymes that help the sperm penetrate the egg. Numerous mitochondria provide ATP for movement of the flagellum, or tail.

In **oogenesis**, meiotic cytokinesis is unequal, producing one large ovum and up to three small haploid polar bodies that disintegrate. At birth, an ovary already contains all its cells that will develop into eggs. Oogenesis happens in stages: primary oocytes enlarge within follicles and enter prophase I between birth and puberty; after puberty, FSH periodically

---

■ **INTERACTIVE QUESTION 42.5**

List the three ways in which oogenesis differs from spermatogenesis.

a.

b.

c.

---

stimulates the first meiotic division to occur within maturing follicles; the secondary oocyte is released during ovulation, and the second meiotic division is triggered by penetration of the egg cell by the sperm.

■ A complex interplay of hormones regulates reproduction

The Male Pattern Androgens are responsible for the male primary (associated with reproduction) and secondary (associated with the voice, hair distribution, and muscle growth) sex characteristics. The most important androgen is testosterone, produced mainly by interstitial cells of the testes. Androgens are also determinants of sexual and other behaviors. Androgen secretion and sperm production are controlled by hormones from the anterior pituitary and hypothalamus.

The Female Pattern Female humans and many primates have **menstrual cycles**, during which the endometrium lining thickens to prepare for the implantation of the embryo and then is shed if fertilization does not occur. This bleeding, called **menstruation**, occurs on a cycle of approximately 28 days in humans. Other mammals have **estrous cycles**, during which the endometrium thickens, but, if fertilization does not occur, it is reabsorbed. Estrous cycles are often coordinated with season or climate, and females are receptive to sexual activity only during **estrus**, or heat, the period surrounding ovulation.

The human menstrual cycle consists of the **menstrual flow phase**, the few days of menstrual bleeding; the **proliferative phase**, the week or two during which the endometrium begins to thicken; and the **secretory phase**, about two weeks during which the endometrium becomes more vascularized and develops glands that secrete a glycogen-rich fluid.

The **ovarian cycle** begins with the **follicular phase**, during which an egg cell enlarges, its follicular cells become multilayered and enclose a fluid-filled cavity, and the large, mature follicle forms a bulge near the ovary surface. Ovulation occurs during the **ovulatory phase** with the rupture of the follicle and adjacent ovary wall. The remaining follicular tissue develops into the hormone-secreting corpus luteum during the **luteal phase**.

Five hormones coordinate the menstrual and ovarian cycles using positive and negative feedback. During the follicular phase, the hypothalamus secretes GnRH (gonadotropin-releasing hormone), which stimulates the anterior pituitary to secrete small amounts of FSH (follicle-stimulating hormone) and LH (luteinizing hormone). FSH stimulates follicular growth, and the cells of the follicle secrete estrogen. When the level of estrogen secretion rises sharply, the hypothalamus is stimulated to increase

GnRH output, which results in a rise in LH and FSH release.

By positive feedback, the increase in LH, caused by increased estrogen secretion from the follicle, induces maturation of the follicle, and ovulation occurs.

In the luteal phase, LH stimulates the transformation of the ruptured follicle and maintains the corpus luteum, which secretes estrogen and progesterone. The rising level of these two hormones exerts negative feedback on the hypothalamus and pituitary, inhibiting secretion of LH and FSH. The corpus luteum, deprived of its LH stimulation, disintegrates, thereby dropping the levels of estrogen and progesterone. This drop releases the inhibition of the hypothalamus and pituitary, and FSH and LH secretion begins again, stimulating growth of new follicles and the start of the next follicular phase.

The hormones of the ovarian cycle synchronize the menstrual cycle and the preparation of the uterus for possible implantation of an embryo. Estrogen causes the endometrium to begin to thicken in the proliferative phase. After ovulation, estrogen and progesterone stimulate increased vascularization and gland development of the endometrium. Thus, the luteal phase of the ovarian cycle corresponds with the secretory phase of the menstrual cycle. The rapid drop of ovarian hormones caused by the disintegration of the corpus luteum reduces blood supply to the endometrium and begins its disintegration, leading to the menstrual phase of the next cycle.

Estrogens are also responsible for female secondary sex characteristics and influence sexual behavior.

Hormones and Sexual Maturation   In humans, the onset of reproductive ability, termed puberty, is a gradual process that is completed around the age of 12 to 14 and includes the onset of menstruation or the production of viable sperm, as well as the development of secondary sex characteristics. (See Interactive Question 42.6, p. 313.)

■ Embryonic and fetal development occur during pregnancy in humans and other placental mammals

From Conception to Birth   The development of one or more **embryos** in the uterus, preceded by **conception** and ending with birth, is called **pregnancy** or **gestation**. The gestation period correlates with body size and the degree of development of the young at birth. Human pregnancy averages 266 days (38 weeks).

Human gestation can be divided into three **trimesters**. About 24 hours after fertilization in the oviduct, the zygote begins to divide; it travels to the uterus in about 3 to 4 days. After about a week of **cleavage**, the zygote develops into a hollow ball of cells called a **blastocyst** and implants in the endometrium in about 5 more days. Tissues grow out of the developing embryo to meet with the endometrium and form the **placenta**, a disk-shaped organ in which gas and nutrient exchange and waste removal take place between the maternal and embryonic circulations. Differentiation leads to **organogenesis**, and, by the eighth week, the embryo has all the rudimentary structures of the adult and is called a **fetus**.

The embryo secretes **human chorionic gonadotropin (HCG)**, which maintains the corpus luteum's secretion of progesterone and estrogen through the first trimester. High progesterone levels initiate growth of the maternal part of the placenta, enlargement of the uterus, and cessation of ovulation and menstrual cycling.

During the second trimester, HCG declines, the corpus luteum degenerates, and the placenta secretes its own progesterone, which maintains the pregnancy. The fetus grows rapidly and is quite active. The third trimester is a period of rapid fetal growth.

High levels of estrogen stimulate the development of oxytocin receptors on the uterus. Oxytocin produced by the fetus and released from the mother's posterior pituitary stimulates contractions and uterine secretion of prostaglandins, which enhance uterine contractions.

Labor consists of a series of strong contractions of the uterus and results in birth, or **parturition**. During the first stage of labor, the cervix dilates. The second stage consists of the contractions that force the fetus out of the uterus and through the vagina. The placenta is delivered in the final stage of labor.

Lactation is unique to mammals. Prolactin secretion stimulates milk production, and oxytocin controls the release of milk during nursing.

Reproductive Immunology   Several hypotheses attempt to explain why a mother does not reject an embryo, which has paternal as well as maternal chemical markers. There is evidence that the trophoblast, a protective layer that develops from the blastocyst and surrounds the embryo, induces the development of white blood cells that suppress other white cells from mounting an immune attack. This suppression may occur only after an initial immune response to the trophoblast. Some researchers suggest that if the initial

■ **INTERACTIVE QUESTION   42.6**

In the diagram of the human menstrual cycle, label the lines indicating the levels of the gonadotropic and the ovarian hormones, and the phases of the ovarian and menstrual cycles.

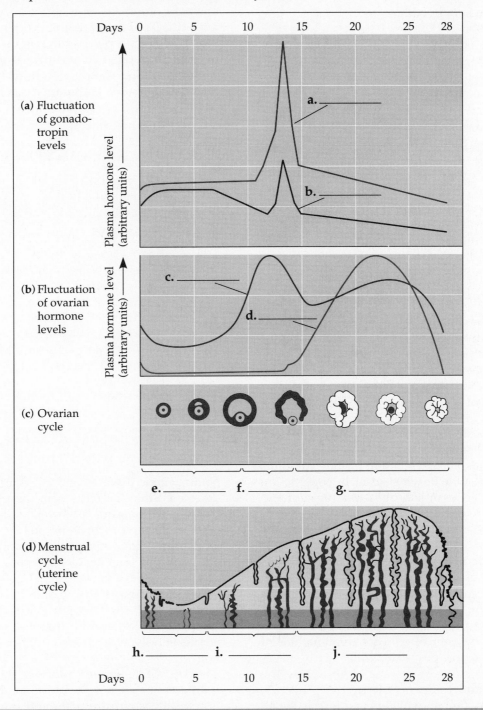

▣ INTERACTIVE QUESTION   42.7

List the functions of the following hormones and where they are secreted:

a. human chorionic gonadotropin

b. progesterone

c. oxytocin

response is too weak, then suppression may not occur, and the continued immunological attack results in spontaneous abortion of the embryo.

▪ Contraception prevents pregnancy

Contraception, the deliberate prevention of pregnancy, can be accomplished by one of several methods: preventing fertilization, preventing implantation, or preventing release of egg or sperm. The **rhythm method** is based on refraining from intercourse during the period that conception is most likely, the few days before and after ovulation. *Coitus interruptus* is an undependable method of contraception. Abstinence is 100 percent effective.

Barrier methods of contraception include **condoms** and diaphragms, which, when used in conjunction with spermicidal foam or jelly, present a physical and chemical barrier to fertilization.

The intrauterine device (IUD) probably prevents implantation by irritating the endometrium. IUDs have a low failure rate but have been associated with harmful side effects.

The release of gametes may be prevented by chemical contraception, as in birth control pills, and by sterilization, as in **tubal ligation** in women or **vasectomy** in men. Birth control pills are combinations of synthetic estrogen and progestin, which act by negative feedback to stop the release of GnRH by the hypothalamus and FSH and LH by the pituitary, resulting in a cessation of ovulation and follicle development. A progestin minipill, which will release progestin for up to 5 years when inserted under the skin, alters the cervical mucus so that it blocks sperm entry

to the uterus. Cardiovascular problems are a potential effect of birth control pills.

Abortion is the termination of a pregnancy. Spontaneous abortion, or miscarriage, occurs in as many as one-third or more of all pregnancies. The drug RU-486 is an analog of progesterone that blocks progesterone receptors in the uterus and can be used to terminate a pregnancy in the first few weeks.

Condoms are the only form of birth control that offer some protection against sexually transmitted diseases.

▣ INTERACTIVE QUESTION   42.8

List the three general types, along with examples, of birth control methods. Which examples are most likely to prevent pregnancy? Which are least likely to do so?

a.

b.

c.

▪ New technologies offer help for reproductive problems

Some genetic diseases and congenital defects can be detected while the fetus is in the uterus. The use of **ultrasound imaging** produces an image of the fetus from the echoes of high-frequency sound waves. In **amniocentesis**, a sample of amniotic fluid is withdrawn with a long needle, and fetal cells are cultured and analyzed for chromosomal defects. In **chorionic villus sampling (CVS)**, a small sample of tissue is removed from the fetal portion of the placenta. CVS can be done earlier and produces faster results than amniocentesis, but carries a higher risk of complications.

In vitro **fertilization** involves the removal of ova from a woman whose oviducts are blocked, fertilization within a culture dish, and implantation of the developing embryo in the uterus.

The development of male chemical contraception is currently under research.

STRUCTURE YOUR KNOWLEDGE

1. Trace the path of a human sperm from the point of production to the point of fertilization, briefly commenting on the functions of both the structures it passes through and the associated glands.

2. Answer the following questions concerning the human menstrual cycle.
   a. What does GnRH do?

   b. What does FSH stimulate?

   c. What causes the spike in LH level?

   d. What does this high level of LH induce?

   e. What does LH maintain during the luteal phase?

   f. What inhibits secretion of LH and FSH?

   g. What allows LH and FSH secretion to begin again?

3. Describe how birth control pills work. How does the French drug RU-486 function?

TEST YOUR KNOWLEDGE

FILL IN THE BLANKS

_____ 1. type of asexual reproduction in which a new individual grows while attached to the parent's body

_____ 2. sequential hermaphroditic in which an organism is first a male and then becomes a female

_____ 3. development of egg without fertilization

_____ 4. individual with functioning male and female reproductive systems

_____ 5. common opening of digestive, excretory, and reproductive systems in non-mammalian vertebrates

_____ 6. type of reproductive cycle in which thickened endometrium is reabsorbed

_____ 7. hormone that prepares the uterus for pregnancy

_____ 8. common duct for urine and semen in mammalian males

_____ 9. filling of a tissue with blood due to increased blood flow

_____ 10. period when reproductive ability begins in humans

MULTIPLE CHOICE: *Choose the one best answer.*

1. Which of the following is an explanation for the periodicity of reproductive cycles in animals?
   a. Reproduction may correspond with periods of increased food supply, during which energy can be invested in gamete formation.
   b. Seasonal cycles may allow offspring to be produced during favorable environmental conditions when chances of survival are highest.
   c. Hormonal control of reproduction may be tied to biological clocks and seasonal cues.
   d. Synchronicity in release of gametes increases probability of fertilization.
   e. All of these may contribute to periodic reproductive activity.

2. Which of the following is least likely to be hermaphroditic?
   a. earthworm
   b. barnacle
   c. tapeworm
   d. honeybee
   e. liver fluke

3. Which of the following is *incorrectly* paired with its function?
   a. seminiferous tubules—add fluid containing mucus, fructose, and prostaglandins to semen
   b. scrotum—encases testes and suspends them below abdominal cavity
   c. epididymis—stores sperm
   d. prostate gland—adds alkaline secretion to semen
   e. vas deferens—transports sperm from epididymis to ejaculatory duct

4. The function of the corpus luteum is to
   a. nourish and protect the egg cell.
   b. produce prolactin in the milk sacs of the mammary gland.
   c. produce progesterone and estrogen.
   d. produce estrogens and disintegrate following ovulation.
   e. maintain pregnancy by production of human chorionic gonadotropin.

5. Which of the following hormones is *incorrectly* paired with its function?
   a. GnRH—controls release of FSH and LH
   b. oxytocin—stimulates uterine contractions during parturition
   c. estrogen—responsible for primary and secondary female sex characteristics
   d. LH—stimulates secretion of estrogen by developing follicle
   e. progesterone—secreted by corpus luteum and stimulates growth of placenta

6. Myotonia is
   a. a congenital birth defect.
   b. the hormone responsible for breast development.
   c. muscle tension.
   d. the rhythmic contractions of reproductive organs during orgasm.
   e. responsible for delivery of the placenta.

7. The secretory phase of the menstrual cycle
   a. is associated with dropping levels of estrogen and progesterone.
   b. is when the endometrium begins to degenerate and menstrual flow occurs.
   c. involves the initial proliferation of the endometrium.
   d. corresponds with the follicular phase of the ovarian cycle.
   e. corresponds with the luteal phase of the ovarian cycle.

8. Examples of birth control methods that prevent release of gametes from the gonads are
   a. sterilization and chemical contraception.
   b. birth control pills and IUDs.
   c. condoms and diaphragms.
   d. abstinence and chemical contraception.
   e. the progestin minipill and RU-486.

9. The ability of a pregnant woman not to reject her "foreign" fetus may be due to
   a. the fact that fetal and maternal blood never mix.
   b. the suppression of her immune response to the paternal chemical markers on fetal tissue.
   c. the protection of the fetus within the trophoblast.
   d. the production of human chorionic gonadotropin that maintains the pregnancy.
   e. the masking of paternal markers by specialized white blood cells.

10. Certain maternal diseases, drugs, alcohol, and radiation are most dangerous to embryonic development
    a. during the first 2 weeks when the embryo has not yet implanted and spontaneous abortion may occur.
    b. during the first 2 months when organogenesis is occurring.
    c. during the first and second trimesters when the embryonic liver is not yet filtering toxins.
    d. during the second trimester when the corpus luteum no longer secretes estrogen and progesterone.
    e. during the third trimester when the most rapid growth is occurring.

CHAPTER 43

ANIMAL DEVELOPMENT

FRAMEWORK

Fertilization initiates physical and molecular changes in the egg cell. Early embryonic development includes cleavage of the fertilized egg, gastrulation, and organogenesis. The amount of yolk in the egg and the evolutionary history of the animal determine how these processes occur in any particular animal group.

Determination and differentiation of cells are the result of the control of gene expression by both cytoplasmic determinants and the positional information a cell receives within a developing structure. Developmental biologists, using transplant experiments and techniques from molecular biology, are gradually unraveling some of the mechanisms and regulatory genes underlying the complex processes of animal development.

CHAPTER REVIEW

Animal development charts the course from a single fertilized egg to an organism made of many differentiated cells organized into specialized tissues and organs. The combination of molecular genetics, classical studies, and the use of well-known research organisms, such as the fruit fly and mouse, are helping developmental biologists to ask and answer basic questions about embryonic development.

■ From egg to organism, an animal's form develops gradually: *the concept of epigenesis*

Preformation, or the belief that the egg or sperm contains a miniature embryo, was once the favored theory of animal development. **Epigenesis** is the now accepted theory that the form of an embryo gradually devel-

ops from an egg. The developmental plan of an embryo is predetermined by the zygote genome and by the distribution of material in the egg cytoplasm. As cell division separates cytoplasmic components, nuclei are exposed to different environments that affect which genes are expressed by different cells. Signal transduction pathways are often involved in conveying developmental messages between cells.

■ Embryonic development involves cell division, differentiation, and morphogenesis

The three key processes of embryonic development are cell division, the production of large numbers of cells; **differentiation**, the formation of specialized cells that make up tissues and organs; and **morphogenesis**, the movement of cells and tissues to produce body shape and form. All three processes must be tightly regulated in order for normal development to occur.

■ Fertilization activates the egg and brings together the nuclei of sperm and egg

Fertilization, the union of egg and sperm, combines the haploid sets of chromosomes of these specialized cells and activates the egg by initiating metabolic reactions that trigger embryonic development.

The Acrosomal Reaction In sea urchin fertilization, the acrosome at the tip of the sperm discharges hydrolytic enzymes when it comes in contact with the jelly coat of an egg. This **acrosomal reaction** allows the acrosomal process to elongate through the jelly coat. Fertilization within the same species is assured when a protein on the surface of the acrosomal process attaches to specific receptor molecules on the vitelline layer of the egg, which is external to the plasma membrane.

In response to the fusion of sperm and egg plasma membranes, ion channels open in the egg membrane, and sodium ions flow into the egg. The resulting depolarization of the membrane prevents other sperm cells from fusing with the egg, providing a **fast block to polyspermy**.

The Cortical Reaction   Membrane fusion also initiates the **cortical reaction**, involving a signal transduction pathway in which a G protein triggers release of calcium ions into the egg cytoplasm. In response to the Ca^{2+} increase, **cortical granules** located in the outer cortex of the egg fuse with the plasma membrane and release their contents into the perivitelline space. The released enzymes and macromolecules cause the vitelline layer to elevate and harden to form the **fertilization membrane**, which functions as a **slow block to polyspermy**.

Activation of the Egg   The release of the second messenger Ca^{2+} activates the egg by increasing the rates of cellular respiration and protein synthesis. Hydrogen ions leave the cell, and the increase in cellular pH may be the trigger for activation.

Parthenogenetic development can be initiated by injecting calcium into an egg. Even an enucleated egg can be activated to begin protein synthesis, showing that inactive mRNA had been stockpiled in the egg.

After the sperm nucleus fuses with the egg nucleus, DNA replication begins in preparation for the cleavage division that begins the development of the embryo.

▓ INTERACTIVE QUESTION   43.1

a. What assures that only sperm of the right species fertilize sea urchin eggs?

b. What assures that only one sperm will fertilize an egg?

Fertilization in Mammals   Following internal fertilization, secretions of the mammalian female reproductive tract enhance motility and alter surface molecules of the sperm. A thus capacitated sperm migrates through the follicle cells released with the egg to the **zona pellucida**, a three-dimensional filament network made of three types of glycoproteins. Complementary binding of a molecule on the sperm head to the glycoprotein ZP3 induces an acrosomal reaction. The hydrolytic enzymes released from the acrosome enable the sperm cell to penetrate the zona pellucida.

Binding of a sperm membrane protein with the egg membrane triggers a depolarization and fast block to polyspermy. Release of granules from the egg cortex during the cortical reaction cause the hardening of the zona pellucida and a slow block to polyspermy.

Microvilli of the egg enclose the entire sperm. The basal body of the sperm flagellum divides to form centrioles that function in cell division. Nuclear envelopes disperse and the first mitotic division occurs.

▪ Cleavage partitions the zygote into many smaller cells

Cleavage is a succession of rapid cell divisions during which the embryo becomes partitioned into many small cells, called **blastomeres**. Cleavage parcels different regions of the cytoplasm, which contain different cytoplasmic components, into cells and sets the stage for later development.

The axis of the egg of many frogs and other animals is defined by the **vegetal pole**, where the stored nutrients in **yolk** are most concentrated, and the opposite end called the **animal pole**, where the polar bodies budded from the egg during meiosis and the anterior part of the embryo usually forms. The animal hemisphere has melanin granules in the outer cytoplasm, whereas the vegetal hemisphere contains the yellow yolk. Fertilization results in a rotation of the outer cytoplasm toward the point of sperm entry, creating an opposite narrow **gray crescent** that serves as a marker of egg polarity.

Yolk impedes cell division, and smaller cells are produced in the animal hemisphere in the frog. In species in which eggs have little yolk, blastomeres are of equal size, although an animal–vegetal axis results from other chemical concentration gradients.

The first two cleavage divisions are vertical, or polar; the third division is equatorial. Deuterotomes—echinoderms and chordates—exhibit radial cleavage, in which each tier of cells aligns directly with the cells of the lower tier. Most protostomes—annelids, arthropods, and mollusks—have spiral cleavage with upper tier cells in the grooves between lower tier cells.

Further cleavage produces a **morula**, a solid ball of cells, followed by the arrangement of cells into a hollow ball, the **blastula**, surrounding a fluid-filled cavity called the **blastocoel**. The frog blastula has its blastocoel restricted to the animal hemisphere due to unequal cell division.

The large amount of yolk in eggs of birds results in **meroblastic cleavage** restricted to the small disc of cytoplasm on top of the yolk. **Holoblastic cleavage** is the complete division of eggs with little or moderate amounts of yolk. In the yolk-rich eggs of insects, the zygote nucleus undergoes repeated divisions and the nuclei migrate to the egg surface, forming a stage called a **syncytium**. Plasma membranes eventually form around the nuclei, creating a blastula made of cells surrounding a mass of yolk.

■ INTERACTIVE QUESTION   43.2

Identify the early embryonic stages diagrammed below and indicate whether they represent a sea urchin, frog, or fruit fly embryo. Label the indicated regions or structures.

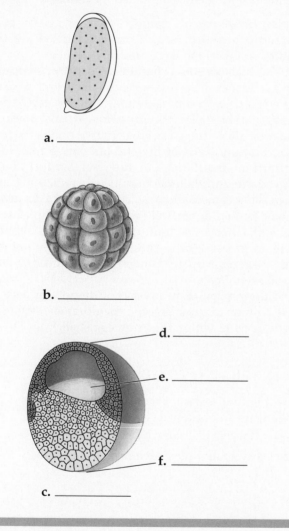

a. _____

b. _____

d. _____

e. _____

f. _____

c. _____

■ Gastrulation rearranges the blastula to form a three-layered embryo with a primitive gut

Changes in cell motility, shape, and adhesion are part of the morphogenetic rearrangements of cells in **gastrulation** that result in a three-layered embryo called the **gastrula**. The gastrula consists of the primary

germ layers—the outer **ectoderm**, the middle **mesoderm**, and the inner **endoderm**—that line the embryonic digestive tract. Different organs develop from these three layers of embryonic tissues. This triploblastic body plan is characteristic of most animal phyla.

In a sea urchin, gastrulation occurs when cells at the vegetal pole flatten into a plate that then buckles inward by a process known as **invagination**. Some cells near the plate detach and move into the blastocoel as migratory mesenchyme cells. The invagination deepens to form a narrow pouch called the **archenteron**, a cavity that will become the digestive tract. The archenteron opening is the **blastopore**, which develops into the anus. Cytoplasmic extensions (filopodia) from mesenchyme cells at the tip of the archenteron adhere to the blastocoel wall and pull the archenteron to the animal pole. The tip fuses with the outer wall to form the mouth. This three-layered gastrula develops into a ciliated larva.

Gastrulation in frog development is more complicated because of the multilayered blastula wall and the large, yolk-filled cells of the vegetal hemisphere. A group of cells burrowing inward begins gastrulation in the region of the gray crescent, forming a tuck that will become the **dorsal lip** of the blastopore. In a process called **involution**, surface cells roll over the dorsal lip and migrate along the roof of the blastocoel. The blastopore eventually forms a circle surrounding a **yolk plug** of large, yolk-filled cells. Gastrulation produces an endoderm-lined archenteron surrounded by mesoderm, with the outer layer of the gastrula composed of ectoderm. (See Interactive Question 34.3, p. 320.)

■ Organogenesis forms the organs of the animal body from the three embryonic germ layers

Rudiments of organs develop from the primary germ layers in a process known as **organogenesis**. Morphogenetic changes include folds, splits, and dense clustering of cells. The **notochord** forms from condensation of dorsal mesoderm along the roof of the archenteron. Ectoderm above the developing notochord forms a neural plate. The edges of the plate fold upward and join to form a hollow **neural tube**, which will develop into the central nervous system. The elongating notochord stretches the anterior–posterior axis of the embryo.

Mesoderm along the sides of the notochord condenses and separates into blocks, called **somites**, which give rise to the segmented vertebrae and muscles associated with the axial skeleton. Lateral to the somites, the mesoderm splits to form the lining of the coelom.

■ INTERACTIVE QUESTION 43.3

Label the parts in these diagrams of a developing sea urchin and frog gastrula.

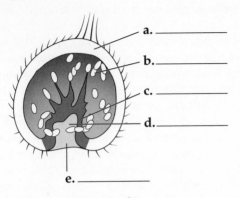

a. _____

b. _____

c. _____

d. _____

e. _____

a. Gastrulation in sea urchin

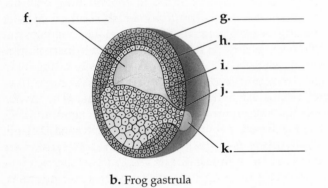

f. _____   g. _____

h. _____

i. _____

j. _____

k. _____

b. Frog gastrula

■ INTERACTIVE QUESTION 43.4

Label the indicated structures in this diagram of organogenesis in a frog embryo.

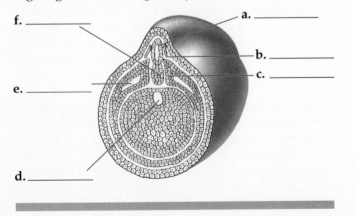

f. _____   a. _____

b. _____

c. _____

e. _____

d. _____

A band of cells called the **neural crest** separates from the meeting margins of the neural tube when it forms. These cells later migrate to form pigment cells in the skin, bones and muscles of the skull, teeth, the adrenal medulla, and components of the peripheral nervous system.

■ Amniote embryos develop in a fluid-filled sac within a shell or uterus

Reptiles, birds, and mammals are **amniotes**, meaning that they create the aqueous environment necessary for embryonic development with a fluid-filled sac surrounded by a membrane called the amnion.

Avian Development   Meroblastic cleavage of the fertilized egg produces a cap of cells called the **blastodisc** on top of the large, undivided yolk.

The blastomeres separate into an upper epiblast and a lower hypoblast layer, forming a cavity between them comparable to a blastocoel. A straight invagination, called the **primitive streak**, marks the anterior–posterior axis of the embryo and is the site of gastrulation. Cells of the epiblast move toward the primitive streak, detach, and ingress, forming the mesoderm and the endoderm. Cells remaining in the epiblast become ectoderm. The borders of the embryonic disc fold down and join, forming a three-layered tube attached by a stalk to the yolk. The hypoblast cells form part of the yolk sac and connecting stalk. Organogenesis proceeds in a fashion similar to that of the frog embryo.

The primary germ layers and hypoblast also give rise to four **extraembryonic membranes**, essential to development within the avian egg shell.

■ INTERACTIVE QUESTION 43.5

List the four extraembryonic membranes found in an avian egg and briefly describe their functions.

a.

b.

c.

d.

Mammalian Development   Fertilization occurs in the oviduct, and embryonic development begins on the journey to the uterus. The egg of placental mammals has little yolk, thus cleavage is holoblastic. Gastrulation and early organogenesis, however, are similar in pattern to those of birds and reptiles.

The **blastocyst** consists of an outer epithelium, the **trophoblast**, surrounding a cavity into which protrudes a cluster of cells, called the **inner cell mass**. These inner cells develop into the embryo and some of the extraembryonic membranes. The trophoblast secretes enzymes that enable the blastocyst to embed in the uterine lining (endometrium) and extends projections that develop, along with mesodermal tissue, into the fetal portion of the placenta.

The inner cell mass forms a flat embryonic disc with an upper epiblast and lower hypoblast, resembling the two-layered blastodisc of birds and reptiles. In gastrulation, cells from the epiblast move through a primitive streak to form the mesoderm and endoderm.

The four extraembryonic membranes are homologous to those of reptiles and birds. The chorion surrounds the embryo and the other membranes. The amnion encloses the embryo in a fluid-filled amniotic cavity. The yolk-sac membrane, enclosing a small fluid-filled cavity, is the site of early formation of blood cells. The allantois forms blood vessels that connect the embryo with the placenta through the umbilical cord.

Organogenesis begins with the formation of the notochord, neural tube, and somites. In the human embryo, all major organs have begun development by the end of the first trimester.

■ **INTERACTIVE QUESTION   43.6**

Label the indicated cell layers in this diagram of a human blastocyst implanted into the endometrium. List the embryonic structures that will develop from each cell layer.

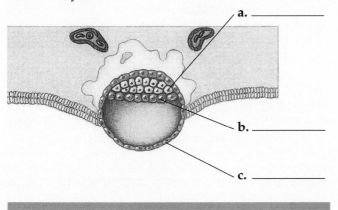

a. _____

b. _____

c. _____

■ The developmental fate of cells depends on cytoplasmic environment, location, and cell–cell interactions

The mechanisms underlying animal development are organized into two basic concepts: heterogeneous cytoplasm in the egg is partitioned through cleavage into different blastomeres, and the interactions between cells affect a cell's developmental fate. Both of these phenomena influence gene expression, which leads to the differentiation of cells.

Polarity and the Basic Body Plan   In mammals, polarity is not established until after cleavage has begun. In many other species, the anterior–posterior, dorsal–ventral, and left–right axes are established before cleavage.

Fate Maps and the Analysis of Cell Lineages   In the 1920s, Vogt developed a **fate map** for amphibian embryos by labeling regions of the blastula with dye to determine where specific cells showed up in later developmental stages. Cell lineage analysis follows the mitotic descendants of individual blastomeres. The developmental history of every cell has been determined for the nematode *Caenorhabditis elegans*.

Programmed cell death, or **apoptosis**, is an essential part of normal development. The genome of *C. elegans* includes two genes coding for toxins that kill the cells that express them and a gene that suppresses these suicide genes in other cells. In vertebrates, programmed cell death provides for normal development of the nervous system, immune system, and body parts such as fingers and toes.

Determination and Restriction of Cellular Potency   Certain unevenly distributed cytoplasmic components, called **cytoplasmic determinants**, fix the developmental fates of different embryonic regions.

Early blastomeres that can be separated and still produce complete embryos are said to be **totipotent**. Even though cytoplasmic determinants are asymmetrically distributed in the frog egg, the first cleavage division bisects the gray crescent and equally separates these determinants. A zygote's characteristic pattern of cleavage influences the developmental fate of cells.

The zygote is the only totipotent cell in many species. Mammalian blastomeres remain totipotent until the inner cell mass forms. The progressive restriction of a cell's developmental potency is called **determination**. Experimental manipulation shows that the developmental fates of cells of late gastrulas have been determined and they will give rise to only one type of cell.

■ INTERACTIVE QUESTION 43.7

If an eight-cell stage of a sea urchin embryo is split vertically into two four-cell groups, both develop into normal larvae. If the division is made horizontally, abnormal larvae develop. Explain this experimental result.

Morphogenetic Movements Changes in cell shape usually involve reorganization of the cytoskeleton. Extension of microfilaments to elongate a cell and then contraction of microfilaments at the apical end create a wedge shape, initiating invaginations and evaginations. Some morphogenesis involves ameboid movement. Cells at the leading edge of migrating tissue may extend filopodia and then drag along neighboring cells. Ameboid cells wander individually from the neural crest to various parts of the embryo.

Extracellular glycoproteins, such as laminin and the fibronectins, help moving cells adhere to the extracellular matrix. Receptor proteins on migrating cells receive directional cues from the environment and transmit that information to the cytoskeleton in order to direct movement. Orientation of extracellular fibronectin fibrils corresponds to the arrangement of contractile microfilaments in the cytoskeleton of migrating cells. **Cell adhesion molecules (CAMs)** on the surface of cells hold the cells of specific tissues and organs together and help regulate morphogenesis.

Induction In **induction**, one group of cells influences the development of another through physical contact or chemical signals that switch on sets of genes. Neural plate formation is induced by the chordamesoderm that forms the notochord. Using transplant experiments, Mangold and Spemann established that the dorsal lip is the primary organizer of the embryo because of its influence in early organogenesis. Peptide growth factors seem to be involved in the induction of chordamesoderm.

Development of the vertebrate eye involves a series of inductions from cells in all three germ layers, progressively determining the fate of epidermal cells.

Differentiation Cells differentiate, developing characteristic structures and functions, as a result of being determined during their developmental history. Differentiation is marked by the production of **tissue-specific proteins**. Even after cells begin expressing different genes, nearly all cells in an organism still possess the same genes and have **genomic equivalence**.

To determine whether genes are irreversibly inactivated during differentiation, Briggs and King transplanted nuclei from embryonic and tadpole cells into enucleated frog egg cells. The ability of the transplanted nucleus to direct normal development was inversely related to its developmental age. These studies indicate that the nucleus of a differentiated cell still contains all the genes necessary for development, but that the genome of a cell undergoes some change during differentiation.

■ INTERACTIVE QUESTION 43.8

What is the difference between determination and differentiation?

Pattern Formation **Pattern formation** is the ordering of cells and tissues into the characteristic structures and in the characteristic locations of the animal. Cells and their offspring differentiate based on **positional information**, molecular cues that indicate their position within a field of cells.

Several developmental biologists have been studying positional information in vertebrate limb development. Mesoderm just under the *apical epidermal ridge* at the tip of the wing bud may provide position assignment along the proximal–distal axis. Position along the dorsal–ventral axis may reflect relative distance from the dorsal and ventral ectoderm of the limb bud. The *zone of polarizing activity (ZPA)*, located at the posterior attachment of the bud, appears to communicate position along the anterior–posterior axis in a developing limb. Gradients of positional signals along three axes provide positional information to developing cells.

What would be the effect of transplanting an extra ZPA to the anterior attachment of a limb bud?

■ Pattern formation in *Drosophila* is controlled by a hierarchy of gene activations

Pattern formation relies on genes that control overall body plan. In fruit flies and other arthropods, their ordered series of segments become progressively distinct in response to regional differences in concentrations of regulatory proteins during development.

Egg Polarity Genes Studies of developmental mutants show the influence of *maternal-effect genes*, maternal genes that influence early embryonic development. **Egg-polarity genes** in follicle cells and nurse cells in the ovary produce and secrete mRNA for proteins that specify head–tail and dorsal–ventral axes of fly embryos. Proteins translated from this mRNA after fertilization diffuse through the embryo's syncytial cytoplasm and create **morphogen** gradients that provide positional information and regulate the activity of zygotic genes. Gradients of bicoid protein, coded for by the maternal gene *bicoid*, and posterior morphogens turn on specific genes in newly formed cells of the blastoderm, determining the basic body axes.

Segmentation Genes These early morphogen gradients turn on three sets of **segmentation genes**, which then set off a cascade of gene expression that determines segment division and differentiation. Products of the **gap genes** first determine the basic subdivisions along the embryo. **Pair-rule genes** establish the pairs of segments, and the **segment-polarity genes** then determine the anterior–posterior axis of each segment. Products of segmentation genes are DNA-binding transcription factors that initiate this hierarchy of gene activation responsible for the segmentation pattern of the embryo.

Homeotic Genes Regulatory genes called **homeotic genes** determine each segment's anatomical fate. Which homeotic genes are activated in each segment is determined by the positional imprints left by seg-

mentation genes. Homeotic genes also encode regulatory proteins that will activate genes that trigger development of segment appendages.

a. What could cause a mutant fruit fly embryo to develop two tail ends but no head?

b. What could cause a mutant embryo to have half the normal number of segments?

■ Comparisons of genes that control development reveal homology in animals as diverse as flies and mammals

A sequence of 180 nucleotides, called a **homeobox**, has been found in each *Drosophila* homeotic gene. The same or similar homeobox nucleotide sequences have now been identified in numerous eukaryotic organisms, including humans. These nucleotide sequences are translated into amino acid sequences, called *homeodomains*. The proteins containing homeodomains are regulatory proteins that control the transcription of other genes by binding to DNA. Thus, homeotic genes code for proteins that may coordinate the timing and location of transcription of developmental genes and thus determine pattern formation.

The similarity of eukaryotic homeodomain amino acid sequences and their relationship to DNA-binding domains of prokaryotic regulatory proteins indicate that these homeodomain sequences must have arisen early and been conserved through evolution as important regulators of gene expression and development.

STRUCTURE YOUR KNOWLEDGE

1. Create a flow chart that shows the sequence of key events in the fertilization of a sea urchin egg and indicate the functions of these events.

2. Fill in the table below, briefly describing the early stages of development for sea urchin, frog, bird, and mammal embryos.

Animal	Cleavage	Blastula	Gastrula
Sea urchin			
Frog			
Bird			
Mammal			

3. Use the following terms to create a concept map that illustrates the relationships among these concepts. Put in additional concepts as needed.

   development

   differentiation

   cytoplasmic determinants

   positional information

   determination

   cell division

   pattern formation

   morphogenesis

TEST YOUR KNOWLEDGE

MULTIPLE CHOICE: *Choose the one best answer.*

1. The vitelline layer
   a. is inside the fertilization membrane.
   b. releases calcium, which initiates the cortical reaction.
   c. has receptor molecules that are specific for proteins on the acrosomal process of sperm.
   d. is hydrolyzed by the release of enzymes by the acrosomal process.
   e. fuses with the sperm plasma membrane, initiating the depolarization of the membrane.

2. The slow block to polyspermy
   a. prevents sperm from other species from fertilizing the egg.
   b. is directly produced by the depolarization of the membrane.
   c. is a result of the formation of the fertilization membrane.

   d. is caused by the expulsion of hydrogen ions.
   e. involves all of the above.

3. The blastocoel
   a. develops into the archenteron or embryonic gut.
   b. is a fluid-filled cavity in the blastula.
   c. opens to the exterior through a blastopore.
   d. forms a hollow chamber during gastrulation.
   e. is lined with mesoderm.

4. Which of the following groups does *not* have eggs with definite polarity?
   a. sea urchin
   b. frog
   c. fruit fly
   d. bird
   e. mammal

5. Cytoplasmic determinants
   a. are unevenly distributed cytoplasmic components that fix the developmental fates of cells.
   b. are involved in the regulation of gene expression.
   c. set the stage for the eventual differentiation of cells and tissues.
   d. are often separated in the first few cleavage divisions.
   e. are all of the above.

6. Which of the following is *incorrectly* paired with its primary germ layer?
   a. muscles—mesoderm
   b. central nervous system—ectoderm
   c. lens of the eye—mesoderm
   d. liver—endoderm
   e. notochord—mesoderm

7. In a frog embryo, gastrulation
   a. is impossible because of the large amount of yolk.
   b. proceeds by involution as cells roll over the dorsal lip of the blastopore.
   c. produces a blastocoel displaced into the animal hemisphere.
   d. occurs along the primitive streak.
   e. involves the formation of the notochord and neural tube.

8. The primitive streak of mammalian embryos is analogous to
   a. the dorsal lip of the frog embryo.
   b. the blastodisc of a bird embryo.
   c. the syncytium of a fruit fly embryo.
   d. the archenteron of a sea urchin embryo.
   e. none of the above.

9. A function of the allantois in birds is to
   a. provide for nutrient exchange between the embryo and yolk sac.
   b. store nitrogenous wastes in the form of urea.
   c. form a respiratory organ in conjunction with the chorion.
   d. provide an aqueous environment for the developing embryo.
   e. produce the blood vessels of the umbilical cord.

10. Somites are
   a. blocks of mesoderm circling the archenteron.
   b. condensations of cells from which the notochord arises.
   c. serially arranged mesoderm blocks that develop into vertebrae and skeletal muscles.
   d. mesodermal derivatives of the notochord that line the coelom.
   e. produced during organogenesis in sea urchins, frogs, birds, and mammals, but not in fruit flies.

11. If the suppression gene for the "suicide genes" in *C. elegans* has a mutation that produces a nonfunctional product,
   a. The programmed cell death of regions of the embryo will not occur.
   b. Toxins may destroy all the cells of the embryo.
   c. The cell lineage analysis will show more cells than normal.

   d. The "suicide genes" will not be activated.
   e. The roundworm produced will have two head ends.

12. In her transplant experiments with frog embryos, Hilde Mangold found that a part of the dorsal lip of the blastopore moved to another location would result in a second gastrulation and even the development of a doubled, face-to-face tadpole. Which of the following is the best explanation for these results?
   a. morphogenetic movements
   b. separation of cytoplasmic determinants
   c. homeobox regulation of gene expression
   d. genomic equivalence of cells
   e. induction by dorsal lip tissue

13. Pattern formation appears to be determined by
   a. positional information a cell receives from gradients of chemical signals called morphogens.
   b. differentiation of cells, which then migrate into developing organs.
   c. the movement of cells along fibronectin fibrils.
   d. the induction of cells from chordamesoderm cells found in the center of organs.
   e. gastrulation and the formation of the three tissue layers.

14. Homeodomains are
   a. coded for by homeobox DNA sequences found in homeotic genes.
   b. part of regulatory proteins that bind to DNA and influence transcription.
   c. similar to DNA-binding domains of prokaryotic regulatory proteins.
   d. highly conserved amino acid sequences that are important in development.
   e. all of the above.

15. A mutation that inactivates a gap segmentation gene of a *Drosophila* would result in
   a. a larva with two head ends or two tail ends.
   b. a larva missing body segments and subsequent changes in the expression of pair-rule and segment-polarity genes.
   c. homeotic mutations producing flies with abnormal placement of body parts.
   d. a change in the expression of bicoid gene.
   e. all of the above because of the cascade effect of gene expression.

CHAPTER 44

NERVOUS SYSTEMS

FRAMEWORK

This chapter describes the structural components of nervous systems and how they functionally integrate, coordinate, transmit, and respond to information from the internal and external environment.

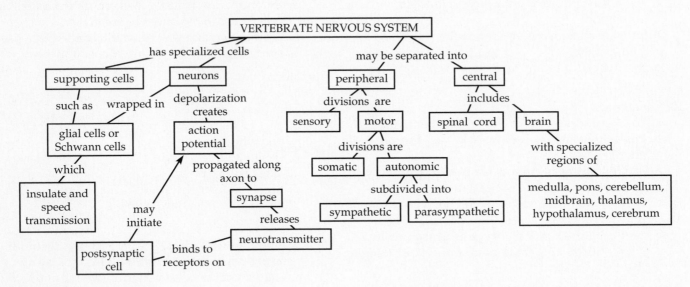

CHAPTER REVIEW

The coordination of behavior and physiology is a product of both the nervous and the endocrine systems. The structural complexity of the nervous system enables an animal to integrate information and respond rapidly to environmental stimuli.

■ **Nervous systems perform the three overlapping functions of sensory input, integration, and motor output:** *an overview*

The three main functions of the nervous system are sensory input, integration, and motor output.

Information from sensory receptors is conducted to integration centers in the brain and spinal cord (**central nervous system** or **CNS**). Motor output sends signals to **effector cells**, muscle or gland cells. The **peripheral nervous system (PNS)** carries sensory and motor information between the body and the central nervous system in **nerves**, bundles of neuron processes wrapped in connective tissue.

■ **The nervous system is composed of neurons and supporting cells**

Neurons   A **neuron** consists of a **cell body**, containing a nucleus and organelles, and fiberlike processes, **axons** and **dendrites**, which conduct impulses. A typical neuron usually has one long axon that transmits signals away from the cell body. Dendrites are the more numerous, shorter, and branched processes that carry signals toward the cell body.

Axons originate from a region of the cell body called the axon hillock. In many vertebrate PNS neurons, axons are wrapped in **Schwann cells** that collectively form an insulating **myelin sheath**. Axons may terminate in branches with numerous specialized endings called **synaptic terminals**, which release neurotransmitters that relay signals across the **synapse** between the axon and another neuron or effector cell.

Sensory neurons transmit information about the internal and external environment from sensory receptors to the CNS. **Motor neurons** carry information from the CNS to effector organs. **Interneurons** integrate information between sensory and motor neurons.

Supporting Cells   The numerous **supporting cells** give structural integrity to the nervous system. **Glial cells**, the supporting cells in the CNS, include astrocytes, which line capillaries in the brain and contribute to the **blood–brain barrier** that restricts the passage of most substances into the brain, and oligodendrocytes, which insulate axons in a myelin sheath. Schwann cells wrap to form concentric membrane layers around axons in the PNS.

■ INTERACTIVE QUESTION   44.1

Label the indicated structures on this diagram of a neuron. Indicate the direction of impulse transmission.

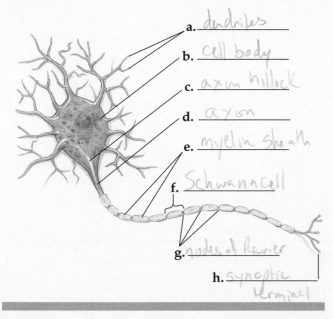

a. dendrites

b. cell body

c. axon hillock

d. axon

e. myelin sheath

f. Schwann cell

g. nodes of Ranvier

h. synaptic terminal

■ Impulses are action potentials, electrical signals transmitted along neuronal membranes

The Origin of Electrical Membrane Potential   A fundamental feature of all cells, an electrical voltage gradient results from the difference in charge between cytoplasm and extracellular fluid. Electrophysiologists can measure the magnitude of the charge separation by placing microelectrodes connected to a voltmeter inside and outside a cell. The **membrane potential** of a typical nontransmitting neuron is about −70 mV.

Membrane potential results from the concentrations of ions on either side of the membrane and on the membrane's selective permeability to these ions. The principal cation outside cells is sodium (Na^+) and the principal anion is chloride (Cl^-). Inside cells, potassium (K^+) is the principal cation. Negatively charged proteins, amino acids, phosphate, sulfate, and other ions are grouped together and symbolized as the principal anion, A^-.

The number and type of ion channels determine a membrane's permeability to different ions. Cells are usually much more permeable to K^+ than to Na^+. The internal anions (A^-) contribute a constant internal negative charge because most are too large to cross the membrane.

The potassium concentration gradient drives K^+ out of the cell until the electrical gradient, created by the nonmoving A^- and the exiting of positive charge, starts to drive K^+ back across the membrane. The resulting equilibrium potential for potassium is about −85 mV. Despite the low permeability of the membrane to Na^+, the concentration gradient and inner negative charge will move some Na^+ into the cell, increasing the internal positive charge and raising the membrane potential of a resting neuron to −70 mV. Sodium-potassium pumps maintain these resting concentration gradients using energy from ATP to move sodium ions back out of the cell and potassium ions into the cell.

Membrane Potential Changes and the Action Potential   Neurons and muscle cells are **excitable cells**; they are able to generate changes in their membrane potentials. A change in their **resting potential** may generate an electrical impulse.

When a neuron receives a stimulus, **gated ion channels** open, causing the cell to change its membrane potential. Should the stimulus open potassium channels, the efflux of K^+ will cause the membrane potential to become more negative, resulting in **hyperpolarization**. When sodium channels open and Na^+ is allowed to flow in, the membrane potential is reduced, **depolarizing** the membrane as the cytoplasm becomes more positively charged. With this type of **graded potential**, the magnitude of a voltage change is proportional to the strength of the stimulus: the stronger the stimulus, the more gated ion channels that open.

Once depolarization of a typical neuron reaches a **threshold potential** of –55 to –50 mV, an **action potential**, or nerve impulse, is triggered. The action potential is an *all or none* event, always creating the same voltage spike, regardless of the intensity of the stimulus once the threshold potential is reached. An action potential causes the membrane potential first to reverse polarity in the depolarizing phase, and then

return to the normal negative resting potential during a repolarizing phase. The membrane potential may temporarily become more negative than the resting potential in a phase called the undershoot.

The **voltage-gated channels** of excitable cells open and close in response to changes in membrane potential. Voltage-gated potassium channels open slowly in response to depolarization. Sodium activation gates open rapidly in response to depolarization. The slower-acting sodium inactivation gates close in response to depolarization. When resting, the inactivation gate is open but the activation gate is closed, and Na⁺ does not enter the neuron. Depolarization opens the activation gate, allowing Na⁺ to enter the cell, which further

---

■ INTERACTIVE QUESTION   44.2

This diagram shows the changes in voltage-gated ion channels during the triggering of an action potential. Label the gates and channels and phases of the action potential. Label the axes of the graph and indicate where the phases are shown by the graph. Describe which ion movements are associated with each phase.

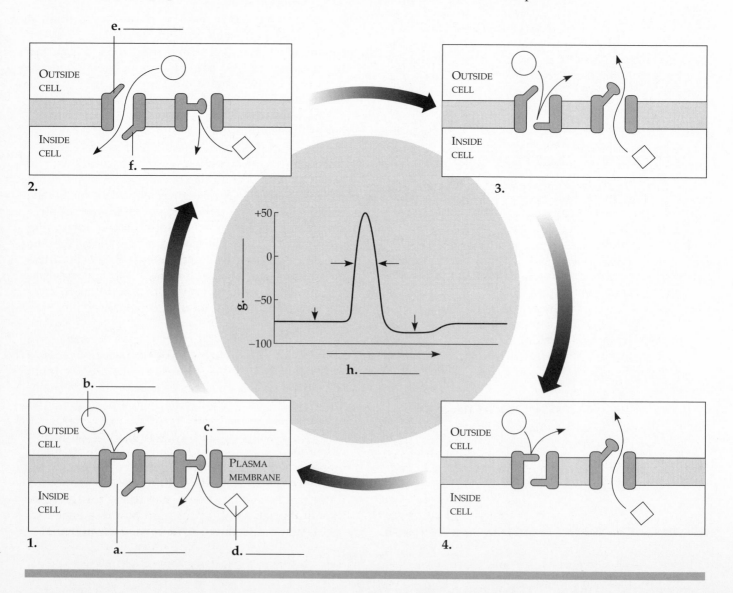

depolarizes the membrane and opens more voltage-gated sodium channels. The sodium channel inactivation gate closes, greatly reducing sodium permeability during the repolarizing phase, and the potassium gates completely open, allowing K⁺ to flow from the cell and helping return the cell to its resting-stage negativity. During the undershoot, the sodium channel inactivation gates have closed, but the potassium channels have not yet closed in response to repolarization; the increased permeability to K⁺ may temporarily hyperpolarize the membrane. During the **refractory period**, which occurs at the end of an action potential before the inactivation gates have reopened, the neuron cannot respond to another stimulus.

Though the amplitude of an action potential is always the same, its frequency varies with the intensity of a stimulus.

Propagation of the Action Potential   As sodium ions move into the cell during the action potential, they depolarize adjacent sections of the membrane, bringing them to the threshold potential. Local depolarizations and action potentials across the membrane result in the propagation of serial action potentials along the length of the neuron. Because of the brief refractory period, the action potential is propagated in only one direction.

Action Potential Transmission Speed   Resistance to current flow is inversely proportional to the cross-sectional area of the conducting "wire." The greater the axon diameter, the faster the action potential is conducted. Some invertebrates, such as squid and lobsters, have giant axons that conduct impulses rapidly.

In vertebrates, voltage-gated ion channels are concentrated in regions called the *nodes of Ranvier*, where there are small gaps in the myelinated membranes and the neuron membrane has contact with extracellular fluid. Action potentials can be generated only at these nodes, and a nerve impulse "jumps" from node to node, resulting in a faster mode of transmission known as **saltatory conduction**.

■ Chemical or electrical communication between cells occurs at synapses

Synapses between neurons conduct impulses from the synaptic terminal of a **presynaptic cell** to a dendrite or cell body of the **postsynaptic cell**.

Electrical Synapses   Electrical synapses allow action potentials to flow directly from presynaptic to postsynaptic cells via gap junctions. Electrical synapses are found in the giant neurons of some crustaceans but

are less common than chemical synapses in vertebrates and most invertebrates.

Chemical Synapses   At a chemical synapse, the electrical message is converted to a chemical message that travels from the presynaptic cell across a **synaptic cleft** to the postsynaptic cell. A synaptic terminal contains numerous **synaptic vesicles**, in which thousands of molecules of **neurotransmitter** are stored. The depolarization of the **presynaptic membrane** opens voltage-gated calcium channels in the membrane. The influx of Ca²⁺ causes the synaptic vesicles to fuse with the presynaptic membrane and release neurotransmitter into the cleft.

The **postsynaptic membrane** contains receptor proteins for different neurotransmitter molecules, which are associated with particular ion channels. These chemically sensitive gates allow ions to cross the membrane, altering the membrane potential by either depolarizing or hyperpolarizing it. Enzymes rapidly break down the neurotransmitter.

■ INTERACTIVE QUESTION   44.3

Identify the components of this chemical synapse that has received an action potential.

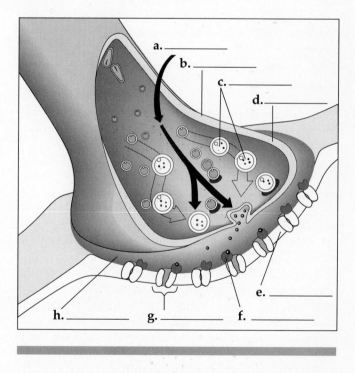

Summation: Neural Integration at the Cellular Level
A neuron must integrate the information it receives from numerous neighboring neurons at thousands of its excitatory and inhibitory synapses. At an excitatory synapse, binding of neurotransmitter to receptors opens a chemically gated channel that allows Na⁺ to

flow into and K$^+$ to flow out of the cell. A net flow of positive charge into the cell depolarizes the membrane, creating an **excitatory postsynaptic potential (EPSP)** and bringing the membrane potential closer to an action potential threshold. At an inhibitory synapse, binding of neurotransmitter opens ion gates to allow K$^+$ to flow out and/or Cl$^-$ to move into the cell, hyperpolarizing the membrane and producing an **inhibitory postsynaptic potential (IPSP).**

EPSPs and IPSPs are graded potentials. Their magnitude depends on the number of neurotransmitter molecules that bind to receptors. **Summation** of several postsynaptic potentials is usually necessary to bring the axon hillock to threshold potential. Summated IPSPs will hyperpolarize the membrane. The membrane potential of the axon hillock at any given time is determined by the sum of all EPSPs and IPSPs.

▨ INTERACTIVE QUESTION   44.4

a. _____ summation occurs with repeated release of neurotransmitters from one or more synaptic terminals before the postsynaptic potential returns to its resting potential.

b. _____ summation occurs when several different presynaptic terminals, usually from different neurons, release neurotransmitter simultaneously.

Neurotransmitters and Receptors   Dozens of molecules have been identified as neurotransmitters so far. The criteria for neurotransmitters are as follows: (1) the compound must be contained in synaptic vesicles and discharged when the presynaptic cell is stimulated, and it must alter the membrane potential of the postsynaptic cell; (2) when experimentally injected into the synapse, it must cause an EPSP or IPSP; and (3) the compound must be rapidly degraded or removed from the synapse.

Acetylcholine is a common neurotransmitter in invertebrates and vertebrates. In vertebrate neuromuscular junctions, acetylcholine released from a motor axon depolarizes the postsynaptic muscle cell. It can also be inhibitory; for example, it may slow down the heart rate of vertebrates and mollusks.

Biogenic amines, neurotransmitters derived from amino acids, usually function within the CNS. **Epinephrine, norepinephrine,** and **dopamine** are derived from tyrosine. **Serotonin,** synthesized from tryptophan, and dopamine affect sleep, mood, attention, and learning. Imbalances of these transmitters have been associated with several disorders.

The amino acids **glycine, glutamate, aspartate,** and **gamma aminobutyric acid (GABA)** function as neurotransmitters in the CNS. GABA is the most common inhibitory transmitter in the brain.

Substance P is a **neuropeptide** that functions in pain perception. **Endorphins** are neuropeptides produced in the brain during physical or emotional stress, which have painkilling and other functions.

Amino acid transmitters bind to receptors on the postsynaptic membrane, affect ion permeability, and create either EPSPs or IPSPs. Biogenic amines and neuropeptides have a longer-lasting effect because they activate a second messenger that affects the metabolism of the postsynaptic cell.

▨ INTERACTIVE QUESTION   44.5

How can acetylcholine have opposite effects on heart and skeletal muscle cells?

Gaseous Signals of the Nervous System   Neurons of the vertebrate CNS and PNS use nitric oxide and carbon monoxide as local regulators. Neurons signal endothelial cells in blood vessels to synthesize and release NO, triggering relaxation of smooth muscle cells and vessel dilation.

Neural Circuits and Clusters   Groups of neurons that interact and carry information along specific pathways are called circuits. In **convergent circuits,** several neurons come together and feed information into a single postsynaptic neuron. **Divergent circuits** spread out information from one neuron to several postsynaptic neurons. **Reverberating circuits** are circular paths in which signals return to their source.

Nerve cell bodies are arranged in functional groups called **ganglia.** Functional clusters of cell bodies in the brain are called **nuclei.**

▪ Invertebrate nervous systems are highly diverse

The cnidarian **nerve net** is a loosely organized system of nerves in which impulses are conducted in both directions by means of electrical synapses. Some centralization is seen in jellyfish, in which clusters of nerve cells around the margin of the bell coordinate swimming movements. Echinoderms have a central nerve ring with radial nerves connected to a nerve net in each arm.

Bilateral animals show **cephalization**—the concentration of sense organs and feeding structures in the head. A brain is an anterior enlargement of one or more **nerve cords**, thick bundles of nerves running longitudinally through the body. This central nervous system processes sensory information and controls motor responses of the PNS.

Flatworms have a simple brain and two or more nerve cords. Annelids and arthropods have a prominent brain and a ventral nerve cord, which may have ganglia within each body segment. Sessile mollusks have little cephalization and only simple sense organs. Cephalopod mollusks, such as the octopus, have large brains, image-forming eyes, and giant axons, all of which contribute to their active predatory life and their ability to learn and remember.

■ **INTERACTIVE QUESTION 44.6**

The study of invertebrates has contributed to neurobiology. Our knowledge of action potentials comes from work with (**a**)_____. Neurobiologists are studying the neuronal basis of learning using (**b**) _____ .

■ The vertebrate nervous system is a hierarchy of structural and functional complexity

The Peripheral Nervous System   The peripheral nervous system consists of the **sensory division** with sensory, or afferent, neurons that bring information to the CNS, and the **motor division** with efferent neurons that carry signals away from the CNS to effector cells. The human PNS contains 12 pairs of cranial nerves and 31 pairs of spinal nerves. Most cranial and all spinal nerves contain both sensory and motor neurons.

The motor division has two parts. The **somatic nervous system** carries signals to skeletal muscles, the movement of which is largely under conscious control. The **autonomic nervous system** maintains involuntary control over smooth and cardiac muscles and various organs.

The autonomic nervous system is subdivided: In general, the **parasympathetic division** carries signals that enhance activities that gain and conserve energy, such as digestion and slowing the heart rate. The **sympathetic division** accelerates the heart and metabolic rate, preparing an organism for action.

The Central Nervous System   The CNS consists of the **spinal cord** and the **brain**. Three protective layers of connective tissue called the **meninges** cover these

bilaterally symmetrical organs. Axons are in bundles or tracts; **white matter** is named for the white color of the myelin sheaths. Neuron cell bodies make up the **gray matter**, which is outside the white matter in the brain but inside it in the spinal cord.

Spaces in the brain called **ventricles** are continuous with the narrow **central canal** of the spinal cord and filled with **cerebrospinal fluid**. This fluid, formed by filtration of the blood, cushions the brain and carries out circulatory functions.

The spinal cord integrates simple responses to some stimuli, usually in the form of an unconscious, programmed response called a **reflex**, and carries information to and from the brain.

Evolution of the Vertebrate Brain   The vertebrate brain evolved as three bulges in the anterior end of the spinal cord. The **hindbrain**, the **midbrain**, and the **forebrain** are present in all vertebrates, although these regions may be subdivided to enhance the capacity for integration of complex activities.

Three evolutionary trends in the vertebrate brain include an increase in relative size, subdivision into areas with specific functions, and additional complexity of the forebrain. More sophisticated behaviors are correlated with increased size of one region of the forebrain, the **cerebrum**. Folding or convolution of the cerebral cortex increases the surface area and integrative capacity of the cerebrum.

■ The human brain is a major research frontier

The **brain stem** is a stalk and caplike swelling formed from the hindbrain and midbrain. The hindbrain has three parts: the **medulla oblongata**, which contains control centers for such homeostatic functions as respiration, heart and blood vessel actions, and digestion; the **pons**, which functions with the medulla in some of these activities and in conducting information between the rest of the brain and the spinal cord; and the **cerebellum**, which coordinates movements. The tracts of motor neurons from the mid- and forebrain cross in the medulla, so that the right side of the brain controls much of the movement of the left side of the body and vice versa. The cerebellum integrates information from the auditory and visual systems with sensory input from the joints and muscles as well as motor pathways from the cerebrum to provide automatic coordination of movements and balance.

The midbrain receives and integrates sensory information and sends this information to specific regions of the forebrain. Fibers involved in hearing pass through or terminate in the **inferior colliculi**. The **superior colliculi** form prominent optic lobes in nonmammalian vertebrates but only coordinate visual reflexes in mammals because vision is integrated in

■ INTERACTIVE QUESTION 44.7

Complete the following concept map to help you understand the organization of the vertebrate nervous system.

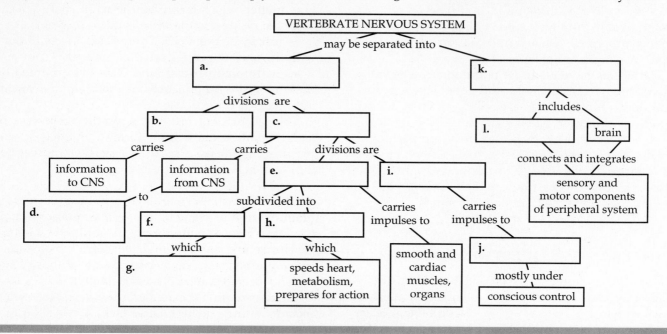

the forebrain. The **reticular formation**, a major group of nuclei localized in the medulla, pons, and midbrain, regulates states of arousal.

The most intricate neural processing occurs in the forebrain. The **diencephalon** contains the thalamus and hypothalamus, and the **telencephalon** consists of the cerebrum.

The **thalamus** is a major relay area for sensory information going to the cerebrum. It receives input from the cerebrum and from parts of the brain regulating emotion and arousal, and it filters information sent to the cerebrum.

The **hypothalamus** is the major site for homeostatic regulation. It produces the posterior pituitary hormones and the releasing hormones that control the anterior pituitary. The hypothalamus contains the regulating centers for many autonomic functions and also plays a role in sexual response, mating behaviors, the alarm response, and pleasure. The **suprachiasmatic nucleus** functions as the biological clock maintaining daily biorhythms.

The cerebrum is divided into right and left **cerebral hemispheres**, each with an outer covering of gray matter, called the cerebral cortex, and internal white matter. The **basal ganglia**, a cluster of nuclei located in the white matter, are important in relaying motor impulses and coordinating motor responses.

The **cerebral cortex** has changed the most during vertebrate evolution. Communication between the two hemispheres travels through the **corpus callosum**, a thick band of fibers.

Found at the boundary between the frontal and parietal lobe of the cerebral cortex, the motor cortex sends signals to skeletal muscles and the somatosensory cortex receives and partially integrates input from touch, pain, pressure, and temperature receptors. The proportion of somatosensory and motor cortex devoted to controlling each part of the body is correlated with the importance of that area. Vision, hearing, smell, and taste are controlled by other cortical regions. Adjacent association areas cooperate in producing sensory perceptions. (See Interactive Question 44.8, p. 333.)

Integration and Higher Brain Functions Nerve impulses are integrated on all levels of the nervous system, from the spinal reflex to the intellectual creations of the cerebral cortex.

Arousal is a state in which an individual is aware of the external world, whereas sleep is a state in which the individual is not conscious of external stimuli. All birds and mammals have a characteristic sleep–wake cycle. Different patterns in the electrical activity of the brain can be recorded by an **electroencephalograph**, or **EEG**. Slow, synchronous alpha waves are produced by a person lying quietly with eyes closed. Faster beta waves are associated with opened eyes or thinking about a complex problem. In early stages of sleep, irregular theta waves predominate. Quite slow and highly synchronized delta waves occur during deeper sleep. Periods of delta waves alternate with periods of a desynchronized EEG and

■ INTERACTIVE QUESTION 44.8

Identify the structures of the human brain. Match the functions to these structures.

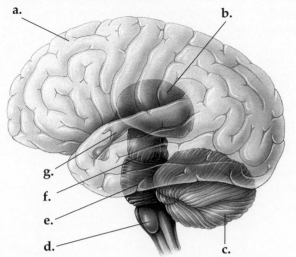

Structure		Function
a.	_____	_____
b.	_____	_____
c.	_____	_____
d.	_____	_____
e.	_____	_____
f.	_____	_____
g.	_____	_____

Functions

1. Coordination, balance, movement
2. Aids medulla in some functions, conducts information between brain and spinal cord
3. Screens and relays information to cerebrum
4. Regulates breathing, heart rate, digestion
5. Integrates sensory and motor information, thinking
6. Produces hormones, homeostatic regulation
7. Sends sensory information to forebrain, contains nuclei involved in hearing and vision

rapid eye movements, called REM sleep, when most dreaming occurs.

The reticular formation filters the sensory information reaching the cortex. The level of arousal relates to the amount of input the cortex receives. Sleep-producing nuclei are located in the pons and medulla, and serotonin may be the neurotransmitter involved in these areas. A center that causes arousal is found in the midbrain.

The association areas of the cerebral cortex are not bilaterally symmetrical. The left hemisphere controls speech, language, and calculation, whereas the right hemisphere controls artistic ability and spatial perception. Much of the information on this **lateralization** of the brain comes from Sperry's work with "split-brain" patients.

A speech association area in the left parietal lobe stores information required for speech content and construction, whereas a speech association area in the frontal lobe provides motor information required for speech production. Different types of aphasia, the inability to speak coherently, occur if one or the other of these regions is damaged.

Human emotions have been mapped to a group of nuclei and interconnecting axon tracts in the forebrain called the **limbic system**. This system includes parts of the thalamus, hypothalamus, and inner parts of the cerebral cortex, including two nuclei, the **amygdala** and **hippocampus**. The prefrontal cortex interacts with the limbic system to evaluate emotional content.

Human memory consists of **short-term memory**, the immediate sensory perception of an object or idea, and **long-term memory**, the recall of these perceptions after the passage of time. Transferring information from short-term to long-term memory is facilitated by rehearsal, a favorable emotional state, and associations with previously learned and stored information.

■ INTERACTIVE QUESTION 44.9

What is the difference between fact memories and skill memories?

Various brain imaging techniques are useful as diagnostic and research tools. Magnetic resonance imaging (MRI) measures the radio signals given out as the nuclei of hydrogen atoms in water molecules are aligned by powerful magnets. MRI is useful for detecting problems in the brain and spinal cord. Computed tomography (CT) produces high-resolution video images of a series of thin X-ray sections taken through the body and is useful for detecting ruptured blood vessels on the brain. Positron-emission tomography (PET) is widely used in brain research. Radioactively labeled glucose injected into the body is taken up by metabolically active cells and is detected by the PET scanner, revealing metabolic hot spots in the brain during various activities.

Neuroscientists use amnesia victims, animals, and PET to determine the complex pathways involved in memory. To develop a fact memory, sensory signals may pass through vision centers to the reticular formation to parts of the hypothalamus and limbic system to the prefrontal cortex for integration and back to cortical vision centers.

Memory seems to be stored with some redundancy in a cortical association areas. The amygdala seems to tie information to be saved to an event or emotion. In the hippocampus, memory storage and learning may result from **long-term potentiation (LTP)**, in which a postsynaptic cell has an enhanced response to a single action potential as a result of strong depolarization by repeated bursts of action potentials from a presynaptic cell. The excitatory neurotransmitter glutamate is released by the presynaptic cell, binding with special receptors in the postsynaptic membrane and opening gated channels for Ca^{2+}. The influx of Ca^{2+} triggers a cascade of changes that contribute to LTP.

STRUCTURE YOUR KNOWLEDGE

1. Develop a flow chart, diagram, or description of the sequence of events in the creation and propagation of an action potential and in the transmission of this potential across a chemical synapse.

2. List the location and functions of these important brain nuclei.

   a. reticular formation

   b. suprachiasmatic nucleus

   c. basal ganglia

   d. limbic system

   e. amygdala and hippocampus

TEST YOUR KNOWLEDGE

MULTIPLE CHOICE: *Choose the one best answer.*

1. Which of the following does *not* distinguish the nervous system from the endocrine system?
   a. Transmission of nervous impulses is more rapid.
   b. The nervous system uses chemical communication.
   c. Nervous system messages are delivered directly to target cells or organs.
   d. The structural complexity of the nervous system allows for integration of more information and responses.
   e. The response of the body to nervous communication is more rapid.

2. Interneurons
   a. may connect sensory and motor neurons.
   b. are confined to the PNS.
   c. are confined to the CNS.
   d. are electrical synapses between neurons.
   e. do not have cell bodies.

3. Nodes of Ranvier are
   a. gaps where Schwann cells abut at which action potentials are generated.
   b. neurotransmitter-containing vesicles located in the synaptic terminals.
   c. the major components of the blood–brain barrier that restrict the passage of substances into the brain.
   d. clusters of receptor proteins located on the postsynaptic membrane.
   e. ganglia adjacent to the spinal cord.

4. Which of the following is *not* true of the resting potential of a typical neuron?
   a. The inside of the cell is more negative than is the outside.
   b. There are concentration gradients with more sodium outside the cell and a higher potassium concentration inside the cell.
   c. It is about −70 mV and can be measured by using microelectrodes placed inside and outside the cell.
   d. It is formed by the sequential opening of voltage-gated channels.
   e. It results from the combined equilibrium potentials of potassium and sodium.

5. After the depolarization of an action potential, the resting potential is restored by
   a. the closing of sodium activation and inactivation gates.
   b. the opening of sodium activation gates.
   c. the refractory period in which the membrane is hyperpolarized.
   d. the delay in the action of the sodium-potassium pump.
   e. the opening of voltage-gated potassium channels and the closing of sodium inactivation gates.

6. The threshold potential of a membrane
   a. is a positive value, often equal to about 35 mV.
   b. opens voltage-gated channels and permits the rapid outflow of sodium ions.
   c. is the depolarization that is needed to generate an action potential.
   d. is a graded potential that is proportional to the strength of a stimulus.
   e. is an all-or-none event.

7. Which of the following is *not* true of chemical synapses?
   a. Neurotransmitter is rapidly degraded in the synaptic cleft.
   b. The influx of calcium when an action potential reaches the presynaptic membrane causes synaptic vesicles to release their neurotransmitter into the cleft.
   c. The binding of neurotransmitter to receptors on the postsynaptic membrane changes the membrane's permeability to certain ions.
   d. An excitatory postsynaptic potential forms when sodium channels open and the membrane potential moves closer to an action potential threshold.
   e. Synaptic terminals at the ends of branching dendrites contain synaptic vesicles, which enclose the neurotransmitter.

8. An inhibitory postsynaptic potential occurs when
   a. Sodium flows into the postsynaptic cell.
   b. Enzymes do not break down the neurotransmitter in the synaptic cleft.
   c. Binding of the neurotransmitter opens ion gates that result in the membrane becoming hyperpolarized.
   d. Acetylcholine is the neurotransmitter.
   e. Both c and d occur.

9. Which of the following does *not* function as a neurotransmitter?
   a. acetylcholine
   b. neuropeptides
   c. biogenic amines
   d. steroids
   e. NO

10. Which of the following animals is mismatched with its nervous system?
    a. sea star (echinoderm)—modified nerve net, central nerve ring with radial nerves
    b. hydra (cnidarian)—ring of ganglia, paired ventral nerve cords
    c. annelid worm—brain, ventral nerve cord with segmental ganglia
    d. vertebrate—dorsal central nervous system of brain and spinal cord
    e. clam (mollusk)—widely separated ganglia, little centralization

11. Which of the following is *not* true of the autonomic nervous system?
    a. It is a subdivision of the somatic nervous system.
    b. It consists of the sympathetic and parasympathetic divisions.
    c. It is part of the peripheral nervous system.
    d. It controls smooth and cardiac muscles.
    e. It is part of the motor division.

12. If you needed to obtain nuclei from the cell bodies of neurons for an experiment, you would want to make preparations of
    a. astrocytes of the brain.
    b. the gray matter of the brain.
    c. the inner portion of the spinal cord.
    d. sympathetic ganglia near the spinal cord.
    e. either b, c, or d.

13. The superior and inferior colliculi
    a. control biorhythms and are found in the thalamus.
    b. are the part of the limbic system found in the midbrain.
    c. are located in the parietal and frontal lobe, respectively, and involved in language and speech.
    d. are nuclei in the midbrain involved in hearing and vision.
    e. are found in the inner cortex and are involved in memory storage.

14. Which of the following structures is *incorrectly* paired with its function?
    a. pons—conducts information between spinal cord and brain
    b. cerebellum—contains the tracts that cross motor neurons from one side of the brain to the other side of the body
    c. thalamus—screens and relays incoming impulses to the cerebrum
    d. corpus callosum—band of fibers connecting left and right hemispheres
    e. hypothalamus—homeostatic regulation, pleasure centers

15. REM sleep
    a. is characterized by a highly synchronized EEG.
    b. is characterized by delta waves.
    c. alternates with periods of beta waves.
    d. is when dreaming occurs.
    e. is characterized by rapid eye movements and desynchronized alpha waves.

16. Long-term potentiation seems to be involved in
    a. memory storage and learning.
    b. enhanced response of a presynaptic cell resulting from strong depolarization from bursts of action potential.
    c. the attachment of emotion to information to be learned.
    d. the release of the excitatory neurotransmitter glutamate that opens sodium-gated channels.
    e. all of the above.

CHAPTER 45

SENSORY AND MOTOR MECHANISMS

FRAMEWORK

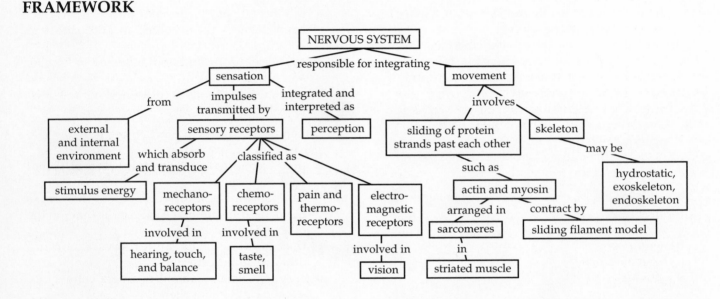

CHAPTER REVIEW

■ **Sensory receptors detect changes in the external and internal environment:** *an overview*

Sensation and Perception Information is transmitted in the nervous system as action potentials. Sensations, or impulses traveling along sensory neurons, are routed to different parts of the brain that interpret them as perceptions.

General Function of Sensory Receptors **Sensory receptors** are usually modified neurons or epithelial cells that occur singly or in groups within sensory organs and that collect and transmit information from environmental stimuli. **Reception** is the absorption of energy from a particular stimulus. The conversion of that energy into a **receptor potential** is called **transduction**. The stimulus energy may need to undergo

amplification, either by accessory structures of sense organs or as part of the transduction process.

Transmission of the receptor potential to the central nervous system may occur either as an action potential, if the receptor is a sensory neuron, or by the release of neurotransmitter from a receptor cell into a synapse with a sensory neuron, which then may translate it into an action potential. The strength of the stimulus and the receptor potential correlates with the frequency of action potentials or the quantity of neurotransmitter released.

Integration of sensory information begins with the summation of graded potentials from receptors. Ongoing stimulation may lead to **sensory adaptation** of the receptor cell in which continued stimulation results in a decline in sensitivity. A receptor cell's threshold for firing may vary with conditions. Sensory information is further integrated by complex receptors and the CNS.

Types of Receptors Sensory receptors may be **exteroreceptors** or **interoreceptors**, which provide information from the external or internal environment, respectively. Receptors may also be categorized by the energy stimulus to which they respond.

Mechanoreceptors respond to the mechanical energy of pressure, touch, motion, and sound. Bending or stretching of the mechanoreceptor cell membrane increases its permeability to sodium and potassium ions, creating a receptor potential. In humans, modified dendrites of sensory neurons found in deep skin layers respond to strong pressure, whereas receptors that detect light touch are closer to the surface.

The length of skeletal muscles is monitored by **muscle spindles**, stretch receptors that are stimulated by stretching of the muscles. **Hair cells**, mechanoreceptors that detect motion, are found in the vertebrate ear and lateral line organs of fishes. When motion produces bending in the cilia or microvilli projecting upward from a hair cell, ion permeabilities either increase or decrease, and the rate of action potential firing changes.

Chemoreceptors include both general receptors that monitor the total solute concentration and specific receptors that respond to individual kinds of important molecules. **Gustatory** (taste) and **olfactory** (smell) **receptors** respond to groups of related chemicals.

Electromagnetic receptors respond to energy of various wavelengths. **Photoreceptors** detect visible light and are often organized into eyes. Some animals have receptors that detect infrared rays, electric currents, and magnetic fields.

Thermoreceptors respond to heat or cold and help to regulate body temperature.

Pain receptors in humans are naked dendrites in the epidermis called **nociceptors**. Different groups of receptors respond to excess heat, pressure, or chemi-cals released by injured cells. Histamines and acids trigger pain receptors, and prostaglandins increase pain by sensitizing receptors.

■ Photoreceptors contain light-absorbing visual pigments: *a closer look*

Invertebrates Photoreceptors containing light-absorbing pigments are found in most invertebrates. The **eye cup** of planaria, which detects light intensity and direction, consists of receptor cells within a cup formed from darkly pigmented cells. The brain compares impulses from the left and right eye cups to help the animal navigate a direct path away from a light source.

The **compound eye** of insects and crustaceans contains up to thousands of light detectors called **ommatidia**, each of which has a cornea and lens. The different intensities of light entering the many ommatidia produce a mosaic image. Compound eyes are adept at detecting movement, partly due to the rapid recovery of the photoreceptors, and may detect color and ultraviolet radiation.

Some jellyfish, spiders, and many mollusks have a **single-lens eye**, in which light is focused through the single lens onto the retina containing light-transducing receptor cells.

Structure and Function of the Vertebrate Eye The eyeball consists of a tough, outer connective tissue layer called the **sclera** and a thin, pigmented inner layer called the **choroid**. The **conjunctiva** is a mucous membrane covering the sclera. At the front of the eye, the sclera becomes the transparent **cornea**, and the choroid forms the colored **iris**, which regulates the amount of light entering through the **pupil**. The **retina**, a layer inside the choroid, contains the photoreceptor cells. The optic nerve attaches to the eye at the optic disc.

The transparent **lens** focuses an image onto the retina. The **ciliary body** produces the **aqueous humor** that fills the anterior eye cavity; jellylike **vitreous humor** fills the posterior cavity. Many fishes focus by moving the lens backward or forward. Mammals focus on close objects by **accommodation**, in which ciliary muscles contract, causing suspensory ligaments to slacken and the elastic lens to become rounder.

Rod cells and **cone cells** are the photoreceptors in the retina. The relative proportion of each of these receptors correlates with the activity pattern of the animal: rods are more light sensitive and enable night vision, whereas cones distinguish colors. In the human eye, rods are concentrated toward the edge of the retina, whereas the center of the visual field, the **fovea**, is filled with cones.

■ INTERACTIVE QUESTION   45.1

List the five functions common to all receptor cells as they absorb stimulus energy and transmit it to the nervous system.

a.

b.

c.

d.

e.

■ INTERACTIVE QUESTION   45.2

Label the parts of the vertebrate eye.

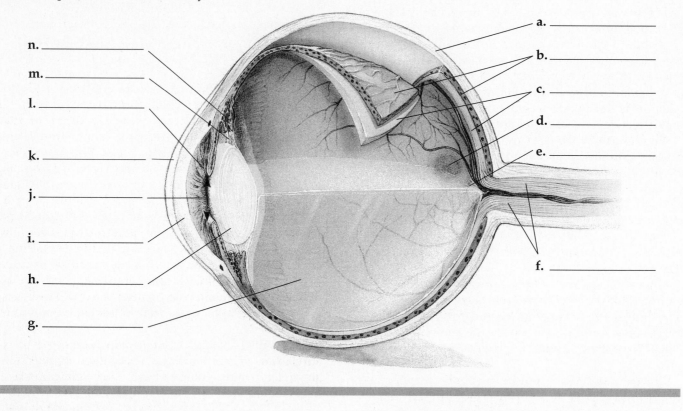

n. _____

m. _____

l. _____

k. _____

j. _____

i. _____

h. _____

g. _____

a. _____

b. _____

c. _____

d. _____

e. _____

f. _____

Signal Transduction in the Eye   Both rods and cones have visual pigments embedded in an outer stack of folded membranes or discs. **Retinal** is the light-absorbing molecule and is bonded to a membrane protein called an **opsin**. Rods contain the visual pigment **rhodopsin**, which changes shape when it absorbs light and activates a membrane-bound G protein called transducin. Transducin activates an effector enzyme that alters the second messenger, cyclic guanosine monophosphate. In the dark, cGMP is bound to sodium ion channels, keeping the rod-cell membrane depolarized and releasing an inhibitory neurotransmitter. The light-triggered rhodopsin signal transduction pathway converts cGMP to GMP, closing sodium channels, hyperpolarizing the membrane, and decreasing its release of inhibitory neurotransmitter. Enzymes convert the "bleached" rhodopsin back to its original form in the dark. In bright light, the rhodopsin remains bleached, and cones are responsible for vision.

The three subclasses of red, green, and blue cones, each with its own type of opsin that binds with retinal to form a **photopsin**, are named for the color of light they are best at absorbing. Signal processing in cones is more complex than in rods.

■ INTERACTIVE QUESTION   45.3

Summarize the effects of light on rod cells and rhodopsin by filling in the following table.

	In the Dark	In the Light
Rhodopsin	a.	b.
cGMP	c.	d.
Sodium channels	e.	f.
Rod-cell membrane	g.	h.
Neuro-transmitter	i.	j.
Postsynap-tic cell	k.	l.

Visual Integration Rods and cones synapse with **bipolar cells** in the retina, which in turn synapse with **ganglion cells**, whose axons transmit action potentials to the brain. **Horizontal cells** and **amacrine cells** help to integrate visual information.

Signals from rods and cones may follow the vertical pathway directly from receptor cells to bipolar cells to ganglion cells. A lateral pathway involves lateral integration by horizontal cells, which carry signals from one receptor cell to others and to several bipolar cells, and by amacrine cells, which relay information from one bipolar cell to several ganglion cells. **Lateral inhibition** of nonilluminated receptors enhances contrast.

Axons of the ganglion cells form the optic nerve. The left and right optic nerves meet at the **optic chiasma** at the base of the cerebral cortex. Information from the left visual field of both eyes travels to the right side of the brain, whereas what is sensed in the right field of view goes to the left side. Most axons of the ganglion cells go to the **lateral geniculate nuclei** of the thalamus, where neurons lead to the **primary visual cortex** in the occipital lobe of the cerebrum. Interneurons also carry information to visual processing and integrating centers in the cortex.

■ **INTERACTIVE QUESTION** 45.4

The receptive field of a ganglion cell includes the rods or cones that supply information to the bipolar cells connected to it.

a. Would a large or small receptive field produce the sharpest image?

b. Where would the smallest receptive fields be found?

■ Hearing and balance are related in most animals: *a closer look*

The Human Ear The mechanoreceptors for hearing and balance are hair cells in fluid-filled canals. The human ear consists of three regions: The external **pinna** and the **auditory canal** make up the **outer ear**. The **tympanic membrane** (eardrum), separating the outer ear from the **middle ear**, transmits sound waves to three ossicles—the **malleus** (hammer), **incus** (anvil), and **stapes** (stirrup)—which conduct the waves to the **inner ear** by way of a membrane called the **oval window**. The **Eustachian tube**, connecting the pharynx and the middle ear, equalizes pressure within the middle ear. The inner ear is a fluid-filled labyrinth of channels within the temporal bone.

The coiled **cochlea** of the inner ear has two large fluid-filled chambers—an upper vestibular canal and a lower tympanic canal—separated by the smaller cochlear duct, which is filled with endolymph. The **organ of Corti**, located on the **basilar membrane** that forms the floor of the cochlear duct, contains receptor hair cells. Their tips extend into the cochlear duct, and some attach to the tectorial membrane, which overhangs the organ of Corti.

Sound waves—transmitted and amplified by the tympanic membrane, the three bones of the middle ear, the oval window, and pressure waves in the cochlear fluid—are transduced into receptor potentials in the cochlea. Pressure waves, traveling from the vestibular canal through the tympanic canal and dissipating when they strike the **round window**, vibrate the basilar membrane. The bending of the hairs of the receptor cells against the tectorial membrane opens ion channels, and K^+ enters the cells. The resulting depolarization increases neurotransmitter release from the hair cells and increases the frequency of action potentials generated in their associated sensory neuron and carried through the auditory nerve to the brain.

Volume is a result of the **amplitude**, or height, of the sound wave; a stronger wave bends the hair cells more and results in more action potentials. **Pitch** is related to the **frequency** of sound waves, usually expressed in hertz (Hz). Different regions of the basi-

■ **INTERACTIVE QUESTION** 45.5

Label the parts in this diagram of the middle and inner ear.

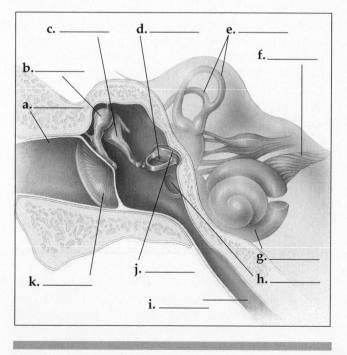

c. _____ d. _____ e. _____

b. _____ f. _____

a. _____

g. _____

j. _____ h. _____

k. _____

i. _____

lar membrane vibrate in response to different frequencies, and neurons associated with the vibrating region transmit action potentials to specific auditory regions of the cortex where the sensation is perceived as a particular pitch.

Balance and Equilibrium in Humans Within the inner ear are two chambers, the **utricle** and **saccule**, and three **semicircular canals**, responsible for balance and equilibrium. Hair cells in the utricle and saccule project into a gelatinous material containing calcium carbonate particles called otoliths. These heavy particles exert a pull on the hairs in the direction of gravity; different head angles stimulate different hair cells, thereby altering the output of neurotransmitters and the resulting action potentials of sensory neurons. Other hair cells in the swellings at the base of the semicircular canals (the ampulla) project into a gelatinous mass called the cupula and are able to sense rotational movements by the pressure of the endolymph in the semicircular canals against the cupula.

Hearing and Equilibrium in Other Vertebrates Mechanoreceptor units called **neuromasts**—clusters of hair cells with sensory hairs embedded in a gelatinous cap or cupula—are contained in the **lateral line system** of fish and aquatic amphibians. Water moving through the tube past these mechanoreceptors bends the cupula and stimulates the hair cells, enabling the fish to perceive its own movement, water currents, sounds of low frequency, and the pressure waves generated by other moving objects.

Fishes also have inner ears, consisting of a saccule, utricle, and semicircular canals, within which sensory hairs are stimulated by the movement of otoliths. Sound-wave vibrations pass through the skeleton of the head to the inner ear. Some fishes have a Weberian apparatus, a series of bones that transmits vibrations from the swim bladder to the inner ear.

In amphibians and reptiles, sound vibrations are conducted by a tympanic membrane on the body surface and a single bone to the inner ear. A cochlea has evolved in birds.

Sensory Organs for Hearing and Balance in Invertebrates Many arthropods sense sounds with body hairs that vibrate in response to sound waves of specific frequencies. Localized "ears" are found on many insects, consisting of a tympanic membrane with attached receptor cells stretched over an internal air chamber.

Most invertebrates have **statocysts**, which often consist of a layer of hair cells around a chamber containing **statoliths**, dense granules or grains of sand. The stimulation of the hair cells under the statoliths provides positional information to the animal.

■ **INTERACTIVE QUESTION   45.6**

What organs in mammals, fishes, and arthropods are used to sense movement or equilibrium? How are these organs similar?

  a. Mammals:

  b. Fishes:

  c. Arthropods:

  d. Similarities:

The interacting senses of taste and smell enable animals to detect many different chemicals: *a closer look*

The sense of taste (gustation) detects chemicals in solution, whereas smell (olfaction) detects airborne chemicals. Sensillae, or sensory hairs containing chemoreceptor taste cells, are found on the feet and mouthparts of insects. Insects also have olfactory sensillae, usually located on their antennae.

In mammals, the chemical senses of taste and smell are produced when a molecule dissolved in liquid binds to a protein in a receptor cell membrane and triggers a membrane depolarization and the release of neurotransmitter. **Taste buds**, which contain groups of modified epithelial cells, are scattered on the tongue and mouth. The four primary taste sensations—sweet, sour, salty, and bitter—are detected in distinct regions of the tongue. The brain integrates the input from the taste buds as they differentially respond to various chemicals, and it creates the perception of a complex flavor.

Olfactory receptor cells are neurons that line the upper part of the nasal cavity; binding of solubilized chemicals to specific receptor molecules on their cilia triggers a signal transduction pathway. The second messenger cAMP opens Na^+ channels, depolarizing the membrane and sending impulses to the olfactory bulb of the brain.

■ **INTERACTIVE QUESTION   45.7**

For what purposes do animals use their chemical senses of gustation and olfaction?

■ Movement is a hallmark of animals

Animals move to obtain food, escape from danger, and find mates. The evolution of various skeletal and muscle designs and body shapes reflects adaptations to overcome the challenges of friction and gravity in a water, land, or air environment.

On a cellular level, all animal movement depends on energy expended to move protein strands past each other, either in microtubules, which are responsible for beating of cilia and flagella, or in microfilaments, which are involved in amoeboid movement and muscle contraction.

■ INTERACTIVE QUESTION   45.8

a. Which mode of **locomotion** is the most energy efficient?

b. Which mode is the most energetically expensive per distance traveled?

c. Which mode consumes the most energy per unit of time?

■ Skeletons support and protect the animal body and are essential to movement

Hydrostatic Skeletons   Fluid under pressure in a closed body compartment creates a **hydrostatic skeleton**. As muscles change the shape of the fluid-filled compartment, the animal moves. Earthworms and other annelids move by **peristalsis** using rhythmic waves of contractions of circular and longitudinal muscles.

Exoskeletons   Typical of mollusks and arthropods, **exoskeletons** are hard coverings deposited on the surface of animals. The **cuticle** of an arthropod contains fibrils of the polysaccharide **chitin** embedded in a protein matrix. Where protection is needed, this flexible support is hardened by cross-linking proteins and addition of calcium salts.

Endoskeletons   The supporting elements of an **endoskeleton** are embedded in the soft tissues of the animal. Sponges have endoskeletons composed of spicules or fibers, and echinoderms have hard plates beneath the skin. The endoskeletons of chordates are composed of cartilage and/or bone. The vertebrate axial skeleton consists of skull, vertebral column, and rib cage; the appendicular skeleton contains the pectoral and pelvic girdles and the limb bones.

■ INTERACTIVE QUESTION   45.9

List an advantage and a disadvantage for each of the three types of skeletons.

   a. hydrostatic skeleton

   b. exoskeleton

   c. endoskeleton

■ Muscles move skeletal parts by contracting

Structure and Function of Vertebrate Skeletal Muscle
Vertebrate skeletal muscle consists of a bundle of multinucleated muscle cells, called fibers, running the length of the muscle. Each fiber is a bundle of myofibrils, each composed of two kinds of myofilaments: thin filaments consist of two strands of actin coiled with a strand of regulatory protein, and thick filaments are made of myosin molecules.

The regular arrangement of myofilaments produces repeating light and dark bands; thus skeletal muscle is also called striated muscle. The repeating units are called **sarcomeres** and are delineated by **Z lines**. Thin filaments attach to the Z line and project toward the center of the sarcomere. Thick filaments lie free in the center, forming the **A band**, with an **H zone** in the center where the thin filaments do not reach. At the edges of the sarcomere, where thick filaments do not extend when the muscle is at rest, is an **I band** of only thin filaments.

The Molecular Mechanism of Muscle Contraction
According to the **sliding filament model** of muscle contraction, the thick and thin filaments do not change length but simply slide past each other, increasing their degree of overlap. The mechanism for the sliding of filaments is the hydrolyzing of ATP by the globular head of a myosin molecule, which then changes to a high-energy configuration that binds to actin and forms a **cross-bridge**. When relaxing to its low-energy configuration, the myosin head bends and

pulls the attached thin filament toward the center of the sarcomere. When a new molecule of ATP binds to the myosin head, it breaks its bond to actin, and the cycle repeats with the high-energy configuration attaching to an actin molecule farther along on the thin filament. Most of the energy for muscle contraction comes from **phosphagens**, such as **creatine phosphate** in vertebrates, which regenerate ATP.

■ INTERACTIVE QUESTION   45.10

Identify the components in this diagram of a section of a skeletal muscle fiber.

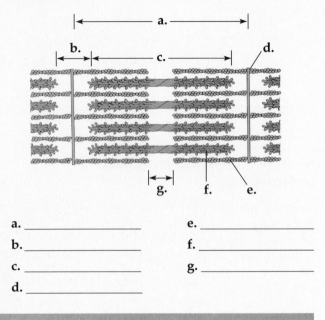

a. _____   e. _____

b. _____   f. _____

c. _____   g. _____

d. _____

The Control of Muscle Contraction   When the muscle is at rest, the myosin-binding sites of the actin molecules are blocked by a strand of the regulatory protein **tropomyosin**, whose position is controlled by another set of regulatory proteins, the **troponin complex**. When troponin binds to calcium ions, the tropomyosin–troponin complex changes shape, and myosin-binding sites on the actin are exposed.

Calcium ions are actively transported into the **sarcoplasmic reticulum** of a muscle cell. When an action potential of a motor neuron causes the release of acetylcholine into the neuromuscular junction, an action potential spreads into **transverse tubules**, infoldings of the muscle cell plasma membrane. The action potential changes the permeability of the sarcoplasmic reticulum, and calcium ions are released into the cytoplasm. Calcium binding with troponin exposes myosin-binding sites and contraction begins. Contraction ends when the sarcoplasmic reticulum pumps Ca^{2+} back out of the cytoplasm, and the tropomyosin–troponin complex again blocks the binding sites.

Graded Contractions of Whole Muscles   A contraction of a single muscle fiber is an all-or-none action, producing a twitch. More graded contraction of a muscle fiber occurs by temporal summation when action potentials arrive rapidly, producing a greater level of tension and a state of smooth and sustained contraction called **tetanus**.

Graded muscular contractions also result from involving additional fibers. A single motor neuron and all the muscle fibers it innervates make up a **motor unit**. The strength of a muscle contraction depends on the size and number of motor units involved. Muscle tension can be increased by the activation of additional motor units, called **recruitment**.

The muscle fatigue of extended tetanus is caused by depletion of ATP, loss of ion gradients required for depolarization, and accumulation of lactate. Some postural support muscles are always in a state of partial contraction, in which different motor units take turns contracting.

Fast and Slow Muscle Fibers   Slow muscle fibers have less sarcoplasmic reticulum; thus calcium remains in the cytoplasm longer and twitches last longer. Slow fibers, common in muscles that must sustain long contractions, have many mitochondria, a good blood supply, and **myoglobin**, which extracts oxygen from the blood and stores it. **Fast muscle fibers** produce rapid and powerful contractions.

Other Types of Muscles   Vertebrate **cardiac muscle** cells are connected by **intercalated discs** through which action potentials spread to all the cells of the heart. The plasma membrane of a cardiac muscle cell can generate action potentials by rhythmic depolarizations. The action potentials of cardiac muscle cells last quite long and have a role in controlling the duration of contraction.

The spiral arrangement of actin and myosin filaments in **smooth muscle** accounts for its nonstriated appearance. Smooth muscle lacks a transverse tubule system and well-developed sarcoplasmic reticulum. Its relatively slow contractions can occur over a greater range of lengths.

Invertebrates have muscle cells similar to the skeletal and smooth muscle cells of vertebrates. The flight muscles of insects are capable of independent and rapid contraction.

STRUCTURE YOUR KNOWLEDGE

1. Describe the path of a light stimulus from where it enters the vertebrate eye to its transmission as an impulse through the optic nerve.

2. Arrange the following structures in an order that enables you to describe the passage of sound waves from the external environment to the brain.

round window	basilar membrane
oval window	cerebral cortex
pinna	vestibular and tympanic canals
auditory nerve	tectorial membrane
cochlea	tympanic membrane
auditory canal	malleus, incus, stapes
cochlear duct	organ of Corti with hair cells

3. Trace the sequence of events in muscle contraction from an action potential in a motor neuron to the relaxation of the muscle.

TEST YOUR KNOWLEDGE

MATCHING: *Match the term with its description.*

_____ 1. oxygen-storing compound in muscles

_____ 2. stretch receptor in muscle, monitors position

_____ 3. stimulus energy converted into receptor potential

_____ 4. light-absorbing molecule

_____ 5. naked dendrites that detect pain

_____ 6. bones that transmit vibrations to inner ear of some fish

_____ 7. invertebrate mechanoreceptors for position; chamber of hair cells

_____ 8. energy-storing compound in muscles

A. accommodation

B. adaptation

C. ampulla

D. creatine phosphate

E. cupula

F. ganglion cells

G. muscle spindle

H. myofibrils

I. myoglobin

J. nociceptors

K. ommatidia

L. rhodopsin

M. statocyst

N. transduction

O. transmission

P. tropomyosin

Q. Weberian apparatus

_____ 9. changing shape of lens to focus

_____ 10. numerous light detectors of compound eye

_____ 11. decrease in sensitivity of receptor during constant stimulation

_____ 12. blocks myosin-binding sites on actin

MULTIPLE CHOICE: *Choose the one best answer.*

1. The absorption of energy from a particular stimulus is called
   a. reception.
   b. transduction.
   c. integration.
   d. transmission.
   e. amplification.

2. Prostaglandins
   a. function as chemoreceptors for histamine.
   b. act like histamines to trigger pain receptors.
   c. are released by nociceptors in response to a pain stimulus.
   d. are released by aspirin and decrease pain.
   e. increase perception of pain by sensitizing pain receptors.

3. The compound eye found in insects and crustaceans
   a. is composed of thousands of prisms that focus light onto a few photoreceptor cells.
   b. cannot sense colors.
   c. contains many ommatidia, each of which consists of a cornea and a lens that focuses light from a tiny portion of the field of view.
   d. may detect infrared and ultraviolet radiation.
   e. do all of the above.

4. Which of the following statements is *not* true?
   a. The iris regulates the amount of light entering the pupil.
   b. The choroid is the thin, pigmented inner layer of the eye that contains the photoreceptor cells.
   c. The ciliary muscle changes the shape of the lens.
   d. Thin aqueous humor fills the anterior eye cavity.
   e. The conjunctiva is a mucous membrane surrounding the white of the eye.

5. Which of the following is *incorrectly* paired with its function?
   a. cones—respond to different light wavelengths producing color vision
   b. horizontal cells—produce lateral inhibition of receptor and bipolar cells
   c. bipolar cells—relay impulses between rods or cones and ganglion cells
   d. ganglion cells—synapse with the optic nerve at the optic disc or blind spot
   e. fovea—center of the visual field, containing only cones

6. The molecule rhodopsin
   a. is involved with color vision.
   b. dissociates when struck by light and is functional only in complete dark.
   c. is recombined by enzymes from retinal and opsin in the light.
   d. is more sensitive to light than are photopsins and is involved in black and white vision in dim light.
   e. activates a signal transduction pathway that increases a rod cell's permeability to sodium.

7. Rotation of the head of a vertebrate is sensed by
   a. the Weberian apparatus located in the utricle and saccule.
   b. changes in the action potentials of hair cells in response to the pull on them by calcium carbonate particles.
   c. bending of hair cells in the ampulla of a semicircular canal in response to the inertia of the endolymph.
   d. hair cells in response to movement of fluid in the coiled cochlea.
   e. statocysts when statoliths stimulate hair cells.

8. Fish can perceive pressure waves from moving objects with their
   a. lateral line system containing mechanoreceptor units called neuromasts.
   b. tectorial membrane as it is stimulated by the bending of sensory hairs embedded in a gelatinous cap.
   c. inner ears, which consist of a saccule, utricle, and semicircular canals, within which sensory hairs are stimulated by the movement of otoliths.
   d. Weberian apparatus, a series of bones that transmits vibrations from the swim bladder to the inner ear.
   e. sensillae collected into ampulla in the lateral line system.

9. The volume of a sound is determined by
   a. the area in the brain that receives the signal.
   b. the section of the basilar membrane that vibrates.
   c. the number of hair cells that are stimulated.
   d. the frequency of the sound wave.
   e. the degree to which hair cells are bent and the resulting increase in the number of action potentials produced in the sensory neuron.

10. Hydrostatic skeletons are used for movement by all of the following *except*
    a. cnidarians.
    b. echinoderms.
    c. nematodes.
    d. annelids.
    e. flatworms.

11. When striated muscle fibers contract,
    a. the Z lines are pulled closer together.
    b. the I band remains the same.
    c. the A band becomes shorter.
    d. the H zone widens slightly.
    e. all of the above occur.

12. Tetanus
    a. is muscle fatigue resulting from a buildup of lactate during anaerobic respiration.
    b. is the all-or-none contraction of a single muscle fiber.
    c. is the result of stimulation of additional motor neurons.
    d. is the result of a volley of action potentials, whose summation produces an increased and sustained contraction.
    e. is the rigidity of muscles following death.

13. What is the role of ATP in muscle contraction?
    a. to form cross-bridges between thick and thin filaments
    b. to break the cross-bridge when it binds to myosin and provide energy to myosin to form its high-energy configuration
    c. to remove the tropomyosin–troponin complex from blocking the binding sites on the actin filament
    d. to bend the cross-bridge and pull the thin filaments toward the center of the sarcomere
    e. to replace the supply of creatine phosphate required for movement of actin past myosin.

14. When an action potential spreads into the transverse tubules,
    a. Calcium is released from the tubules into the sarcoplasmic reticulum.

b. Acetylcholine is released into the neuromuscular junction.

c. Smooth muscle fibers begin their slow and sustained contraction.

d. Calcium binds with troponin and the actin binding sites are exposed.

e. The sarcoplasmic reticular membrane becomes permeable and calcium ions are released.

15. Most of the readily available energy for muscle contraction in vertebrates comes from
a. phosphagens.
b. myoglobin.
c. glycogen.
d. anaerobic respiration.
e. both b and c.

16. A motor unit is
a. a sarcomere that extends from one Z line to the next Z line.
b. a motor neuron and the muscle fibers it innervates.
c. the myofibrils of a fiber and the transverse tubules that connect them.
d. the muscle fibers that make up a muscle.

e. the antagonistic set of muscles that flex and extend a body part.

17. Which of the following is *not* a characteristic of cardiac muscle?
a. spiral arrangement of actin and myosin filaments
b. intercalated discs that spread action potentials between cells
c. action potentials that last a long time
d. ability to generate action potentials without nervous input
e. striations

18. Smooth muscle contracts relatively slowly because
a. The only ATP available is supplied by fermentation.
b. Its contraction is stimulated by hormones, not motor neurons.
c. It does not have a well-developed sarcoplasmic reticulum, and calcium enters the cell through the plasma membrane during an action potential.
d. It is not striated.
e. It is composed exclusively of slow muscle fibers.

CHAPTER **46**

AN INTRODUCTION TO ECOLOGY: DISTRIBUTION AND ADAPTATIONS OF ORGANISMS

FRAMEWORK

This chapter describes the organizational levels at which ecological questions are asked, the abiotic factors to which organisms have adapted in both an ecological and an evolutionary time frame, and the major world communities, or biomes, in which adaptations to climate and abiotic factors have produced similar and characteristic life forms.

CHAPTER REVIEW

■ Ecology is the scientific study of the interactions between organisms and their environments

Ecology is a multidisciplinary science, incorporating many fields of biology and the physical sciences. Ecology is a challenging science because of the complexity and scale of ecological questions. Organisms are affected by and, in turn, affect both the **abiotic** and **biotic** components of their environments.

■ Basic ecology provides a scientific context for evaluating environmental issues

The science of ecology is not synonymous with the growing "ecological" consciousness about current environmental problems, but it helps us to understand these problems and their possible solutions.

■ Ecological research ranges from the adaptations of organisms to the dynamics of ecosystems

The Questions of Ecology **Organismal ecology** considers behavioral, morphological, and physiological responses of an organism to its abiotic environment. Population ecology is concerned with the factors that control the size of **populations**, which are groups of individuals of the same species in an area. The **community** includes all the populations of organisms in an area; ecology on this level looks at interactions such as predation and competition. **Ecosystem ecology** considers abiotic factors as well as the biological community, and it addresses such topics as the flow of energy and chemical cycling.

Ecology as an Experimental Science   Ecology had its foundation in natural history, but experimental approaches in both the laboratory and the field are increasingly used to investigate ecological questions. Tremendous creativity is often required to control variables and manipulate communities in field experiments. Some ecologists develop mathematical models and computer simulations to study the interactions of variables and gain insight into complex ecological questions.

Ecology and Evolution   Interactions between organisms and their environments occur within ecological time. The cumulative effects of these interactions are realized on the scale of evolutionary time. The distribution and abundance of organisms are the result of both evolutionary history and present interactions with the environment.

■ INTERACTIVE QUESTION   46.1

List the four levels of ecological study and give examples of the focus of inquiry at each level.

a.

b.

c.

d.

■ Climate and other abiotic factors are important determinants of the biosphere's distribution of organisms

The **biosphere** is the relatively thin layer of Earth inhabited by life, including the waters of seas, lakes, and streams, the land to a few meters below the soil surface, and the atmosphere up to a few kilometers. **Biomes** are characteristic communities found in broad geographical regions as a result of patterns in climate and other abiotic factors.

Important Abiotic Factors   Temperature is an important environmental factor because of its effects on metabolism and enzyme activity. Most organisms can-

not maintain body temperatures that vary more than a few degrees from ambient temperature.

The availability of water in different habitats can vary greatly. In order to maintain homeostasis, organisms must regulate their osmolarity, compensating for different osmolarities in aquatic environments and avoiding desiccation in terrestrial habitats.

Light energy drives almost all ecosystems. The intensity and quality of light are limiting factors in aquatic environments. Many plants and animals are sensitive to photoperiod, which serves as a reliable indicator of seasonal changes.

Wind increases the rate of heat and water loss in organisms. Soils, which vary in their physical structure, pH, and mineral composition, affect the distribution of plants and, in turn, the distribution of animals. Substrate composition in aquatic environments influences water chemistry and the types of organisms that can inhabit intertidal and benthic zones.

Fire, hurricanes, and other periodic disturbances can drastically affect biological communities. Many plants have evolved adaptations to periodic fires.

Climate and the Distribution of Organisms   The **climate**, or prevailing weather conditions of a locality, is determined by temperature, water, light, and wind. A climograph plots annual mean temperature and rainfall for a region; generally these values correlate with the distribution of various biomes. Overlaps of biomes on a climograph indicate the importance of seasonal patterns of variation in rainfall and temperatures.

Global Climate Patterns   The absorption of solar radiation heats the atmosphere, land, and water, setting patterns for temperature variations, air circulation, and water evaporation that cause latitudinal variations in climate. The shape of the Earth and the tilting of its axis create seasonal variations in day length and temperature that increase with latitude. The **tropics** receive the greatest amount of and least variation in solar radiation.

The global circulation of air begins as intense solar radiation near the equator. It causes warm, moist air to rise, producing the characteristic wet tropical climate, the arid conditions around 30° north and south as dry air descends, the fairly wet though cool climate about 60° latitude as air rises again, and the cold and rainless climates of the arctic and antarctic. Air flowing in the lower levels of the three major air circulation cells on either side of the equator produces predictable global wind patterns.

Local and Seasonal Effects of Climate   Regional climatic patchiness is influenced by proximity to water and topographical features. Coastal areas are generally more moist, and large bodies of water moderate climate.

Seasonal changes affect local climate. Seasonal changes in wind patterns affect ocean currents, sometimes causing upwellings of cold, nutrient-rich water. Seasonal temperature changes produce the biannual **turnover** of waters in ponds and lakes that brings oxygenated water to the bottom and nutrient-rich water to the surface.

Climate also varies on a very small scale. Microclimates within an area have differences in abiotic features that affect the distributions of organisms.

▓ INTERACTIVE QUESTION 46.2

Mountains affect local climate. Describe their influence in the following three areas:

a. solar radiation:

b. temperature:

c. rainfall:

■ The costs and benefits of homeostasis affect an organism's responses to environmental variation

The geographical distribution of a species is largely determined by its ability to tolerate a combination of abiotic factors.

Regulators and Conformers Plants and animals that use behavioral and physiological mechanisms to maintain a steady internal environment are called **regulators.** **Conformers,** whose internal conditions vary with environmental changes, often live in relatively stable environments.

The Principle of Allocation Regulating mechanisms require an expenditure of energy. The **principle of allocation** is used to analyze how an organism divides its limited energy among reproduction, eating, growing, escaping from predators, and homeostasis. The maintenance costs of endotherms are significantly higher than those of ectotherms.

The distribution of organisms is related to different systems of energy allocation. In stable environments, organisms can allot more energy to growth and reproduction. Organisms that put more of their energy into

dealing with environmental fluctuation, however, can successfully live in a wider range of habitats.

■ The response mechanisms of organisms are related to environmental grain and the time scale of environmental variation

Environmental Grain Size and activity level influence the ways an organism relates to spatial and temporal variations within an environment, called **environmental grain.** A *coarse-grained environment* has patches so large in scale relative to the size and activity of the organism that the organism can choose from different environmental patches. Patches in a *fine-grained environment* are so small relative to the organism that it does not choose between patches. Temporal environmental variations can also be described as fine or coarse grained, with seasonal shifts in climate considered coarse grained even for large and long-lived organisms.

Behavioral Responses Behavioral responses to environmental variations may include relocation to a more favorable local or distant environment. Cooperative social behavior can help organisms respond to unfavorable conditions.

Physiological Responses Optimal environmental conditions and tolerance curves for an organism can be determined experimentally by varying abiotic factors. Tolerance limits largely determine the distribution of organisms. Physiological adjustments to environmental change may include fairly rapid changes in the rates of processes or more gradual and substantial changes that shift the tolerance curve and are called **acclimation.**

Morphological Responses Morphological responses to environmental changes may involve either seasonal (reversible) changes that are considered acclimation or changes in growth and differentiation. The latter irreversible changes are most common in plants and provide these rooted organisms with a means of adapting to changing environments.

Adaptation over Evolutionary Time Behavioral, physiological, and morphological responses occur within an ecological time scale but are based on adaptations that have developed by natural selection over an evolutionary time span. The adaptation of organisms to localized environments may restrict their geographic distribution and often their ability to respond to environmental change. In order for a species to exist in a particular location, it must first have reached that location, and second, the conditions there must be within its tolerance limits.

▦ INTERACTIVE QUESTION   46.3

a. What are the advantages and disadvantages of being a regulator?

b. How does the principle of allocation relate to a conformer?

c. What is acclimation?

■ The geographic distribution of terrestrial biomes is based mainly on regional variations in climate

The geographic distribution of the world's major biomes is related to abiotic factors—in particular, the prevailing climate. Biomes are usually named for their predominant vegetation, and the general appearance of a biome may be similar over large areas, even though species composition varies locally. Often, convergent evolution has produced superficial resemblance of unrelated "ecological equivalents."

Biomes grade into each other. If they do so over a large area, a separate biome, or ecotone, may be recognized.

The predominant ecological communities in a biome result from natural succession; patchiness in biomes results from natural or human disturbances. Urban and agricultural biomes cover a large portion of the Earth's land mass.

Tropical Forest   **Tropical forests**, within 23° latitude of the equator, vary little in temperature or day length. Variations in rainfall result in **tropical dry forests**, lowlands where rainfall is scarce; **tropical deciduous forests**, where trees and shrubs drop their leaves during a long dry season that alternates with monsoons; and **tropical rain forests**, where rainfall is abundant.

The tropical rain forest has the largest species diversity of all communities. The vegetation can be divided into five layers: trees that emerge above the canopy, the high upper **canopy**, the low-tree layer, the shrub understory, and the herbaceous plant and fern ground layer. Many of the diverse animal species have mutualistic interactions with plants when they disperse pollen, fruits, and seeds while foraging.

The soil is generally poor and thin, and most populations of species have low densities. Human destruction of the tropical rain forest is reducing biological diversity and may cause large-scale climate changes.

Savanna   **Savannas** are tropical and subtropical grasslands with only scattered trees and typically have three distinct seasons: cool and dry, hot and dry, and warm and wet. Frequent fires and large grazing mammals restrict vegetation to grasses and forbs, small broad-leaved plants. Large herbivores and burrowing animals are common, although the dominant herbivores are insects. Areas where forest and grassland intergrade are also referred to as savanna.

Desert   Characterized by low and unpredictable precipitation, **deserts** may be hot or cold, depending on location. Perennial vegetation, in deserts where there is enough precipitation to sustain it, consists of widely scattered drought-resistant shrubs or succulents. Rainy periods are marked by rapid blooms of annual plants. Seed eaters and their reptilian predators are common animals. Desert animals have physiological and behavioral adaptations to dry conditions and extreme temperatures.

Chaparral   **Chaparral** is common along coastlines in mid latitudes that have mild, rainy winters and hot, dry summers. The dominant vegetation, dense, spiny evergreen shrubs and annuals, is maintained by and adapted to periodic fires; many species have storage roots that permit quick regeneration or seeds that germinate only after a fire. Animals are typically browsers, fruit-eating birds, seed-eating rodents, ants, and reptiles.

Temperate Grassland   Somewhat similar to savannas but found in regions with relatively cold winters, **temperate grasslands** are maintained by fire, seasonal drought, and grazing. Soils are deep and rich in nutrients, and mulch maintains moisture. Large grazing mammals are typical, but underground invertebrates are the most significant herbivores.

Temperate Deciduous Forest   Characterized by broad-leaved deciduous trees, **temperate deciduous forests** grow in midlatitude regions that have adequate moisture to support the growth of large trees. Hot summers alternate with cold winters, during which trees drop their leaves and become dormant. Forbs, shrubs, and one or two strata of trees are typical; a rich diversity of animal life is associated with this variety of food and habitat. Human activity has greatly reduced these forests worldwide.

Taiga (Coniferous Forest)   A large biome found in northern latitudes and at high elevations in lower latitudes, the **taiga** is characterized by harsh winters and short summers. Soil is usually thin and acidic. Coniferous trees grow in dense, uniform stands, shading out undergrowth. Heavy snowfall insulates the soil and protects small mammals. Animals include seed-eaters, insect herbivores, larger browsers, and predators.

Tundra   The northernmost limit of plant growth is in the **arctic tundra** and is characterized by dwarfed, shrubby, or matlike vegetation. The **alpine tundra**, found at all latitudes on high mountains above the tree line, has similar flora and fauna, but day length may be more even and plant growth steadier. The **permafrost** of the arctic tundra prevents roots from penetrating very far into the continually saturated soil. The arctic tundra has long, dark winters and brief summers marked by nearly continuous daylight and rapid plant growth.

Animal species diversity is low. Insects are common during the arctic summer and provide food for migratory birds. Some large herbivores are found, as well as predators and some small, cyclically abundant herbivores.

■ INTERACTIVE QUESTION   46.4

Temperature and precipitation are two of the key factors that influence the vegetation found in a biome. On the climograph shown below, label the North American biomes (arctic and alpine tundra, coniferous forest, desert, grassland, temperate forest, and tropical forest) represented by each area of temperature and precipitation.

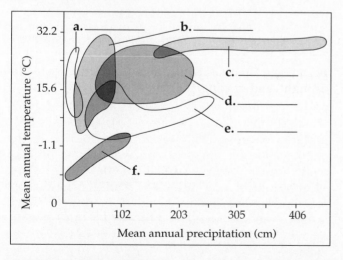

■ Aquatic ecosystems, consisting of freshwater and marine biomes, occupy the largest part of the biosphere

One of the chemical differences in aquatic habitats is salt concentration—less than 1 percent for freshwater biomes versus an average of 3 percent for marine biomes. Freshwater biomes are closely linked with and shaped by the surrounding terrestrial biomes. Three-fourths of Earth is covered by oceans, which influence global rainfall, climate, and wind patterns. Algae of the marine biomes produce a large portion of the world's oxygen and consume enormous amounts of carbon dioxide.

Ponds and Lakes   Small bodies of standing fresh water are called **ponds**; larger ones are **lakes**. Ponds and lakes are stratified in availability of light, temperature, and community structure. The **photic zone** receives sufficient light for photosynthesis, whereas little light penetrates into the lower **aphotic zone**. A narrow **thermocline** separates surface waters from the cold bottom layer.

The shallow and warm waters close to shore, called the **littoral zone**, support a diverse community of rooted and floating aquatic plants, attached algae, mollusks, insects, crustaceans, fish, and amphibians. The open surface waters of the **limnetic zone** are occupied by a variety of phytoplankton and zooplankton, supporting a food chain of larger organisms. **Detritus**, dead organic matter, settles into the deep, aphotic **profundal zone** where it is decomposed by microbes and other organisms, causing these cold, dark waters to become oxygen depleted but nutrient rich. The biannual turnover in temperate zones exchanges oxygen and nutrients between the layers.

Oligotrophic lakes are deep, clear, nutrient poor, and fairly nonproductive. The shallower, nutrient-rich waters of **eutrophic** lakes support large, productive phytoplankton communities. Runoff carrying nutrients and sediment may gradually convert oligotrophic lakes into eutrophic lakes. Cultural eutrophication results when municipal wastes and fertilizer-enriched runoff contribute excessive nitrogen and phosphorus, leading to explosive algal growth, high production of detritus, and depletion of oxygen.

Streams and Rivers   **Streams** and **rivers** are freshwater, flowing habitats, whose physical and chemical characteristics vary from their source to their point of entry into oceans or lakes. Overhanging vegetation contributes to nutrient content. Oxygen levels are high in turbulently flowing water and low in murky, warm waters. Seasonal changes occur in flow and oxygen content.

Food chains are built on the photosynthesis of attached algae and rooted plants or on the addition of organic matter and nutrients from the land. Animal communities vary with changes in the physical environment along the length of streams and rivers. Organisms show evolutionary adaptations that allow them to exploit these habitats. Many insects filter their food from the moving waters. Human impact on streams and rivers has included pollution with wastes, stream channelization, and dams.

Wetlands   Defined as areas covered with water and supporting hydrophytes ("water plants"), **wetlands** range from marshes to swamps to bogs. Topography results in basin wetlands in shallow basins, riverine wetlands along shallow and periodically flooded rivers and streams, and fringe wetlands with rising and falling water levels along coasts of large lakes and seas. These richly diverse biomes are important to flood control and water quality. Wetlands are now being protected and regulated.

Estuaries   Where a freshwater river or stream meets the ocean, an **estuary** is formed. These highly productive areas are often bordered by wetlands called mudflats and saltmarshes. Salinity varies both spatially and daily with the rise and fall of tides. Saltmarsh grasses, algae, and phytoplankton are the major producers. Estuaries serve as feeding and breeding areas for marine invertebrates, fish, and waterfowl. Little undisturbed estuary habitat remains, as these areas have been developed for commercial or residential use and destroyed by the pollution of their streams or rivers.

Introduction to Marine Zones   The marine environment can be classified into a photic zone, in which phytoplankton, zooplankton, and many fishes are found, and an aphotic zone, into which light does not penetrate. The **intertidal zone** is the shallow area along the shore. The **neritic zone** is the shallow region over the continental shelf. The **oceanic zone** reaches great depths. Open water of any depth is the **pelagic zone**, the bottom surface of which is the **benthic zone**.

The Intertidal Zones   The daily cycle of tides exposes the shoreline to variations in water, nutrients, and temperature, and to the mechanical force of wave action. Rocky intertidal communities are vertically stratified, with organisms adapted to firmly attach to the hard substrate and to withstand variations in exposure and salinity. A diverse array of algae, seaweeds, invertebrates, and fishes is usually present. Sandy or mudflat intertidal zones are less stratified, have few large algae or plants, and are home to bur-

■ **INTERACTIVE QUESTION   46.5**

Indicate with a + or - whether the following are relatively high or low in oxygen level, nutrient content, and productivity.

Biome	Oxygen Level	Nutrient Content	Productivity
Oligotrophic lake	a.		
Eutrophic lake	b.		
Headwater of stream	c.		
Turbid river	d.		
Estuary	e.		

rowing worms, clams, and crustaceans. Recreational use and intertidal pollutants such as oil have severely reduced species diversity and population numbers.

Coral Reefs   Found in tropical waters in the neritic zone, **coral reefs** are highly diverse and productive biomes. The structure of the reef is produced by the calcium carbonate skeletons of the coral and serves as a substrate for other corals, sponges, and algae. Coral feed on microscopic organisms and organic debris and are nourished by symbiotic photosynthetic dinoflagellates. These communities support a huge variety of herbivores and carnivores. Coral reefs are easily damaged by pollution, development, native and introduced predators, and souvenir hunters.

The Oceanic Pelagic Biome   The generally cold water of the **oceanic pelagic biome** is typically nutrient poor because detritus sinks to the benthic zone. Seasonal mixing of temperate oceans stimulates phytoplankton growth. Phytoplankton flourish in the photic region and are grazed on by numerous types of zooplankton, invertebrate larvae, and fishes. Morphological adaptations of plankton help them maintain buoyancy. Free-swimming animals, called nekton, include squid, fishes, turtles, and marine mammals. Many pelagic birds feed from the surface waters.

Benthos   The communities that occupy the sea bottom of the neritic and pelagic zones are collectively called **benthos**. Nutrients reach the benthic zone as detritus falling from the waters above. Neritic benthic communities are very diverse and productive. Various invertebrates and fishes inhabit the **abyssal zone**, the deep benthic region where light does not penetrate, and are adapted to darkness, cold, and high water pressure. Chemosynthetic bacteria form the basis of a collection of organisms adapted to the hot, anoxic environment surrounding deep-sea hydrothermal vents.

■ INTERACTIVE QUESTION   46.6

Different marine environments can be classified on the basis of light penetration, distance from shore, and open water or bottom. Match the following zones to their corresponding number on the diagram below:

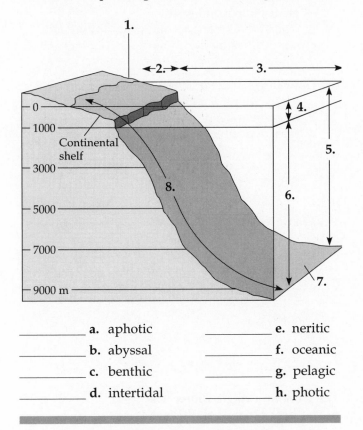

_____	**a.** aphotic	_____	**e.** neritic
_____	**b.** abyssal	_____	**f.** oceanic
_____	**c.** benthic	_____	**g.** pelagic
_____	**d.** intertidal	_____	**h.** photic

STRUCTURE YOUR KNOWLEDGE

1. **a.** Define *ecology.*
   b. What methods are used to answer ecological questions?
   c. What theory guides the interpretation of data?

2. **a.** What are biomes?
   b. What accounts for the similarities in life forms found in the same type of biome in geographically separated areas?

TEST YOUR KNOWLEDGE

MULTIPLE CHOICE: *Choose the one best answer.*

1. Which level of ecology considers energy flow and chemical cycling?
   a. community
   b. ecosystem
   c. organismal
   d. population
   e. abiotic

2. Which of the following would be *least* true of a regulator?
   a. It can live in a variable climate because of its homeostatic mechanisms.
   b. It may have a larger geographic range than a conformer.
   c. Much of its energy budget can be allocated to reproduction.
   d. It can increase its tolerance limits through acclimation.
   e. It has behavioral mechanisms for responding to changing conditions.

3. Ecologists use mathematical models and computer simulations because
   a. Ecological experiments are too broad in scope to be performed.
   b. Most of them are mathematicians.
   c. Ecology is becoming a more descriptive science.
   d. These approaches allow them to study the interactions of multiple variables and simulate large-scale experiments.
   e. Variables can be manipulated with computers but not in field experiments.

4. A fast-running stream with a rocky bottom would be a coarse-grained environment to
   a. benthic insect nymphs.
   b. trout.
   c. phytoplankton.
   d. rooted plants.
   e. a snake.

5. Acclimation
   a. is a morphological response to environmental change.
   b. can extend the tolerance limit of an organism.
   c. involves behavioral responses to environmental change.
   d. is an irreversible physiological change to environmental change.
   e. is an evolutionary change in the range of a species.

6. Which of the following is *incorrectly* paired with its description?
   a. neritic zone—shallow area over continental shelf
   b. abyssal zone—benthic region where light does not penetrate
   c. littoral zone—area of open water
   d. intertidal zone—shallow area at edge of water
   e. profundal zone—deep, aphotic region of lakes

7. A conformer is most likely to be successful in a(n)
   a. intertidal zone.
   b. coral reef.
   c. taiga.
   d. chaparral.
   e. estuary.

8. Two communities have the same mean temperature and rainfall but very different compositions and characteristics. The best explanation for this phenomenon is that the two
   a. are found at different altitudes.
   b. are composed of species that have very low dispersal rates.
   c. are found on different continents.
   d. receive different amounts of sunlight.
   e. have a different range of temperatures and pattern of rainfall throughout the year.

9. Phytoplankton are the basis of the food chain in
   a. streams.
   b. wetlands.
   c. the oceanic pelagic biome.
   d. rocky intertidal zones.
   e. deep-sea thermal vents.

10. The ample rainfall of the tropics and the arid areas around 30° north and south latitudes are caused by
    a. ocean currents that flow clockwise in the northern hemisphere and counterclockwise in the southern.
    b. the global circulation of air initiated by intense solar radiation near the equator producing wet and warm air.
    c. the tilting of the earth on its axis and the resulting seasonal changes in climate.
    d. the heavier rain on the windward side of mountain ranges and the "rainshadow" on the leeward side.
    e. the location of tropical rain forests and deserts.

11. The permafrost of the arctic tundra
    a. prevents plants from getting established and growing.
    b. protects small animals during the long winters.
    c. prevents plant roots from penetrating very far into the soil.
    d. helps to keep the soil from getting wet since water cannot soak in.
    e. both c and d.

12. Many plant species have adaptations for dealing with the periodic fires typical of a
    a. savanna.
    b. chaparral.
    c. temperate grassland.
    d. temperate deciduous forest.
    e. a, b, or c.

13. Track athletes may train at high altitudes in order to
    a. take advantage of the better weather.
    b. produce more red blood cells through acclimation.
    c. get used to running hills.
    d. extend their tolerance limits for altitude.
    e. adapt physiologically and morphologically to cold and wind.

14. Upwellings in the ocean
    a. are locations of reef communities.
    b. occur over deep-sea hydrothermal vents.
    c. are responsible for ocean currents.
    d. bring nutrient-rich water to the surface.
    e. are most common in tropical waters, where they bring oxygen-rich water to the surface.

MATCHING: *Match the biotic description with its biome.*

Biome	Biotic Description
_____ 1. chaparral	A. broad-leaved deciduous trees
_____ 2. desert	B. lush growth, vertical layers
_____ 3. savanna	C. evergreen shrubs, fire adapted vegetation
_____ 4. taiga	D. tropical grasses and forbs
_____ 5. temperate forest	E. coniferous forests
_____ 6. temperate grassland	F. low shrubby or matlike vegetation
_____ 7. tropical rain forest	G. grasses in relatively cool regions
_____ 8. tundra	H. widely scattered shrubs, cacti, succulents

CHAPTER 47

POPULATION ECOLOGY

FRAMEWORK

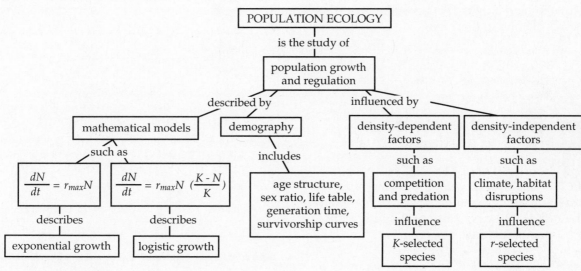

CHAPTER REVIEW

The continuing growth of the human population in the face of limited resources is a critical biological phenomenon. Population ecology is the study of the fluctuations in and regulation of population size and composition. A **population** is a group of individuals of the same species that occupy the same area, use the same resources, and respond to similar environmental factors. The characteristics of a population are determined by interactions with the environment on both ecological and evolutionary time scales.

■ Two important characteristics of any population are its density and the spacing of individuals

Every population has geographic boundaries; ecologists use boundaries based upon the type of organism and the research question being asked. The number of individuals per unit area or volume is the population **density**; the spacing of those individuals within the population is referred to as **dispersion**.

Measuring Density   Population density is often measured by using one of a variety of sampling techniques to count and estimate population size. Indirect indicators, such as burrows or nests, also may be used. In the **mark–recapture method**, animals are trapped, marked, released, and trapped again. The number of marked recaptured animals of the total second trapping indicates the percent of the population that was marked; dividing the initial number marked by this percent estimates population size.

Patterns of Dispersion   Individuals may be dispersed in the population's **geographical range** in several patterns. **Clumping** may indicate a heterogeneous environment, with organisms congregating in suitable microenvironments; clumping also may be related to social interactions between individuals. **Uniform** distribution is usually related to competition for resources, resulting in interactions that produce territories or spaces between individuals. **Random** spacing, indicating the absence of strong attractions or repulsions between individuals, is not very common.

Populations of a species may also show dispersion patterns within their geographical range. Biogeography considers the factors involved in the distribution of a species.

■ **INTERACTIVE QUESTION   47.1**

In a mark–recapture study, an ecologist traps, marks, and releases 25 voles in a small wooded area. Three nights later she resets her traps and captures 30 voles, 10 of which were marked. What is her estimate of the population size of voles in that area?

■ **Demography is the study of factors that affect birth and death rates in a population**

Population size changes in response to the relative rates of birth and immigration versus death and emigration. **Demography** is the study of the vital statistics that affect population size.

Age Structure and Sex Ratio   The **age structure** of a population is the relative number of individuals of each age and may be shown with an age pyramid. Each age group in a population has a characteristic **birth rate**, or **fecundity**, and **death rate**. Populations with the greatest proportion of members of reproductive age or slightly younger grow the fastest. A shorter **generation time**, the average span between the birth of individuals and birth of their offspring, results in faster population growth. Generation time generally increases with body size.

Sex ratio is the proportion of each sex in a population. The number of females in the population may have a greater effect on birth rate if males of the species typically mate with many females.

Life Tables and Survivorship Curves   A **life table** can be constructed either by following a **cohort** of newborn organisms as they reproduce and die or by knowing the age-specific mortality and birth rates in a population during a period of time. A typical life table shows age-specific mortality and birth rates and the proportion of individuals that survive at a given age.

A **survivorship curve** shows the number of members of a cohort still alive at each age. There are three general types of survivorship curves. Type I, with low mortality during early and middle age and a rapid increase with old age, is typical of populations that produce relatively few offspring and provide parental care. In a Type II curve, death rate is relatively constant throughout the life span. A Type III curve is typical of populations that produce many offspring, most of which die off rapidly. The few that survive are likely to reach adulthood. Many species show intermediate or more complex survivorship patterns.

■ **INTERACTIVE QUESTION   47.2**

Identify the types of survivorship curves shown below and give examples of groups that exhibit each curve.

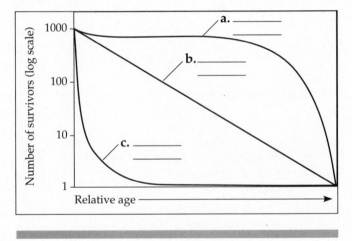

■ **The traits that affect an organism's schedule of reproduction and death make up its life history**

Variation in Life Histories   The **life history** of an organism—its pattern of reproduction and death—affects population growth in ecological time but evolved as a result of natural selection operating in evolutionary time. Life history factors may vary in conjunction with environmental factors, such as increasing clutch size with increasing latitude. Life history traits are often related; high fecundity often correlates with high mortality.

Allocation of Limited Resources   Darwinian fitness is measured by how many offspring survive to produce their own offspring. Because organisms have a finite energy budget, they cannot maximize all life history factors simultaneously. Evolution has fashioned how individuals in a population "choose" how often to breed, at what age to begin reproduction, and how many offspring to produce at a time.

Number of Reproductive Episodes per Lifetime
Some organisms use a strategy called **semelparity**, expending all the organism's energy budget in a single large reproductive effort. Other organisms are **iteroparous**, making a smaller reproductive effort over many seasons and dividing their energy into maintenance, growth, and reproduction. Population mathematical models relate survival probability of both adult and young to the evolution of semelparity or iteroparity. Semelparity is expected when adult survival is uncertain. Plants and animals that typically have longer life spans are usually iteroparous.

Most organisms that live more than one or two years are iteroparous. After growing for several seasons, however, some organisms reproduce with a "big bang," investing all their resources in reproduction and then dying.

Clutch Size   The number of offspring produced is related to the adult's chance of survival to the next breeding season—a trade-off of current fecundity versus adult survival and future fecundity. Clutch size usually is inversely related to the size of offspring. Organisms with a Type III survivorship curve usually produce large numbers of small eggs or offspring. Fewer, larger offspring may have a better chance of surviving to adulthood, typical of Type I and II curves. Large numbers of offspring are related to the selective pressures of high mortality rates of young in uncertain environments or from intense predation. Parental investment in size of offspring, incubation, and/or parental care increases survival chances of the young.

Age at First Reproduction   Delaying first reproduction can result in a maximum lifetime reproductive output if older females produce larger clutches and if survivorship to an older age is likely. Birds with a higher chance of surviving from one year to the next have an older age of sexual maturity.

■ A mathematical model for exponential growth describes an idealized population in an unlimited environment

Observations, experiments, and mathematical modeling are used to determine rates of population growth, to study variables affecting growth, and to predict population sizes.

■ INTERACTIVE QUESTION   47.3

Fecundity, mortality, age at first reproduction, clutch size, and parental investment are usually interrelated. On the following graphs, sketch the relationship you would predict between the two variables.

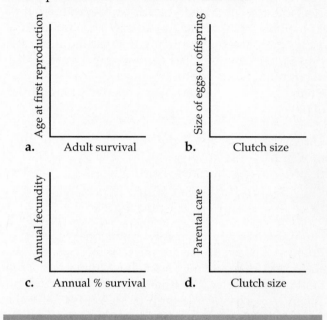

a.   Adult survival

b.   Clutch size

c.   Annual % survival

d.   Clutch size

Growth of a small population in a very favorable environment will be restricted only by the physiological ability of that species to reproduce. Ignoring immigration and emigration, the change in population size during a specific time period is equal to the number of births minus deaths. Births and deaths can be expressed in terms of a per capita birth and death rate; the change in population size per unit of time can be signified as $\Delta N/\Delta t = bN - dN$. The difference in per capita birth and death rates is the per capita population growth rate, symbolized by r; $r = b - d$. **Zero population growth (ZPG)** occurs when $r = 0$. The formula describing change in the population at any one instant uses the notation of differential calculus and is written as $dN/dt = rN$.

The **intrinsic rate of increase** (r_{max}) is the fastest growth rate possible for a population reproducing under ideal conditions. This **exponential population growth**, expressed as $dN/dt = r_{max} N$, produces a J-shaped growth curve when graphed. The larger the population (N) becomes, the faster the population grows. The size of r_{max} determines a population's growth rate and is related to life history features. Periods of exponential growth may occur when populations exploit an unfilled environment or rebound from a catastrophic event.

■ A logistic model of population growth incorporates the concept of carrying capacity

A population may grow exponentially for only a short time before its increased density limits the resources available for continued growth and reproduction of its members. Crowding and resource limitation may lead to decreased per capita birth rates and increased per capita death rates. The **carrying capacity** (*K*) is the maximum stable population size that a particular environment can support over a relatively long period of time.

The Logistic Growth Equation   The mathematical model of **logistic population growth** ($dN/dt = r_{max}N(K-N)/K$) includes the term $(K-N)/K$ to reflect the impact of the increasing population size on *r* as the population approaches the carrying capacity.

When *N* is small, the $(K-N)/K$ term is close to 1, and growth is approximately exponential ($r_{max}N$). As population size approaches the carrying capacity, the $(K-N)/K$ term becomes a small fraction, and exponential growth ($r_{max}N$) is reduced by that fraction. When *N* reaches *K*, the term $(K-N)/K$ is 0, and the growth rate (*r*) is 0. At this point, birth rate equals death rate, and the size of the population does not increase. According to the logistic model, population growth is **density dependent**. The logistic model produces an S-shaped growth curve, and maximum increase in population numbers occurs when *N* is intermediate.

How Well Does the Logistic Equation Actually Describe the Growth of Populations? Some laboratory populations of small animals and microorganisms show logistic growth, and some studies show rebounding populations or organisms introduced into new habitats growing to a carrying capacity.

The logistic model makes the assumption that any increase in population numbers will have a negative effect on population growth. The **Allee effect** is seen in some populations, however, when individuals benefit as the population grows, either from physical support, as in plants, or from social interactions important to reproduction.

The logistic model also assumes that populations approach their carrying capacity smoothly, but many populations overshoot and then oscillate above and below a general carrying capacity.

There are populations for which population density is not an important factor. These populations are often reduced by environmental conditions before resources have a chance to become limiting.

Population Growth Models and Life Histories
Populations that tend to remain at high density, close to their carrying capacity, may exhibit the *K*-selected traits of competitive ability, efficient resource utilization, and the related life history traits of late maturation, low mortality rate, small clutch size with large offspring and extensive parental care, and iteroparity. In variable environments, population densities fluctuate, and populations may have the *r*-selected traits of increased fecundity, early maturity, and high rates of reproduction. Most populations, however, appear to show a mix of traditional *r*- and *K*-selected traits.

■ INTERACTIVE QUESTION   47.4

Label the exponential and logistic growth curves, and show the equation associated with each curve. What is *K* for the population shown with curve **b**?

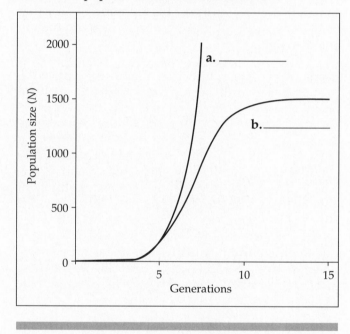

■ Both density-dependent and density-independent factors can affect population growth

Density-Dependent Factors   According to the logistic model, an increasing population ultimately leads to resource limitations that curb population growth. **Intraspecific competition** by members of the same species for the same resources increases with population size, causing growth rate (*r*) to decline. Density-dependent factors decrease reproduction or increase mortality and determine the carrying capacity of the environment.

Limited food supply often limits reproductive output. Food and other resources may be protected in defended areas by animals that exhibit **territoriality**, and the availability of territorial space then may become the limiting resource. Increased population densities may affect health and survivorship. The accumulation of toxic metabolic wastes may also be a limiting factor.

Predation may be a density-dependent factor when a predator shows switching behavior and feeds preferentially on a prey population that has reached a high density.

Intrinsic factors may also regulate population size. Studies of mice have shown that even when food or shelter is not limiting, population size stabilizes at high densities as mice develop a stress syndrome of hormonal changes that inhibit reproduction.

Density-Independent Factors **Density-independent factors**, such as weather or habitat disruption, affect a population regardless of its size. Density-independent factors may reduce population size before resources or other density-dependent factors become limiting. Many insect populations appear to be regulated by density-independent factors.

Interaction of Regulating Factors Many populations remain fairly stable in size, apparently influenced by density-dependent factors that maintain the populations close to their carrying capacity. Short-term or seasonal fluctuations in population size, however, may be due to density-independent factors. Most populations are probably regulated by some mix of density-dependent and density-independent factors.

Population Cycles Some populations of birds, mammals, and insects show regular density fluctuations. One hypothesis for the cause of these cycles is that crowding may produce stress that affects hormonal balance, reducing fertility, increasing aggressiveness, or inducing mass migrations. The time lag in a population's response to density-dependent factors is another explanation for the regular oscillation about the carrying capacity. Cycles in the snowshoe hare population were once thought to be caused by the predation of lynx on the increasing density of prey, but these cycles have been found in hare populations where lynx are absent. The current thinking is that an increasing herbivore population deteriorates its food supply, which may regulate population size more than increasing predator pressure. Prey cycles, however, may be responsible for the population cycles of their predators.

▪ INTERACTIVE QUESTION   47.5

a. List some density-dependent factors that may limit population growth.

b. List some density-independent factors that may limit growth.

▪ The human population has been growing exponentially for centuries, but will not be able to do so indefinitely

From 1650, it took 200 years for the global population to double to 1 billion. In the next 80 years, it doubled to 2 billion; it doubled again in the next 45 years and is projected to reach 8 billion by the year 2017 if the present growth rate is maintained. Growth since the Industrial Revolution has resulted mainly from a drop in death rates, especially among infants. Severe environmental problems are caused by this explosive population growth.

The ultimate carrying capacity of Earth may be determined by food supplies, space, nonrenewable resources, or degradation of the environment.

Worldwide, population growth rates vary by country. High birth rates are more common in developing countries. The age structure of a population influences growth rate; a large proportion of individuals of reproductive age or younger results in more rapid growth.

Human population growth is unique in that it can be consciously controlled by voluntary contraception or government-supported birth control programs. In many cultures, women are delaying marriage and reproduction, thus slowing population growth.

When and how human population growth will level off is an issue of great ecological consequence.

STRUCTURE YOUR KNOWLEDGE

1. Create a concept map to organize your understanding of the exponential and logistic equations—the mathematical models of population growth.

2. Explain how the demographic factors of age structure, age-specific fecundity, sex ratio, generation time, and survivorship influence population growth rates.

TEST YOUR KNOWLEDGE

MULTIPLE CHOICE: *Choose the one best answer.*

1. In a range with a heterogeneous distribution of suitable habitats, the dispersion pattern of a population probably would be
   a. clumped.
   b. uniform.
   c. random.
   d. unpredictable.
   e. dense.

2. The age structure of a population influences population growth because
   a. Younger females have more offspring than do older females.
   b. Populations with shorter generation times grow more rapidly.
   c. Different age groups have different reproductive capabilities.
   d. Life tables show that mortality rates change with age.
   e. The more individuals that are immature, the faster the population will grow.

3. A Type I survivorship curve is level at first, with a rapid increase in mortality in old age. This type of curve is
   a. typical of many invertebrates that produce large numbers of offspring.
   b. typical of humans and other large mammals.
   c. found most often in *r*-selected species.
   d. almost never found in nature.
   e. typical of all species of birds.

4. The middle of the S growth curve in the logistic growth model
   a. shows that at middle densities, individuals of a population do not affect each other.
   b. is best described by the term $r_{max}N$.
   c. shows that reproduction will occur only until the population size reaches K and dN/dt becomes 0.
   d. is the period when competition for resources is highest.
   e. is the period when the population is increasing the fastest.

5. A Type III survivorship curve would be more likely to be found in
   a. a semelparous species.
   b. a *K*-selected species.
   c. a species that undergoes periodic molting.
   d. a species that is territorial.
   e. a population that is regulated by density-dependent factors.

6. A few members of a population have reached a favorable habitat with few predators and unlimited resources, but their population growth rate is slower than that of the parent population. What is a possible explanation for this situation?
   a. The genetic makeup of these founders may be less favorable than that of the parent population.
   b. The parent population may still be in an exponential part of its growth curve and not yet limited by density-dependent factors.
   c. The Allee effect may be operating; there are not enough population members present for successful reproduction.
   d. a, b, and c may apply.
   e. This scenario would not happen.

7. The term $(K–N)/K$
   a. is the carrying capacity for a population.
   b. is greatest when K is very large.
   c. is zero when population size equals carrying capacity.
   d. increases in value as N approaches K.
   e. accounts for the overshoot of carrying capacity.

8. Density-independent factors
   a. tend to maintain a population around the carrying capacity.
   b. are involved in the population cycles seen in some mammals.
   c. are important in the regulation of *K*-selected populations.
   d. include climatic events and habitat disruptions.
   e. affect a higher proportion of a small population.

9. Which of the following is *not* a characteristic of a *K*-selected species?
   a. usually one reproductive episode per lifetime with little parental care
   b. extensive homeostatic capability to deal with environmental fluctuations
   c. long maturation time
   d. usually low mortality
   e. large offspring or eggs

10. Which of the following is *not* true? A population with a large r_{max} value
    a. probably has a short generation time.
    b. probably has large clutch sizes.
    c. is most likely found in variable environments.
    d. is most likely to be regulated by density-independent factors.
    e. is most likely to have a large body size.

11. The human population is growing at such an alarmingly fast rate because
    a. Technology has increased our carrying capacity and, thus, density-dependent factors have not slowed reproduction.
    b. The death rate has greatly decreased since the Industrial Revolution.
    c. The age structure of many countries is highly skewed toward younger ages.
    d. Infant mortality has decreased.
    e. All of the above are true.

Questions 12–18. Use the following choices to indicate how these life history characteristics would be affected by the described changes.
    a. increase
    b. decrease
    c. stay the same
    d. no relationship or unable to predict

12. How would r_{max} be expected to vary with an increase in generation time?

13. How would generation time be expected to vary with an increase in body size?

14. For a population regulated by density-dependent factors, how might clutch size change with increased population density?

15. How would clutch size be expected to vary with a decrease in size of eggs or offspring?

16. For a population in a particular environment, how would K be expected to change with an increase in N?

17. In a population showing exponential growth, how would dN/dt be expected to change with an increase in N?

18. How would generation time be expected to vary with an increase in K?

CHAPTER **48**

COMMUNITY ECOLOGY

FRAMEWORK

Communities are composed of populations of various species that may interact. The structure of a community—including its species diversity, trophic relations, and stability—is determined by these interactions, the past history of the community, and its successional stage. Disturbances and chance events contribute to the dynamic equilibrium (or nonequilibrium) of a community.

Island biogeography permits the study of community structure and diversity within a relatively simple system and provides guidance for establishing nature preserves.

CHAPTER REVIEW

The aggregate of different species living close enough to allow for interaction is called a **community**. Community ecology studies the factors involved in determining species composition and the absolute and relative abundance of species in a community.

■ **The interactive and individualistic hypotheses pose alternative explanations of community structure:** *science as a process*

Ecologists in the 1920s and 1930s developed two views on community composition based upon plant distributions. Gleason advanced the **individualistic hypothesis** that saw communities as chance groupings of species found in the same area because of similar abiotic requirements. Clements advocated the **interactive hypothesis** of a community functioning as an integrated unit.

The individualistic hypothesis predicts that species have independent distributions along environmental gradients and that boundaries between communities are indistinct. The interactive hypothesis, in contrast, predicts that species are clustered into discrete communities. In most plant communities, species appear to vary on a continuum, with each species independently distributed along environmental gradients according to its tolerance range. The distribution of animal species may be more dependent on the presence of other species. Community composition is most likely determined by both abiotic gradients and biological interactions.

■ **INTERACTIVE QUESTION   48.1**

Most plant communities appear to be individualistic. What may explain the occasional occurrence of sharp delineations in species composition between communities?

■ **Community interactions can provide strong selection factors in evolution**

Both abiotic and biotic components of the environment can act as selective factors for evolution. **Coevolution** refers to reciprocal evolutionary adaptations of two species, in which a change in one acts as a selective agent on the other species, whose adaptations in turn act as selective forces on the first.

A complex example of coevolution is found in the interactions of passion-flower vines, the butterfly *Heliconius*, whose larvae are able to feed on these vines despite their toxic chemicals, and ants and wasps, which are attracted to the yellow nectaries on some vines and prey on the *Heliconius* eggs and larvae. Whereas it is often difficult to determine whether intertwined adaptations among several species have served as selective factors for each other, it is generally accepted that interactions between species in ecological time are linked to adaptations over evolutionary time.

■ Interspecific interactions may have positive, negative, or neutral effects on a population's density: *an overview*

Interspecific interactions occur among populations of different species living in the same community.

■ Predation and parasitism are (+ -) interactions: *a closer look*

Predators eat **prey**. Interactions that are (+ -) also include **parasitism**, in which predators live on or in their hosts; **parasitoidism**, in which insects lay eggs on hosts and their larvae feed on them; and **herbivory**, in which animals eat plants.

Predation   Adaptations to increase success in predation include acute senses, speed and agility, and physical structures such as claws, fangs, teeth, and stingers.

Plants may defend themselves with mechanical devices, such as thorns or microscopic spines, or chemical compounds. Distasteful or toxic chemicals, called *secondary compounds* because they form as metabolic by-products, include such well-known compounds as strychnine, morphine, nicotine, digitoxin, various spices, and tannins. Counteradaptations often enable herbivore populations to overcome plant defenses and even use the compounds in their own defense.

Animals can defend against predation by hiding, fleeing, or defending. Alarm calls may bring in reinforcements that mob the predator. Distraction displays may divert a predator's attention. Potential prey may use camouflage in the form of **cryptic coloration** or body form to blend in with the background. Deceptive markings may confuse predators. Mechanical and chemical defenses discourage predation. From the food they eat, some animals passively accumulate compounds that would be toxic to their predators. Bright, conspicuous, **aposematic coloration** warns predators not to eat animals with these defenses.

Mimicry may be used by prey to exploit the aposematic coloration of other species. Predators may use mimicry to "bait" their prey.

■ INTERACTIVE QUESTION   48.2

Name the following two types of mimicry:

a. a harmless species resembling a poisonous or distasteful species:

b. mutual imitation by two or more distasteful species:

Parasitism   Parasites harm but usually do not kill the host on which they feed. Parasites that live within a host are **endoparasites**; those that feed briefly on the surface of a host are **ectoparasites**. Parasites best adapted to find and feed on hosts and hosts best able to defend against parasites are favored by natural selection. Examples of host defenses are secondary plant chemicals and the immune system of vertebrates. Coevolution tends to stabilize host–parasite relationships.

Brood parasitism is another (+ -) interaction; in this type, a host's parental behavior is exploited. Coevolution is evident in the defense strategies of some host birds and in the egg mimicry of the parasite species.

■ Interspecific competitions are (- -) interactions: *a closer look*

If two species use the same limiting resource, **interspecific competition** may affect the density of both species. Competition for resources that involves fighting is called **interference competition**, whereas simply the use of the same resources is called **exploitative competition**.

The Competitive Exclusion Principle   Lotka and Volterra modified the logistic equation to include the effects of interspecific competition and predicted that two species that relied on the same limited resource could not coexist in the same community. Gause's laboratory experiments with *Paramecium* tested and supported this **competitive exclusion principle**.

Ecological Niches   An organism's **ecological niche** is described as its place in an ecosystem—its habitat and use of biotic and abiotic resources. The **fundamental niche** includes resources an organism theoretically could use; the **realized niche** is the resources it actually does use as determined by constraints such as competition, predation, or the absence of a potential resource. The competitive exclusion principle holds that two species with the identical niche cannot coexist in a community.

Evidence for Competition in Nature   It is difficult to evaluate the role of competition in nature, since it is presumed not to be able to operate very long before a species either becomes extinct or adapts to a slightly different niche. Populations in the same geographic area can interact and are called *sympatric. Allopatric* populations are found in different areas.

Resource partitioning, slight variations in niche that allow ecologically similar species to coexist, provides circumstantial evidence that competition was a major factor in natural selection. **Character displacement** of some morphological or behavior trait allows closely related sympatric species to avoid competition. When these same species are allopatric, the differences between them may be much less. In addition to the accumulated circumstantial evidence, controlled field studies have documented the competitive exclusion of species in areas of overlap in fundamental niche.

■ Commensalism and mutualism are (+ 0) and (+ +) interactions, respectively: *a closer look*

Close association between a **symbiont** and its **host** is called **symbiosis**. In **parasitism**, the host is harmed; in **commensalism**, one individual is benefited and the other unaffected; and in **mutualism**, both symbionts benefit from the relationship.

Commensalism (+ 0)   It is difficult to establish that a relationship is truly commensal, that one member is completely unaffected by the other. Evolutionary

adaptations in a commensal relationship are likely to involve only the species that benefits.

Mutualism (+ +)   Coevolution in mutualistic relationships affects both species, as changes in one alter the fitness of the other. Mutualistic interactions may have evolved when organisms became able to derive some benefit from their predator or parasite.

■ A community's structure is defined by the activities and abundances of its diverse organisms

Feeding Relationships Within Communities   The feeding relationships, or **trophic structure**, of a community include plant–herbivore, host–parasite, predator–prey, and competitor interactions. The study of trophic structure can lump species into functional groups, such as producers or consumers, or can analyze food web connections among species in the community.

Species Richness, Relative Abundance, and Diversity   **Species richness**, the number of different species found in a community, and **relative abundance**, the relative numbers of individuals in each species, are both components of the ecological concept of **species diversity**. Human impact on most communities is to reduce species diversity.

Disturbances and Community Stability   The ability of a community to reach and maintain equilibrium and to return to its original makeup following a disturbance is known as **stability. Community resilience** is the persistence of a community's major structural features in the face of disturbance, even though not all populations within it remain stable.

■ INTERACTIVE QUESTION   48.3

Name and give examples of the interspecific interactions symbolized in the table.

	Interaction	Examples
+ +	a.	
+ 0	b.	
+ -	c.	
- -	d.	

■ INTERACTIVE QUESTION   48.4

Community A has six different species; 90 percent of the population members belong to one of those species. Community B has five different species; 20 percent of the population belongs to each of those species.

  a. The community with the greatest species richness is _____.

  b. The community with the greatest species diversity is _____.

■ The factors that structure communities include competition, predation, and environmental patchiness

The Role of Competition Field studies have documented the importance of competition in some communities. Many introduced, or **exotic**, species have changed community structure by outcompeting native species. Interspecific competition appears to influence relative abundance and richness in many communities, but it does not always lead to competitive exclusion. Competition is not thought to influence herbivorous populations, which tend not to be limited by their food supply but to be regulated by density-independent factors. Competition is seen more frequently among carnivores and plants. Competition cannot be the major factor structuring communities unless population sizes are near carrying capacity.

The Role of Predation The introduction of exotic species to aquatic communities has shown that predation may reduce diversity. The more common effect, however, may be to reduce competition among prey species and help to stabilize species diversity. Paine's study of a predatory sea star in an intertidal community introduced the concept of a **keystone predator** and its role in maintaining species diversity by reducing the density of a highly competitive prey species.

The Role of Environmental Patchiness A spatially and temporally heterogeneous habitat provides more ecological niches and can support a more diverse community. The vegetative structure of a community largely determines the resident animals, and more complex vegetation provides more diverse habitats. Environmental patchiness facilitates resource partitioning and provides varying local conditions to which different species may be better adapted. Temporal use of habitats also adds to the species diversity of a community.

Evaluating Causative Factors Interspecific interactions, environmental heterogeneity, or an interaction of these factors may all affect community structure and diversity. Which factors are most important may vary from community to community. Determining the evolutionary factors that have led to today's ecological relationships is particularly difficult.

■ Succession is the sequence of changes in a community after a disturbance

Transition in species composition in a community, usually following some disturbance, is known as eco-

logical succession. In traditional views, this succession follows a definite sequence of predictable transitional stages, leading to a characteristic, stable **climax community**. If no soil was originally present, the process is called **primary succession**. **Secondary succession** occurs when an existing community is disrupted by fire, logging, or farming, but the soil remains intact.

Causes of Succession Early successional stages are dominated by *r*-selected species that disperse readily and grow rapidly. Species whose tolerances best fit the specific abiotic environment of an area are the most successful colonizers. As resource availability changes, different species compete more successfully.

Successional changes may be influenced by the biotic community. Through *inhibition*, early colonizers may retard the growth of other species by exploitative or interference competition. In *facilitation*, one stage of vegetation alters the environment, perhaps by changing soil pH or increasing organic matter, making the environment more suitable for species in the next stage.

Factors that periodically disturb communities may prevent succession to a typical climax community. There is also evidence that climax communities may not remain as stable over long periods of time as previously thought.

Natural and Human Disturbance Disturbances change resource availability and create opportunities for new species. Natural disturbances include fire, drought, wind, and moving water. Animals can also create major disturbances. Human activities have altered the structure and succession of communities all over the world, disrupting mature communities and restarting successional growth. The disappearance of the tropical rain forests and the creation of vast barren areas in Africa are the results of human disturbances.

Small-scale natural disturbances may produce patches of different successional stages and help maintain species diversity in a community. They may also prevent large-scale disturbances, such as the destructive fires in Yellowstone Park in 1988.

Community Equilibrium, Disturbance, and Species Diversity   During succession, *K*-selected species tend to replace *r*-selected species, species diversity generally increases, and predation, competition, and symbiosis become more extensive. According to the equilibrial model, which stresses population interactions, succession reaches a climax when interactions are so intricate that no new niches are available for additional species.

The nonequilibrial model views communities as being in continual flux, with the number and identity of species changing even in the climax stage. A mature community is described as an unpredictable combination of patches at different successional stages. Chance events such as dispersal and disturbance are thought to be major determinants of community composition and species diversity. According to the **intermediate disturbance hypothesis**, species diversity is greatest when disturbances are intermediate in severity and frequency, allowing organisms from different successional stages to be represented in the community.

■ **INTERACTIVE QUESTION**   48.6

How might a community exposed to severe and frequent disturbances differ in composition from one that experiences mild and infrequent disturbances?

■ Biogeography complements community ecology in the analysis of species distribution

Biogeography is the study of past and present distributions of species and communities. Historical biogeography considers how the distributions of species reflect both past history and present interactions with the biotic and abiotic environment. At the intersection of biogeography and community ecology, the following three areas are studied.

Limits of Species Ranges   A species may be limited to a particular range for one of three reasons: it never dispersed beyond that range; pioneers to areas outside the range failed to tolerate the physical environment or to compete successfully with resident species; or it retracted from a former larger range in the course of evolutionary time. Transplant experiments can determine which of the first two reasons has operated to limit a range. Paleontology and historical biogeography can provide evidence of species for which the third explanation applies.

Global Clines in Species Diversity   Many taxonomic groups increase in diversity along a global climatic cline from northern latitudes toward the equator. Many hypotheses try to explain this pattern: the tropics rarely experience major natural disturbances, and their resulting age has allowed complex interactions to coevolve; they generally experience intermediate levels of disturbance and have more environmental patchiness; the stability of the climate may allow organisms to specialize, resulting in a finer level of resource partitioning; increased solar radiation results in increased photosynthesis and a large resource base; the structural complexity provides a great variety of microhabitats; and the existing diversity is self-propagating, preventing any populations from establishing dominance. A complex combination of these factors may be involved in latitudinal gradients and the high diversity of the tropics.

The increase in marine animal diversity with increasing depth may relate to the long-term stability of deep-sea benthic communities. All of these explanations are difficult to address experimentally.

Island Biogeography   Any habitat surrounded by a significantly different habitat is considered an island and allows ecologists to study interactions that are difficult to isolate in more complex systems. In the 1960s, MacArthur and Wilson developed a general theory of island biogeography, stating that the size of the island and its closeness to the mainland (or source of dispersing species) are important variables directly correlated with species diversity. Larger islands closer to the mainland will have a higher species diversity than smaller or more distant islands. Following a period of initial immigration and colonization, fewer new species become established, and the extinction rate of present species increases. When immigration and extinction rates are equal, an equilibrium in species diversity develops, although species composition may continue to change. Over long periods, evolutionary changes in species and speciation on the island may determine community structure and composition.

Observations and experiments, including the work of Simberloff and Wilson with arthropod diversity, provide support for the island biogeography theory. These experiments have shown that, whereas species diversity correlates with size and closeness of islands, chance events such as dispersal affect the actual species composition of communities.

■ INTERACTIVE QUESTION 48.7

Many biogeographical studies have found that large islands have greater species richness than small islands. Label the lines on the following graph that show how immigration rate and extinction rate vary with the number of species on large and small islands. Indicate the location of the equilibrium number on the *x* axis for a small and large island.

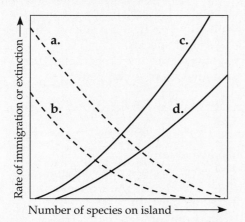

■ Lessons from community ecology and biogeography can help us preserve biodiversity

Efforts to conserve biological diversity in the face of increasing human encroachment on natural communities are based on our esthetic regard for other life forms, the awareness that products useful to humans may be derived from as yet undiscovered species, and the recognition that natural environments maintain the proper functioning of the biosphere. Continuing population growth and widespread economic difficulties complicate conservation efforts.

Information from community ecology and biogeography can be applied to efforts to preserve biodiversity and conserve ecosystems. Island biogeography can be applied to help determine the optimal size of nature preserves. The Lovejoy study in the Amazon region has found that the smaller "islands" have experienced high levels of species extinction. The heterogeneity of the environment may determine the preferability of one large preserve to a collection of smaller, diverse preserves. Continued study to create well-designed nature areas may help to preserve diversity.

STRUCTURE YOUR KNOWLEDGE

1. Complete this concept map to organize your understanding of the important factors that structure a community.

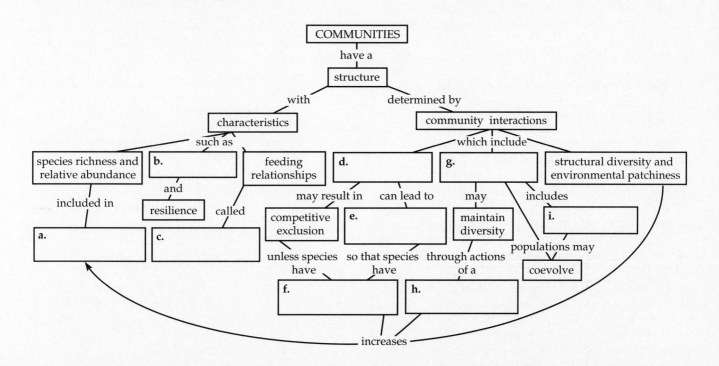

2. Describe succession and the factors that contribute to this process.

TEST YOUR KNOWLEDGE

MULTIPLE CHOICE: *Choose the one best answer.*

1. Which of the following is *not* part of Gleason's individualistic concept of communities?
   a. Communities are chance collections of species that are in the same area because of similar environmental requirements.
   b. There should be no distinct boundaries between communities.
   c. The consistent composition of a community is based on interactions that cause it to function as an integrated unit.
   d. Species are distributed independently along environmental gradients.
   e. Most communities studied meet the predictions made by this concept.

2. Two species, A and B, occupy adjoining environmental patches that differ in several abiotic factors. When species A is experimentally removed from a portion of its patch, species B colonizes the vacated area and thrives. When species B is experimentally removed from a portion of its patch, species A does not successfully colonize the area. From this, one might conclude that
   a. Both species A and species B are limited to their range by abiotic factors.
   b. Species A is limited to its range by competition and species B is limited by abiotic factors.
   c. Both species are limited to their range by competition.
   d. Species A is limited to its range by abiotic factors and species B is limited to its range because it cannot compete with species A.
   e. Species B is *K*-selected and species A is *r*-selected.

3. The species richness of a community refers to
   a. the relative numbers of individuals in each species.
   b. the number of different species found in a community.
   c. the feeding relationships or trophic structure within the community.
   d. the species diversity that is characteristic of that community.
   e. the ability to persist through disturbances.

4. Which of the following is *not* an example of coevolution?
   a. adaptations of flowers and their exclusive pollinators
   b. passion-flower vines and the butterfly *Heliconius*
   c. a parasite that is specific for one host
   d. a brood parasite's eggs that mimic the host species' eggs
   e. aposematic coloration of monarch butterflies and predators that learn not to eat them

5. Through resource partitioning,
   a. Two species can compete for the same prey item.
   b. Slight variations in niche allow closely related species to coexist in the same habitat.
   c. Two species can share the same realized niche in a habitat.
   d. Competitive exclusion results in the success of the superior species.
   e. Two species undergo character displacement that allows them to compete.

6. A species may be restricted to a particular range because
   a. It cannot tolerate environmental conditions outside that range.
   b. It has never dispersed beyond that range.
   c. It has retracted from a former range due to local extinctions.
   d. It would outcompete native species if it were transplanted to their habitat.
   e. a, b, and c are all true.

7. The ability of some herbivores to eat plants that have toxic secondary compounds may be an example of
   a. coevolution.
   b. mutualism.
   c. commensalism.
   d. niche partitioning.
   e. parasitism.

8. A palatable (good-tasting) prey species may defend against predation by
   a. Müllerian mimicry.
   b. Batesian mimicry.
   c. secondary compounds.
   d. aposematic coloration.
   e. either a or b.

9. When one species was removed from a tidepool, the species richness became significantly reduced. The removed species was probably
   a. a strong competitor.
   b. a potent parasite.
   c. a resource partitioner.
   d. a keystone predator.
   e. the species with the highest relative abundance.

10. A highly successful parasite
    a. will not harm its host.
    b. may benefit its host.
    c. will be able to feed without killing its host.
    d. will kill its host fairly rapidly.
    e. will have coevolved into a commensalistic interaction with its host.

11. The most important factor(s) in determining community structure
    a. may change from one community to another.
    b. is predation.
    c. is competition.
    d. is history.
    e. are structural diversity and environmental patchiness.

12. During succession, inhibition
    a. may prevent the achievement of a climax community.
    b. is evidence for the equilibrial theory of succession.
    c. is one of the factors that determines the most tolerant species in an area.
    d. interferes with the successful colonization of other species.
    e. may involve changes in soil pH or accelerated accumulation of humus.

13. According to the nonequilibrial model of succession,
    a. Chance events such as dispersal and disturbance play major roles in succession, and species composition remains in flux.
    b. Species diversity is greatest in the climax community.
    c. When succession reaches a climax community, only extinctions make room for new colonists.

d. The communities with the greatest diversity have the greatest resiliency and resistance to change.
   e. Early colonizers are *r*-selected and later community members are *K*-selected.

14. The increasing diversity of the benthic community with depth is an example of
    a. the diversity of an area correlating with its productivity.
    b. a cline that may be related to increasing environmental stability.
    c. the equilibrium model of community structure.
    d. a community of intermediate disturbance reaching a higher species diversity.
    e. the environmental patchiness and structural diversity of a habitat supporting diversity.

15. An island that is small and far from the mainland, as compared to a large island close to the mainland,
    a. would be expected to have a low species diversity.
    b. would be expected to be in an early successional stage.
    c. would have a small species diversity but a large abundance of organisms.
    d. would have a high rate of colonization but a high rate of extinction.
    e. would have a low rate of colonization and a low rate of extinction.

16. The island recolonization experiment of Simberloff and Wilson showed that
    a. Species diversity returns very slowly to an island after a disturbance.
    b. The species diversity was highest when disturbances were intermediate in frequency and severity.
    c. Whereas the same numbers of species of arthropods returned to each island, the species composition was different, indicating the importance of chance events.
    d. Islands closest to the mainland had the greatest numbers of arthropods recolonize, and their community composition and diversity were the same as prior to fumigation.
    e. The largest islands had the greatest species richness but the least species diversity.

CHAPTER 49

ECOSYSTEMS

FRAMEWORK

This chapter describes energy flow and chemical cycling through ecosystems. Producers convert light energy into chemical energy, which is then passed, with a loss of energy at each level, from consumer to consumer and ultimately to decomposer. Energy makes a one-way trip through ecosystems. Chemical elements, such as carbon, oxygen, nitrogen, and phosphorus, are cycled in the ecosystem from reservoirs in the atmosphere, oceans, or soil through producers, consumers, and decomposers, and back to the reservoirs.

Burning fossil fuels, deforestation, dumping of toxic chemicals into the environment, and washing excess nutrients into lakes are human activities that have altered nutrient cycles and reduced biodiversity. Solving ecological problems requires extending our understanding of ecosystems.

CHAPTER REVIEW

An **ecosystem** is a community and its physical environment; it includes all the biotic and abiotic components of an area. Most ecosystems are powered by energy from sunlight, which is transformed to chemical energy by autotrophs, passed to a series of heterotrophs in the organic compounds of food, and continually dissipated in the form of heat. Chemical elements, such as carbon and nitrogen, are cycled between the abiotic and biotic components of the ecosystem as autotrophs incorporate them into organic compounds and the processes of metabolism return them to the soil, air, and water.

■ Trophic structure determines an ecosystem's routes of energy flow and chemical cycling

The **trophic structure** of an ecosystem determines how energy and chemicals move through different **trophic levels**. Most **primary producers**, or autotrophs, use light energy to photosynthesize sugars for use as fuel in respiration and as building materials for other organic compounds. The **primary consumers** are herbivores. **Secondary consumers** are carnivores that eat herbivores; **tertiary consumers** eat other carnivores. **Decomposers**, or detritivores, consume detritus, which is organic wastes, fallen leaves, and dead organisms.

A **food chain** shows the transfer of food between trophic levels. **Omnivores** eat producers as well as consumers of various levels. Most ecosystems have complex, branching feeding relationships called **food webs**.

Producers Plants are the main producers in terrestrial ecosystems; phytoplankton form the basis for most marine ecosystems. Chemosynthetic bacteria found around thermal vents deep in the seas depend on chemical rather than solar energy.

Consumers Terrestrial herbivores include insects, snails, grazing mammals, and birds and mammals that eat fruit and seeds. In aquatic ecosystems, zooplankton are the main primary consumers. Terrestrial secondary consumers include spiders, frogs, insect-eating birds, and carnivorous mammals. Secondary consumers in aquatic habitats include fish and larger invertebrates.

Decomposers Fungi and bacteria are the most important decomposers in most ecosystems. Earthworms and such scavengers as crayfish, cockroaches, and bald eagles are also decomposers. Decomposers that feed on plant remains often form a link between producers and secondary and tertiary consumers.

■ **INTERACTIVE QUESTION   49.1**

Label the levels of this food chain. How does a food chain differ from a food web?

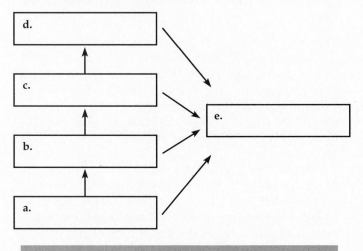

---

large area, account for much of Earth's overall productivity. Estuaries and coral reefs also have high rates of productivity; deserts, tundra, and oceans have low rates. The oceans, however, contribute the most primary productivity due to their huge area.

Productivity in terrestrial ecosystems may be limited by precipitation, heat, and light intensity, as well as by nutrients. A **limiting nutrient** is one whose inadequate supply restricts productivity. Nitrogen or phosphorus is the limiting nutrient in many ecosystems.

Productivity in the seas is greatest in shallow waters and along coral reefs. The open ocean generally has low productivity because mineral nutrients are limited near the surface where light is available for photosynthesis. In freshwater ecosystems, light intensity and availability of minerals affect productivity.

■ **INTERACTIVE QUESTION   49.2**

Antarctic seas are often more productive than most tropical seas, even though they are colder and receive lower light intensity. Explain.

■ An ecosystem's energy budget depends on primary productivity

The Global Energy Budget   The intensity of solar energy striking the earth varies by latitude and by region, depending on cloud cover and dust in the air. Only a small portion of incoming solar radiation strikes algae and plants, and, of that, only 1 to 2 percent is converted to chemical energy by photosynthesis. Nevertheless, worldwide photosynthetic production is about 170 billion tons of organic material per year.

Primary Productivity   The rate of conversion of light to chemical energy is called the **primary productivity** of an ecosystem. **Net primary productivity (NPP)** is the **gross primary productivity (GPP)** minus the energy used by plants in their own cellular respiration (Rs). NPP = GPP − Rs. For most plants, 50 to 90 percent of gross primary productivity remains as net primary productivity, which is the stored chemical energy available to consumers.

Primary productivity can be expressed as energy per unit area per unit time $(J/m^2/yr)$ or as **biomass** measured in terms of dry weight of organic material $(g/m^2/yr)$. **Standing crop biomass** is the total biomass of plants in an ecosystem. Tropical rain forests are very productive ecosystems and, because of their

---

■ As energy flows through an ecosystem, much is lost at each trophic level

Secondary Productivity   The rate at which consumers in an ecosystem produce new biomass from their food is called **secondary productivity**. Each trophic level sees a decline in productivity, partly because of a loss of energy in the form of heat as each consumer converts organic fuel into its own molecules or into energy in cellular respiration. Only the chemical energy stored as growth or in offspring is available as food to higher trophic levels.

Herbivores consume only a fraction of plant material produced; they cannot digest all they eat, and much of the energy they do absorb is used for cellular respiration. Carnivores can digest and absorb more of the organic compounds in their food, but often they use more of it for cellular respiration.

Ecological Efficiency and Ecological Pyramids   **Ecological efficiency** is the percentage of the energy of one trophic level that makes it to the next level, usually ranging from 5 to 20 percent. A **pyramid of productivity** shows this multiplicative loss of energy. A **biomass pyramid** illustrates the standing crop biomass of organisms at each trophic level. This pyramid usually narrows rapidly from producers to the top trophic level. Some aquatic ecosystems have inverted biomass pyramids where zooplankton (consumers) outlive and outweigh the highly productive, but heavily consumed, phytoplankton. Phytoplankton have a short **turnover time**, determined by dividing standing crop biomass by productivity. The productivity pyramid for this ecosystem, however, is normal in shape.

The **pyramid of numbers** illustrates that higher trophic levels contain small numbers of individuals, resulting from the larger size of these animals and the greatly decreased energy availability illustrated by the pyramid of productivity.

▓ INTERACTIVE QUESTION   49.3

a. What are the ultimate energy source and sink for most ecosystems?

b. Approximately what proportion of the chemical energy produced in photosynthesis makes it to a tertiary consumer?

■ Matter cycles within and between ecosystems

Chemical elements are passed between abiotic and biotic components of ecosystems through **biogeochemical cycles**. Plants and other autotrophs use inorganic nutrients to build new organic matter, which is passed through the food chain. Atoms of organisms are returned to the atmosphere, water, or soil through respiration and the action of decomposers.

The route of a biogeochemical cycle depends on the element involved. Gaseous carbon, oxygen, sulfur, and nitrogen have global cycles involving atmospheric reservoirs. Less mobile elements, such as phosphorus, potassium, calcium, and the trace elements, have a more localized cycle in which soil is the main abiotic reservoir. Elements are found in four types of reservoirs or compartments: organic material in living organisms, available to other organisms; unavailable organic material in fossilized deposits; available inorganic elements, ions, and molecules in water, soil, or air; and unavailable elements in rocks. Matter may leave the unavailable reservoirs through weathering of rock, erosion, or burning of fossil fuels.

The actual movement of elements through biogeochemical cycles is quite complex, with influx and loss of nutrients from ecosystems occurring in many ways. Ecologists have studied this movement by adding radioactive tracers to chemical elements in several ecosystems.

■ A combination of biological and geological processes drives chemical cycles

The Water Cycle   Water is essential to living organisms and is the force for the transfer of materials in several cycles. The water cycle involves evaporation, precipitation, and transpiration, with a net flow of water evaporated by solar energy from the oceans, moved as water vapor to the land where it precipitates, and returned to the oceans through runoff and groundwater.

The Carbon Cycle   Carbon generally cycles at a fast rate, as plants take CO_2 from the atmosphere for photosynthesis and organisms release it in cellular respiration. Wood, coal, and petroleum can store carbon for long periods of time. The relatively low concentration of CO_2 in the atmosphere (0.03 percent) fluctuates slightly with seasonal changes in photosynthetic activity in the northern hemisphere. Fossil fuel combustion is increasing the amount of atmospheric CO_2.

In aquatic environments, CO_2 interacts with water and limestone to reversibly form bicarbonate, which acts as a carbon reservoir. The ocean reservoirs contain 50 times as much carbon as does the atmosphere and may continue to absorb some of the CO_2 from fossil fuel combustion.

The Nitrogen Cycle   Plants require nitrogen in the form of NH_4^+ or NO_3^-; they cannot assimilate atmospheric nitrogen. A small amount of nitrogen enters the ecosystem through atmospheric deposition, in rain or the settling of particulates. Most nitrogen enters through **nitrogen fixation**: soil bacteria and symbiotic bacteria in root nodules fix nitrogen in terrestrial ecosystems, and some cyanobacteria do so in aquatic ecosystems. Fertilizers also add a significant amount of nitrogen to agricultural soils and associated waters.

The excess ammonia released by these bacteria is protonated to ammonium in the slightly acidic soil. Evaporation of ammonia back to the atmosphere from neutral soils then raises NH_4^+ concentrations in rainfall.

In a process called **nitrification**, some soil bacteria oxidize ammonium to NO_2^- (nitrite) and NO_3^- (nitrate). Plants primarily absorb nitrate. Animals obtain nitrogen in organic form from plants or other animals. Some bacteria convert nitrate to N_2 to obtain their oxygen. This **denitrification** returns N_2 to the atmosphere.

Decomposers, in a process called **ammonification**, break down nitrogenous compounds from organic wastes and dead organisms and return ammonia to the soil.

Most nitrogen cycling involves nitrates recycled from organic compounds in soil and water by ammonification and nitrification. Nitrogen fixation and denitrification of atmospheric N_2 are minor processes in the nitrogen cycle.

The Phosphorus Cycle Weathering of rock adds phosphorus to the soil in the form of PO_4^{3-}, which is absorbed by plants. Organic phosphate is transferred from plants to consumers and returned to the soil through the action of decomposers or by animal excretion. Humus and soil particles usually bind phosphate, keeping it available locally for recycling. The phosphorus that leaches into water eventually travels to the sea, where sediments become incorporated into rocks. These rocks may eventually return to terrestrial ecosystems as a result of geological processes occurring within geological time. The local cycle between soil, plants, and consumers operates in ecological time. Phosphorus in runoff from agricultural

lands and sewage may overstimulate productivity in aquatic habitats.

Variations in Nutrient Cycling Time Temperature and the availability of water and O_2 influence decomposition rates, thus nutrient cycling times vary in different ecosystems. Nutrients cycle rapidly in a tropical rain forest; the soil contains only about 10 percent of the ecosystem's nutrients. In temperate forests where decomposition is slower, 50 percent of organic nutrients are stored in the detritus and soil. Decomposition rates are slow in aquatic ecosystems, and sediments constitute a nutrient sink.

■ Field experiments reveal how vegetation regulates chemical cycling: *science as a process*

Research groups are conducting **long-term ecological research (LTER)** to follow ecosystem dynamics over relatively long time periods.

A team of scientists has looked at nutrient cycling in one forest ecosystem since 1963. The mineral budget for each of six valleys was determined by measuring the influx of key nutrients in rainfall and their outflow through the creek that drained each water-

■ **INTERACTIVE QUESTION 49.4**

Label the organisms and compounds indicated in the nitrogen cycle.

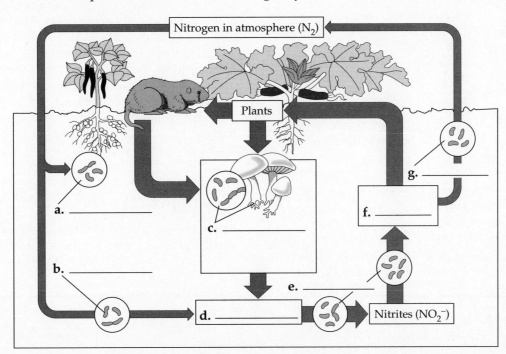

shed. About 60 percent of the precipitation exited through the stream; the rest was lost by transpiration and evaporation. Most minerals were recycled within the forest ecosystem.

The effect of deforestation on nutrient cycling was measured for 3 years in a valley that was completely logged. Compared with a control, water runoff from the deforested valley increased 30 to 40 percent; net loss of minerals, such as Ca^{2+}, K^+, and nitrate, was large. The sixtyfold increase in nitrate in the creek created unsafe drinking water.

■ INTERACTIVE QUESTION 49.5

a. In which natural ecosystem do nutrients cycle the fastest? Why?

b. In which natural ecosystem do nutrients cycle the slowest? Why?

c. What is the effect of loss of vegetation on nutrient cycling?

■ The human population is disrupting chemical cycles throughout the biosphere

Agricultural Effects on Nutrient Cycling The harvesting of crops removes nutrients that would otherwise recycle into the soil. After depleting the organic and inorganic reserves of nutrients, crops require the addition of expensive synthetic fertilizers. The addition of nitrogen fertilizers now equals the amount of nitrogen contributed by all nitrogen-fixing bacteria worldwide. With the balance between denitrification and nitrogen fixation upset, more N_2 is lost to the atmosphere, and nitrates are leached into groundwater or lost in water runoff. Excess nutrients in aquatic systems stimulate algal growth.

Accelerated Eutrophication of Lakes Sewage, factory wastes, and runoff of animal wastes and fertilizers from agricultural lands can lead to *cultural eutrophication* of lakes. The rapid increase in nutrients can cause an explosive increase in producers. Oxygen shortages, due to algal respiration at night and the metabolism of decomposers that work on the accumulating organic material, kill off many fish and other lake organisms.

Poisons in Food Chains Many toxic chemicals dumped into ecosystems are nonbiodegradable; some may become more harmful as they react with other environmental factors. Organisms absorb these toxins from food and water and may retain them within their tissues. In a process known as **biological magnification**, the concentration of such compounds increases in each successive link of the food chain. The magnification of DDT in birds of prey interferes with successful eggshell production. Chlorinated hydrocarbons, such as DDT, and polychlorinated biphenols, or PCBs, have been implicated in endocrine system problems in many animal species.

Intrusions in the Atmosphere: Carbon Dioxide Emissions and the Greenhouse Effect The concentration of CO_2 in the atmosphere has been increasing since the Industrial Revolution as a result of the combustion of fossil fuel and wood. If C_3 plants become able to outcompete C_4 plants with the increase in CO_2, species composition in natural and agricultural communities may be altered.

Through a phenomenon known as the **greenhouse effect**, CO_2 and water vapor in the atmosphere absorb infrared radiation reflected from Earth and rereflect it back to Earth, causing an increase in temperature. The continued increase in atmospheric CO_2 concentration may have far-reaching consequences.

Scientists have used mathematical models to estimate the extent and consequences of increasing CO_2 levels. **Paleoecologists** study the ecology of fossil communities. From dated core samples from lake sediments, they can study pollen and determine which species were present. Species composition allows them to infer the climate from that time period. The effects on vegetation of previous global warming trends can be used to predict the effects of increasing temperatures. A treaty that set up a plan for stabilizing CO_2 emissions was signed at the Earth Summit in 1992, but changing the level of combustion of fossil fuel and wood in increasingly industrialized societies is a huge challenge.

■ INTERACTIVE QUESTION 49.6

List some of the potential consequences of global warming.

Depletion of Atmospheric Ozone   A layer of ozone molecules (O_3) in the lower stratosphere absorbs damaging ultraviolet radiation. This layer has been gradually thinning since 1975, largely as a result of the accumulation of breakdown products of chlorofluorocarbons in the atmosphere. The dangers of ozone depletion may include increased incidence of skin cancer and cataracts and unpredictable effects on phytoplankton, crops, and natural ecosystems.

■ Human activities are altering species distribution and reducing biodiversity

The Introduction of Exotic Species   Exotic species have been introduced into communities both accidentally by human travels and intentionally for agricultural or ornamental purposes. Some exotic species displace native species or alter community structure. Examples of introduced species that have had a negative impact include the northern march of the destructive fire ants in the southern United States and the introduction of the Nile perch into Lake Victoria in East Africa.

The release of genetically engineered organisms into natural environments has potential dangers. For example, insect- or herbicide-resistant crops may become pests if they escape from agricultural fields or interbreed with noncrop species and invade new areas. Genetically engineered animals could also interbreed with or outcompete native species.

Habitat Destruction and the Biodiversity Crisis   Natural ecosystems are cleared for agricultural, industrial, and residential development, and clear-cutting of timber eliminates huge areas of forests. Nearly all large tracts of tropical forest will be gone in a decade or two if present practices continue. Wars also have severe ecological impacts. Destruction of habitat is involved with 73 percent of species listed as extinct, endangered, vulnerable, and rare by the International Union for Conservation of Nature and Natural Resources (IUCN).

The extent of the biodiversity crisis is difficult to measure because only a fraction of species have been identified and described. Preserving endangered species is complicated by lack of knowledge about minimum viable population sizes and habitat size. Migratory species that range over several countries require international cooperation in conservation measures.

Although extinction is a natural process, the current rate is about 50 times greater than at any time in the past 100,000 years. In the tropical rain forests, this rate may be 1000 to 10,000 times greater than the normal rate of one in every million species per year.

■ **INTERACTIVE QUESTION   49.7**

a. What is the major cause of the current high rate of extinction?

b. Why does the loss of biodiversity threaten human welfare?

■ The Sustainable Biosphere Initiative is reorienting ecological research

The Ecological Society of America has endorsed a new research agenda, the **Sustainable Biosphere Initiative**, to encourage studies of global change, biological diversity, and maintenance of the productivity of natural and artificial ecosystems. The goal of this huge research initiative is to develop the ecological knowledge needed to make intelligent decisions on the development, management, and conservation of Earth's resources.

STRUCTURE YOUR KNOWLEDGE

1. Two processes that emerge at the ecosystem level of organization are energy flow and chemical cycling. Develop a concept map that explains, compares, and contrasts these two processes.

2. Describe four or five human intrusions in ecosystem dynamics that have detrimental effects.

TEST YOUR KNOWLEDGE

MULTIPLE CHOICE: *Choose the one best answer.*

1. Which of the following organisms and trophic levels is mismatched?
   a. algae—producer
   b. phytoplankton—primary consumer
   c. earthworm—decomposer
   d. bobcat—secondary consumer
   e. eagle—tertiary consumer

2. Chemosynthetic bacteria found around deep-sea vents are examples of
   a. producers.
   b. decomposers.
   c. chemical cycling.
   d. secondary productivity.
   e. upwellings that make nutrients available.

3. In an ecosystem,
   a. Energy is recycled through the trophic structure.
   b. Energy is usually captured from sunlight by primary producers, passed to secondary producers in the form of organic compounds, and lost to decomposers in the form of heat.
   c. Chemicals are recycled between the biotic and abiotic sectors, whereas energy makes a one-way trip through the food web.
   d. There is a continuous process by which energy is lost as heat, and chemical elements leave the ecosystem through runoff.
   e. A food chain shows that all trophic levels may feed off each other.

4. Primary productivity
   a. is equal to the standing crop of an ecosystem.
   b. is greatest in freshwater ecosystems.
   c. is the rate of conversion of light to chemical energy in an ecosystem.
   d. is inverted in some aquatic ecosystems.
   e. is all of the above.

5. The open ocean and tropical rain forest are the two largest contributors to Earth's net primary productivity because
   a. Both have high rates of net primary productivity.
   b. Both cover large surface areas of the Earth.
   c. Nutrients cycle fastest in these two ecosystems.
   d. The ocean covers a large surface area and the tropical rain forest has a high rate of productivity.
   e. Both a and b are correct.

6. Productivity in terrestrial ecosystems is affected by
   a. temperature.
   b. light intensity.
   c. availability of nutrients.
   d. availability of water.
   e. all of the above.

7. Secondary productivity
   a. is measured by the standing crop.
   b. is the rate of biomass production in consumers.
   c. is greater than primary productivity.
   d. is 10 percent less than primary productivity.
   e. is the gross primary productivity minus the energy used for respiration.

8. Which of the following is *not* true of a pyramid of productivity?
   a. Only about 10 percent of the energy in one trophic level is passed into the next level.
   b. Because of the loss of energy at each trophic level, most food chains are limited to three to five steps.

   c. The pyramid of productivity of some aquatic ecosystems is inverted because of the large zooplankton primary-consumer level.
   d. Eating grain-fed beef is an inefficient means of obtaining the energy trapped by photosynthesis.
   e. A pyramid of productivity is the same shape as a pyramid of numbers.

9. Biogeochemical cycles are global for elements
   a. that are found in the atmosphere.
   b. that are found mainly in the soil.
   c. such as carbon, nitrogen, and phosphorus.
   d. that are dissolved in water.
   e. in the nonavailable reservoirs.

10. Which of these processes is *incorrectly* paired with its description?
   a. nitrification—oxidation of ammonium in the soil to nitrite and nitrate
   b. nitrogen fixation—reduction of atmospheric nitrogen into ammonia
   c. denitrification—removal of nitrogen from organic compounds
   d. ammonification—decomposition of organic compounds into ammonia
   e. evaporation of ammonia—loss of ammonia to the atmosphere from nonacidic soils

11. Carbon cycles relatively rapidly except when it is
   a. dissolved in freshwater ecosystems.
   b. released by respiration.
   c. converted into sugars.
   d. stored in petroleum, coal, or wood.
   e. part of the bicarbonate reservoir in oceans.

12. The ecological time scale of phosphorus cycling involves
   a. the uptake of phosphate from drinking water.
   b. the weathering of rock to add PO_4^{3-} to the soil.
   c. sedimentation to form rocks in the sea bed.
   d. the incorporation of phosphorus into organisms that become fossils.
   e. both a and b.

13. Which of the following was *not* shown by the Hubbard Brook Forest study?
   a. Most minerals recycle within a forest ecosystem.
   b. Deforestation results in a large increase in water runoff.
   c. Mineral losses from a valley were great following deforestation.
   d. Nitrate was the mineral that showed the greatest loss.
   e. Nutrient loss returned to normal in less than two years after deforestation.

14. The finding of harmful levels of DDD in grebes (fish-eating birds) in Clear Lake, California, following years of trying to eliminate bothersome gnat populations, is an example of
    a. eutrophication.
    b. biological magnification.
    c. the biomass pyramid.
    d. chemical cycling.
    e. increasing resistance to pesticides.

15. The greenhouse effect
    a. could change global weather and lead to the flooding of coastal areas.
    b. could result in more C_4 plants in plant communities that were previously dominated by C_3 plants.
    c. causes an increase in temperature when CO_2 absorbs more sunlight entering the atmosphere.
    d. could increase precipitation in central continental areas.
    e. could do all of the above.

CHAPTER 50

BEHAVIOR

FRAMEWORK

This chapter introduces the complex and fascinating subject of animal behavior. The study of behavior integrates biochemistry, genetics, physiology, evolutionary theory, and ecology.

Behaviors can range from simple fixed-action patterns in response to specific stimuli to insight learning in novel situations. Behaviors result from interactions among environmental stimuli, experience, and individual genetic makeup. The parameters of behavior are controlled by genetics and thus are acted upon by natural selection. Behavioral ecology focuses on the ultimate cause of reproductive fitness, which can be used to interpret foraging behavior, agonistic interactions, mating patterns, and altruistic behavior. Sociobiology extends evolutionary interpretations to human social interactions.

CHAPTER REVIEW

■ Behavior is what an animal does and how it does it

Animal **behaviors** are observable movements or actions as well as unobservable physiological or neural changes in response to a stimulus.

■ Behavioral ecology emphasizes evolutionary hypotheses: *science as a process*

Behavioral ecology looks to evolution for explanations of animal behavior, with the underlying expectation that animals behave in ways that maximize their Darwinian fitness. This expectation of optimal behavior relies on the fact that behavior is influenced by genetics and therefore subject to natural selection. Behavioral ecology allows researchers to frame questions about animal behavior in an evolutionary context, develop hypotheses, and make testable predictions.

■ A behavior has both an ultimate and a proximate cause

The **ultimate causation** of a particular animal behavior relates to the evolutionary basis of the behavior—why it has been favored by natural selection. **Proximate causation** explains the immediate cause of a behavior in terms of the cues or stimuli that trigger it and the internal mechanisms that produce it. Animal behaviorists must consider what they call *umvelt*, the ways in which a species perceives the physical world and the constraints of its physiology, when they frame behavioral questions. Proximate causes produce behaviors that evolve because they result in increased fitness, which is the ultimate cause of behaviors.

■ **INTERACTIVE QUESTION   50.1**

Many animals breed in the spring or early summer.

a. What is a probable proximate cause of this behavior?

b. What is the probable ultimate cause of this behavior?

■ Certain stimuli trigger innate behaviors called fixed-action patterns

Ethology, the study of how animals behave in their natural environment, originated in the 1930s. The work of Lorenz, Tinbergen, and von Frisch provided evidence for the innate components of behavior.

Fixed-Action Patterns A **fixed-action pattern (FAP)** is a highly stereotyped, innate behavior that, once begun, is usually carried through to completion. A FAP is triggered by a **sign stimulus** or **releaser**— some external sensory stimulus that, when perceived by the animal, initiates or "releases" a specific behavior.

The sign stimulus may be a feature of individuals of the same or another species. The releaser for attack behavior in stickleback fish is the red belly of the intruder. FAPs often are associated with interactions between parents and young, such as the begging and feeding behavior of many bird species. A limited subset of available sensory information usually elicits behavior in animals. Human behavior is usually based on more diverse, integrated information, which leads to intelligent action.

FAPs are adaptive responses that have evolved due to selection pressures. Finer discrimination among sign stimuli has evolved in situations where there has been strong selective pressure, as in the ability in certain species of birds to recognize foreign cuckoo eggs in their nests. Intelligence, the ability to learn from novel stimuli and adjust behavior based on this learning, may not have evolved in many animal groups due to the costs involved with the development and maintenance of neural tissue and the extension of juvenile phases with higher parental investment per offspring.

The mechanistic aspect of FAPs is illustrated by the repeated stereotyped behavior seen in some animals, such as a digger wasp performing its "drag and inspect" subroutine over and over when the paralyzed cricket it placed outside its nest is experimentally moved.

The Nature of Sign Stimuli The experiments of ethologists have shown that sign stimuli usually involve just one or two simple characteristics of the object or organism. The pecking behavior of a herring gull chick, for example, is released by the red spot on its parent's moving beak.

An animal's level of sensitivity to general stimuli correlates closely with the type of sign stimuli to which it responds. Frogs are quite good at detecting movement, and it is the movement of objects that releases the frog's feeding behavior. Experiments with *supernormal stimuli*, such as giant models of eggs, illustrate that stronger responses often are elicited by exaggerated sign stimuli.

Ethologists have suggested that the use of simple cues in the animal's environment to release pre-programmed behavior (FAPs) ensures behavioral responses that require the least amount of processing and integrating of input.

■ **INTERACTIVE QUESTION 50.2**

a. Give an example of a FAP in a human infant and the sign stimulus that illicits it.

b. The baby bird with the largest gape usually gets fed first. What is this an example of?

■ Learning is experience-based modification of behavior

Learning is the modification of behavior as a result of experience and can affect even innate behaviors.

Nature Versus Nurture Early European ethologists were studying innate aspects of animal behavior, while North American psychologists focused on how learning could influence behavior. The "nature versus nurture" debate concerns the relative importance of instinct and learning to behavior. Human language and bird songs are examples of learned behavioral variations of an innate ability.

Learning Versus Maturation Improved performance of innate behaviors may result from neuromuscular development called **maturation**. Birds prevented from flying until older are able, upon release, to fly without practice or learning.

Habituation A simple type of learning called **habituation** is the loss of sensitivity to unimportant stimuli or to stimuli not associated with appropriate feedback.

Imprinting Learning has a specific innate component in the phenomenon of **imprinting**. Mother–offspring bonding is critical for survival of offspring and for the reproductive fitness of the parent. The *imprinting stimulus* for newly hatched geese is the movement of an object, whether it be their mother, a ticking box, or Konrad Lorenz. The return of salmon to their home stream to spawn is an example of olfactory imprinting.

Imprinting is characterized by the irreversibility of learning and a limited **critical period** during which some type of imprinting may occur. The timing and duration of the critical period of sexual imprinting may vary.

The complex songs of male birds may advertise for mates, warn male competitors, announce group membership, or initiate group behaviors. Learning and genetic programming closely interact in vocal learning in young birds. In studies with experimentally raised white-crowned sparrows, a male bird raised in isolation developed a crude song called the template, a neural basis for the full adult song A bird learns to sing normally (i.e., to modify its template to the appropriate species' song) only if it can hear the song of its species during a critical period and then compare its own singing with the memory of that song. If a bird hears a taped song of only a related species during its critical period, it does not learn that song.

However, innate programming can be changed by experience. When sparrows older than 50 days can interact with singing adult birds of another species, the length of their critical period for learning songs is longer, and they are able to learn the song of the other species. People also have a critical period for learning vocalizations; learning foreign languages is easier up until the teen years.

Different species of birds show a great deal of diversity in their ability to learn songs. Some can develop nearly normal songs even when raised in isolation or deafened early in life.

Nottebohm's research has identified the forebrain region responsible for song learning in canaries. Changes in the size of this area correlate with the learning of new songs during the breeding season and the size of a song repertoire.

The diverse patterns of song acquisition among different groups of birds indicate that bird song is a fairly flexible behavioral system that has been shaped by natural selection and improves reproductive fitness.

Classical Conditioning   In **associative learning**, animals learn to associate one stimulus with another. **Classical conditioning**, described through the work of Pavlov in the 1900s with salivating dogs, illustrates the association of an irrelevant stimulus with a fixed physiological response.

Operant Conditioning   **Operant conditioning** refers to trial-and-error learning through which an animal associates a behavior with a reward or punishment. Skinner's work with rats is the best-known laboratory study of operant conditioning. The association of good or bad tastes with food items is probably a common form of operant conditioning in nature. A food's nutritive value may lead to the innate programming of good or bad taste through natural selection.

Observational Learning   Many vertebrates learn by observing the behavior of others. Song development in birds illustrates this type of learning.

Play   Species that engage in **play** often have a highly social lifestyle. Despite its use of energy and the risk of injury, play behavior is common because it may confer adaptive advantages. Play that simulates stalking and attacking may allow animals to practice these behaviors. Play also promotes muscular and cardiovascular fitness.

Insight   **Insight learning**, sometimes called reasoning or innovation, involves the ability to perform correctly in a novel situation. Insight learning is most often observed in mammals, especially in primates, although such learning has been documented in some bird species.

■ **INTERACTIVE QUESTION   50.3**

Indicate the type of learning illustrated by the following examples:

a. Ewes will adopt and nurse a lamb shortly after they give birth to their own lamb but will butt and reject a lamb introduced a day or two later.

b. A dog, whose early "accidents" were cleaned up with paper towels accompanied with harsh discipline, hides under the bed any time a paper towel is used in the household.

c. Ducklings eventually ignore a cardboard silhouette of a hawk that is repeatedly flown over them.

d. Kittens stalk and pounce on each other, biting and kicking as they roll around together.

e. Chickadees that join large winter flocks learn flock-specific songs.

f. Ravens develop novel strategies to retrieve a food reward attached to a string.

g. In Pavlov's experiments, the ringing of a bell caused a dog to salivate.

Animal Cognition   **Cognition** refers to an animal's ability to be aware of and make judgments about its environment. Because it is impossible to know what goes on inside the minds of animals, many researchers have taken a mechanistic approach to the study of animal behavior. Recently, Griffin and others have argued that conscious thinking is an inherent part of the behavior of many nonhuman, and even nonprimate, animals. According to **cognitive ethology**, cognitive ability forms an evolutionary continuum through many animal groups and has adaptive advantages that have been acted on by natural selection.

■ Rhythmic behaviors synchronize an animal's activities to daily and seasonal changes in the environment

Feeding, sleeping, reproducing, and migrating are all regularly repeated behaviors that show a temporal rhythm. The safe and profitable execution of such behaviors is the explanation for the ultimate causation of these rhythms. The proximate mechanisms are the subject of much study.

Animals, as well as plants, show circadian rhythms of roughly 24 hours. To determine whether these rhythms are based on *exogenous* (external) cues or *endogenous* (internal) timers, researchers have placed animals into environments with no external cues and observed that their rhythmic behaviors continue. An endogenous *biological clock* uses an exogenous cue, sometimes called a *Zeitgeber*, to keep the behavior timed with the real world. Humans living in free-running conditions with no external time cues have a biological clock with a period of about 25 hours.

The role of endogenous factors in **circannual behaviors**, such as reproduction, hibernation, and migration, is not well understood. These behaviors are based in part on physiological and hormonal changes linked with changes in day length. One of the few studies of endogenous control of these long-term behaviors has shown that fat deposition in ground squirrels, associated with hibernation, occurs regularly, even in a constant daylight cycle.

Some anatomical structure may serve as a pacemaker that controls behavioral rhythms. In mammals, the suprachiasmatic nucleus (SCN) in the hypothalamus receives nervous input from the retina and may produce proteins that regulate a variety of physiological functions involved in behavior.

■ Environmental cues guide animal movement

The ultimate cause for oriented movements is to bring animals to environments for which they are adapted. Much research has focused on the proximate causes or mechanisms that animals use to detect and respond to environmental cues.

Kinesis and Taxis   A **kinesis** is a change in activity rate in response to a stimulus. Although kinetic movements are randomly directed, they tend to maintain organisms in favorable environments. A **taxis** is an oriented movement toward or away from a stimulus.

Migration Behavior   **Migrations** are long-distance, regular movements, often involving a round trip each year. Three mechanisms may be used in migration: Traveling from one familiar landmark to another is called *piloting*. *Orientation* occurs when an animal detects compass directions and moves in a straight line. *Navigation* involves the ability to detect compass direction and to determine present location relative to destination (a "map sense").

Some species of birds and other animals orient by the heavens, using the sun or stars as directional cues. Many migrating animals use internal timing mechanisms to adjust to the changing positions of the sun and stars, both throughout the day and along the migratory route.

Many bird species are able to continue migrating under clouds or through fog, possibly because of their ability to detect and orient to Earth's magnetic field. Magnetite has been found in the heads of some birds, in the abdomens of bees, and in certain bacteria, although its role in sensing the earth's magnetic field has not been experimentally established.

■ INTERACTIVE QUESTION   50.4

a. What is the likely Zeitgeber for circadian rhythms?

b. What is the likely Zeitgeber for circannual behaviors?

c. How has the importance of endogenous mechanisms in rhythmic behavior been established?

■ INTERACTIVE QUESTION   50.5

Sow bugs are placed in experimental chambers that have light and dark areas and are either humid or dry. In the humid chamber, the sow bugs move into the dark area and stop moving. In the dry chamber, they move into the dark area and continue to move about in that area. Explain these experimental results.

■ Behavioral ecologists are using cost/benefit analysis to study foraging behavior

Some animals are generalists, feeding on a large variety of items, whereas others are specialists with quite restricted diets. Specialists usually have morphological and behavioral adaptations specific for their food, which makes them extremely efficient at foraging.

Generalists often concentrate on an abundant food item, developing a **search image** for it. Search images allow for short-term specialization while maintaining the advantages of being a generalist.

Behavioral ecologists study **foraging**, with the assumption that natural selection favors *optimal foraging*, choosing foods that maximize energy intake over expenditure. Factors involved in an optimal foraging strategy include the distance to a food item; the time and energy required to pursue, catch, and handle the item; the amount of usable energy and nutrients contained in it; and the risk of being preyed upon during feeding. Studies such as those on prey selection by bluegill sunfish have indicated that animals can modify their foraging behavior in ways that tend to maximize overall energy intake. This ability appears to be innate, although experience and physical maturation are thought to increase foraging efficiency.

■ INTERACTIVE QUESTION 50.6

What similarities and differences would you expect in the optimal foraging strategies of a generalist and a specialist?

■ Sociobiology places social behavior in an evolutionary context

Social behavior involves interaction between two or more animals, usually of the same species. Growing from the 1975 publication of E. O. Wilson's *Sociobiology*, the discipline of **sociobiology** uses evolutionary theory as the basis for studies of social behavior.

■ Competitive social behaviors often represent contests for resources

Agonistic Behavior **Agonistic behavior** involves a contest to determine which competitor gains access to a resource, such as food or a mate. The encounter may include a test of strength or, more commonly, threat displays and **ritual**, which serve to avoid actual physical conflicts. A display of submission or appeasement by one of the competitors inhibits further aggressive activity. The ending of a conflict without a violent combat avoids reducing the reproductive fitness of the winner as well as the loser. Scarcity of resources may result in more intense combat.

Dominance Hierarchies **Dominance hierarchies**, as illustrated by the "pecking order" of hens, prevent continual combat by establishing which animals get first access to resources. In wolf packs, the alpha female controls mating of the others, allowing them to mate if there is an abundance of food and preventing it otherwise.

Territoriality A **territory** is an area established for feeding, mating, and/or rearing young from which conspecifics are excluded. Territory size varies with species, function, and resource abundance. Agonistic behavior is used to establish and defend territories. A home range is the area in which an animal may roam, but a territory is the area that the animal defends. Vocal displays, scent marks, and patrolling may be used to continually proclaim ownership.

Dominance systems and territoriality help to stabilize population density by ensuring that at least some individuals have a sufficient supply of a limited resource in order to reproduce.

■ INTERACTIVE QUESTION 50.7

a. Why are many interactions between members of the same species agonistic?

b. What mechanisms reduce violent encounters between conspecifics?

■ Mating behavior relates directly to an animal's fitness

Courtship Most animals tend to view conspecifics as threatening competitors to be driven off. Complex courtship interactions, unique to each species, assure that individuals are not a threat and that they are of the proper species, sex, and physiological mating condition. These complex rituals often consist of a series of fixed-action patterns, each released in sequence by the reciprocal behavior of the other individual involved.

Courtship behavior may also function as a basis for mate selection. Females, who usually have more **parental investment** in each offspring due to greater time and resource allocation, usually show more discrimination in choosing mates than do males. Males often compete with each other for access to females. Intense courtship displays and secondary sexual characteristics may be used to attract females. If males provide parental care, a female's *assessment* of a specific mate may be based on the competence he displays. Vigorous courtship displays and extreme secondary sexual characteristics also may indicate good health. Especially if a male's only contribution to offspring is genetic, it is most beneficial to a female to choose a mate whose "performance" may indicate his genetic fitness.

Courtship and agonistic ritualized acts probably evolved from actions that originally had a more direct meaning in ancestral species.

Mating Systems   Many species have **promiscuous** mating, with no strong pair bonds forming. Longer lasting relationships may be **monogamous** or **polygamous**. Polygamous relationships are most often **polygynous** (one male and many females), although a few are **polyandrous**.

The needs of young offspring are often an ultimate factor for the reproductive pattern of the parents. If young require more food than one parent can supply, a male may increase his reproductive fitness by helping to care for offspring rather than going off in search of more mates. With mammals, the female often provides all the food, and males are often promiscuous or polygynous.

Certainty of paternity also influences mating systems and parental care. With internal fertilization, the acts of mating and egg laying or birth are separated, and paternity is less certain than when eggs are fertilized externally. Natural selection has resulted in male parental care being much more frequent in species with external fertilization, where the male's genetic contribution to the offspring is more certain.

Molecular biology can determine paternity using DNA fingerprinting. Several studies have shown that extrapair copulations are common in many bird species that were previously believed to form monogamous pairs.

■ INTERACTIVE QUESTION   50.8

a. Which sex usually shows more discrimination in choosing potential mates?

b. Why might the sexes show differences in discrimination in mate selection?

■ Social interactions depend on diverse modes of communication

Communication is the intentional transmission of information between individuals using special behaviors called *displays*. A change in behavior of the "receiver" is an indication that communication has occurred. Whereas ethologists assumed that communication systems evolved in ways to maximize information transfer, behavioral ecologists believe that communications develop due to the benefits for the "sender." Deceptive communications are adaptive for the sender but not the receiver.

Communication may involve visual, auditory, chemical, tactile, and electrical signals, depending on the lifestyle and sensory specializations of a species. **Pheromones** are chemical signals commonly used by mammals and insects in reproductive behavior to attract mates and to release specific courtship behaviors.

Social or hive bees have highly complex communication systems, using ritualized dances to communicate the location of food sources. Round dances are used when the food source is relatively close to the hive. Regurgitated nectar provides a scent to direct other bees to the food. The location of more distant food is communicated by waggle dances, in which the duration and vertical orientation of the dance on the comb indicate the distance from the hive and the direction to the food relative to the horizontal angle to the sun. Additional research indicates that sounds and odors given off by the dancing bee may communicate information about the food resource.

■ INTERACTIVE QUESTION   50.9

a. Give an example of a deceptive communication.

b. Why is most communication among mammals olfactory and auditory, whereas communication among birds is visual and auditory?

■ The concept of inclusive fitness can account for most altruistic behavior

Many social behaviors are selfish, benefiting an individual at the expense of others. **Altruism** is behavior that reduces an individual's fitness while increasing the fitness of the recipient. In the 1960s, Hamilton explained altruistic behavior in terms of **inclusive fitness**, the ability of an individual to pass on its genes either by producing its own offspring or by helping close relatives produce their offspring. The *coefficient of relatedness* is a measure of the proportion of genes shared by two related individuals. **Kin selection** suggests that altruistic behavior is strongest between close relatives and less common as genetic relatedness decreases. Studies show that most cases of altruistic behavior involve close relatives and thus improve the individual's inclusive fitness.

When altruistic behavior involves nonrelated animals, the explanation offered is **reciprocal altruism;** there is no immediate benefit for the altruistic individual, but some future benefit may occur when the helped animal may "return the favor." Reciprocal altruism often is used to explain altruism in humans.

■ INTERACTIVE QUESTION   50.10

a. According to kin selection, would an individual be more likely to exhibit altruistic behavior toward a grandparent, a sibling, or a first (full) cousin?

b. Explain your answer in terms of the coefficient of relatedness.

■ Human sociobiology connects biology to the humanities and social sciences

In *Sociobiology*, Wilson speculated on the evolutionary basis of certain social behaviors of humans. The sociobiology debate continues the nature-versus-nurture controversy. Some sociobiologists explain that cultural and genetic components of social behavior are linked together in a cycle of reinforcement. The development of cultural regulations of innate behaviors serves as an additional factor for the natural selection of that behavior. According to this view, human behavior represents an integration of genes and culture.

The parameters of human social behavior may be set by genetics, but the environment undoubtedly shapes behavioral traits just as it influences the expression of physical traits. Due to our capacity for learning and integration, human behavior appears to be quite plastic. Our structured societies, with their regulations on behaviors, including behaviors that would enhance an individual's fitness, may be the one unique characteristic separating humans and other animals.

STRUCTURE YOUR KNOWLEDGE

1. How does the nature-versus-nurture controversy apply to behavior?

2. How does the concept of Darwinian fitness apply to behavior?

TEST YOUR KNOWLEDGE

MULTIPLE CHOICE: *Choose the one best answer.*

1. Behavioral ecology is the
   a. mechanistic study of the behavior of animals, focusing on stimulus and response.
   b. application of human emotions and thoughts to other animals.
   c. study of animal cognition.
   d. study of animal behavior from an evolutionary perspective of Darwinian fitness.
   e. consideration of an animal's *umvelt* when considering the ultimate cause of behavior.

2. Proximate causes
   a. explain the evolutionary significance of a behavior.
   b. are immediate causes of behavior such as environmental stimuli.
   c. indicate that much of animal behavior is innate.
   d. are endogenous, although they may be set by exogenous cues.
   e. show that nature is more important than nurture.

3. Supernormal stimuli
   a. may elicit stronger responses or FAPs.
   b. are innate releasing mechanisms between individuals of the same species.
   c. are illustrated by the removal of the cricket from near a digger wasp's nest.
   d. are illustrated by the bobbing of a stick with a red spot past herring gull chicks.
   e. are illustrated by the imprinting of goslings on Konrad Lorenz.

4. Which of the following is an example of the evolution of a fixed-action pattern with finer discrimination among sign stimuli?
   a. a bluegill sunfish feeding on larger *Daphnia* when prey are abundant
   b. a chick pecking at the red spot on a parent's moving beak
   c. a bird recognizing and expelling an egg of a cuckoo from its nest
   d. a songbird learning its song after listening to a taped song of its species
   e. a bird learning to avoid monarch butterflies

5. A change in behavior as a result of experience is called
   a. habituation.
   b. imprinting.
   c. insight.
   d. learning.
   e. maturation.

6. A critical period
   a. is the time right after birth when sexual identity is developed.
   b. usually follows the receiving of a sign stimulus.
   c. is a limited time during which imprinting can occur.
   d. is the period during which birds can learn to fly.
   e. is the time during which social animals play.

7. In classical conditioning,
   a. An animal associates a behavior with a reward or punishment.
   b. An animal learns as a result of trial and error.
   c. Sensitivity to unimportant or repetitive stimuli occurs.
   d. A bird can learn the song of a related species if it hears only that song.
   e. An irrelevant stimulus can elicit a response because of its association with a normal stimulus.

8. Circannual behaviors
   a. are often linked to changes in day length.
   b. rely solely on endogenous cues.
   c. involve foraging, reproduction, and migration.
   d. do not occur in free-running conditions.
   e. have a rhythm of 24 hours that is based on exogenous cues and endogenous timers.

9. A kinesis
   a. is a randomly directed movement that is not caused by external stimuli.
   b. is a movement that is directed toward or away from a stimulus.
   c. is a change in activity rate in response to a stimulus.
   d. is illustrated by trout swimming upstream.
   e. often involves piloting but not orientation or navigation.

10. A dominance hierarchy
    a. may be established by agonistic behavior.
    b. determines which animals get first access to resources.
    c. helps to avoid potential injury of competitors.
    d. may help to stabilize population density.
    e. applies to all of the above.

11. An animal's territory
    a. may be larger than its home range.
    b. may decrease in size if resources dwindle and expand if resources become more plentiful.
    c. excludes both conspecifics and members of other species.
    d. may be proclaimed by scent marks, vocal displays, and patrolling.
    e. applies to all of the above.

12. In a species in which females provide all the needed food and protection for the young,
    a. Males are likely to be promiscuous.
    b. Mating systems are likely to be monogamous.
    c. Mating systems are likely to be polyandrous.
    d. Males most likely will show sexual selection.
    e. Females will have a higher Darwinian fitness than males.

13. The ability of honeybees to fly directly to a food source, after having to wait several hours from the time of the waggle dance, indicates that
    a. The waggle dance provided directions relative only to the position of the hive.
    b. Bees have an internal clock that compensates for the movement of the sun during the elapsed time.
    c. The bees must have been to that food source before.
    d. Bees are directed more by olfactory cues than by directional cues.
    e. The intensity of the dance provided directional cues.

14. The concept of inclusive fitness explains
    a. optimal feeding behavior.
    b. sexual selection.
    c. kin selection and altruistic behavior.
    d. monogamous mating systems.
    e. the coefficient of relatedness.

15. Sociobiology
    a. explains the evolutionary basis of behavioral characteristics within animal societies.
    b. applies evolutionary explanations to human social behaviors.
    c. studies the roles of culture and genetics in human social behavior.
    d. considers communication and mating systems from the viewpoint of Darwinian fitness.
    e. does all of the above.

CHAPTER 2
THE CHEMICAL CONTEXT OF LIFE

■ INTERACTIVE QUESTIONS

2.1 a. Mass is a measure of the amount of matter within an object.
   b. Weight is how strongly that mass is pulled upon by gravity.

2.2 oxygen, carbon, hydrogen, nitrogen, calcium, phosphorus, potassium, sulfur, sodium, chlorine, magnesium

2.3 neutrons; 15, 15, 16, 31

2.4 absorb; released

2.5 a. Nitrogen $_7N$ **b.** Phosphorus $_{15}P$

   c. Oxygen $_8O$ **d.** Chlorine $_{17}Cl$

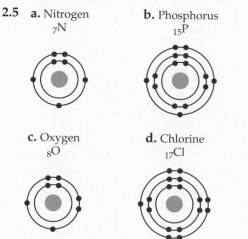

N and P; both elements have five electrons in their valence shells.

2.6 a. protons
   b. atomic number
   c. element
   d. neutrons
   e. mass number or atomic weight
   f. isotopes
   g. electrons
   h. electron shells or energy levels
   i. valence shell

2.7 a. 1 **b.** 2 **c.** 3 **d.** 4

2.8 a. $CaCl_2$ **b.** Ca^{2+} is the cation.

2.9

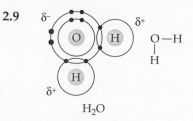

H_2O

2.10 6, 6, 6

Suggested Answers to Structure Your Knowledge

1.

Particle	Charge	Mass	Location
Proton	+1	1 dalton	nucleus
Neutron	0	1 dalton	nucleus
Electron	-1	negligible	orbitals in energy shells

2. a. The atoms of each element have a characteristic number of protons in their nuclei, referred to as the **atomic number.** In a neutral atom, the atomic number also indicates the number of electrons.

The **mass number** is an indication of the approximate mass of an atom and is equal to the number of protons and neutrons in the nucleus.

The **atomic weight** refers to the atomic mass of an atom. It is equal to the mass number and is measured in the atomic mass unit of daltons. Protons and neutrons both have a mass of approximately 1 dalton.

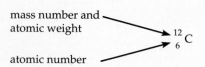

b. The **valence,** an indication of bonding capacity, is the number of bonds that must be formed for an atom to complete its valence shell with eight electrons.

c. The valence of an atom is most related to the chemical behavior of an atom because it is an indication of the number of bonds the atom will make, or the number of electrons the atom must share in order to reach a filled valence shell.

3. Ionic and nonpolar covalent bonds represent the two ends of a continuum of electron sharing between atoms in a molecule. In ionic bonds the electrons are completely pulled away from one atom by the other, creating negatively and positively charged ions (anions and cations). In nonpolar covalent bonds the electrons are equally shared between two atoms. Polar covalent bonds form when a more electronegative atom pulls the shared electrons closer to it, producing a partial negative charge associated with that portion of the molecule and a partial positive charge associated with the atom from which the electrons are pulled.

Answers to Test Your Knowledge

Multiple Choice:

1. b	**5. a**	**9. d**	**13. c**
2. d	**6. e**	**10. c**	**14. b**
3. a	**7. d**	**11. a**	**15. e**
4. e	**8. b**	**12. c**	**16. c**

CHAPTER 3
WATER AND THE FITNESS OF THE ENVIRONMENT

■ INTERACTIVE QUESTIONS

3.1

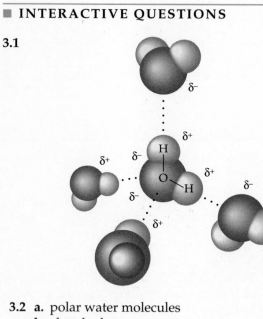

3.2 a. polar water molecules
b. absorbed
c. heat of vaporization
d. evaporative cooling
e. solar heat
f. released
g. specific heat
h. rain
i. ice forms

3.3 a. The molecular weight of $C_3H_6O_3$ is 90 d, the combined atomic weights of its atoms. A mole of lactic acid = 90 g. A 0.5 M solution would require $1/2$ mol, or 45 g.

b. 228 grams of NaCl

3.4

carbonic acid   bicarbonate   hydrogen ion

$$H_2CO_3 \rightleftharpoons HCO_3^- + H^+$$

H^+ donor H^+ acceptor

a. Bicarbonate acts as a base to accept excess H^+ ions when the pH starts to fall; the reaction moves to the left.
b. When the pH rises, H^+ ions are donated by carbonic acid, and the reaction shifts to the right.

Suggested Answers to Structure Your Knowledge

1. **a.** cohesion, adhesion
   b. A water column is pulled up through plant vessels.
   c. Heat is absorbed or released when hydrogen bonds break or form.
   d. high heat of vaporization
   e. Solar heat is absorbed by tropical seas.
   f. evaporative cooling
   g. Evaporation of water cools surfaces of plants and animals.
   h. Hydrogen bonds in ice space water molecules apart, making ice less dense.

i. Floating ice insulates bodies of water so they don't freeze solid.

j. versatile solvent

k. Polar water molecules cause ions and polar solutes to dissolve.

2.

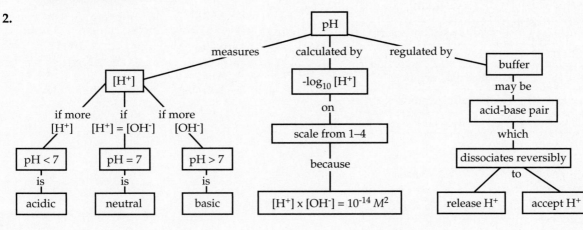

Answers to Test Your Knowledge

Multiple Choice:

1. c	**4.** e	**7.** e	**10.** d	**13.** c
2. a	**5.** b	**8.** b	**11.** b	**14.** c
3. e	**6.** d	**9.** b	**12.** d	**15.** e

Fill in the Blanks:

$[H^+]$	$[OH^-]$	pH	Acidic, Basic, or Neutral?
10^{-3}	10^{-11}	3	acidic
10^{-8}	10^{-6}	8	basic
10^{-12}	10^{-2}	12	basic
10^{-7}	10^{-7}	7	neutral
10^{-1}	10^{-13}	1	acidic

CHAPTER 4
CARBON AND THE MOLECULAR DIVERSITY OF LIFE

■ INTERACTIVE QUESTIONS

4.1

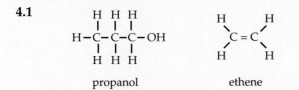

propanol   ethene

4.2 Ethanol and dimethyl ether, structural isomers, have the same number and kinds of atoms but a different bonding sequence and very different properties. Maleic acid and fumaric acid are geometric isomers whose double bonds fix the spatial arrangement of the molecule. Although the enantiomers *l*- and *d*-lactic acid look similar in a flat representation of their structures, they are not superimposable.

Suggested Answers to Structure Your Knowledge

1.

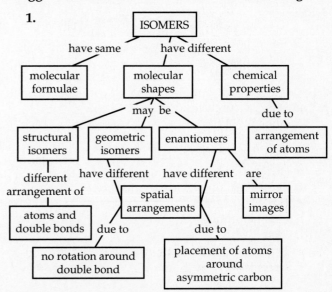

2.

Functional Group	Molecular Formula	Names and Characteristics of Organic Compounds Containing Functional Group
Hydroxyl	–OH	Alcohols; polar group
Carbonyl	$\geq C=O$	Aldehyde or ketone; polar
Carboxyl	–COOH	Carboxylic acid; release H^+
Amino	$-NH_2$	Amines; basic, accept H^+
Sulfhydryl	–SH	Thiols; stabilize proteins
Phosphate	$-OPO_3^{-2}$	Used in energy transfers

Answers to Test Your Knowledge

Multiple Choice:

1. c	4. a	7. d
2. b	5. d	8. c
3. e	6. a	

Matching:

1. a, c	4. e	7. a, d, g	10. b, f
2. b, f	5. e	8. e	11. e
3. d, g	6. a, c, d, g	9. a, c	12. c

CHAPTER 5
THE STRUCTURE AND FUNCTION OF MACROMOLECULES

■ INTERACTIVE QUESTIONS

5.1 a. monosaccharides
   b. $(CH_2O)_n$
   c. energy compounds
   d. carbon skeletons, monomers
   e. glycosidic bonds
   f. disaccharides
   g. polysaccharides
   h. glycogen
   i. animals
   j. starch
   k. cellulose
   l. chitin

5.2

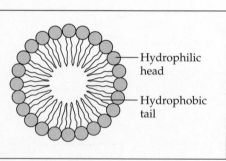

Hydrophilic head

Hydrophobic tail

(a) Micelle

5.3 a. fats, triacylglycerides
   b. phospholipids
   c. glycerol
   d. fatty acids
   e. unsaturated: C=C bonds
   f. saturated: all C-H bonds, no C=C
   g. phosphate group
   h. cell membrane
   i. steroids
   j. hormones, cell membrane component (cholesterol)

5.4 a. hydrogen bond
   b. hydrophobic interaction
   c. disulfide bridge
   d. ionic bond
These interactions produce tertiary structure.

5.5 a. A change in pH alters the availability of H^+, OH^-, or other ions, thereby disrupting the hydrogen bonding and ionic bonds that maintain protein shape.
   b. An organic solvent would disrupt hydrophobic interactions, and the protein would turn inside out as the hydrophilic regions became clustered on the inside of the molecule.
   c. The return to its functional shape indicates that a protein's conformation is intrinsically determined by its primary structure—the sequence of its amino acids.

5.6

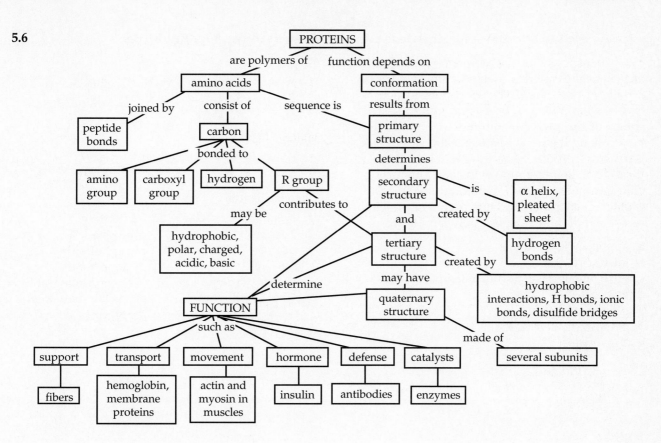

5.7   DNA $\rightarrow$ RNA $\rightarrow$ Protein

5.8

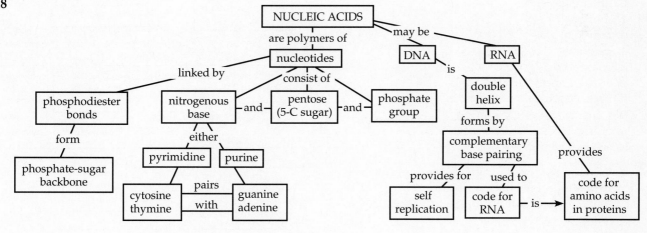

Suggested Answers to Structure Your Knowledge

1. The primary structure of a protein is the specific, genetically coded sequence of amino acids in a polypeptide chain. The secondary structure involves the coiling (α helix) or folding (pleated sheet) of the protein, stabilized by hydrogen bonds along the polypeptide backbone. The tertiary structure involves interactions between the side chains of amino acids and produces a characteristic three-dimensional shape for a protein. Quaternary structure occurs in proteins composed of more than one polypeptide chain.

2. **a.** amino acid (glycine)
   b. fatty acid
   c. nitrogenous base, purine (adenine)
   d. glycerol
   e. phosphate group
   f. pentose (deoxyribose)
   g. sugar (triose)

1. b, d	3. c, e, f	5. c	7. e, f
2. a	4. f, g	6. a	

Answers to Test Your Knowledge

Matching:

1. A	4. C	7. B	10. A
2. B	5. C	8. C	
3. D	6. D	9. A	

Multiple Choice:

1. e	3. a	5. c	7. c	9. d
2. c	4. e	6. b	8. a	10. c

Fill in the Blanks:

1. F. Sanger
2. guanine
3. purines
4. nucleotide
5. primary structure or amino acid sequence
6. quaternary
7. glycogen
8. phospholipids
9. peptide bonds
10. X-ray crystallography, computer modeling, graphics

CHAPTER 6
AN INTRODUCTION TO METABOLISM

■ INTERACTIVE QUESTIONS

6.1 a. capacity to do work
  b. kinetic
  c. motion
  d. potential
  e. position
  f. conserved
  g. created nor destroyed
  h. first
  i. transformed or transferred
  j. entropy
  k. second

6.2 a. When a system is high in free energy, it is unstable, can change spontaneously, will move toward equilibrium, and has the potential to do work.
  b. When a system has low free energy, it is more stable, won't change spontaneously, is at equilibrium, and has less capacity to perform work.

6.3

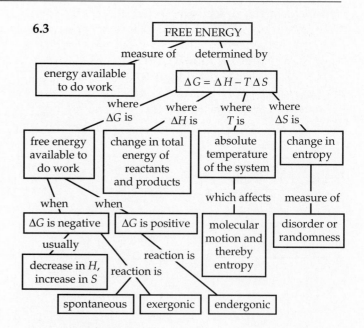

6.4 a. adenine
  b. ribose
  c. phosphates
  d. Circle the phosphate portion of the molecule and indicate that the terminal phosphate bond is most likely to break.

e. The negatively charged phosphate groups are crowded together, and their mutual repulsion makes the bonds unstable. A hydrolysis reaction breaks the terminal phosphate bond and releases a molecule of inorganic phosphate.

f. The products of hydrolysis (ADP and P_i) are more stable than ATP. The chemical change to a more stable state releases energy.

6.5 a. free energy
b. transition state
c. E_A (free energy of activation) with enzyme
d. E_A without enzyme
e. ΔG of reaction

6.6

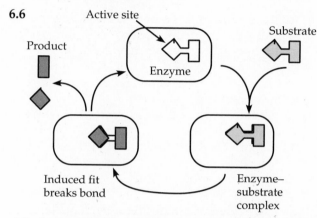

6.7 A competitive inhibitor would mimic the shape of the substrate and compete with the substrate for the active site. A noncompetitive inhibitor would be a shape that could bind to another portion of the enzyme molecule and would thus change the conformation of the active site such that the substrate couldn't bind.

Suggested Answers to Structure Your Knowledge

1. Metabolism is the totality of chemical reactions that take place in living organisms. To create and maintain the structural order required for life requires an input of free energy—from sunlight for photosynthetic organisms and from energy-rich food molecules for other organisms. A cell couples catabolic, exergonic reactions ($-\Delta G$) with anabolic, endergonic reactions ($+\Delta G$), using ATP as the primary energy shuttle between the two.

2. Enzymes are essential for metabolism because they lower the free energy of activation of the specific reactions they catalyze and allow those reactions to occur extremely rapidly at a temperature conducive to life. By regulating the enzymes it produces, a cell can regulate which of the myriad of possible chemical reactions take place at any given time. Metabolic control also occurs through feedback inhibition. The compartmentalization of the cell can order metabolic pathways in time and space.

Answers to Test Your Knowledge

Multiple Choice:

1. c	4. e	7. a	10. c
2. b	5. c	8. e	11. b
3. a	6. c	9. e	12. c

Fill in the Blanks:

1. metabolism	6. free energy of activation
2. anabolic	7. competitive
3. kinetic	8. coenzymes
4. allosteric	9. feedback inhibition
5. entropy	10. phosphorylated intermediates

CHAPTER 7
A TOUR OF THE CELL

■ INTERACTIVE QUESTIONS

7.1 a. the study of cell structure
b. the internal ultrastructure of cells
c. the three-dimensional surface topography of a specimen
d. LM enables study of living cells and does not introduce the artifacts that may be introduced by TEM and SEM.

7.2 a. 10^2, or 100 times the surface area. **b.** 10^3, or 1000 times the volume.

7.3 The genetic instructions for specific proteins are transcribed from DNA into messenger RNA (mRNA), which then passes into the cytoplasm to complex with ribosomes where it is translated into the primary structure of proteins.

7.4 a. Golgi apparatus—processes products of ER; makes polysaccharides, packages products in vesicles targeted to specific locations
b. transport vesicle—carries products of ER to various locations
c. nuclear envelope—double membrane that encloses nucleus; pores regulate passage of materials
d. rough ER—attached ribosomes produce proteins that enter cisternae, and makes secretory proteins and membranes

e. smooth ER—houses enzymes that synthesize lipids, metabolize carbohydrates, detoxify drugs and alcohol; stores and releases calcium ions in muscle cells

f. transport vesicle—fuses with plasma membrane, secreting contents and adding to membrane

g. plasma membrane—selective barrier that regulates passage of materials into and out of cell

h. lysosome—houses hydrolytic enzymes to digest macromolecules

7.5 Peroxisomes do not bud from the endomembrane system, but grow by incorporating proteins and lipids from the cytosol; they increase in number by dividing.

7.6

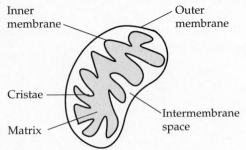

Mitochondrion

Inner membrane
Outer membrane
Cristae
Matrix
Intermembrane space

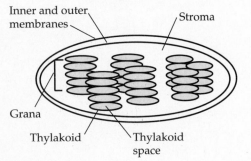

Chloroplast

Inner and outer membranes
Stroma
Grana
Thylakoid
Thylakoid space

7.7 a. hollow tube, helix of α- and β-tubulin dimers; 25-nm diameter

b. cell shape and support, tracks for moving organelles, chromosome movement, beating of cilia and flagella

c. two twisted chains of actin molecules: 7-nm diameter

d. muscle contraction, maintain and change cell shape, pseudopod movement, cytoplasmic streaming

e. supercoiled fibrous proteins of keratin family; 8–12 nm

f. reinforce cell shape, anchor nucleus

7.8

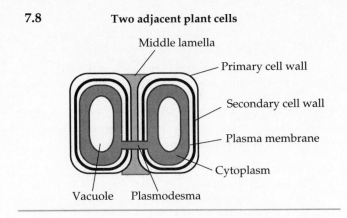

Two adjacent plant cells

Middle lamella
Primary cell wall
Secondary cell wall
Plasma membrane
Cytoplasm
Vacuole   Plasmodesma

Suggested Answers to Structure Your Knowledge

1. a. nucleus, chromosomes, centrioles, microtubules (spindle), microfilaments (actin-myosin aggregates pinch apart cell)

b. nucleus, chromosomes, DNA $\longrightarrow$ mRNA $\longrightarrow$ ribosomes $\longrightarrow$ enzymes and other proteins

c. mitochondria

d. ribosomes, rough and smooth ER, Golgi apparatus, vesicles

e. smooth ER (peroxisomes also detoxify substances)

f. lysosomes, food vacuoles

g. peroxisomes

h. cytoskeleton: microtubules, microfilaments, intermediate filaments, extracellular matrix

i. cilia and flagella (microtubules), microfilaments (actin) in muscles and pseudopodia

j. plasma membrane, vesicles

k. desmosomes, tight and gap junctions, ECM

2. a. structural support, middle lamella glues cells together

b. storage, digestion, growth of cell by water absorption

c. photosynthesis, production of carbohydrates

d. starch storage

e. cytoplasmic connections between cells

3. a. peroxisome
b. microfilament
c. mitochondrion
d. flagellum
e. nuclear envelope
f. chromatin
g. nucleus
h. nucleolus
i. ribosomes
j. rough endoplasmic reticulum
k. plasma membrane
l. smooth endoplasmic reticulum
m. Golgi apparatus
n lysosome
o. microtubule
p. centriole

4.

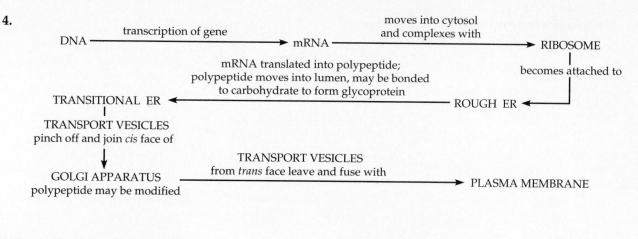

Answers to Test Your Knowledge

Multiple Choice:

1. c	**4.** e	**7.** a	**10.** d	**13.** b
2. b	**5.** a	**8.** c	**11.** e	**14.** c
3. d	**6.** b	**9.** b	**12.** b	**15.** e

Fill in the Blanks:

1. transport vesicles
2. cristae
3. extracellular matrix
4. peroxisomes
5. grana
6. basal body
7. cytosol
8. cytoskeleton
9. tight junction
10. tonoplast

CHAPTER 8
MEMBRANE STRUCTURE AND FUNCTION

■ INTERACTIVE QUESTIONS

8.1 a. phospholipid bilayer
    b. hydrocarbon tail—hydrophobic
    c. phosphate head—hydrophilic
    d. hydrophobic region of protein
    e. hydrophilic region of protein

8.2 a. In hybrid human/mouse cells, membrane proteins intermingle.
    b. The proportion of unsaturated phospholipids may increase.

8.3 Ions and large polar molecules, such as glucose, are impeded by the hydrophobic center of the plasma membrane's lipid bilayer.

8.4 Side A is hypertonic; side B is hypotonic; water will move from B to A.

8.5 membranes that are less permeable to water and contractile vacuoles that expel excess water

8.6 a. isotonic
    b. hypotonic

8.7 Although it may speed diffusion, facilitated diffusion is still passive transport because the solute is moving down its concentration gradient.

8.8 Three Na^+ ions are pumped out of the cell for every two K^+ ions pumped in, resulting in a net movement of positive charge from the cytoplasm to the extracellular fluid.

8.9 a. Human cells use receptor-mediated endocytosis to take in cholesterol.
    b. In familial hypercholesteremia, LDL receptor proteins in the plasma membrane are defective and low-density lipoproteins cannot bind and be transported into the cell.

8.10 a. first messenger
    b. receptor protein
    c. relay protein
    d. effector protein (enzyme)
    e. second messenger
    f. cellular responses

Suggested Answers to Structure Your Knowledge

1.

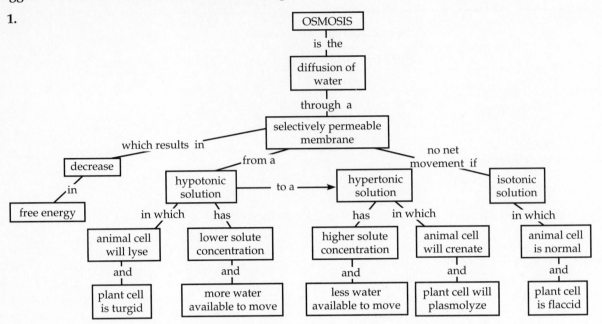

2. **a.** II represents facilitated diffusion because the solute is moving through a transport protein and down its concentration gradient as it crosses the membrane. The cell does not expend energy in this transport. Polar molecules and ions may be moved by facilitated diffusion.

   b. III represents active transport because the solute is clearly moving against its concentration gradient. Energy from the cell must be expended to drive this transport.

   c. I illustrates diffusion through the lipid bilayer. The solute molecules must be nonpolar or small polar molecules.

   d. Both diffusion and facilitated diffusion are considered passive transport because the solute moves down its concentration gradient and the cell does not expend energy in the transport.

Answers to Test Your Knowledge

Multiple Choice:

1. c	5. d	9. b	13. b	17. a
2. b	6. a	10. a	14. c	18. e
3. a	7. e	11. e	15. b	
4. e	8. c	12. a	16. c	

Explanation for answer to question 17: This problem involves both osmosis and diffusion. Although the solution is initially isotonic, glucose will diffuse down its concentration gradient until it reaches dynamic equilibrium with a 1.5 M concentration on both sides. The increasing solute concentration on side A will cause water to move into this hypertonic side, and the water level will rise.

CHAPTER 9
CELLULAR RESPIRATION: HARVESTING CHEMICAL ENERGY

■ INTERACTIVE QUESTIONS

9.1 $C_6H_{12}O_6$, 6 CO_2, energy (ATP + heat)

9.2 a. oxidized
   b. oxidizing agent
   c. reduced

9.3 a. oxygen
   b. glucose
   c. some stored in ATP and some released as heat

9.4 a. an electron acceptor or oxidizing agent
   b. NADH

9.5 a. glycolysis (glucose $\longrightarrow$ pyruvate)
   b. Krebs cycle
   c. electron transport chain and oxidative phosphorylation
   d. substrate-level phosphorylation
   e. substrate-level phosphorylation

f. oxidative phosphorylation

Top two arrows show flow of electrons via NADH to the electron transport chain.

9.6 a. pyruvate (from glycolysis)
  b. CO_2 removed
  c. CoA
  d. NADH
  e. acetyl CoA
  f. citrate
  g. α-ketoglutarate
  h. ATP
  i. succinate
  j. $FADH_2$
  k. malate
  l. oxaloacetate

9.7 a. intermembrane space
  b. inner mitochondrial membrane
  c. mitochondrial matrix
  d. electron transport chain
  e. $NADH + H^+$
  f. NAD^+
  g. nH^+
  h. $2 H^+ + 1/2 O_2$

i. H_2O
j. ATP synthase
k. $ADP + ℗_i$
l. ATP

9.8 a. –2
  b. 4
  c. Krebs cycle
  d. 32
  e. 36
  f. 8
  g. 2
  h. 2
  i. 2

9.9 By oxidizing pyruvate to CO_2 and passing electrons from NADH (and $FADH_2$) through the electron transport chain, respiration can produce approximately 36 ATP compared to the 2 net ATP that are produced by fermentation. In order to use the electron transport chain, O_2 must be present to act as the final electron acceptor.

Suggested Answers to Structure Your Knowledge

1. No Answer for 1.

2.

Process	Main Function	Inputs	Outputs
Glycolysis	Oxidation of glucose to 2 pyruvate, 2 ATP net	glucose 2 ATP $2 NAD^+$ $4 ADP + ℗_i$	2 pyruvate 4 ATP (2 net) 2 NADH
Pyruvate to acetyl CoA	Oxidation of pyruvate to acetyl CoA, which then enters Krebs cycle	2 pyruvate 2 CoA $2 NAD^+$	2 acetyl CoA $2 CO_2$ 2 NADH
Krebs cycle	Redox reactions produce NADH and $FADH_2$, ATP by substrate-level phosphorylation; CO_2 released.	2 acetyl CoA 2 oxaloacetate $2 ADP + ℗_i$ $6 NAD^+$ 2 FAD	2 CoA $4 CO_2$ 2 ATP 6 NADH $2 FADH_2$
Electron transport chain and oxidative phosphorylation	NADH (from glycolysis and Krebs cycle) and $FADH_2$ (from Krebs) transfer electrons to carrier molecules in mitochondrial membrane. In a series of redox reactions, H^+ is pumped into intermembrane space, and electrons are delivered to $1/2 O_2$. By chemiosmosis, proton motive force drives H^+ back through ATP synthase to make ATP.	NADH $FADH_2$ H^+ $\frac{1}{2}O_2$ $ADP + ℗_i$	NAD^+ FAD H_2O ATP
Fermentation	Regenerates NAD^+ so glycolysis can continue. Pyruvate is either reduced to ethyl alcohol and CO_2 or to lactate.	pyruvate NADH or lactate	NAD^+ ethanol and CO_2 or lactate

3.

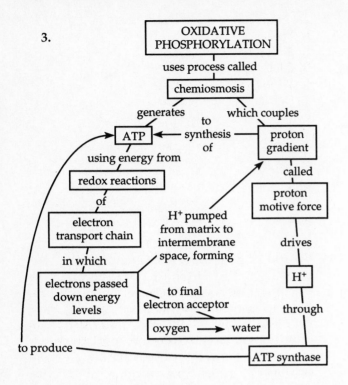

Answers to Test Your Knowledge

1. b	5. e	9. b	13. b	17. b
2. a	6. d	10. b	14. b	18. d
3. c	7. e	11. a	15. c	19. c
4. a	8. a	12. e	16. c	20. e

CHAPTER 10
PHOTOSYNTHESIS

■ INTERACTIVE QUESTIONS

10.1 **a.** outer membrane
   b. granum
   c. inner membrane
   d. thylakoid compartment
   e. thylakoid
   f. stroma

10.2 **a.** light
   b. H_2O
   c. light reactions taking place in grana
   d. O_2
   e. ATP
   f. NADPH
   g. CO_2
   h. Calvin cycle taking place in stroma
   i. CH_2O (sugar)

10.3 The solid line is the absorption spectrum; the dotted line is the action spectrum. Some wavelengths of light, particularly in the yellow and orange range, result in a higher rate of photosynthesis than would be indicated by the absorption of those wavelengths by chlorophyll *a*. These differences are partially accounted for by accessory pigments, such as chlorophyll *b* and the carotenoids, that absorb light energy from different wavelengths and pass that energy on to chlorophyll *a*.

10.4 All these chlorophyll *a* molecules are identical. The differences between reaction-center molecules and other chlorophyll *a* molecules is their location next to a primary electron acceptor. P700 and P680 have different electron distributions, resulting from the specific proteins to which they are bound in the reaction center, and thus slightly different absorbances.

10.5 **a.** photosystem II
   b. P680
   c. electrons
   d. oxygen ($1/2\ O_2$)
   e. water (H_2O)
   f. primary electron acceptor
   g. electron transport chain
   h. photophosphorylation by chemiosmosis
   i. ATP
   j. photosystem I
   k. P700
   l. primary electron acceptor
   m. $NADP^+$ reductase
   n. NADPH

ATP and NADPH provide the chemical energy and reducing power for the Calvin cycle.

10.6 a. Electrons from P700 in photosystem I are transferred from the primary electron acceptor to ferredoxin. NADP⁺ reductase transfers the electrons to NADP⁺ to form NADPH.

b. Ferredoxin passes the electrons to the electron transport chain, from which they return to P700. The energy released through the electron transport chain drives cyclic photophosphorylation.

c. Electrons from P680 are not passed to P700. Without the oxidizing agent P680, water is not split.

10.7 a. in the thylakoid compartment (pH 5)

b. (1) transport of protons into the thylakoid compartment by the electron transport chain; (2) protons from the splitting of water remain in the thylakoid compartment; (3) removal of hydrogen in the stroma during the reduction of NADP⁺.

10.8 a. carbon fixation
b. reduction
c. regeneration of RuBP
d. $3 CO_2$
e. ribulose bisphosphate (RuBP)

f. rubisco
g. 3-phosphoglycerate
h. $6 ATP \longrightarrow 6 ADP + 6 \textcircled{P}_i$
i. 1,3-bisphosphoglycerate
j. $6 NADPH \longrightarrow 6 NADP^+$
k. glyceraldehyde 3-phosphate (G3P)
l. G3P
m. glucose and other carbohydrates
n. $3 ATP \longrightarrow 3 ADP + 3 \textcircled{P}_i$

10.9 Photorespiration may be an evolutionary relic from the time when there was little O_2 in the atmosphere and the ability of rubisco to distinguish between O_2 and CO_2 was not critical. Now, in our oxygen-rich atmosphere, photorespiration seems to be an agricultural liability.

10.10 a. in the bundle-sheath cells
b. Carbon is initially fixed into a four-carbon compound in the mesophyll cells by PEP carboxylase. When this compound is broken down in the bundle sheath cells, CO_2 is maintained at a high enough concentration that rubisco does not accept O_2 and cause photorespiration.

Suggested Answers to Structure Your Knowledge

1.

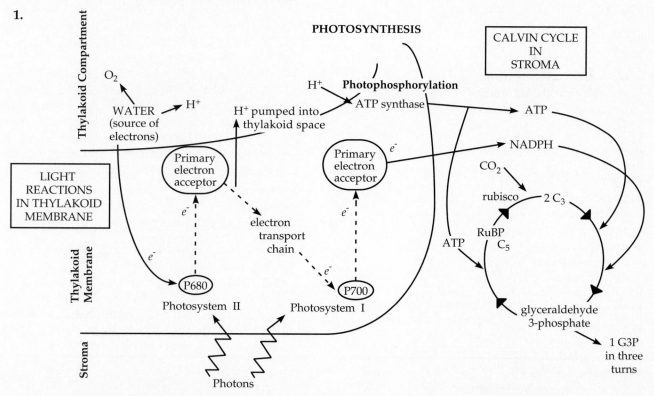

2.

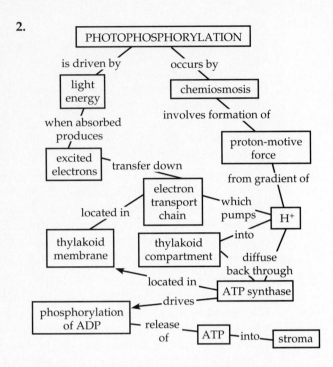

PHOTOPHOSPHORYLATION

is driven by — light energy — when absorbed produces — excited electrons — transfer down — electron transport chain — located in — thylakoid membrane

occurs by — chemiosmosis — involves formation of — proton-motive force — from gradient of — H⁺ — which pumps — into — thylakoid compartment — located in — ATP synthase

diffuse back through — ATP synthase — drives — phosphorylation of ADP — release of — ATP — into — stroma

Answers to Test Your Knowledge

Multiple Choice:

1. d	6. b	11. a	16. b	21. a
2. a	7. e	12. b	17. c	22. b
3. b	8. c	13. d	18. c	23. d
4. a	9. a	14. d	19. a	24. a
5. b	10. e	15. c	20. b	25. c

CHAPTER 11
THE REPRODUCTION OF CELLS

■ INTERACTIVE QUESTIONS

11.1 genome

11.2 a. 46
b. 23

11.3 a. Growth—most organelles and cell components are produced continuously throughout these subphases.
b. DNA synthesis

11.4 Refer to Campbell Fig. 11.6 for chromosome diagrams.
a. G₂ of interphase
b. prophase
c. prometaphase
d. metaphase
e. anaphase
f. telophase and cytokinesis
g. centrosomes (with centrioles)
h. aster
i. chromatin
j. nuclear envelope
k. nucleolus
l. early mitotic spindle
m. pair of centrioles
n. nonkinetochore microtubules
o. kinetochore microtubules
p. metaphase plate
q. spindle

r. cleavage furrow
s. nuclear envelope forming

11.5 a. a complex of cyclin and Cdk that orchestrates the steps of mitosis by phosphorylating proteins and other kinases
b. The Cdk level is constant through the cell cycle, but the level of cyclin varies because active MPF phosphorylates an enzyme that degrades cyclin. Thus MPF regulates its own level as it initiates mitosis and then becomes inactive until sufficient cyclin accumulates again during the next interphase.

Suggested Answers to Structure Your Knowledge

1. **Interphase:** 90 percent of cell cycle; growth, metabolism, DNA replication.
   • G₁ Phase: The chromosome, consisting of chromatin fibers made of DNA and associated proteins, is diffuse throughout the nucleus. RNA molecules are being transcribed from genes that are switched on.
   • S Phase—synthesis of DNA: The chromosome is replicated; two exact copies, called sister chromatids, are produced and held together at a region called the centromere. Growth and metabolic activities continue.

- G_2 Phase: Growth and metabolic activities of the cell continue.

Mitosis: cell division
- Prophase: The chromosome, consisting of two sister chromatids attached at the centromere, becomes tightly coiled and folded.
- Prometaphase: Kinetochore fibers from opposite ends of the mitotic spindle attach to the kinetochores of the sister chromatids; the chromosome moves toward midline.
- Metaphase: The centromere of the chromosome is aligned at the metaphase plate along with the centromeres of the other chromosomes.
- Anaphase: The sister chromatids separate (now considered to be individual chromosomes) and move to opposite poles.
- Telophase: Chromatin fiber of chromosome uncoils and is surrounded by re-forming nuclear membrane.

2.

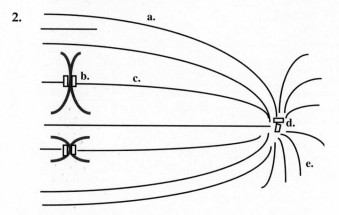

a. nonkinetochore microtubules: push poles apart by sliding past microtubules from the opposite pole
b. kinetochore: region of centromere where kinetochore microtubules attach
c. kinetochore microtubules: move chromosomes to metaphase plate and separate chromosomes as the microtubules disassemble
d. centrosome and centrioles: region of initiation of mitotic spindle formation
e. aster: radiating spindle fibers (in animal cells)

3. a. anaphase
   b. interphase
   c. late telophase
   d. metaphase

Answers to Test Your Knowledge

Fill in the Blank:

1. G_0
2. anaphase
3. prophase
4. telophase
5. S phase
6. metaphase
7. G_2
8. prophase
9. prometaphase
10. G_1 phase

Multiple Choice:

1. d
2. c
3. e
4. b
5. c
6. c
7. c
8. b
9. a
10. c

CHAPTER 12
MEIOSIS AND SEXUAL LIFE CYCLES

■ INTERACTIVE QUESTIONS

12.1 a. 14
     b. 7
     c. 28
     d. 1

12.2 Area of each life cycle from fertilization to meiosis should be shaded in as diploid.
     a. meiosis
     b. fertilization
     c. zygote
     d. gametes
     e. fertilization
     f. $2n$
     g. zygote
     h. meiosis
     i. meiosis
     j. spores
     k. mitosis
     l. gametes
     m. fertilization
     n. zygote

12.3 a. metaphase II
     b. prophase I
     c. anaphase I
     d. interphase
     e. metaphase I
     f. anaphase II

     Proper sequence: **d. b. e. c. a. f.**

12.4 2^{23}, approximately 8 million

Suggested Answers to Structure Your Knowledge

1. **a.** Chromosome replication, sister chromatids attached at centromere.

 b. Synapsis of homologous pairs, crossing over at chiasmata, spindle forms.

 c. Homologous pairs line up at metaphase plate, with centromeres attached to spindle fibers from opposite poles.

 d. Homologous pairs of chromosomes separate and move toward opposite poles, centromeres do not separate (sister chromatids remain attached).

 e. Haploid set of chromosomes, each consisting of two sister chromatids, align at metaphase plate.

 f. Centromeres separate and single-stranded chromosomes move to opposite poles.

2.

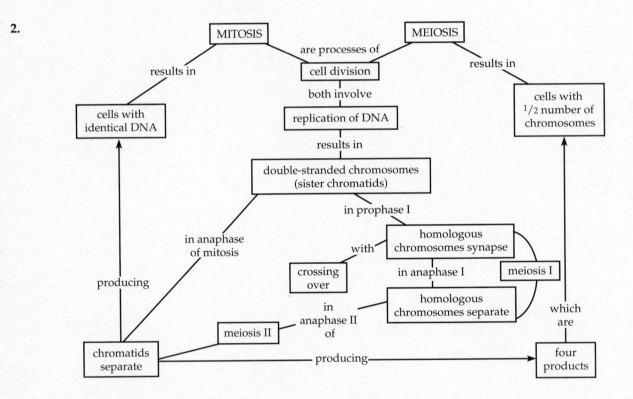

Answers to Test Your Knowledge

Multiple Choice:

1. b	4. a	7. c	10. b	13. c
2. d	5. e	8. c	11. a	14. d
3. e	6. b	9. b	12. b	15. e

CHAPTER 13
MENDEL AND THE GENE IDEA

■ INTERACTIVE QUESTIONS

13.1 a. $\textcircled{R}$

b. $\textcircled{r}$

c. F$_1$ Generation

d. *Rr*

e. F$_2$ Generation

f. $\textcircled{R}$

g. $\textcircled{r}$

h. ⬭ *Rr*

i. ⬭ *Rr*

j. ⬬ *rr*

k. 3 round:1 wrinkled

l. 1 *RR*:2 *Rr*:1 *rr*

13.2 a. all *Tt* tall plants

b. 1:1 tall (*Tt*) to dwarf (*tt*)

13.3 a. tall purple plants

b. *TtPp*

c. *TP, Tp, tP, tp*

d.

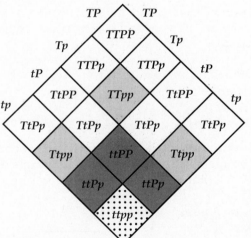

e. 9 tall purple:3 tall white:3 dwarf purple:1 dwarf white

13.4 You could determine this in two ways. The probability of getting both recessive alleles in a gamete is 1/4, and for two such gametes to join is 1/4 × 1/4, or 1/16. An easier way is to treat each gene involved as a monohybrid cross. The probability of obtaining a homozygous recessive offspring in a monohybrid cross is 1/4. The probability of the offspring being homozygous recessive for both genes is 1/4 × 1/4, or 1/16.

13.5 a. Consider the outcome for each gene as a monohybrid cross. The probability that a cross of *Aa* × *Aa* will produce an *A_* offspring is 3/4. The probability that a cross of *Bb* × *bb* will produce a *B_* offspring is 1/2. The probability that a cross of *cc* × *CC* will produce a *C_* offspring is 1. To have all of these events occur simultaneously, multiply their probabilities: 3/4 × 1/2 × 1 = 3/8 .

b. Offspring could be *A_bbC_*, *aaB_C_*, or *A_B_C_*. The genotype *A_B_cc* is not possible. Probability of *A_bbC_* = 3/4 × 1/2 × 1 = 3/8 Probability of *aaB_C_* = 1/4 × 1/2 × 1 = 1/8 Probability of *A_B_C_* = 3/4 × 1/2 × 1 = 3/8 Probability of offspring showing at least two dominant traits is the total of these independent probabilities, or 7/8.

13.6 a. A: $I^A I^A$ and $I^A i$

b. B: $I^B I^B$ and $I^B i$

c. AB: $I^A I^B$

d. O: *ii*

13.7 The ratio of offspring would be 9:3:4, a common ratio when one gene is epistatic to another.

Phenotype	Genotype	Ratio
Black	*M_B_*	3/4 × 3/4 = 9/16
Brown	*M_bb*	3/4 × 1/4 = 3/16
White	*mm__*	1/4 × 1 = 4/16

13.8 a. The parental cross produced 25-cm tall F$_1$ plants, all *AaBbCc* plants with 3 units of 5 cm added to the base height of 10 cm.

b. Of the 64 possible combinations of gametes in the F$_2$, there will be 7 different phenotypic classes, varying from 6 dominant alleles (40 cm), 5 dominant (35 cm), 4 dominant (30 cm), and so on, to all 6 recessive alleles (10 cm).

13.9 a. This trait is recessive. If it were dominant, albinism would be present in every generation, and it would be impossible to have albino children with nonalbino (homozygous recessive) parents.

b. father = *Aa*; mother = *Aa*

c. mate 1 = *AA* (probably); mate 2 = *Aa*; grandson 4 = *Aa*

d. The genotype of son 3 could be *AA* or *Aa*. If his wife is *AA*, then he could be *Aa* (both his parents are carriers) and the recessive allele never would be expressed in his offspring.

Even if he and his wife were both carriers (heterozygotes), there would be a $243/1024$ or 24% chance that all five children would be normally pigmented. ($3/4 \times 3/4 \times 3/4 \times 3/4 \times 3/4$)

13.10 a. $1/4$

b. $2/3$. There is a probability that $3/4$ of the offspring will have a normal phenotype. Of these, $2/3$ would be predicted to be heterozygotes and, thus, carriers of the recessive allele.

13.11 Both sets of prospective grandparents must have been carriers. The parents do not have the disorder so they are not homozygous recessive. Thus each has a $2/3$ chance of being a heterozygote carrier. The probability that both parents are carriers is $2/3 \times 2/3 = 4/9$; the chance that two heterozygotes will have a recessive homozygous child is $1/4$. The overall chance that a child will inherit the disease is $4/9 \times 1/4 = 1/9$. Should this couple have a baby that has the disease, this would establish that they are both carriers, and the chance that a subsequent child would have the disease is $1/4$.

Suggested Answers to Structure Your Knowledge

1. Mendel's law of segregation occurs in anaphase I, when alleles segregate as homologous chromosomes move to opposite poles of the cell. The two cells formed from this division have one-half the number of chromosomes and one copy of each gene. Mendel's law of independent assortment relates to the lining-up of synapsed chromosomes at the equatorial plate in a random fashion during metaphase I. Genes on different chromosomes will assort independently into gametes.

2. One allele does not "dominate" another. All alleles operate independently within a cell, coding for their gene products, usually enzymes, as specified by their particular sequence of DNA nucleotides. If both alleles code for functional enzymes, then the alleles may be codominant with both traits expressed in the heterozygote, as illustrated by MN or AB blood types. If the recessive allele codes for a nonfunctional enzyme, then the resulting recessive trait may be obvious only when homozygous (such as O blood type in which there are no carbohydrate molecules on the cell membrane). If one copy of an allele produces sufficient product that the dominant trait is expressed, then we have complete dominance where the heterozygote phenotype is indistinguishable from the homozygote dominant (for example, smooth peas in which the enzyme converts sugar to starch). With incomplete dominance, the heterozygote is distinguishable from homozygous dominants and recessives. The functional allele is insufficient to produce the full trait (as with the pink color of a heterozygote in carnations). Genotypes translate into phenotypes by way of biochemical pathways, which are controlled by enzymes coded for by genes. Thus molecular processes produce physical, physiological, and behavioral results.

Answers to Genetics Problems

1. White alleles are dominant to yellow alleles. If yellow were dominant, then you should be able to get white squash from a cross of two yellow heterozygotes.

2. **a.** $1/4$ [$1/2$ (to get AA) $\times 1/2$ (bb)]
   b. $1/8$ [$1/4$ (aa) $\times 1/2$ (BB)]
   c. $1/2$ [1 (Aa) $\times 1/2$ (Bb) $\times 1$ (Cc)]
   d. $1/32$ [$1/4$ (aa) $\times 1/4$ (bb) $\times 1/2$ (cc)]

3. Since flower color shows incomplete dominance, there will be six phenotypic classes in the F_2 instead of the normal four classes found in a 9:3:3:1 ratio. You could find the answer with a Punnett square, but multiplying the probabilities of the monohybrid crosses is more efficient.

Tall red	$T\_RR = 3/4 \times 1/4$	$3/16$
Tall pink	$T\_Rr = 3/4 \times 1/2$	$3/8$ or $6/16$
Tall white	$T\_rr = 3/4 \times 1/4$	$3/16$
Dwarf red	$ttRR = 1/4 \times 1/4$	$1/16$
Dwarf pink	$ttRr = 1/4 \times 1/2$	$1/8$ or $2/16$
Dwarf white	$ttrr = 1/4 \times 1/4$	$1/16$

4. Determine the possible genotypes of the mother and child. Then find the blood groups for the father that could not have resulted in a child with the indicated blood group.
   a. no groups exonerated
   b. A or O
   c. A or O
   d. AB
   e. B or O

5. The parents are $CcBb$ and $Ccbb$. Right away you know that all cc offspring will die and no BB black offspring are possible because one parent is bb. Only four phenotypic classes are possible. Determine the proportion of each type by applying the law of multiplication.
   Normal brown ($CCBb$) $= 1/4 \times 1/2 = 1/8$
   Normal white ($CCbb$) $= 1/4 \times 1/2 = 1/8$
   Deformed brown ($CcBb$) $= 1/2 \times 1/2 = 1/4$
   Deformed white ($Ccbb$) $= 1/2 \times 1/2 = 1/4$

6. Father's genotype must be *Pp* since polydactyly is dominant and he has had one normal child. Mother's genotype is *pp*. The chance of the next child having normal digits is 1/2 or 50% because the mother can only donate a *p* allele and there is a 50% chance that the father will donate a *p* allele.

7. The genotypes of the puppies were 3/8 *B_S_*, 3/8 *B_ss*, 1/8 *bbS_*, and 1/8 *bbss*. Because recessive traits show up in the offspring, both parents had to have had at least one recessive allele for both genes. Black occurs in a 6:2 or 3:1 ratio, indicating a heterozygous cross. Solid occurs in a 4:4 or 1:1 ratio, indicating a cross between a heterozygote and a homozygous recessive. Parental genotypes were *BbSs* × *Bbss*.

8. The 1:1 ratio of the first cross indicates a cross between a heterozygote and a homozygote. The second cross indicates that the hairless hamsters cannot have had a homozygous genotype, because then only hairless hamsters should be produced. Let us say that hairless hamsters are *Hh* and normal-haired are *HH*. The second cross of heterozygotes should yield a 1:2:1 genotype ratio. The 1:2 ratio indicates that the homozygous-recessive genotype may be lethal, with embryos that are *hh* never developing.

9. **a.** A general rule for the number of alleles involved in a cross is one less than the number of phenotypic classes. Here there are five phenotypic classes, thus four alleles or two gene pairs. Each dominant gene adds 5 cm in tail length. The 15-cm pigs are double recessives; the 35-cm pigs are double dominants.

b. A cross of a 15-cm and a 30-cm pig would be equivalent to *aabb* × *AABb*. Offspring would have either one or two dominant alleles and would have tail lengths of 20 or 25 cm.

10. **a.** The black parent would be *CC^{ch}*. The Himalayan parent could be either *C^hC^h* or *C^hc*. First list the alleles in order of dominance: *C, C^{ch}, C^h, c*. To approach this problem you should write down as much of the genotype as you can be sure about for each phenotype involved. The parents were black (*C_*) and Himalayan (*C^h_*). The offspring genotypes would be black (*C_*) and Chinchilla (*C^{ch}_*). Right away you can tell that the black parent must have been *CC^{ch}* because half of the offspring are Chinchilla; *C^{ch}* is dominant over *C^h* and *c* and could not have been present in the Himalayan parent. The genotype of this parent could be either *C^hC^h* or *C^hc*. A testcross would be necessary to determine this genotype.

b. You cannot definitely determine the genotype of the parents in the second cross. You can, however, eliminate some genotypes. The black parent could not be *CC* or *CC^{ch}*, because both the *C* and *C^{ch}* alleles are dominant to *C^h*, and Himalayan offspring were produced. The Chinchilla parent could not be *C^{ch}C^{ch}* for the same reason. One or both parents had to have *C^h* as their second allele; one, but not both, might have *c* as their second allele.

11. **a.** First figure out possible genotypes: _ _E_ = golden (any *B* combination with at least 1 *E*), *B_ee* = black, *bbee* = brown. All you know of first parents at start is _ _E_ × _ _E_. Since you see black and brown offspring, you know that both had to be heterozygous for *Ee*, and at least one was heterozygous for *Bb* (to get a black dog). Since black and brown are in a 1:1 ratio (like a testcross ratio), then one dog was *Bb* and the other was *bb*. You need to consider the results from the second cross to know which dog was *Bb*.

b. For the second cross, you now know that Dog 2 is *?bEe* and dog 3 must be *B?ee*. A ratio of approximately 3:1 black to brown looks like a cross of heterozygotes, so both parents must be *Bb*. So Dog 3 is *Bbee*, and Dog 2 must be *BbEe*. That means that Dog 1 must be *bbEe*.

12.

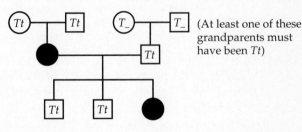

Answers to Test Your Knowledge

Matching:

1. I	3. F	5. L	7. H	9. C
2. A	4. E	6. J	8. M	10. K

Multiple Choice:

1. c	3. e	5. b	7. c	9. b
2. a	4. c	6. d	8. d	10. b

CHAPTER 14
THE CHROMOSOMAL BASIS OF INHERITANCE

■ INTERACTIVE QUESTIONS

14.1

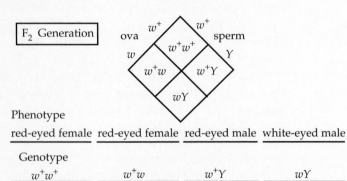

Phenotype

red-eyed female	red-eyed female	red-eyed male	white-eyed male

Genotype

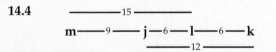

w^+w^+	w^+w	w^+Y	wY

14.2 **a.** tall, purple-flowered and dwarf, white-flowered

b. tall, white-flowered and dwarf, purple-flowered

14.3 Since linked genes have their loci on the same chromosome, they travel together during meiosis and you tend to have more parental offspring produced. Recombinants are the result of crossing over between homologous chromosomes, an event that is less frequent than 50% for genes located close to each other.

14.4

$$\text{m} \underset{\underset{\displaystyle 12}{\rule{3cm}{0.4pt}}}{\overset{\overset{\displaystyle 15}{\rule{4cm}{0.4pt}}}{\rule{0pt}{0pt}}} \text{m}\!-\!\!9\!\!-\!\text{j}\!-\!6\!-\!\text{l}\!-\!6\!-\!\text{k}$$

14.5 The gene is sex-linked, so a good notation is X^C, X^c, and Y so that you will remember that the Y does not carry the gene. Capital C indicates normal sight. Genotypes are:

1. X^CY
2. X^CX^c
3. X^CX^C or X^CX^c
4. X^CX^c
5. X^CY
6. X^cY
7. X^CX^c

14.6 **a.** A trisomic organism ($2n + 1$) has an extra copy of one chromosome, usually caused by a nondisjunction during meiosis. A triploid organism ($3n$) has an extra set of chromosomes, possibly caused by a total nondisjunction in gamete formation.

b. The genetic balance of a trisomic organism would be more disrupted than that of an organism with a complete extra set of chromosomes.

14.7 translocation, deletion, and inversion

14.8 Aneuploidies of sex chromosomes appear to upset genetic balance less, perhaps because relatively few genes are located on the Y chromosome and extra X chromosomes are inactivated as Barr bodies.

Suggested Answers to Structure Your Knowledge

1. Genes that are not linked assort independently, and the ratio of offspring from a testcross with a dihybrid heterozygote should be 1:1:1:1 (*AaBb* × *aabb* gives *AaBb, Aabb, aaBb, aabb* offspring). Genes that are linked and do not cross over should produce a 1:1 ratio in this testcross (*AaBb* and *aabb*). If crossovers occur 50% of the time, the heterozygote will produce equal quantities of *AB, Ab, aB,* and *ab* gametes, and the genotype ratio of offspring will be 1:1:1:1, the same as it is for unlinked genes. Because each 1% of crossovers is equal to 1 map unit, this measurement ceases to be meaningful at relative distances of 50 or more map units. However, crosses with intermediate genes on the chromosome could establish both that the genes *A* and *B* are on the same chromosome and that they are a certain distance apart.

2. If the gene for the mutant is sex-linked, you could assume it was on the X chromosome. A cross between a mutant female fly and a normal male should produce all normal females (who get a wild-type allele from their father) and mutant male flies (who get the mutant allele on the X chromosome from their mother). $X^mX^m × X^+Y$ produces X^+X^m and X^mY offspring (normal females and mutant males).

3. The serious phenotypic effects that are associated with these chromosomal alterations indicate that normal development and functioning is dependent on genetic balance. Most genes appear to be vital to an organism's existence, and extra copies of genes upset genetic balance. Inversions and translocations, which do not disrupt the balance of genes, can alter phenotype because of effects on gene functioning from neighboring genes.

Answers to Genetics Problems

1. **a.** The trait is sex-linked and recessive.
   b. Using the symbols X^T for the dominant allele and X^t for the recessive:
      1. X^tY 5. X^TX^t
      2. X^TX^t 6. X^TX^t or X^TX^T
      3. X^TX^t 7. X^TX^t
      4. X^TY
   c. #6 has a brother who has the trait, so her mother must be a carrier of the trait, meaning there is a 1/2 probability that #6 is a carrier. Since she is mated to a phenotypically normal male, none of her daughters will show the trait (0 probability). Her sons have 1/2 chance of having the trait if she is a carrier. $1/2 \times 1/2 = 1/4$ probability that her sons will have the trait. There is a 1/2 chance that a child would be male, so the probability of an affected child is $1/2 \times 1/4$, or 1/8.

2. e   a   c   b   d

3. The genes appear to be linked because the parental types appear most frequently in the offspring. Recombinant offspring represent 10 out of 40 total offspring for a recombination frequency of 25 percent, indicating that the genes are 25 map units apart.

4. One of the mother's X chromosomes carries the recessive lethal allele. One-half of male fetuses would be expected to inherit that chromosome and spontaneously abort. Assuming an equal sex ratio at conception, the ratio of girl to boy children would be 2:1, or 6 girls and 3 boys.

5. **a.** For female chicks to be black, they must have received a recessive allele from the male parent. If all female chicks are black, the male parent must have been Z^bZ^b. If the male parent was homozygous recessive, then all male offspring will receive a recessive allele, and the female parent would have to be Z^BW to produce all barred males.
   b. For female chicks to be both black and barred, the male parent must have been Z^BZ^b. If the female parent were Z^BW, only barred male chicks would be produced. To get an equal number of black and barred male chicks, the female parent must have been Z^bW.

Answers to Test Your Knowledge

Multiple Choice:

1. e	5. a	9. a
2. a	6. e	10. a
3. b	7. b	11. d
4. d	8. c	12. b

CHAPTER 15
THE MOLECULAR BASIS OF INHERITANCE

■ INTERACTIVE QUESTIONS

15.1 a. Griffith injected mice with heat-killed S and live R bacteria. R cells were transformed by heat-stable genetic material of S cells.
   b. Avery purified chemicals from heat-killed S cells. DNA is a transforming agent.
   c. Hershey and Chase radioactively labeled protein and DNA of T2 phage, then infected bacterial samples. DNA of T2 is injected into bacteria and is hereditary material.
   d. Chargaff compared DNA of several organisms, found DNA composition is species specific and A=T and G=C.

15.2 a. sugar-phosphate backbone
   b. complementary base pair
   c. adenine
   d. pyrimidine bases
   e. guanine
   f. thymine
   g. purine bases
   h. hydrogen bonds
   i. cytosine
   j. nucleotide
   k. deoxyribose
   l. phosphate
   m. 3.4 nm
   n. 0.34 nm
   o. 2 nm

15.3

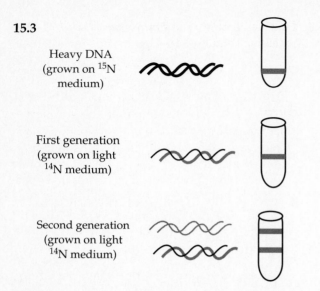

Heavy DNA (grown on ^{15}N medium)

First generation (grown on light ^{14}N medium)

Second generation (grown on light ^{14}N medium)

15.4 The phosphate end of each strand is the 5′ end, and the hydroxyl group extending from the 3′ carbon of the sugar marks the 3′ end.

15.5
a. helicase
b. single-strand binding protein
c. DNA polymerase
d. leading strand
e. lagging strand
f. DNA ligase
g. Okazaki fragment
h. RNA primer
i. primase
j. replication fork
k. 3′ end of parental strand
l. 5′ end

Suggested Answers to Structure Your Knowledge

1. Watson and Crick used the X-ray crystallography data from Franklin to deduce that DNA was a helix 2 nm wide, with nitrogenous bases stacked 0.34 nm apart, and making a full turn every 3.4 nm. Using molecular models of wire, they experimented with various arrangements and finally placed the sugar-phosphate backbones on the outside of the helix with the bases extending inside. Pairing of a purine base with a pyrimidine base produced the proper diameter. Specificity of base pairing (A with T and C with G) is assured by hydrogen bonds.

2. Replication bubbles form where proteins recognize specific base sequences and open up the two strands. **Helicase,** an enzyme that works at the replication fork, untwists the helix and separates the strands. **Single-strand binding proteins** support the separated strands while replication takes place. **Primase** lays down about 10 RNA bases to start the new strand. After a proper base pairs up on the exposed template, **DNA polymerase** joins the nucleotide to the 3′ end of the new strand. On the lagging strand, short Okazaki fragments are formed by primase and polymerase (again moving 5′ → 3′). Ligase joins the 3′ end of one fragment to the 5′ end of its neighbor. Proofreading enzymes check for mispaired bases, and excision enzymes and other enzymes repair damage or mismatches.

Answers to Test Your Knowledge

Multiple Choice:

1. c	5. b	9. d	13. d
2. a	6. a	10. a	14. b
3. b	7. b	11. e	15. c
4. d	8. e	12. a	

CHAPTER 16
FROM GENE TO PROTEIN

■ INTERACTIVE QUESTIONS

16.1 transcription translation
DNA → RNA → protein

16.2 Met Pro Asp Phe Lys stop

16.3
a. Initiation: Transcription factors bind to promoter; polymerase II binds to promoter and separates DNA strands at initiation site.
b. Elongation: Polymerase II moves along DNA strand, connecting RNA nucleotides that have paired to the DNA template to the 3′ end of the growing RNA strand.
c. Termination: Upon reaching the termination site, polymerase II releases the RNA.

16.4

DNA triplet	mRNA codon	Anticodon	Amino acid
TAC	AUG	UAC	methionine
GGA	CCU	GGA	proline
TTC	AAG	UUC	lysine
ATC	UAG	AUC	stop

16.5 1. Codon recognition: An elongation factor helps an aminoacyl-tRNA into the A site where its anticodon hydrogen-bonds to the mRNA codon; one GTP is used.

2. Peptide bond formation: Ribosome catalyzes peptide bond formation between new amino acid and polypeptide held in P site.

3. Translocation: The tRNA in the P site is released; the tRNA now holding the polypeptide moves from the A to the P site; one GTP used.

4. Termination: Release factor binds to termination codon in the A site. Free polypeptide and tRNA are released from the P site. Ribosomal subunits separate.
  a. amino end of growing polypeptide
  b. amino acid–tRNA complex (aminoacyl-tRNA)
  c. large subunit of ribosome
  d. A site
  e. small subunit of ribosome
  f. 5′ end of mRNA
  g. peptide bond formation
  h. P site
  i. release factor
  j. termination codon
  k. free polypeptide

16.6 Whether or not it is bound to an mRNA that codes for a signal sequence.

16.7 "All biological catalysts are enzymatic proteins" is no longer a valid generalization due to the discovery of ribozymes. "All genes consist of DNA" is no longer true due to the recognition of RNA viruses.

16.8 a. A base-pair substitution in the third nucleotide of a codon that still codes for the same amino acid.
  b. A base-pair substitution that results in a codon for a different amino acid.
  c. A base-pair substitution or frameshift mutation that creates a stop codon and prematurely terminates translation.
  d. An insertion or deletion of one, two, or more than three nucleotides that disrupts the reading frame and creates extensive missense and nonsense mutations.

Suggested Answers to Structure Your Knowledge

1.

	Transcription	Translation
Template	DNA	RNA
Location	nucleus (cytoplasm in prokaryotes)	cytoplasm; ribosomes can be free or attached to ER
Molecules involved	RNA nucleotides, DNA template strand, RNA polymerase, transcription factors	amino acids, tRNA, mRNA, ribosomes, ATP, GTP, enzymes, initiation, elongation, and release factors
Enzymes involved	RNA polymerases, RNA processing enzymes, ribozymes	aminoacyl-tRNA synthetase, ribosomal enzymes, ribozymes
Control—start and stop	transcription factors locate promoter region with TATA box, terminator sequence	initiation factors, initiation sequence (AUG), stop codons, release factor
Product	mRNA	protein
Product processing	RNA processing: 5′ cap and poly-A tail, splicing of hnRNA—introns removed by snRNPs in spliceosomes	spontaneous folding, disulfide bridges, signal sequence removed, cleaving, quaternary structure, modification with sugars, etc.
Energy source	ribonucleoside triphosphate	ATP and GTP

2. The genetic code includes the sequence of nucleotides on DNA that is transcribed into the codons found on mRNA and translated into their corresponding amino acids. There are 64 possible codons created from the four nucleotides used in the triplet code (4^3). Redundancy of the code refers to the fact that several triplets may code for the same amino acid. Often these triplets differ only in the third nucleotide. The wobble phenomenon explains the fact that there are only about 45 different tRNA molecules that pair with the 64 possible codons. The third nucleotide of many tRNAs can pair with more than one base. Because of the redundancy of the genetic code, these wobble tRNAs still place the correct amino acid in position.

The genetic code was thought to be universal, that each codon coded for exactly the same amino acid in all organisms. Recently, some exceptions to this universality have been found in a few ciliates and in the DNA of mitochondria and chloroplasts.

3.

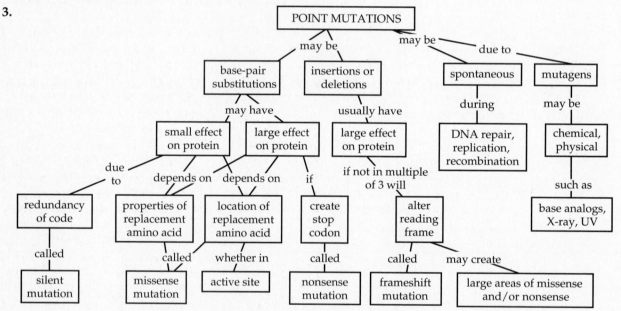

Answers to Test Your Knowledge

Multiple Choice:

1. d	5. b	9. b	13. d
2. a	6. e	10. b	14. e
3. b	7. a	11. b	15. c
4. e	8. a	12. d	16. b

CHAPTER 17
MICROBIAL MODELS: THE GENETICS OF VIRUSES AND BACTERIA

▦ INTERACTIVE QUESTIONS

17.1 1. Phage attaches to host cell and injects DNA.
2. Phage DNA forms circle.
3. New phage DNA and proteins are synthesized and assembled into phages.
4. Bacterium lyses, releasing phages.
5. Phage DNA integrates into bacterial chromosome.
6. Bacterium reproduces, passing prophage to daughter cells.
7. Colony of infected bacteria forms.
8. Occasionally, prophage exits bacterial chromosome and begins lytic cycle.
   a. phage DNA
   b. bacterial chromosome
   c. new phages
   d. prophage
   e. replicated bacterial chromosome with prophage

17.2 RNA ⟶ DNA ⟶ RNA; viral reverse transcriptase, host RNA polymerase

17.3 Oncogenes generally code for cellular growth factors or the receptor proteins for growth factors.

17.4 a. DNA
   b. RNA
   c. protein capsid
   d. host
   e. bacterium
   f. lytic or lysogenic cycles
   g. animal
   h. oncogenes
   i. viral envelope
   j. retrovirus
   k. reverse transcriptase
   l. plant
   m. viroids
   n. plasmids
   o. transposons

17.5 a. virulent phage reproducing by lytic cycle
   b. temperate phage that then leaves lysogenic cycle and enters lytic cycle

17.6 a. circular chromosome
   b. F plasmid, R plasmid
   c. mutation
   d. transformation
   e. naked DNA
   f. transduction
   g. phage
   h. conjugation
   i. F$^+$ or Hfr, and F$^-$ cells
   j. transposons
   k. antibiotic-resistant genes
   l. adaptation to environment/evolution

17.7 a. regulatory gene **f.** RNA polymerase
   b. promoter   **g.** active repressor
   c. operator   **h.** inducer metabolite
   d. structural genes **i.** mRNA for enzymes
   e. operon   of pathway

17.8 a. anabolic; corepressor; on; inactive
   b. catabolic; inducers; off; active

Suggested Answers to Structure Your Knowledge

1.

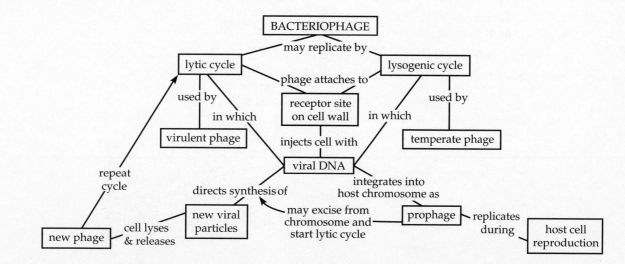

2.

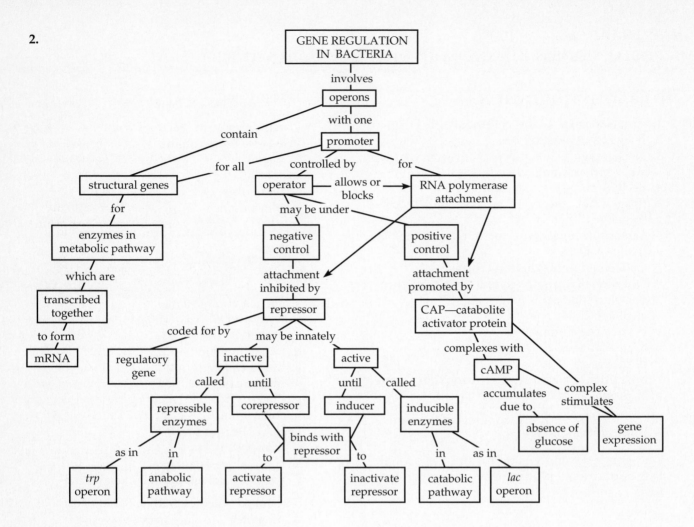

Answers to Test Your Knowledge

Multiple Choice:

1. b	5. a	9. c	13. c	17. a
2. d	6. d	10. e	14. e	18. b
3. a	7. b	11. a	15. b	
4. e	8. b	12. e	16. c	

Matching:

1. C	3. G	5. F	7. A
2. D	4. E	6. B	8. H

CHAPTER 18
GENOME ORGANIZATION AND EXPRESSION IN EUKARYOTES

■ INTERACTIVE QUESTIONS

18.1 a. nucleosome; 30-nm chromatin fiber; looped domains; coiling and folding of looped domains
b. Barr body—compacted *X* chromosome in cells of female

18.2 fragile *X* syndrome and Huntington's disease

18.3 a. transcription factor
b. enhancer
c. transcription factors
d. RNA polymerase

18.4 a. express different genes
b. available or unavailable for expression
c. regulation of gene expression
d. binding of transcription factors to enhancer sequences and promoters
e. hormones and other chemical signals

18.5 Different combinations of the multiple variable and constant regions come together during cellular differentiation.

Suggested Answers to Structure Your Knowledge

1. a. DNA packing and methylation (heterochromatin in Barr body); gene amplification or loss (multiple copies of rRNA genes in ovum); gene rearrangements (transposons, immunoglobulin genes)
b. Transcription factors that bind with enhancer and promoter regions; steroid hormones or other chemical messages may interact with receptor proteins and influence transcription.
c. RNA processing and export (introns removed, 5' cap and poly-A tail added); mRNA degradation
d. Repressor proteins may prevent ribosome binding so mRNA can be stockpiled (awaiting fertilization in ovum); initiation factors necessary for mRNA and ribosomes to join.
e. Protein processing by cleavage or modification; signal sequence targets proteins to proper location; selective degradation of proteins.

2. a. Proto-oncogenes are key genes that control cellular growth, differentiation, or adhesion. When such genes mutate to form a more active product, become amplified, or have changes in their normal control mechanisms, they may become oncogenes and produce a tumor.
b. Tumor-suppressor genes code for proteins that regulate cell division. Loss or mutation of both alleles of a tumor-suppressor gene may allow tumors to develop. Usually mutations or other changes must occur in oncogenes and suppressor genes for cancer to develop.

Answers to Test Your Knowledge

Multiple Choice:

1. b	4. b	7. c	10. e
2. a	5. a	8. d	11. c
3. e	6. e	9. c	12. e

CHAPTER 19
DNA TECHNOLOGY

19.1 a. The third sequence, because it has a symmetrical sequence running in opposite directions.

   b. Bacterial plasmids, bacteriophage, yeast plasmids, retroviruses.

   c. Their fast generation time (rapid growth rate), their ability to take up plasmids and be infected by phage, their plasmids that replicate and may form multiple copies.

19.2 a. plasmid
   b. antibiotic-resistant gene
   c. *lacZ* gene
   d. same restriction sites
   e. human gene of interest
   f. sticky ends
   g. human gene
   h. nonfunctional *lacZ* gene
   i. recombinant plasmid
   j. plate with antibiotic and X-gal
   1. Plasmid and gene of interest are isolated and cut with same restriction enzyme (disrupts *lacZ* gene of plasmid).
   2. Fragments are mixed and some foreign fragments base-pair with plasmid.
   3. DNA ligase seals ends.
   4. Recombinant plasmid transforms bacteria.
   5. Cells containing recombinant plasmid are identified by their ability to grow in presence of the antibiotic and by their white color. Blue colonies contain plasmids that resealed and thus have a functioning *lacZ* gene.

19.3 Use cDNA that has no introns; attach promoters and proper transcription sequences to vector DNA; attach gene to commonly produced bacterial gene.

19.4 The sequence of nucleotides in the original fragment would be complementary to the sequence shown by the order of fragments. It would read, from the bottom up, A A C A G C T T C A G T C.

19.5 a. restriction enzyme treatment
   b. gel electrophoresis
   c. Southern blotting
   d. addition of labeled probe
   e. autoradiography
   Crime samples contain blood from the victim and, presumably, the perpetrator of the crime. After identifying the bands from the victim, the remaining fragments match those of Suspect 2's DNA fingerprint.

19.6 a. genetic (linkage) mapping
   b. physical mapping of chromosomes
   c. nucleotide sequencing
   d. analysis of genomes of other species

19.7 a. Before, members of the extended family had to be tested to determine the variant of the RFLP marker to which the disease-causing allele was linked. Then RFLP analysis was done with the patient's blood to determine which marker, and thus which allele, he or she had inherited. Now the cloned gene can be used as a probe, and RFLP analysis can determine whether the individual carries the abnormal allele for the disease.

   b. The major difficulty is to assure that proper control mechanisms are present so that the gene is expressed at the proper time, in the proper place, and to the proper degree. If genes can be inserted in cells that will continue to divide, the therapy will last longer. Ethical considerations include how available such expensive treatments will be and whether genetic engineering should be done on germ cells, thus influencing the genetic makeup of future generations.

19.8 The antisense gene would produce mRNA that is complementary to the mRNA for a ripening enzyme. The two mRNAs would hybridize by base-pairing, thus preventing the translation of the ripening enzyme.

Suggested Answers to Structure Your Knowledge

1. a. Bacterial enzymes that cut DNA at restriction sites, creating "sticky ends" that can base-pair with other fragments. Use: make recombinant DNA, form restriction fragments used for many other techniques

   b. Mixture of molecules applied to slab of gel in electric field; molecules separate, moving at different rates due to charge, size, and other physical properties. Use: separate restriction fragments into pattern of distinct bands, fragments can be removed from gel and retain activity or can be identified with probes

   c. mRNA isolated from cell is treated with reverse transcriptase to produce a complementary DNA strand, which then produces a double-stranded DNA gene, minus introns and control regions. Use: creates genes that are easier to clone in bacteria, produces library of genes that are active in cell

d. Radioactively labeled single-stranded DNA or mRNA used to base-pair with complementary sequence of DNA or RNA. Use: locate gene in clone of bacteria, identify similar nucleic acid sequences, make cytological map of genes with *in situ* hybridization

e. DNA fragments separated by gel electrophoresis, transferred by Southern blotting onto filter or membrane, labeled probe added, rinsed, autoradiography. Use: analyze DNA for homologous sequences

f. Single-stranded DNA fragments are incubated with four nucleotides, DNA polymerase, and one of four dideoxy nucleotides that interrupt synthesis, samples separated by high-resolution gel electrophoresis. Use: sequence of nucleotides is read from the four sets of bands on the gel (or from sequence of fluorescent tags)

g. DNA is repeatedly melted and mixed with DNA polymerase, nucleotides, and primers having complementary sequences for targeted DNA section. Use: rapidly produce multiple copies of a gene or section of DNA *in vitro*

h. Restriction fragments separated by gel electrophoresis, Southern blotting transfers single-stranded fragments to filter, radioactive probe added, and autoradiography reveals band pattern. Use: DNA fingerprints for forensic use, map chromosomes using RFLP markers, diagnose genetic diseases

2. Agricultural applications: (1) production of vaccines, hormones, and antibodies, which will improve the health or productivity of livestock, (2) improvement of the genomes of agricultural plants and animals, (3) improvements in the nitrogen-fixing capacity of bacteria or plants, (4) development of plant varieties that have genes for resistance to diseases and herbicides

Medical applications: (1) vaccines developed, either through the production of virus subunits or attenuated viruses, (2) diagnosis of genetic diseases by probing for a potentially defective gene with a normal gene or RFLP analysis, (3) treatment of genetic disorders through the replacement of a defective gene with a normal one, (4) production of insulin, human growth factor, EPO, TPA (tissue plasminogen activator), and other useful products

Answers to Test Your Knowledge

Multiple Choice:

1. c	5. a	9. a	13. d
2. b	6. c	10. b	14. a
3. e	7. e	11. c	15. c
4. e	8. a	12. b	16. d

CHAPTER 20
DESCENT WITH MODIFICATION: A DARWINIAN VIEW OF LIFE

■ INTERACTIVE QUESTIONS

20.1 a. 1. F e   3. B g   5. D c, f, h   7. E h

   2. A b   4. C d   6. G a   8. H f

b. 3, 6, 1, 2, 5, 8, 4, 7

20.2 The excessive production of offspring sets up the struggle for existence; only a small proportion can live to leave offspring of their own. Natural selection is the differential reproductive success of individuals within a population that are best suited to the environment, which leads over generations to greater adaptation of organisms to the environment.

20.3 a. biogeography
   b. endemic island species and mainland species or neighboring species in different habitats
   c. fossil record
   d. ancestral and transitional forms
   e. comparative anatomy
   f. vestigial structures
   g. homologous structures
   h. comparative embryology
   i. ontogeny
   j. molecular biology
   k. DNA and proteins
   l. descent from a common ancestor

Suggested Answers to Structure Your Knowledge

1. The two major components of Darwin's evolutionary theory are that all life has descended from a common ancestral form and that the modification of that form has been the result of natural selection by the reproductive success of the individuals best adapted to the environment. The theory of natural selection is based on several key observations and inferences. The overproduction of offspring in conjunction with limited resources leads to a struggle for existence and the differential reproductive success of those organisms best suited to the local environment. The unequal survival and reproduction of the most fit individuals in a population leads to the gradual accumulation of adaptive characteristics in a population within a particular environment.

Answers to Test Your Knowledge

Multiple Choice:

1. b   3. e   5. c   7. e   9. d
2. c   4. a   6. a   8. d

CHAPTER 21
THE EVOLUTION OF POPULATIONS

■ INTERACTIVE QUESTIONS

21.1 a. The 98 *BB* mice contribute 196 *B* alleles, and the 84 *Bb* mice contribute 84 *B* alleles to the gene pool. These 84 *Bb* mice also contribute 84 *b* alleles, and the 18 *bb* mice contribute 36 *b* alleles. Of a total of 400 alleles, 280 are *B* and 120 are *b*. Allele frequencies are 0.7 *B* and 0.3 *b*.

b. The frequencies of genotypes are 0.49 *BB*, 0.42 *Bb*, and 0.09 *bb*.

21.2 0.7; 0.3; 0.49; 0.42; 0.09

21.3 a. 0.36; 0.48; 0.16. Plug *p* (0.6) and *q* (0.4) in the expanded binomial: $p^2 + 2pq + q^2$

b. 0.6; 0.4. Add the frequency of the homozygous dominant genotype to 1/2 the frequency of the heterozygote for the frequency of *p*. For *q*, add the homozygous recessive frequency and 1/2 the heterozygote frequency. Alternatively, to determine *q*, take the square root of the homozygous recessive frequency if you are sure the population is in Hardy-Weinberg equilibrium. The frequency of *p* is then 1−*q*.

c. 0.9; 0.1. Determine the genotype frequencies by dividing the number of each genotype by the total number counted; 65/80 = 0.81, 14/80 = 0.18, 1/80 = 0.01. Then determine allele frequencies as above. Another approach is to determine the number of al-leles in the population (160) and the number of *S* and *s* alleles contributed by each genotype; *S* is 144/160 = 0.9 and *s* is 15/160 = 0.1.

21.4 a. genetic drift
b. small population
c. bottleneck effect
d. founder effect
e. gene flow
f. migration between populations
g. mutations
h. nonrandom mating
i. inbreeding
j. assortative mating
k. natural selection
l. better reproductive success

21.5 a. mutation
b. recombination
c. Bacteria and microorganisms have very short generation times, and a new beneficial mutation can increase in frequency rapidly in an asexually reproducing bacterial population. Although mutation is the source of new alleles, they are so infrequent that their contribution to genetic variation in a large, diploid population is minimal. However, crossing over and independent assortment produce gametes with a great deal of genetic variation, and each zygote receives a unique assortment of genes from two parents.

21.6 a. 1. The relative fitness of the most fit genotype is 1. The two given fitness values are less than one, so *Bb* must produce the most offspring.

b. The relative fitness for *Bb* of 1 is twice the fitness of 0.5 and would produce twice the number of offspring, or 200. If *Bb* produced 200 offspring, *bb* with a fitness of 0.25 would produce 1/4 as many offspring or 50.

c. The higher relative fitness of *Bb* may be due to heterozygote advantage or hybrid vigor.

Suggested Answers to Structure Your Knowledge

1. **a.** The Hardy-Weinberg theorem states that allele and genotype frequencies within a population will remain constant from one generation to the next as long as the population is large, mutation and migration are negligible, mating is random, and no selective pressure operates. This Hardy-Weinberg equilibrium provides a null hypothesis to allow population geneticists to test for microevolution.

   b. $p^2 + 2pq + q^2 = 1$. In the Hardy-Weinberg equation, p and q refer to the frequencies of two alleles in the gene pool. The frequency of homozygous offspring is $(p \times p)$ or p^2 and $(q \times q)$ or q^2. Heterozygous individuals can be formed in two ways, depending on whether the ovum or sperm carries the p or q allele, and their frequency is equal to $2pq$.

2. Genetic variation is retained within a population by diploidy and balanced polymorphism. Diploidy masks recessive alleles from selection when they occur in the heterozygote. Thus, less adaptive or even harmful alleles are maintained in the gene pool and are available should selection pressures change and they become relatively better adapted to the new environment. Balanced polymorphism results in several alleles at a gene locus being maintained in a population. In situations in which there is heterozygote advantage, two alleles may be simultaneously selected for and retained in roughly equal frequencies within the gene pool. A patchy environment can favor different morphs within the geographic range of a species. Frequency-dependent selection, in which morphs present in higher numbers are selected against by predators or other factors, is another cause of balanced polymorphism.

3. The allele frequencies estimated from this sample are $B = 0.6$ and $b = 0.4$. Assuming this sample truly represents the individuals in the larger population, then we must say that the population is not in Hardy-Weinberg equilibrium. If it were, we would expect to find 36 *BB*, 48 *Bb*, and 16 *bb* individuals. The larger numbers of both homozygous genotypes indicates that assortative mating may be affecting the genotype frequencies in the population. Other possible explanations for these data are incorrect sampling techniques or the action of some selection agent on speckled butterflies that reduces the numbers of heterozygotes in the population.

Answers to Test Your Knowledge

1. e	5. d	9. b	13. d
2. a	6. d	10. e	14. a
3. c	7. e	11. e	15. b
4. c	8. d	12. c	16. b

CHAPTER 22
THE ORIGIN OF SPECIES

■ INTERACTIVE QUESTIONS

22.1 a. asexual species
   b. fossil species
   c. populations that are geographically separated
   d. subspecies that have very limited or circuitous genetic exchange

22.2 a. reduced hybrid fertility   **b.** post-
   c. gametic isolation   **d.** pre-
   e. mechanical isolation   **f.** pre-
   g. temporal isolation   **h.** pre-
   i. hybrid breakdown   **j.** post-
   k. behavioral isolation   **l.** pre-
   m. reduced hybrid viability   **n.** post-
   o. habitat isolation   **p.** pre-

22.3 The Hawaiian Archipelago is a series of relatively young, isolated, and physically diverse islands whose thousands of endemic species are examples of adaptive radiation resulting from multiple invasions and allopatric speciations.

22.4 a. 20 chromosomes   **b.** 24 chromosomes

22.5 a. morphological
   b. biological
   c. recognition
   d. cohesion

22.6 a. In the gradual tempo theory of evolution, small changes accumulate within populations as a result of chance events and natural selection, leading to the gradual evolution of new life forms. The punctuated equilibrium theory holds that evolution occurs in spurts of relatively rapid change interspersed with long periods of stasis.

b. The fossil record seems to indicate that new species appear fairly rapidly, with greatest change early in their existence, and then remain relatively unchanged during their duration on Earth. Sheldon's study of trilobite fossils provides evidence for gradual evolution.

tion. If the makeup of the gene pool changes enough, microevolution may lead to speciation.

2. An environmental change results in a new adaptive landscape with new adaptive peaks to which a population may be pushed by natural selection. A peak shift may occur within one adaptive landscape when some chance, non-adaptive change pushes a population off its adaptive peak and natural selection pushes the population to a new adaptive peak that may now be in reach.

Suggested Answers to Structure Your Knowledge

1. Speciation, by which a new species evolves from its predecessor, is part of the process of evolution and the increase in biological diversity. Microevolution is the process by which changes occur within the gene pool of a population as a result of either chance events or natural selec-

Answers to Test Your Knowledge

1. e	5. c	9. b	13. a
2. c	6. e	10. b	14. b
3. d	7. a	11. b	
4. c	8. d	12. c	

CHAPTER 23
TRACING PHYLOGENY: MACROEVOLUTION, THE FOSSIL RECORD, AND SYSTEMATICS

■ INTERACTIVE QUESTIONS

23.1 Organisms that were abundant, widespread, existed over a long period of time, had hard shells or skeletons, and lived in regions in which the geology favored fossil formation are most likely overrepresented in the fossil record.

23.2 a. 16,800 years. Carbon-14 has a half-life of 5600 years. In 5600 years the ratio of C-14 to C-12 would be reduced by $1/2$; in 11,200 years it would be reduced by $1/4$; and in 16,800 years it would be reduced by $1/8$.
b. 2.6 billion years. The fossil has $1/4$ the amount of potassium-40. It would take two half-lives to decay that amount.

23.3 Feathers and wings first developed as structures for capturing prey or for social displays; light, honeycombed bones may have increased agility of bipedal dinosaurs.

23.4 The continuation of growth of the human brain for several years longer than the chimpanzee brain allows for the development of human culture.

23.5 A trend caused by gradual phyletic evolution could occur if directional selection gradually resulted in populations evolving into species of increased size. If the trend resulted from species selection, a series of speciation events would

have occurred, with the larger species persisting longer and speciating more often, leading to a trend of increasing size.

23.6 Marsupials probably migrated to Australia through South America and Antarctica while the continents were still joined. When Pangaea broke up, marsupials adaptively radiated on isolated Australia. Mammals evolved and diversified on the other continents.

23.7 Mass extinctions empty many adaptive zones, which may then be exploited by species that survived extinction.

23.8 Adaptations to different environments may lead to large morphologic differences within related groups, and convergent evolution can produce similar structures in unrelated organisms in response to similar selective pressures.

23.9 DNA comparisons made by DNA–DNA hybridization, restriction mapping, and DNA sequencing help to elucidate phylogenetic relationships in the field of systematics. Molecular clocks date and sequence branch points and taxonomic origins. DNA comparisons help to determine human genealogy.

23.10 a. P, C, CE   **e.** CE
b. P   **f.** C
c. C   **g.** P, CE
d. C

Suggested Answers to Structure Your Knowledge

1. **a.** Preexisting structures may be co-opted and adapted for new functions.

   b. Slight genetic changes can result in large morphologic differences: allometric growth, paedomorphosis, heterochrony, and homeosis.

   c. Evolutionary trends may be due to which species persist the longest and speciate the most often.

   d. Formation and breakup of Pangaea caused major climatic changes, brought together species that had evolved separately, then separated and species evolved independently, and accounts for biogeography.

   e. Large climatic changes during Permian and Cretaceous resulted in mass extinctions, opening up adaptive zones for other species.

   f. Development of novel adaptations or mass extinctions may open adaptive zones that new species can exploit.

2.

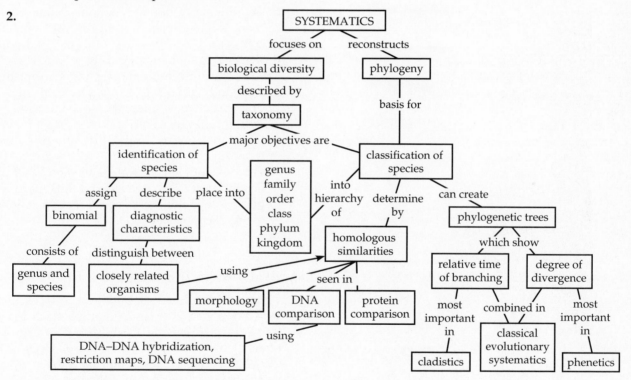

3. Microevolution consists of changes in gene frequencies that occur within a population over generations due to chance events (such as genetic drift and gene flow) and to differential reproductive success as a result of natural selection. New species may form as the result of microevolution or as the result of geographical isolation or mutation. Macroevolution includes the major events and trends that have occurred in the history of life as taxa higher than species evolve. The table in question 1 summarizes some of the mechanisms of macroevolution. Microevolution and macroevolution both involve natural selection and the element of chance. Natural selection appears to play a larger role in microevolution, while historical contingency may play the larger role in macroevolution.

Answers to Test Your Knowledge

Multiple Choice:

1. c	4. e	7. a	10. d	13. b
2. a	5. a	8. e	11. a	14. e
3. b	6. b	9. e	12. a	

True or False:

1. True
2. False—change the *more* to the *less* they vary.
3. True
4. False—change *phenetics* to *cladistics.*
5. True
6. False—change to: because *the half-life of carbon-14 is too short to date fossils that are that old.*
7. False—change to: indicative of *an asteroid having crashed into the Earth.*
8. True

CHAPTER 24
EARLY EARTH AND THE ORIGIN OF LIFE

■ INTERACTIVE QUESTIONS

24.1

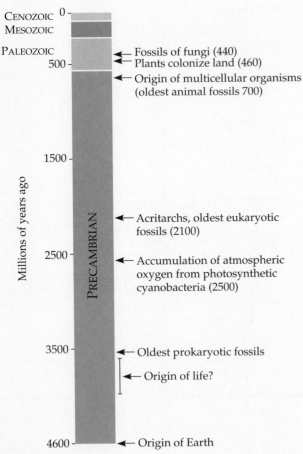

24.2 a. The abiotic synthesis and accumulation of organic molecules such as amino acids, simple sugars, and nucleotides.

   b. The linking of organic monomers into polymers, such as proteins, lipids, and nucleic acids.

   c. The aggregation of abiotically produced molecules into protobionts, separated from the surroundings by selectively permeable membranes that permitted the concentration of critical molecules, and the emerging properties of metabolism, excitability, and primitive reproduction.

   d. The development of RNA and then DNA as genetic material that facilitated the reproduction and evolution of successful aggregations of organic molecules.

Suggested Answers to Structure Your Knowledge

1. The primitive Earth is believed to have had seas, volcanoes, and large amounts of ultraviolet radiation passing through a thin atmosphere, which probably consisted of H_2O, CO, CO_2, N_2, CH_4, and NH_3. Life was able to evolve in this environment because the reducing (electron-adding) atmosphere and the availability of energy from lightning and UV radiation facilitated the abiotic synthesis of organic molecules, and hot lava rocks, clay, or pyrite may have aided the polymerization of these monomers.

2.

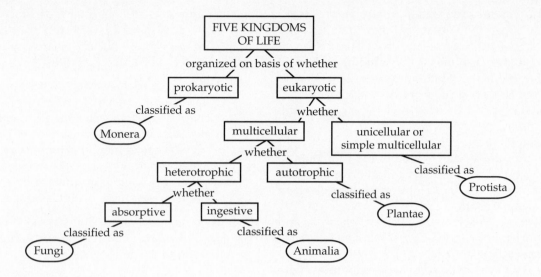

Answers to Test Your Knowledge

Multiple Choice:

1. b	**4. b**	**7. c**	**10. b**
2. d	**5. c**	**8. a**	**11. d**
3. e	**6. b**	**9. e**	

Matching:

1. F	**3. E**	**5. A**
2. D	**4. B**	

CHAPTER 25
PROKARYOTES AND THE ORIGINS OF METABOLIC DIVERSITY

■ INTERACTIVE QUESTIONS

25.1 a. Bacteria (eubacteria)
   b. Archaea (archaebacteria)
   c. Eucarya (eukaryotes)

25.2 a. spheres (cocci), rods (bacilli), or helices (spirilla)
   b. 1–5 μm in diameter, one-tenth the size of eukaryotic cells
   c. cell wall made of peptidoglycan; gram-negative bacteria also have outer lipopolysaccharide membrane; archaebacteria have no peptidoglycan in walls; sticky capsule for adherence; pili for attachment or conjugation
   d. flagella; filaments that spiral around cell and cause cell to cork-screw; or glide in secreted slime; may show taxis to stimuli
   e. some infoldings of plasma membrane used in respiration; thylakoid membranes in cyanobacteria for photosynthesis
   f. circular DNA molecule (genophore) with little associated protein found in nucleoid region; may have plasmids with antibiotic resistance and other genes
   g. binary fission; rapid geometric growth into colony; some genetic recombination in transformation, transduction, or conjugation; most variation from mutation

25.3 a. light
   b. phototroph
   c. chemicals
   d. CO_2
   e. organic nutrients
   f. heterotroph
   g. photoautotroph
   h. photoheterotroph
   i. chemoheterotroph

25.4 a. When free ATP supply diminished, breakdown of abiotically synthesized organic compounds for energy led to *glycolysis* and formation of ATP by substrate-level phosphorylation.

b. Transmembrane *proton pumps* (using ATP) developed to maintain pH of cell.

c. *Electron transport chain* linked oxidation of organic compounds with pumping of H$^+$ out of cell. Efficient electron transport chain could have formed proton gradient so that H$^+$ diffusing back in would reverse proton pump and generate ATP (*chemiosmosis*).

d. *Photosynthesis* evolved with the use of light energy absorbed by membrane pigments to pump H$^+$ out of the cell and to drive electrons from H$_2$S to NADP$^+$ by co-opting electron transport chain.

e. Water was used as the source of electrons and hydrogen for fixing CO$_2$ (*photosynthesis evolving oxygen*).

f. O$_2$ that was now present in the atmosphere was used to pull electrons from organic molecules down existing electron transport chain (*aerobic respiration*).

25.5 a. endospore; thick-walled, resistant cells that can withstand harsh conditions for long periods of time, breaking dormancy when conditions are favorable

b. heterocyst; specialized for nitrogen fixation

Suggested Answers to Structure Your Knowledge

1. Molecular evidence, especially comparison of rRNA signature sequences, indicates that eubacteria and archaebacteria diverged very early in evolutionary history and that archaebacteria may be more closely related to (more recently diverged from) eukaryotes.

2. Decomposers—recycle global nutrients, used for sewage treatment, clean up oil spills.
Nitrogen fixers—provide nitrogen to the soil in nodules on legume roots.
Symbionts—enteric bacteria, some produce vitamins.
Biotechnology—production of antibiotics, hormones, and many useful products.

3. Refer to the time line answer for question 24.1. O$_2$ began accumulating in the atmosphere around 2.5 billion years ago.

Answers to Test Your Knowledge

Multiple Choice:

1. c	**4.** c	**7.** a	**10.** c
2. e	**5.** b	**8.** d	**11.** e
3. b	**6.** d	**9.** a	

Fill in the Blanks:

1. cocci	**7.** endospore
2. nucleoid region	**8.** obligate anaerobe
3. Gram stain	**9.** saprobes, decomposers
4. pili	**10.** exotoxins
5. taxis	**11.** nodules
6. chemoheterotroph	**12.** cyanobacteria

CHAPTER 26
THE ORIGINS OF EUKARYOTIC DIVERSITY

■ **INTERACTIVE QUESTIONS**

26.1 a. from invaginations of the plasma membrane

b. mitochondria and chloroplasts: size, membrane enzymes and transport systems, division process, circular DNA molecule with little associated protein, ribosomal RNA sequences, and the existence of endosymbiotic relationships in modern organisms

26.2 a. acritarchs, ruptured coats of cysts, 2.1 billion years old

b. freshwater, marine, moist terrestrial habitats, symbiotically in hosts

c. photoautotrophs, heterotrophs, or mixotrophs

d. most have flagella or cilia at some point in life cycle

e. asexual by mitosis (process may differ); some have sexual reproduction and alternation of generations; resistant cysts

26.3 *Paramecium* is a ciliate, phylum Ciliophora.

a. cilia
b. oral groove
c. food vacuoles
d. contractile vacuole
e. anal pore
f. macronucleus
g. micronucleus

26.4 a. diploid
b. coenocytic mass, plasmodium
c. amoeboid or flagellated cells
d. haploid
e. solitary amoeboid cells until aggregate to fruiting body

f. two amoebas fuse to form resistant zygote

g. diploid

h. coenocytic hyphae

i. larger egg cell, sperm nucleus

26.5 **a.** diploid generation that produces spores by meiosis

b. haploid generation that produces gametes by mitosis

c. gametophyte and sporophyte generations differ in appearance

d. similar gametes

e. larger, nonmotile egg and flagellated sperm

26.6 Archaebacteria, Eubacteria, Archezoa, Protista (Protozoa), Chromista, Plantae, Fungi, Animalia

---

Suggested Answers to Structure Your Knowledge

1. Unicellular—and a few colonial and multicellular—eukaryotic organisms, most of which use aerobic respiration, have flagella or cilia at some point in their life, reproduce asexually (many have sexual reproduction), require water or moist habitats (which may be inside the bodies of hosts), form cysts to withstand harsh conditions, may be photoautotrophic, heterotrophic, or mixotrophic.

2. Originally, Whittaker assigned only unicellular eukaryotes to the kingdom Protista. The expanded kingdom includes some colonial and multicellular eukaryotes that appear to be more closely related to their unicellular ancestors than to fungi, plants, or animals. However, serial endosymbiosis likely occurred several times to produce the different lineages of protists, and thus, kingdom Protista is polyphyletic. Some systematists propose that the oldest lineage, including the diplomonads that lack mitochondria and have dual haploid nuclei, be placed in a separate kingdom Archezoa; the brown algae and related phyla deserve their own kingdom Chromista; and the green algae, as the ancestors of plants, and the red algae be placed in the kingdom Plantae.

3. Refer to the time line in the answer for question 24.1. The oldest known fossils of eukaryotes are acritarchs, which date from 2.1 billion years ago.

Answers to Test Your Knowledge

Matching:

1. M	4. J	7. L	10. N	13. O
2. A	5. I	8. F	11. G	14. C
3. B	6. E	9. D	12. K	15. H

Multiple Choice:

1. c	4. a	7. b
2. e	5. d	8. d
3. c	6. c	

CHAPTER 27
PLANTS AND THE COLONIZATION OF LAND

■ INTERACTIVE QUESTIONS

27.1 **a.** multicellular, photoautotrophic eukaryotes

b. mostly terrestrial

c. cuticle to protect from desiccation

d. stomata for gas exchange

e. chlorophylls *a* and *b* and carotenoids

f. cellulose cell wall

g. starch as storage compound

h. gametangia with protective cell layer

i. embryophytes; fertilized egg retained

j. alternation of generations that are heteromorphic

27.2 In an ancestral charophyte, a zygote that was enclosed by nonreproductive cells of the haploid thallus may have first divided into a mass of diploid cells and then meiosis occurred, producing an increased number of haploid offspring from the sexual union of egg and sperm.

27.3 **a.** gametophyte

b. archegonia

c. antheridia

d. swim

e. sporophyte

f. meiosis

g. sporangium

h. gametophytes

27.4 **a.** Hollow columns of xylem cells carry water and minerals up from the roots; their lignified walls also provide support.

b. The living cells of phloem form tubes that distribute organic nutrients throughout the plant.

27.5 **a.** meiosis

b. spores

c. gametophyte

d. antheridium

e. sperm

f. archegonium

g. egg

h. fertilization

i. zygote
j. gametophyte
k. young sporophyte
l. mature sporophyte
m. sporangia (in sori)

The gametophyte (upper portion of the diagram) is haploid; the sporophyte (lower portion) is diploid.

27.6 a. The diploid sporophyte would provide spare alleles should an allele be damaged by the more intense, mutagenic radiation on land.

b. The gametophyte generation would allow the screening out of harmful mutations that would be expressed in the haploid condition.

27.7 a. ovulate cone   **e.** megaspore
b. nucellus   **f.** female gametophyte
c. pollen cone   **g.** egg in archegonium
d. pollen grain   **h.** embryo

Meiosis occurs within the sporangia on the microsporophyll to produce microspores that develop into pollen grains, and within the nucellus when the megaspore mother cell gives rise to a megaspore that develops into the female gametophyte. Pollination occurs when pollen enters the micropyle. Fertilization occurs, usually more than a year later, when a sperm nucleus discharged from the pollen tube fertilizes an egg cell contained in an archegonium. Gametophyte structures in the pine life cycle include pollen grains and the multicellular female gametophyte enclosed within integuments of the nucellus.

27.8 a. Allows for greater genetic recombination.

b. May coordinate development of food storage in the seed with the production of an embryo.

27.9 a. Male gametophyte—pollen grain with two haploid cells; female gametophyte—embryo sac contained in the ovule that has eight haploid nuclei in seven cells.

b. Mature ovule with a seed coat derived from the integuments of the ovule containing an embryo and endosperm.

c. Mature ovary that may be modified to help disperse seeds.

Suggested Answers to Structure Your Knowledge

1. a. origin of plants
b. radiation of vascular plants
c. first seed plants
d. radiation of flower plants
e. bryophytes (mosses, liverworts, hornworts)
f. seedless vascular plants (whiskferns, lycopods, horsetails, ferns)

g. gymnosperms (cycads, ginkgo, conifers)
h. angiosperms (flowering plants—monocots, dicots)

During the Carboniferous period, much of the land was covered by swamps, and giant lycopods, horsetails, and ferns dominated the Earth. Formation of the supercontinent Pangaea during the Permian period may have led to a warmer and drier climate, and the great coal forests were replaced by cycads and conifers. With the cooler climates of the Cretaceous period, many groups (including the dinosaurs) disappeared, and flowering plants became the dominant plants.

2. a. cuticle, stomata, jacketed gametangia, protected embryo

b. vascular tissue for transport and support, sporophyte generation dominant

c. protected gametophyte retained on sporophyte plant; nonswimming pollen; vascular fibers for support; seed for protection, dispersal, and nourishment of embryo

d. flowers and fruit, animal pollinators, protected seeds

3. Refer to the time line in the answer for question 24.1. Plants first colonized land around 460 million years ago.

Answers to Test Your Knowledge

Multiple Choice:

1. c	**5.** d	**9.** c
2. e	**6.** a	
3. d	**7.** b	
4. a	**8.** b	

True or False:

1. False—add *or algae.*

2. False—change *heteromorphic* to *heterosporous,* or change *produce...* to *have gametophyte and sporophyte that are morphologically distinct.*

3. True

4. False—change *naked seed* to *seedless,* or change *club mosses and horsetails* to *gymnosperms.*

5. True

6. False—change *fruit* to *seed,* or change *an embryo...* to *a mature ovary.*

7. False—change *tracheids* to *vessel elements.*

8. True

9. False—change *angiosperm* to *gymnosperm,* or change *haploid cells...* to *an embryo sac with seven haploid cells.*

10. True

CHAPTER 28
FUNGI

■ INTERACTIVE QUESTIONS

28.1 a. network of hyphae, typical body form
b. cross-walls between cells, with pores
c. multinucleated hyphae, aseptate
d. two nuclei from different sources exist together in hyphae
e. fusion of hyphal cytoplasm
f. fusion of nuclei

28.2 a. conidia (asexual spores)
b. ascocarp
c. ascospore
d. ascus
e. ascomycete
f. sporangia
g. zygosporangium
h. zygomycete
i. basidiocarp
j. basidium
k. basidiospores
l. basidiomycete

28.3 no known sexual stage

28.4 The flagellated spores of chytrids were the basis for including them with protists. Their similarities with fungi include an absorptive nutrition, chitinous cell walls, hyphae, and common enzymes and metabolic pathways. Recent molecular comparisons of proteins and nucleic acids support a close relationship between chytrids and fungi.

Suggested Answers to Structure Your Knowledge

1. a. zygote fungi, black bread mold, mycorrhizae
   b. coenocytic hyphae
   c. spores in sporangium on aerial hyphae
   d. hyphae fuse, karyogamy, resistant zygosporangia
   e. sac and cup fungi, yeasts, truffles, mycorrhizae
   f. septate hyphae, some unicellular
   g. spores called conidia in chains or clusters
   h. dikaryotic hyphae in ascocarp, karyogamy in asci, ascospores
   i. club fungi, mushrooms, rusts, mycorrhizae
   j. extensive mycelium, dikaryotic hyphae
   k. conidia formed sometimes

l. basidiocarps, karyogamy in basidia, basidiospores
m. imperfect fungi, *Penicillium*
n. may be molds or yeasts
o. form conidia on conidiophores
p. no known
q. lichens (blue-green bacteria or green alga in symbiotic relationship with ascomycete)
r. flat growing thalli
s. soredia or fragments, algae reproduce asexually
t. fungus may form ascocarps

2. The role of fungi as saprobes and decomposers of organic material is central to the recycling of chemicals between living organisms and their physical environment. Decomposers also prevent the accumulation of litter, feces, and dead organisms. Parasitic fungi have economic and health effects on humans. Plant pathogens cause extensive loss of food crops and trees. Mycorrhizae are found on the roots of most plants. These mutualistic associations greatly benefit the plant.

3. Protein and ribosomal RNA comparisons indicate that fungi are more closely related to animals than to plants or modern protists. Molecular systematics indicates that both fungi and animals evolved from a protistan ancestor, probably a choanoflagellate.

4. Refer to the time line in the answer for question 24.1. The first undisputed fossils of fungi appeared 440 million years ago in the Silurian period. The first terrestrial plant fossils have petrified mycorrhizae.

Answers to Test Your Knowledge

Fill in the Blanks:

1. septum	5. basidium	9. coenocytic
2. dikaryon	6. asci	10. Chytridiomycota
3. chitin	7. mycorrhizae	
4. conidia	8. zygospore	

Multiple Choice:

1. a	3. c	5. e	7. a	9. c
2. c	4. d	6. b	8. a	

CHAPTER 29
INVERTEBRATES AND THE ORIGIN OF ANIMAL DIVERSITY

■ INTERACTIVE QUESTIONS

29.1 a. Parazoa; no true tissues.
    b. Eumetazoa; true tissues.
    c. Radiata; radial symmetry, diploblastic.
    d. Bilateria; bilateral symmetry, triploblastic.
    e. Acoelomates; no body cavities or blood vascular system.
    f. Body cavities and blood vascular system.
    g. Pseudocoelomates; body cavity not enclosed by mesoderm.
    h. Coelomates; body cavity enclosed by mesoderm.
    i. Protostomes; spiral, determinate cleavage, mouth from blastopore, schizocoelous.
    j. Deuterostomes; radial, indeterminate cleavage, mouth second opening, enterocoelous.

29.2 a. line body cavity (spongocoel); create water current, trap and phagocytize food particles, may produce gametes
    b. wander through mesohyl; digest food and distribute nutrients, produce skeletal fibers, may produce gametes

29.3 a. polyp
    b. medusa
    c. tentacle
    d. mouth/anus
    e. gastrovascular cavity

29.4 a. Planarians have an extensively branched gastrovascular cavity; food is drawn in and undigested wastes are expelled through the mouth at the tip of a muscular pharynx.
    b. Tapeworms are bathed in predigested food in their host's intestine.

29.5 a. development of unfertilized eggs
    b. Rotifera
    c. pinworms, hookworms, *Trichinella*, which causes trichinosis

29.6 a. 1. muscular foot for movement
    2. visceral mass containing internal organs
    3. mantle that may secrete shell and encloses visceral mass to form mantle cavity that houses gills
    b. 1. Snails creep slowly, using their radula to rasp up algae.
    2. Clams are sessile suspension-feeders that use siphons to draw water over the gills, where food is trapped in mucus.
    3. Squid are active predators that capture and crush prey with jaws.

29.7 a. septum
    b. intestine
    c. metanephridium
    d. setae
    e. ventral nerve cord
    f. dorsal and ventral blood vessels
    g. longitudinal and circular muscles
    h. cuticle

29.8 a. Anterior appendages modified as pincers or fangs.
    b. Many spiders trap their prey in webs they have spun from abdominal gland secretions. Spiders kill prey with poison-equipped, fang-like chelicerae. After masticating and adding digestive juices, they suck up liquid, partially digested food.

29.9 a. The head has fused segments and has one pair of antennae, a pair of compound eyes, and several pairs of mouthparts modified for various types of ingestion.
    b. The segmented thorax has three pairs of walking legs and may have one or two pairs of wings.
    c. The segmented abdomen has no appendages.

29.10 a. mostly terrestrial
    b. mostly aquatic
    c. fly, walk, burrow
    d. swim, float
    e. tracheal system
    f. gills or body surface
    g. Malpighian tubules
    h. diffusion across cuticle, glands regulate salt balance
    i. one pair
    j. two pairs
    k. unbranched; mandibles and modified mouthparts for chewing, piercing, or sucking; thorax has three pairs of walking legs, may have wings (extensions of cuticle)
    l. branched; mandibles and mouthparts, walking legs on thorax, abdominal appendages

29.11 a. radial symmetry, usually five spokes; sessile or slow-moving
    b. endoskeleton of calcareous plates
    c. water vascular system with tube feet for locomotion, feeding, respiration

Suggested Answers to Structure Your Knowledge

1.

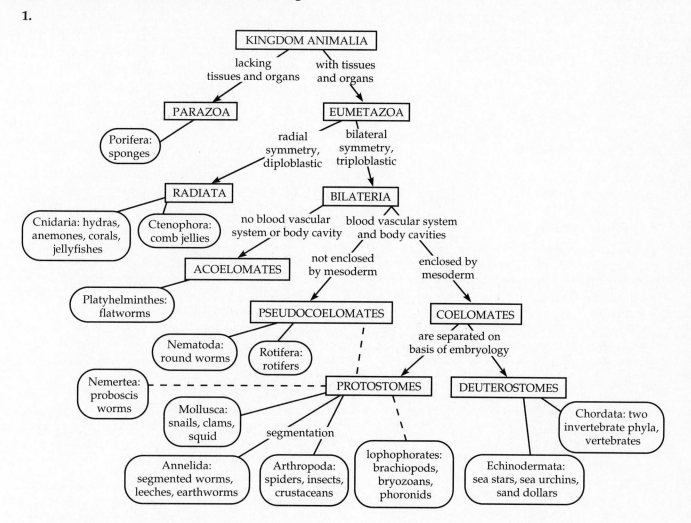

2. a. Fossils from the first 5–10 million years of the Cambrian (about 540 mya) include nearly all modern phyla. The appearance of so many complex animals in such a short time has been called the Cambrian explosion.

   b. The following hypotheses are put forward to explain this explosion: adaptive radiation made possible as the first animals radiated to fill ecological niches, diverse adaptations triggered by emerging predator–prey relationships, higher levels of atmospheric oxygen which allowed for more active lifestyles, development of a mesoderm tissue layer that allowed increasingly diverse body plans, and evolution and then altered expressions of developmental genes that control pattern formation, segmentation, and placement of appendages.

3. The time line in the answer to question 24.1 should show the oldest animal fossils at 700 million years ago.

Answers to Test Your Knowledge

Matching:

1. C l	5. H d	9. D f
2. B e	6. B j	10. C i
3. E g	7. E b	11. H m
4. A h	8. B a	12. E c

Multiple Choice:

1. b	5. e	9. c	13. b
2. c	6. c	10. c	14. a
3. d	7. a	11. b	15. c
4. e	8. a	12. a	16. d

CHAPTER 30
THE VERTEBRATE GENEALOGY

■ INTERACTIVE QUESTIONS

30.1 lancelet, subphylum Cephalochordata
- **a.** mouth
- **b.** *pharyngeal slits
- **c.** intestine
- **d.** segmental muscles
- **e.** anus
- **f.** *postanal tail
- **g.** *dorsal, hollow nerve cord
- **h.** *notochord
* Chordate characteristics

30.2 a. cephalization with sensory organs and enlarged brain to aid navigation
- **b.** cranium and vertebral column to protect brain and provide strong attachments for muscles
- **c.** respiratory, circulatory, feeding, and digestive adaptations to support an active metabolism

30.3 a. chordates
- **b.** vertebrates
- **c.** tetrapods
- **d.** amniotes
- **e.** Urochordata (tunicates)
- **f.** Cephalochordata (lancelets)
- **g.** vertebrae
- **h.** Agnatha (jawless fishes)
- **i.** jaws, paired fins
- **j.** Placodermi (extinct jawed fishes)
- **k.** skeletal elements in appendages
- **l.** Chondrichthyes (sharks and rays)
- **m.** lungs or lung derivatives
- **n.** Osteichthyes (bony fishes)
- **o.** legs
- **p.** Amphibia (salamanders and frogs)
- **q.** amniotic egg
- **r.** Reptilia (snakes, lizards, turtles, crocodilians)
- **s.** feathers
- **t.** Aves (birds)
- **u.** hair
- **v.** Mammalia

30.4 paired fins and hinged jaws

30.5 a. Increase the area for absorption and slow the passage of food through the short shark intestine.
- **b.** Common chamber through which the reproductive tract, excretory system and digestive tract open to the outside.

30.6 a. The common fusiform body shape of fast-swimming sharks, bony fish, and aquatic mammals is an example of convergent evolution, an adaptation to reduce drag while swimming.

30.7 a. lobe-finned fishes, in particular the rhipidistians
- **b.** the Carboniferous period, from 360 to 286 mya

30.8 a. amnion—encases embryo in fluid, prevents dehydration and cushions shocks
- **b.** allantois—stores metabolic wastes, functions with chorion in gas exchange
- **c.** chorion—functions in gas exchange
- **d.** yolk sac—expands over yolk and transports stored nutrients into embryo

30.9 a. sauropsids
- **b.** synapsids
- **c.** anapsids
- **d.** diapsids
- **e.** therapsids

The class Reptilia does not include its descendants, the birds and mammals. Birds are the modern animals that share the most recent ancestor with the dinosaurs.

30.10 Light, strong skeleton, no teeth and absence of some organs to reduce weight; feathers that shape wings into airfoil; strong keel on sternum to which flight muscles attach; high metabolic rate supported by efficient circulatory system and endothermy; well-developed visual and motor areas of brain.

30.11 a. Hair; milk produced by mammary glands; endothermic with active metabolism; diaphragm to ventilate lungs; four-chambered heart; larger brains, increased learning capacity; differentiated teeth and remodeled jaw.

30.12 a. 6 to 8 million years ago
- **b.** *Australopithecus ramidus*, dates from 4.4 million years ago

Suggested Answers to Structure Your Knowledge

1. Agnatha, Chondrichthyes, Osteichthyes, √Amphibia, √*Reptilia, √*Aves, √*Mammalia

2. (1) Pharyngeal slits used for suspension feeding became adapted for gas exchange; (2) skeletal supports for gills became adapted for use as hinged jaws; (3) lungs of bony fish were transformed to air bladders, aiding in buoyancy; (4) fleshy fins supported by skeletal elements developed into limbs for terrestrial animals; (5) feathers, probably originally used for insulation, came to be used for flight; (6) dexterous hands important for an arboreal life evolved to manipulate tools.

3. Amphibians remained in damp habitats, burrowing in mud during droughts; some secrete foamy protection for eggs laid on land. Reptiles developed scaly, waterproof skin, an amniote egg that provided an aquatic environment for developing embryo, and behavioral adaptations to modulate changing temperatures. All terrestrial groups are tetrapods, using limbs for locomotion on land.

4. According to the multiregional model, modern humans evolved in several areas from Neanderthals and other regional archaic *Homo sapiens*, from between 1 and 2 million years ago. According to the monogenesis model, all but the African archaic *H. sapiens* were evolutionary dead ends, and a relatively recent dispersal (100,000 years ago) from Africa gave rise to human geographic diversity.

Some paleoanthropologists claim that the fossil records best support the multiregional model. Molecular data from mtDNA and nuclear DNA indicate a more recent divergence of human groups, supporting the monogenesis model.

Answers to Test Your Knowledge

Fill in the Blanks:

1. Urochordata	7. diapsids
2. Cephalochordata	8. dinosaurs
3. somites	9. monotremes
4. ostracoderms	10. diaphragm
5. operculum	11. Anthropoidea
6. amniotic egg	12. Australopithecus

Multiple Choice:

1. a	5. d	9. b	13. d
2. b	6. d	10. e	14. d
3. e	7. a	11. c	
4. d	8. c	12. b	

CHAPTER 31
PLANT STRUCTURE AND GROWTH

■ INTERACTIVE QUESTIONS

31.1 a. point of stem at which leaf is attached
   b. at the node in the angle between leaf and stem
   c.

Monocots	Dicots
one cotyledon	two cotyledons
fibrous root system	taproot
veins parallel	veins netlike
may lack petiole	have petioles
may have pith inside stele in root	xylem spokes in stele with phloem in between
vascular bundles in stem scattered	vascular bundles in stem in ring, pith inside
no secondary growth	vascular and cork cambiums may produce secondary growth

31.2 a. sclerenchyma (fibers and sclereids), tracheids, and vessel elements
   b. the cells listed in **a** and sieve-tube members

31.3 Primary growth continues to add to the tips of roots and shoots, while secondary growth thickens and strengthens older regions of a woody plant.

31.4 a. epidermis—root hairs function in absorption
   b. cortex—uptake of minerals and water, storage
   c. stele—central vascular cylinder
   d. endodermis—regulates water movement into stele
   e. pericycle—origin of lateral roots
   f. pith—central parenchyma cells
   g. xylem—water and mineral transport
   h. phloem—nutrient transport

31.5 a. cuticle
   b. upper epidermis
   c. palisade parenchyma

d. spongy mesophyll
e. guard cells
f. xylem
g. phloem
h. stomata

31.6 See table in question 31.1.

31.7 H, F, I, A, B, G, D, C, E

Suggested Answers to Structure Your Knowledge

1. **a.** protoderm
   b. ground meristem
   c. cortex
   d. cork cells
   e. procambium
   f. primary phloem
   g. vascular cambium
   h. secondary phloem
   i. secondary xylem

2. This drawing is a dicot stem, indicated by the ring of vascular bundles and the central pith.
   a. xylem   **e.** pith
   b. phloem   **f.** cortex
   c. epidermis   **g.** ground tissue
   d. vascular bundle   connecting pith to cortex

Answers to Test Your Knowledge

Matching:

1. H	**5.** I	**9.** J
2. F	**6.** A	**10.** G
3. B	**7.** C	
4. E	**8.** D	

Multiple Choice:

1. a	**5.** b	**9.** e
2. e	**6.** a	**10.** d
3. b	**7.** e	
4. c	**8.** c	

CHAPTER 32
TRANSPORT IN PLANTS

▣ INTERACTIVE QUESTIONS

32.1 a. $\Psi_P = 0$, $\Psi_S = -0.6$, $\Psi = -0.6$

b. $\Psi_P = 0.6$, $\Psi_S = -0.6$, $\Psi = 0$. The turgor pressure of the cell that develops with movement of water into the cell finally offsets the osmotic potential of the hypertonic cell. No net osmosis occurs, and the water potentials of both cell and solution are 0.

c. $\Psi_P = 0$, $\Psi_S = -0.8$, $\Psi = -0.8$. The cell would plasmolyze, losing water until the osmotic potential of the cell would equal that of the bathing solution. There would be no turgor pressure.

32.2 a. apoplastic   **f.** endodermis
b. symplastic   **g.** stele
c. root hair   **h.** Casparian strip
d. epidermis   **i.** xylem vessel
e. cortex

32.3 a. The evaporation of water from the air spaces through the stomata creates a tension on the remaining water film.

b. Water molecules hold together due to hydrogen bonding, and a pull on one water molecule is transmitted throughout the column of water.

c. Water molecules adhere to the hydrophilic walls of xylem cells, which helps to support the column of water against gravity.

d. The transpiration of water from the leaf creates a tension within the xylem, lowering the water potential and contributing to the bulk flow of water.

32.4 a. The illumination of blue-light receptors activates proton pumps, and the resulting membrane potential facilitates the movement of potassium across the membrane. Photosynthesis in the guard cell produces the ATP needed for the proton pumps.

b. The beginning of photosynthesis in the mesophyll lowers the CO_2 concentration, which stimulates stomata to open.

c. A circadian rhythm contributes to the opening of stomata at dawn.

32.5 A storage organ, such as a root or tuber, stores carbohydrates during the summer and is a sugar sink, but when its starch is broken down to sugar in the spring, it becomes a sugar source.

Suggested Answers to Structure Your Knowledge

1. Solutes may move across membranes by passive transport when they move down their concentration gradient. Transport proteins, which speed this passive transport, may be specific carrier proteins or selective channels. Active transport usually involves a proton pump that creates a membrane potential as positively charged hydrogen ions are moved out of the cell. Cations may now move down their electrochemical gradient. In cotransport, a solute is moved by a transport protein driven by the inward diffusion of H^+.

2. The transpiration-cohesion-tension mechanism explains the ascent of xylem sap. The lower water potential of air compared with that of the soil solution results in the passive movement of water down its water potential gradient, against gravity, in a continuous column of water in xylem vessels. The cohesion of water molecules transmits the pull resulting from transpiration throughout the column, and the adhesion of water to the hydrophilic walls of the xylem vessels aids the flow. Tension created by the evaporation of water from the menisci that form in the leaf's air spaces sets up this bulk flow of water.

A pressure-flow mechanism also explains the movement of phloem sap, but the plant must expend energy to create the differences in water potential that drive translocation. The active accumulation of sugars in sieve-tube members greatly decreases water potential at the source end, resulting in an inflow of water. The removal of sugar from the sink end of a phloem tube results in the osmotic loss of water from the cell. The resulting difference in hydrostatic pressure between the source and sink end of the phloem tube causes the bulk flow of water with the accompanying transport of sugar.

Answers to Test Your Knowledge

1. b	4. a	7. d	10. c
2. e	5. c	8. d	11. a
3. d	6. b	9. a	12. c

CHAPTER 33
PLANT NUTRITION

■ INTERACTIVE QUESTIONS

33.1 a. component of chlorophyll, activates many enzymes
b. component of nucleic acids, phospholipids, ATP, some coenzymes
c. component of cytochromes, activates some enzymes

33.2 in the younger, growing parts of a plant

33.3 The continuous growth of roots exposes them to new sources of mineral nutrients.

33.4 A variety of conservation-minded, environmentally safe, and profitable farming methods that permit the use of soil as a renewable resource by utilizing appropriate soil and water management techniques.

33.5 a. nitrogen-fixing bacteria
b. ammonifying bacteria
c. nitrifying bacteria
d. denitrifying bacteria
e. amino acids and other organic compounds

Suggested Answers to Structure Your Knowledge

1.

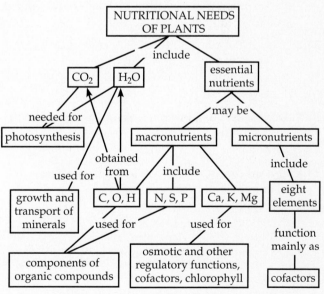

2. The key properties of a fertile soil include:
   - *Texture:* mixture of sand, silt, and clay so that air spaces containing oxygen are provided as well as sufficient surface area for the binding of water and minerals.
   - *Humus:* prevents clay from packing, so soil retains water and has air spaces; provides reservoir of mineral nutrients.
   - *Minerals:* adequate supply of nitrogen, phosphorus, and potassium, as well as other minerals.
   - *pH:* proper pH level so that minerals are in a form that can be absorbed and cation exchange can take place.
   - *Living organisms:* maintain physical and chemical soil properties.

3. Root nodules are mutualistic associations between legumes and nitrogen-fixing bacteria of the genus *Rhizobium.* These bacteria receive nourishment from the plant and provide the plant with ammonium. Mycorrhizae are mutualistic associations between the roots of most plants and various species of fungi. The fungi stimulate root growth, increase surface area for absorption, and secrete acids that increase the solubility of some minerals. In return, the fungi receive nourishment from the plant.

Answers to Test Your Knowledge

1. b	3. c	5. a	7. e	9. c
2. a	4. b	6. e	8. d	10. b

CHAPTER 34
PLANT REPRODUCTION AND DEVELOPMENT

■ INTERACTIVE QUESTIONS

34.1
a. stamen	**f.** stigma
b. anther	**g.** style
c. filament	**h.** ovary
d. petal	**i.** sepal
e. carpel	**j.** ovule

Pollen is formed in the anther. Pollination occurs when pollen lands on the stigma. Fertilization occurs within the embryo sac in the ovule.

34.2 a. a pollen grain with a thick coat surrounding a generative cell and a tube cell
b. the embryo sac with seven cells and eight nuclei

34.3 Double fertilization may conserve resources by coordinating nutrient development with fertilization.

34.4
a. seed coat	**g.** cotyledon (scutellum)
b. epicotyl	**h.** endosperm
c. hypocotyl	**i.** radicle
d. cotyledons	**j.** coleorhiza
e. radicle	**k.** plumule
f. plumule	**l.** coleoptile

34.5 Hormonal interactions induce softening of the pulp, a change in color, and an increase in sugar content.

34.6 The hypocotyl hook continues to elongate but does not straighten. Light is the cue that causes the hook to straighten and thus pull the cotyledons out of the ground.

34.7 The dispersal and dormancy advantages of seeds.

34.8 Harvesting is much easier because plants grow at the same rate and fruits ripen in unison. Crops are more uniform.

34.9 Grow; the indeterminate growth of plants, which retain embryonic root and shoot regions that allow for continuing growth, morphogenesis, and differentiation.

34.10 Microtubules of the preprophase band leave behind ordered actin microfilaments that determine the orientation of the nucleus and the location of the cell plate during cytokinesis. Microtubules may constrain the movement of cellulose-producing enzymes and thus orient the cellulose microfibrils in the cell walls that will determine the direction in which cells can expand.

34.11 a. The outer two zones have their normal gene expression and form sepals and petals. In the third and fourth zones, the lack of gene C activity allows gene A to be expressed, resulting in petal and sepal formation, respectively.
b. A double mutant lacking B and C gene activity would result in gene A being expressed alone in all four zones and a mutant flower of four whorls of sepals.

Suggested Answers to Structure Your Knowledge

1.

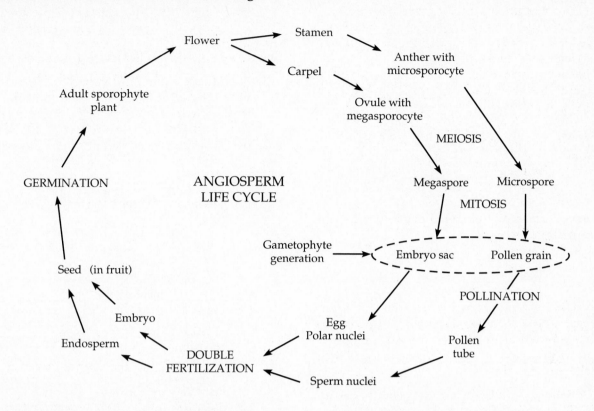

2. *Sexual:* Advantages include increased genetic variability, which provides the potential to adapt to changing conditions, and the dispersal and dormancy capabilities provided by seeds. Disadvantages are that the seedling stage is very vulnerable, sexual reproduction is very energy intensive, and genetic recombination may separate adaptive traits.

Asexual: Advantages include the hardiness of vegetative reproduction and the maintenance of well-adapted plants in a given environment. Disadvantages relate to the advantages of sexual reproduction: there is no genetic variability from which to choose should conditions change, and there are no seeds for dispersal.

Answers to Test Your Knowledge

Fill in the Blanks:

1. preprophase band
2. sporophyte
3. monoecious
4. embryo sac
5. radicle
6. epicotyl
7. hypocotyl or epicotyl hook
8. scion
9. aggregate
10. callus

Multiple Choice:

1. c	5. e	9. b	13. e
2. a	6. a	10. c	14. b
3. d	7. c	11. d	15. b
4. d	8. e	12. a	

CHAPTER 35
CONTROL SYSTEMS IN PLANTS

■ INTERACTIVE QUESTIONS

35.1 Both seedlings would bend to the left, but seedling *a* would show a sharper curve. In the initial seedling illuminated from the right, auxin would move into the agar block in greater concentration on the side away from the light. This higher concentration in the left side of the block would produce a stronger bend when placed on seedling *a*.

35.2

auxin	**D**	**b**
cytokinins	**B**	**e**
gibberellins	**E**	**a**
abscisic acid	**C**	**d**
ethylene	**A**	**c**

35.3 a. positive phototropism and negative gravitropism

    b. mechanical stimulation causing unequal growth rates of cells on opposite sides of the tendril

35.4 Free-running periods are circadian rhythms that have been found to vary from exactly 24 hours when an organism is kept in a constant environment without environmental cues.

35.5

a. no		**g.** yes	
b. yes		**h.** no	
c. no		**i.** yes	
d. yes		**j.** no	
e. no		**k.** yes	
f. yes		**l.** no	

35.6 a. Reduce transpiration water loss by stomatal closing and wilting; deep roots continue to grow in moister soil; heat-shock proteins produced to stabilize plant proteins.

    b. Change membrane lipid composition by increasing unsaturated fatty acids; air tubes may develop in cortex of root.

    c. A systemic response to infection by a plant, created when salicylic acid is released from cells in infected tissue and triggers the production of phytoalexin (an antibiotic) and other defenses throughout the plant.

Suggested Answers to Structure Your Knowledge

1. A synthetic auxin, 2,4-D, used as a broad-leaved herbicide; cytokinin spray used to keep flowers fresh; ethylene used to ripen stored fruit; synthetic auxins induce seedless fruit set; gibberellins sprayed on Thompson seedless grapes.

2.

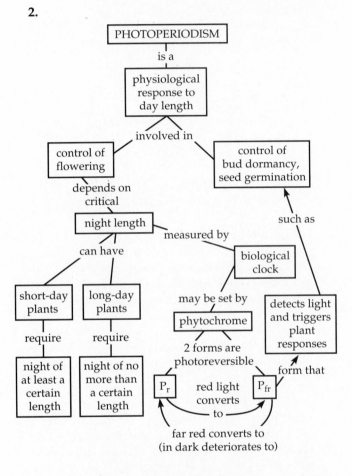

3. In the signal-transduction pathway, a signal is received when a hormone binds to a specific receptor molecule or when a pigment molecule absorbs a certain wavelength of light. The transduction of the signal involves the release of a second messenger, often calcium ions, that serves as a chemical amplifier of the signal. Ca^{2+} binds with calmodulin, and this complex then activates various proteins or other molecules within the cell. These activated molecules then induce specific changes within a cell, which may include fairly rapid responses or alterations of gene expression that may produce longer-term changes. This cascade of events allows a minute concentration of a hormone to produce a widespread cellular response.

Answers to Test Your Knowledge

True or False:

1. False—change *abscisic acid* to *gibberellin,* or say *The removal of abscisic acid.*
2. True
3. False—change *gibberellins* to *cytokinins.*
4. False—change *bolting* to *fruit ripening;* or change to read *Bolting is gibberellin-induced elongation of a flower stalk.*
5. True
6. False—change *Abscisic acid* to *Ethylene;* or change to read *Abscisic acid maintains dormancy and helps plants cope with stresses.*
7. True
8. False—change *thigmotropism* to *turgor movements;* or change to read *Thigmotropism is the change in the growth of a plant in response to touch.*
9. False—change *circadian rhythm* to *photoperiodism;* or change *day or night length* to *on a 24-hour cycle.*
10. True

Multiple Choice:

1. b	4. b	7. e	10. b
2. c	5. a	8. c	11. d
3. e	6. b	9. c	12. b

CHAPTER 36

AN INTRODUCTION TO ANIMAL STRUCTURE AND FUNCTION

▓ INTERACTIVE QUESTIONS

36.1 a. Stratified squamous; outer layer of skin; thick layer is protective, and new cells are produced near basement membrane to replace those sloughed off.
   b. Simple columnar; lining of digestive tract; columnar cells with large cytoplasmic volumes are specialized for secretion or absorption; single layer is better for absorption purposes.

36.2 a. Haversian system
   b. central canal
   c. lacuna
   d. bone
   e. macrophage
   f. fibroblast
   g. collagenous fiber
   h. elastic fiber
   i. reticular fiber
   j. loose connective tissue
   k. white blood cells
   l. red blood cells
   m. platelet
   n. blood

36.3 a. Cardiac; dark bands are intercalated discs that relay electrical signals from cell to cell during heartbeat.
   b. smooth muscle

36.4 digestive, circulatory, respiratory, immune and lymphatic, excretory, endocrine, reproductive, nervous, integumentary, skeletal, muscular

36.5 a. rabbit
   b. frog
   c. rabbit
   d. bear

36.6 a. sweat glands
   b. hypothalamus
   c. set point
   d. hypothalamus

Suggested Answers to Structure Your Knowledge

1.

Tissue	Structural Characteristics	General Functions	Specific Examples
Epithelial	Tightly packed cells; basement membrane; cuboidal, columnar, squamous shapes; simple or stratified	Protection, absorption, secretion; lines body surfaces	Mucous membrane, may be ciliated; skin; lining of blood vessels
Connective	Few cells that secrete extracellular matrix of protein fibers in liquid, gel, or solid ground substance	Connect and support other tissues	Loose connective; adipose; fibrous connective; bone; cartilage; blood
Muscle	Long cells (fibers) with myosin and many microfilaments of actin	Contraction, movement	Skeletal (voluntary); smooth (involuntary); cardiac
Nervous	Neurons with cell bodies, axons, and dendrites	Sense stimuli, conduct impulses	Nerves, brain

2. All organs are covered with epithelial tissue, and internal lumens, ducts, and blood vessels are lined with epithelium. Most organs contain some types of connective tissue, usually serving to connect other tissues and supply blood to the nonvascularized epithelium. Muscle tissue would be present if the organ must contract. Nervous tissue innervates most organs. The small intestine is an example of an organ composed of many tissues. An epithelial mucosa lines the lumen; the submucosa contains blood vessels and loose connective tissue; the muscularis, a layer of circular and longitudinal smooth muscle, comes next; and the outer serosa is an epithelium.

Answers to Test Your Knowledge

Multiple Choice:

1. b	3. d	5. b	7. b	9. b
2. a	4. e	6. d	8. a	10. e

CHAPTER 37
ANIMAL NUTRITION

▪ INTERACTIVE QUESTIONS

37.1 The proteins, fats, and carbohydrates that make up food are too large to pass through cell membranes. Monomers also must be reorganized to form the macromolecules specific to each animal.

37.2 **a.** pharynx—sucks food through mouth
b. esophagus—passes food to crop
c. crop—stores and moistens food
d. gizzard—grinds food
e. intestine—digests and absorbs
The typhlosole increases the surface area for absorption.

37.3 **a.** Polysaccharides (starch and glycogen) broken down in the mouth by salivary amylase to smaller polysaccharides and maltose; hydrolysis continues until salivary amylase is inactivated by the low pH and pepsin in the stomach.
b. Proteins in the stomach are broken down by pepsin to small polypeptides.

37.4 **a.** pancreatic amylases, disaccharidases*
b. trypsin, chymotrypsin, aminopeptidase*, carboxypeptidase*, and dipeptidases*
c. nucleases, nucleotidases*, nucleosidases*
d. lipase (aided by bile salts)
* in brush-border

37.5 **a.** salivary glands
b. oral cavity
c. pharynx
d. esophagus
e. stomach
f. small intestine
g. large intestine (colon)
h. anus
i. rectum
j. pancreas
k. pyloric sphincter
l. gallbladder
m. liver

37.6 Carnivores are characterized by sharp incisors, fanglike canines, and jagged premolars and molars; herbivores have broad molars for grinding plant material. The intestine of an herbivore is longer to facilitate digestion of plant material, and, unlike carnivores, its cecum may contain cellulose-digesting microorganisms.

37.7 Vegetarians must eat a combination of plant foods, complementary in amino acids and consumed at the same meal, to prevent protein deficiencies, because the body cannot store amino acids. A deficiency of a single essential amino acid may prevent protein synthesis and limit the use of other amino acids.

Suggested Answers to Structure Your Knowledge

1. **a.** energy/fuel
   b. carbon skeletons/organic raw materials
   c. essential nutrients
   d. undernourished
   e. biosynthesis

f. liver
g. malnourishment
h. essential amino acids
i. vitamins
j. coenzymes or parts of coenzymes
k. minerals

Answers to Test Your Knowledge

Matching:

1. M	**5.** F	**9.** N
2. I	**6.** G	**10.** E
3. D	**7.** J	
4. B	**8.** H	

Multiple Choice:

1. d	**4.** a	**7.** a	**10.** e	**13.** c
2. c	**5.** d	**8.** c	**11.** b	**14.** c
3. c	**6.** d	**9.** b	**12.** d	**15.** b

CHAPTER 38
CIRCULATION AND GAS EXCHANGE

■ INTERACTIVE QUESTIONS

38.1 The heart pumps hemolymph through vessels into sinuses, and body movements squeeze the sinuses, forcing hemolymph through the body and back to the heart.

38.2 a. capillaries of head and forelimbs
   b. left pulmonary artery
   c. aorta
   d. capillaries of left lung
   e. left pulmonary vein
   f. left atrium
   g. aorta
   h. capillaries of abdomen and hind limbs
   i. left ventricle
   j. right ventricle
   k. posterior vena cava
   l. right atrium
   m. right pulmonary vein
   n. capillaries of right lung
   o. right pulmonary artery
   p. anterior vena cava
   The pulmonary veins and aorta and arteries leading to the systemic capillary beds carry oxygen-rich blood. See text figure 38.4a for numbers.

38.3 a. atrioventricular
   b. semilunar
   c. SA (sinoatrial) node
   d. AV (atrioventricular) node

38.4 a. peripheral resistance
   b. artery walls
   c. contraction of ventricle
   d. diastole
   e. systole
   f. lower number
   g. higher number
   h. contract
   i. relax
   j. nerves, hormones

38.5 a. edema
   b. lack of blood proteins, which lowers the osmotic pressure of the blood

38.6 a. plasma
   b. electrolytes
   c. osmotic balance, buffering, muscle and nerve functioning
   d. proteins
   e. buffers, osmotic factors, lipid escorts, antibodies, clotting
   f. nutrients, wastes, respiratory gases, hormones
   g. erythrocytes
   h. transport O_2
   i. leukocytes
   j. phagocytes
   k. defense and immunity
   l. platelets
   m. blood clotting

38.7 Smoking, not exercising, obesity, and eating a fat-rich diet

38.8 a. water
    b. gill
    c. pump water by moving jaws and operculum
    d. air
    e. trachea
    f. body movements
    g. air
    h. lungs and skin
    i. positive pressure breathing: fill mouth, close nostrils, raise jaw, and blow up lungs
    j. air
    k. lungs
    l. negative pressure breathing: lower diaphragm, raise ribs, thus increasing volume and decreasing pressure of lungs; air flows in

38.9 The dissociation curve for hemoglobin is shifted to the right at a lower pH, meaning that at any given partial pressure of O_2, hemoglobin is less saturated with oxygen. A rapidly metabolizing tissue produces more CO_2, which lowers pH, and hemoglobin will unload more of its O_2 to that tissue.

Suggested Answers to Structure Your Knowledge

1. a. aorta   **g.** inferior vena cava
    b. pulmonary artery   **h.** atrioventricular valve
    c. pulmonary vein   **i.** semilunar valve
    d. left atrium   **j.** right atrium
    e. left ventricle   **k.** superior vena cava
    f. right ventricle

Blood flow from vena cavae $\longrightarrow$ right atrium $\longrightarrow$ right ventricle $\longrightarrow$ pulmonary artery $\longrightarrow$ pulmonary vein $\longrightarrow$ left atrium $\longrightarrow$ left ventricle $\longrightarrow$ aorta

2. Nasal cavity $\longrightarrow$ pharynx $\longrightarrow$ through glottis to larynx $\longrightarrow$ trachea $\longrightarrow$ bronchus $\longrightarrow$ bronchiole $\longrightarrow$ alveoli $\longrightarrow$ interstitial fluid $\longrightarrow$ alveolar capillary $\longrightarrow$ hemoglobin molecule $\longrightarrow$ venule $\longrightarrow$ pulmonary vein $\longrightarrow$ left atrium $\longrightarrow$ left ventricle $\longrightarrow$ aorta $\longrightarrow$ renal artery $\longrightarrow$ arteriole $\longrightarrow$ capillary in kidney

Answers to Test Your Knowledge

Multiple Choice:

1. c	6. b	11. c	16. d	21. a
2. c	7. a	12. e	17. b	22. b
3. d	8. b	13. c	18. c	
4. e	9. a	14. a	19. c	
5. a	10. d	15. a	20. b	

CHAPTER 39
THE BODY'S DEFENSES

■ INTERACTIVE QUESTIONS

39.1 a. Short-lived white blood cells (WBCs) that phagocytize microbes.
    b. WBCs that migrate from blood to tissues, where they develop into macrophages.
    c. Long-lived, large amoeboid WBCs that engulf microbes.
    d. WBCs that attack large parasites with enzymes.
    e. WBCs that attack membranes of the body's infected or aberrant cells.
    f. Release histamine to initiate inflammatory response.
    g. Causes vasodilation and increased permeability of blood vessels.
    h. Proteins released by virus-infected cells that stimulate neighboring cells to produce proteins that inhibit viral replication.
    i. Set of at least 20 proteins that cause lysis of microbes and otherwise cooperate with nonspecific and specific defenses.
    j. Enzyme in tears, saliva, and mucus that attacks bacterial cell walls.
    k. Released by damaged tissue cells and WBCs; promote blood flow.
    l. Chemical released by WBCs that sets body temperature higher, initiating a fever.

39.2 a. Humoral immunity involves antibodies that are produced by B cells and act against freely circulating bacteria and viruses.
    b. Cell-mediated immunity involves T cells that act against body cells that have been infected by bacteria or viruses, against fungi, protozoa, and worms, and against transplanted tissues and cancerous cells. Both B and T cells have membrane-bound receptors that bind to specific antigens and activate the cell to form a clone of effector cells. Plasma cells produce antibodies; cytotoxic T cells kill infected or cancerous cells; and helper T cells secrete cytokines to stimulate B and T cells.

39.3 B or T cells would be activated when they contact class II MHC markers on macrophages and B and T cells, and, through clonal selection, would produce millions of effector cells that would destroy the body's immune system.

39.4 **a.** virus with T-dependent antigens
  b. class II MHC protein
  c. macrophage (APC)
  d. virus antigen fragment
  e. helper T cell
  f. T-cell receptor
  g. B cell
  h. cytokines
  i. virus antigen fragment
  j. class II MHC protein
  k. clone of memory B cells
  l. clone of plasma cells
  m. secreted antibodies

39.5 **a.** neutralization—blocks viral binding sites and coats bacterial toxins
  b. agglutination of particulate antigens
  c. precipitation of soluble antigens
  d. activation of complement to form lesion in cell membrane
  The first three mechanisms tag antigens for phagocytosis.

39.6 **a.** CD4
  b. CD8
  c. interleukin-2 and other cytokines that activate T_H, T_C, and B cells
  d. perforin that lyses the target cell

39.7 **a.** pathogen's membrane
  b. antibody
  c. antigen
  d. complement proteins
  e. membrane attack complex
  f. lesion in cell membrane
  1. Antibodies attach to antigens on pathogen's plasma membrane.
  2. Complement proteins form a bridge between two antibodies.
  3. Complement proteins form a membrane attack complex and attach to the pathogen's membrane.
  4. An area of the target cell's membrane is lysed, and the pathogen is destroyed.

39.8

Blood Type	Antigens on RBCs	Antibodies in Plasma	Can Receive Blood From	Can Donate Blood To
A	A	anti-B	A, O	A, AB
B	B	anti-A	B, O	B, AB
AB	AB	none	A, B, AB, O	AB
O	none	anti-A + B	O	A, B, AB, O

39.9 **a.** AIDS destroys helper T cells, thus crippling the humoral and cell-mediated immune systems and leaving the body unable to fight the AIDS virus and other opportunistic diseases such as Kaposi's sarcoma and *Pneumocystis* pneumonia. While modern medicine has numerous antibiotics for fighting bacterial infections, antiviral drugs are not well developed.
  b. HIV can remain hidden as a provirus for years, and infected individuals can unknowingly pass the virus on through unscreened blood donations, unprotected sex, and needle sharing. The antigenic changes of HIV during replication impair the ability of the immune system to mount an effective defense and greatly complicate the development of a vaccine against HIV. Newly developed drugs may slow the development of the virus or treat the opportunistic infections, but no cure has been found.

Suggested Answers to Structure Your Knowledge

  1. **a.** Set of proteins that circulate in blood.
    b. Cooperates with specific defense mechanisms in classical pathway to form membrane attack complex that lyses cells; cooperates with nonspecific defenses in various alternative pathways, including opsonization, histamine release, immune adherence, and chemotaxis.
    c. Immunoglobulins made by plasma cells derived from B cells, circulate in blood and lymph, form antigen receptors on B cells, attach as receptors to mast cells and basophils.
    d. Bind to specific antigen, mark foreign cells and molecules for destruction, neutralize virus or toxin.
    e. Cytokines secreted by antigen-presenting macrophage (APC) after binding with helper T cell.
    f. Stimulate helper T cell to release interleukin-2.
    g. Cytokine released by activated T_H cell.

h. Activates T_H, T_C, and B cells.

i. Released by T_C cells when attached to cells displaying class I MHC–antigen complexes.

j. Forms hole in target cell's membrane, causing it to lyse.

k. Major histocompatibility complex, set of glycoproteins on plasma membrane of nucleated body cells.

l. Presents fragments of antigens, recognized by T_C cells, CD8 helps interaction.

m. MHC molecules found on macrophages, B and activated T cells.

n. Presents fragments of antigens, recognized by T_H-cell receptor, CD4 helps interaction.

2.

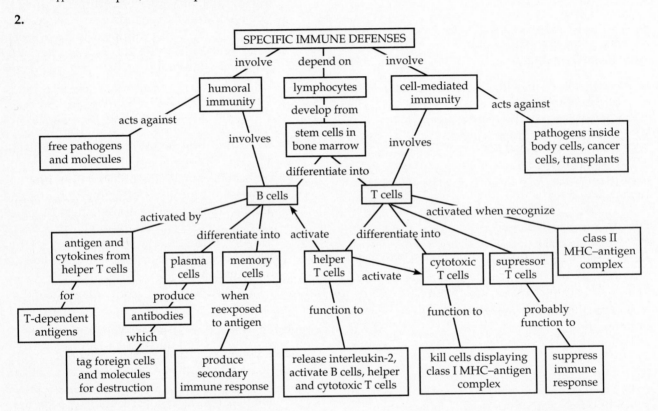

3. An antibody is a protein that typically consists of two identical light polypeptide chains and two identical heavy chains, held together by disulfide bonds in a Y-shaped molecule. The amino acid sequences in the variable sections of the light and heavy chains in the arms of the Y account for the specificity in binding between antibodies and antigenic determinants. The constant region of the antibody determines its effector function: five types of these constant regions correspond to five classes of antibodies. As an example, IgE immunoglobulins are attached to mast cells, which create allergic responses, and IgA is found in body secretions and is the major antibody in colostrum, the first breast milk produced.

4. According to the clonal selection theory, T and B cells become fixed early in their development to produce a specific antigen receptor or antibody. The body's ability to respond to a great variety of antigens depends on a lymphocyte population with a huge diversity of receptor specificities. When a T or B cell encounters its antigen, it is selectively activated to proliferate and produce a clone of lymphocytes, all having the same antigenic specificity.

Answers to Test Your Knowledge

Multiple Choice:

1. d	4. e	7. a	10. b	13. d
2. c	5. b	8. c	11. a	14. c
3. e	6. e	9. a	12. c	15. e

CHAPTER 40
CONTROLLING THE INTERNAL ENVIRONMENT

■ INTERACTIVE QUESTIONS

40.1
  a. conformer   **b.** isosmotic
  c. regulator   **d.** slightly hyperosmotic
  e. regulator   **f.** hypoosmotic
  g. regulator   **h.** hyperosmotic
  i. regulator   **j.** hyperosmotic
  k. regulator   **l.** hyperosmotic

40.2
  a. osmoregulation; remove excess water
  b. both; remove excess water
  c. both; conserve water

40.3
  a. water, salts, nitrogenous wastes, glucose, vitamins, and other small molecules
  b. blood cells and large molecules such as plasma proteins

40.4
  a. Bowman's capsule
  b. glomerulus
  c. proximal tubule
  d. descending limb, loop of Henle
  e. ascending limb, loop of Henle
  f. distal tubule
  g. collecting duct

- HCO_3^-, NaCl*, H_2O, nutrients*, and K^+ are reabsorbed from the proximal tubule. H^{+*} and NH_3 are secreted into the proximal tubule.
- Water is reabsorbed from the descending limb of the loop of Henle.
- NaCl diffuses from the thin segment of the ascending limb and is pumped out* of the thick segment.
- NaCl*, H_2O, and HCO_3^{-*} are reabsorbed from the distal tubule. K^{+*} and H^{+*} are secreted into the tubule.
- Urea and H_2O diffuse out of the collecting tubule, NaCl* is reabsorbed.
  (* indicates active transport.)

40.5
  a. high blood osmolarity
  b. ADH, antidiuretic hormone
  c. distal tubules and collecting ducts
  d. JGA, juxtaglomerular apparatus
  e. renin
  f. NaCl and water reabsorption
  g. adrenal glands
  h. aldosterone
  i. NaCl and water reabsorption
  j. increase in blood pressure or volume
  k. ANP, atrial natriuretic protein

40.6
  a. Juxtamedullary nephrons; the longer the loop of Henle, the more hyperosmotic the urine.
  b. Mammals and birds; loops of Henle are especially long in those taxa inhabiting dry habitats.

40.7 *Uric acid* can be stored safely within the egg while the embryo develops. *Urea* can be removed by the placental blood.

40.8 Internal temperature control mechanisms may be changed. Increased production of enzymes may offset lower enzyme activity at nonoptimal temperatures. Variants of enzymes with different optimal temperatures may also be produced. Changes in the proportions of saturated and unsaturated lipids keep membranes fluid at lower temperatures. Stress-induced proteins may maintain the integrity of other proteins.

Suggested Answers to Structure Your Knowledge

1.
  a. Isosmotic to seawater, osmoconformer; diffusion of ammonia through body wall; ectothermic.
  b. Drink water to compensate for loss to hyperosmotic seawater, salt glands in gills pump out salt; ammonia diffuses through gills, little urine excreted; ectothermic, may have countercurrent heat exchanger.
  c. Copious dilute urine to compensate for osmotic gain, may pump salts in through gills; ammonia lost across epithelium of gills; ectothermic.
  d. Protonephridia, flame-bulb system, dilute fluid excreted; metabolic wastes lost through body wall or excreted into gastrovascular cavity; ectothermic.
  e. Metanephridia, excrete dilute urine to offset osmosis; metanephridia remove wastes from coelomic fluid; ectothermic.
  f. Malpighian tubules, salt and most water reabsorbed; Malpighian tubules remove wastes from hemolymph, uric acid conserves water; ectothermic, behavioral responses, some endothermic.
  g. Nephron in kidney, no loop of Henle, uric acid conserves water, related to shelled egg; ectothermic, behavioral responses.

h. Nephron in kidney, nasal salt glands in sea birds, uric acid conserves water, related to shelled egg; juxtamedullary nephrons; endothermic.

i. Nephron in kidney, concentration of urine related to habitat; juxtamedullary nephrons allow concentration of urine, urea; endothermic

2. a. water, salts, glucose, amino acids, vitamins, urea, other small molecules

   b. ammonia, drugs and poisons, hydrogen ions, K^+

   c. water, glucose, amino acids, vitamins, K^+, NaCl, bicarbonate

Answers to Test Your Knowledge

Multiple Choice:

1. a	5. d	9. d	13. c
2. c	6. a	10. b	14. a
3. c	7. c	11. e	15. e
4. b	8. e	12. d	

CHAPTER 41
CHEMICAL SIGNALS IN ANIMALS

■ INTERACTIVE QUESTIONS

41.1 a. Chemical messages produced by endocrine or neurosecretory cells that usually travel through the bloodstream to target cells.

   b. Volatile chemical signals that communicate between individuals of the same species.

   c. Chemical messages, such as neurotransmitters, histamines, NO, growth factors, and prostaglandins, that communicate between cells.

   d. Local regulator that is a transmitter released by a neuron into a synapse with a target cell.

   e. Local regulators released by a cell into the interstitial fluid that affect nearby cells.

41.2 a. hormone (first messenger)
   b. receptor protein
   c. G protein (relay)
   d. adenylyl cyclase (effector)
   e. GTP
   f. ATP
   g. cAMP (second messenger)

41.3 a. hormone (first messenger)
   b. receptor protein
   c. G protein (relay)
   d. phospholipase C (effector)
   e. diacylglycerol (second messenger)
   f. protein kinase C
   g. GTP
   h. IP_3 (second messenger)
   i. endoplasmic reticulum
   j. Ca^{2+} (third messenger)
   k. Ca^{2+} calmodulin

41.4 Oxytocin and ADH are synthesized by neurosecretory cells and transported down the axons to the posterior pituitary. Releasing and inhibiting hormones are secreted by neurosecretory cells into capillaries that drain through portal vessels to a capillary bed in the anterior pituitary.

41.5 a. stimulates contraction of uterus and milk release

   b. promotes reabsorption of water by kidney collecting ducts

   c. promotes growth and the release of growth factors

   d. various effects, depending on species, including stimulation of milk production and secretion in mammals

   e. stimulates thyroid gland to release hormones

   f. stimulates gonads, production of ova and sperm

   g. stimulates adrenal cortex to produce and secrete corticosteroids

   h. regulates skin pigment cells in amphibians

   i. reduce perception of pain

41.6 a. insulin
   b. glucagon
   c. movement of glucose
   d. glycogen hydrolysis in liver
   e. islets of Langerhans in pancreas
   f. glycogen hydrolysis in liver
   g. blood glucose level

41.7 a. Nervous stimulation of the adrenal medulla causes the secretion of the catecholamine hormones epinephrine and norepinephrine, which increase blood pressure, rate and volume of heart beat, breathing rate, and metabolic rate, stimulate glycogen and fatty acid hydrolysis, and change blood flow patterns.

b. ACTH released from the pituitary stimulates the adrenal cortex to release corticosteroids. Glucocorticoids increase blood glucose through conversion of proteins and fats. Mineralocorticoids increase blood volume and pressure by stimulating the kidney to reabsorb sodium ions and water.

Suggested Answers to Structure Your Knowledge

1.

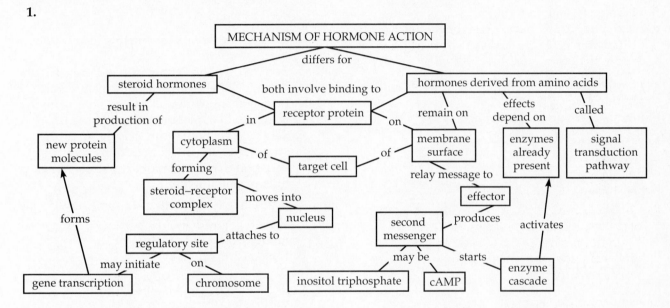

2. Calcitonin is secreted by the thyroid in response to a rise in Ca^{2+} levels above a set point. Its effects are to stimulate Ca^{2+} deposit in bones and reduce Ca^{2+} uptake in the intestines and kidneys. As Ca^{2+} levels in the blood fall, PTH (parathyroid hormone) is secreted by the parathyroid glands and reverses the effects of calcitonin.

Answers to Test Your Knowledge

Matching:

1. I f	**5. A** g	**9. C** c, released from **g**
2. G a	**6. E** b	**10. J** c, released from **g**
3. F d	**7. K** j	
4. B h	**8. D** j	

Multiple Choice:

1. e	**4.** b	**7.** c	**10.** b	**13.** a
2. d	**5.** d	**8.** b	**11.** b	**14.** e
3. d	**6.** a	**9.** e	**12.** c	**15.** c

CHAPTER 42
ANIMAL REPRODUCTION

■ INTERACTIVE QUESTIONS

42.1 a. Rapid colonization of new areas and perpetuation of successful genotypes in stable habitats; eliminates need to locate mate.
b. Varying genotypes and phenotypes of offspring may enhance reproductive success of parents in fluctuating environments.

42.2 a. environmental signals
b. pheromonal communication
c. courtship behaviors that trigger the release of gametes and may provide a means for mate selection

42.3 a. mouth, anus, urethra for both excretory and reproductive systems
b. mouth, anus, urethra, and vagina

42.4 a. vas deferens
b. erectile tissue of penis
c. urethra
d. glans penis
e. prepuce
f. scrotum
g. testis
h. epididymis
i. bulbourethral gland
j. prostate gland
k. seminal vesicle
l. oviduct
m. eggs within follicles
n. corpus luteum
o. uterine wall (muscular layer)
p. endometrium (lining of uterus)
q. cervix
r. vagina
s. ovary

42.5 a. Spermatogonia continue to divide by mitosis and spermatogenesis occurs continuously, whereas a female's supply of primary oocytes is present at birth.
b. Each meiotic division produces four sperm, but only one ovum (and polar bodies that disintegrate).
c. Spermatogenesis is an uninterrupted process. Oogenesis occurs in stages: prophase I between birth and puberty, meiosis I in a maturing follicle just before ovulation, and meiosis II (in humans) when a sperm cell penetrates the egg.

42.6 a. LH
b. FSH
c. estrogens
d. progesterone
e. follicular phase
f. ovulatory phase
g. luteal phase
h. menstrual flow phase
i. proliferative phase
j. secretory phase

42.7 a. Secreted by embryo, maintains corpus luteum in first trimester.
b. Secreted by corpus luteum and later by placenta, regulates growth and maintenance of placenta, formation of mucus plug in cervix, cessation of ovulation, growth of uterus.
c. Secreted by fetus and posterior pituitary, stimulates uterine contractions and uterine secretion of prostaglandins; controls release of milk during nursing.

42.8 a. Prevent fertilization—abstinence, rhythm method, condom, diaphragm, progestin minipill, *coitus interruptus.*
b. Prevent implantation of embryo—IUD, RU-486.
c. Prevent release of gametes from gonads—sterilization, birth control pills.
The most effective methods are abstinence, sterilization, chemical contraception. Least effective are the rhythm method and *coitus interruptus.*

Suggested Answers to Structure Your Knowledge

1. Path of Sperm from Formation to Fertilization

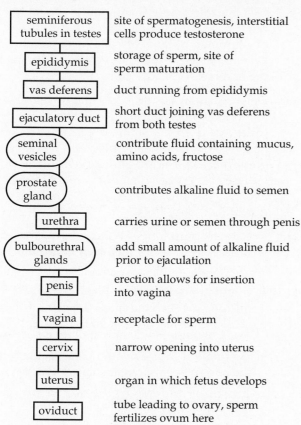

seminiferous tubules in testes	site of spermatogenesis, interstitial cells produce testosterone
epididymis	storage of sperm, site of sperm maturation
vas deferens	duct running from epididymis
ejaculatory duct	short duct joining vas deferens from both testes
seminal vesicles	contribute fluid containing mucus, amino acids, fructose
prostate gland	contributes alkaline fluid to semen
urethra	carries urine or semen through penis
bulbourethral glands	add small amount of alkaline fluid prior to ejaculation
penis	erection allows for insertion into vagina
vagina	receptacle for sperm
cervix	narrow opening into uterus
uterus	organ in which fetus develops
oviduct	tube leading to ovary, sperm fertilizes ovum here

2. a. GnRH is a releasing hormone of the hypothalamus that stimulates the pituitary to secrete FSH and LH.
   b. FSH stimulates growth of follicles.
   c. The spike in LH level is caused by an increase in GnRH production that resulted from increasing estrogen levels produced by the developing follicle.

d. The high level of LH induces the maturation of the follicle and ovulation, and it transforms the ruptured follicle to corpus luteum.
e. LH maintains the corpus luteum, which secretes estrogen and progesterone.
f. High levels of estrogen and progesterone act on the hypothalamus and pituitary to inhibit the secretion of LH and FSH.
g. The corpus luteum disintegrates without LH to maintain it. Estrogen and progesterone production by the corpus luteum ceases, which allows LH and FSH secretion to begin again.

3. Birth control pills, which contain a combination of synthetic estrogen and progestin, act by negative feedback to prevent the release of GnRH by the hypothalamus and FSH and LH by the pituitary, thus preventing the development of follicles and ovulation. RU-486 is an analog of progesterone that blocks progesterone receptors in the uterus, thus preventing progesterone from maintaining pregnancy.

Answers to Test Your Knowledge

Fill in the Blanks:

1. budding
2. protandrous
3. parthenogenesis
4. hermaphrodite
5. cloaca
6. estrous cycle
7. progesterone
8. urethra
9. vasocongestion
10. puberty

Multiple Choice:

1. e	3. a	5. d	7. e	9. b
2. d	4. c	6. c	8. a	10. b

CHAPTER 43
ANIMAL DEVELOPMENT

▧ INTERACTIVE QUESTIONS

43.1 a. Specific binding between receptors on vitelline layer and protein on acrosomal process.
   b. Fast block to polyspermy caused by membrane depolarization during acrosomal reaction and slow block to polyspermy caused by formation of the fertilization membrane during cortical reaction.

43.2 a. syncytium of fruit fly
   b. morula of sea urchin

   c. blastula of frog
   d. animal pole
   e. blastocoel
   f. vegetal pole

43.3 a. ectoderm
   b. mesenchyme
   c. endoderm
   d. archenteron
   e. blastopore
   f. archenteron
   g. ectoderm

h. mesoderm
i. endoderm
j. dorsal lip
k. yolk plug

43.4 a. neural tube
b. neural crest
c. somite
d. archenteron
e. coelom
f. notochord

43.5 a. The **yolk sac** grows to enclose the yolk and develops blood vessels to carry nutrients to the embryo.
b. The **amnion** encloses a fluid-filled sac, which provides an aqueous environment for development and acts as a shock absorber.
c. The **allantois,** growing out of the hindgut, serves as a receptacle for the nitrogenous waste, uric acid.
d. The vascularized allantois, in conjunction with the **chorion,** provides gas exchange for the developing embryo.

43.6 a. epiblast—three primary germ layers, amnion, some mesoderm becomes part of placenta, allantois
b. hypoblast—yolk sac

c. trophoblast—chorion, fetal portion of placenta

43.7 Cytoplasmic determinants in the sea urchin egg must have an unequal polar distribution. The developmental fates of cells in the animal and vegetal halves of the embryo are determined by the first horizontal division.

43.8 Determination occurs when the developmental fate of an embryonic cell is set as a result of the influence of cytoplasmic determinants on the genome. Developmental potential becomes progressively determined until a cell finally differentiates by producing tissue-specific proteins and developing characteristic structures and functions.

43.9 The developing limb would be a mirror image with posterior digits facing both forward and backward.

43.10 a. A mutation in the *bicoid* gene so that no bicoid mRNA was secreted into the egg and thus no bicoid protein was available to establish the anterior end.
b. A mutation in a pair-rule gene so that either the odd or even segments failed to develop.

Suggested Answers to Structure Your Knowledge

1.

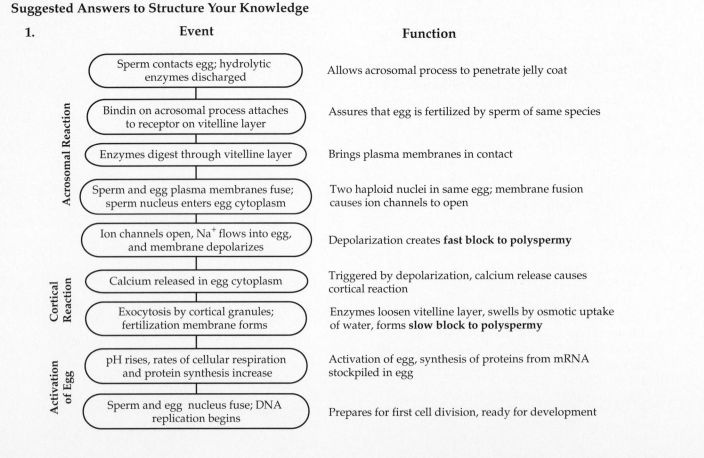

Event	Function
Sperm contacts egg; hydrolytic enzymes discharged	Allows acrosomal process to penetrate jelly coat
Bindin on acrosomal process attaches to receptor on vitelline layer	Assures that egg is fertilized by sperm of same species
Enzymes digest through vitelline layer	Brings plasma membranes in contact
Sperm and egg plasma membranes fuse; sperm nucleus enters egg cytoplasm	Two haploid nuclei in same egg; membrane fusion causes ion channels to open
Ion channels open, Na+ flows into egg, and membrane depolarizes	Depolarization creates **fast block to polyspermy**
Calcium released in egg cytoplasm	Triggered by depolarization, calcium release causes cortical reaction
Exocytosis by cortical granules; fertilization membrane forms	Enzymes loosen vitelline layer, swells by osmotic uptake of water, forms **slow block to polyspermy**
pH rises, rates of cellular respiration and protein synthesis increase	Activation of egg, synthesis of proteins from mRNA stockpiled in egg
Sperm and egg nucleus fuse; DNA replication begins	Prepares for first cell division, ready for development

Acrosomal Reaction

Cortical Reaction

Activation of Egg

2.

	Cleavage	Blastula	Gastrula
Sea urchin	Radial, holoblastic; equal blastomeres	Hollow ball of cells, one cell thick, surrounding blastocoel	Invagination through blastopore to form endoderm of archenteron, migrating mesenchyme forms mesoderm
Frog	Radial, holoblastic, but yolk impedes divisions; unequal blastomeres	Blastocoel in animal hemisphere; walls several cells thick	Involution of cells at dorsal lip of blastopore, migrating cells form mesoderm and endoderm around archenteron
Bird	Meroblastic, only in cytoplasmic disc on top of yolk	Blastodisc; blastocoel cavity between epiblast and hypoblast	Ingression of epiblast cells at primitive streak forms mesoderm and endoderm, lateral folds join to form archenteron
Mammal	Holoblastic, no polarity to egg, blastomeres equal	Embryonic disc forms from inner cell mass with epiblast and hypoblast; trophoblast surrounds embryo	Migration of epiblast cells through primitive streak forms mesoderm and endoderm

3.

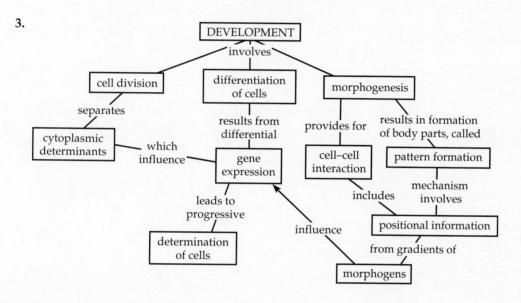

Answers to Test Your Knowledge

Multiple Choice:

1. c	5. e	9. c	13. a
2. c	6. c	10. c	14. e
3. b	7. b	11. b	15. b
4. e	8. a	12. e	

CHAPTER 44

NERVOUS SYSTEMS

■ INTERACTIVE QUESTIONS

44.1 a. dendrites
b. cell body
c. axon hillock
d. axon
e. myelin sheath
f. Schwann cell
g. nodes of Ranvier
h. synaptic terminal
The impulse moves from the dendrites to the cell body and out the axon.

44.2 a. sodium channel
b. Na$^+$
c. potassium channel
d. K$^+$

e. sodium activation gate
f. sodium inactivation gate
g. membrane potential (mV)
h. time

1. Resting state: sodium activation gates and potassium channels are closed, sodium inactivation gates are open.
2. Depolarizing phase: sodium activation gates open and Na$^+$ enters cell, membrane is depolarized; voltage-gated K$^+$ channels are slowly beginning to open.
3. Repolarizing phase: sodium inactivation gates close, potassium gates completely open, potassium leaves cell and membrane repolarizes.
4. Undershoot: both gates of the sodium channels are closed, potassium channels still open, and membrane becomes hyperpolarized.

44.3 **a.** Ca^{2+}
b. synaptic terminal
c. synaptic vesicle
d. presynaptic membrane
e. postsynaptic membrane
f. receptor with bound neurotransmitter
g. ion channel
h. synaptic cleft

44.4 **a.** temporal
b. spatial

44.5 The type of postsynaptic receptor and its mode of action determine neurotransmitter function. Binding of acetylcholine to receptors in heart muscle activates a signal transduction pathway that makes it harder to generate an action potential, thus slowing the strength and rate of contraction. The acetylcholine receptor in skeletal muscle cells opens ion channels and has a stimulatory effect.

44.6 **a.** giant axons of squid
b. the marine snail *Aplysia*

44.7 **a.** peripheral nervous system
b. sensory/afferent
c. motor/efferent
d. effector muscles, glands, and organs
e. autonomic
f. parasympathetic
g. slows heart, conserves energy
h. sympathetic
i. somatic
j. skeletal muscles
k. central nervous system
l. spinal cord

44.8 **a.** cerebral cortex   5
b. thalamus   3
c. cerebellum   1
d. medulla oblongata   4
e. pons   2
f. midbrain   7
g. hypothalamus   6

44.9 Fact memories can be consciously recalled; skill memories are formed by repetition of motor activities and do not require conscious recall of specific information.

Suggested Answers to Structure Your Knowledge

1.

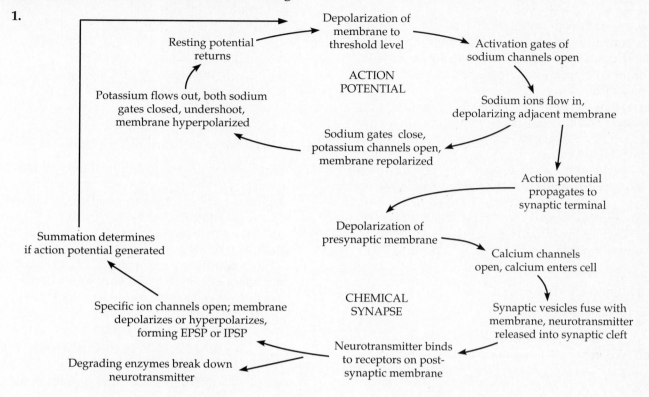

Depolarization of
membrane to
threshold level

Resting potential
returns

Activation gates of
sodium channels open

ACTION
POTENTIAL

Potassium flows out, both sodium
gates closed, undershoot,
membrane hyperpolarized

Sodium ions flow in,
depolarizing adjacent membrane

Sodium gates close,
potassium channels open,
membrane repolarized

Action potential
propagates to
synaptic terminal

Summation determines
if action potential generated

Depolarization of
presynaptic membrane

Calcium channels
open, calcium enters cell

CHEMICAL
SYNAPSE

Specific ion channels open; membrane
depolarizes or hyperpolarizes,
forming EPSP or IPSP

Synaptic vesicles fuse with
membrane, neurotransmitter
released into synaptic cleft

Degrading enzymes break down
neurotransmitter

Neurotransmitter binds
to receptors on post-
synaptic membrane

2. a. Nuclei found in medulla, pons, and mid-brain; regulates state of arousal; filters all sensory information going to cortex.

b. In hypothalamus; maintains biorhythms, functions as biological clock.

c. Cluster of nuclei deep in white matter of cerebrum; centers for motor coordination and motor signal relay.

d. Nuclei and connecting axon tracts in forebrain, also in parts of thalamus, hypothalamus, and inner cerebral cortex; centers of human emotions.

e. Two nuclei in inner cerebral cortex; involved in limbic system and memory; amygdala links information to be stored to an event or emotion; hippocampus may use long-term potentiation for learning and memory storage.

Answers to Test Your Knowledge

Multiple Choice:

1. b	5. e	9. d	13. d
2. a	6. c	10. b	14. b
3. a	7. e	11. a	15. d
4. d	8. c	12. e	16. a

CHAPTER 45
SENSORY AND MOTOR MECHANISMS

■ INTERACTIVE QUESTIONS

45.1 a. reception
b. transduction
c. amplification
d. transmission
e. integration

45.2 a. sclera
b. choroid
c. retina

d. fovea
e. optic disc (blind spot)
f. optic nerve
g. vitreous humor
h. lens
i. aqueous humor
j. pupil
k. cornea
l. iris
m. suspensory ligament
n. ciliary body

45.3
a. inactive
b. active
c. bound to sodium channels
d. converted by signal transduction to GMP, dissociates from sodium channels
e. open
f. closed
g. depolarized
h. hyperpolarized
i. inhibitory, released
j. not released
k. inhibited from firing
l. not inhibited from firing

45.4
a. small
b. fovea

45.5
a. auditory canal
b. malleus
c. incus
d. stapes
e. semicircular canals
f. auditory nerve
g. cochlea
h. round window
i. eustachian tube
j. oval window
k. tympanic membrane

45.6
a. utricle and saccule for position, semicircular canals for movement
b. lateral line system
c. statocysts
d. The balance organs of mammals have otoliths and those of invertebrates contain statoliths. These granules settle by gravity and stimulate hair cells to signal position and movement. Sensory hairs on the hair cells in the neuromasts of fishes are embedded in a gelatinous cap which is bent by moving water and stimulates the hair cells. All of these are mechanoreceptors.

45.7 find mates, recognize territory, navigate, communicate, and locate and taste food

45.8
a. swimming
b. running
c. flying

45.9
a. No extra materials needed, good shock absorber; not much protection or support for lifting animal off the ground.
b. Quite protective; in arthropods, must be molted in order to grow.
c. Various types of joints allow for flexible movement in vertebrates, good structural support; not very protective.

45.10
a. sarcomere
b. I band
c. A band
d. Z line
e. thin filament (actin)
f. thick filament (myosin)
g. H zone

Suggested Answers to Structure Your Knowledge

1. Light enters the eye through the pupil and is focused by the lens onto the retina. When a rhodopsin or photopsin (photopigments in rods and cones) absorbs light energy, it sets off a signal transduction pathway that decreases the receptor cell's permeability to sodium and hyperpolarizes the membrane. The reduction in the release of inhibitory neurotransmitter by a rod or cone cell causes connected bipolar cells to change their release of neurotransmitter, which in turn generates action potentials in ganglion cells—the sensory neurons that form the optic nerve. Horizontal and amacrine cells provide lateral integration of visual information.

2. Sound waves are collected by the *pinna* and travel down the *auditory canal* to the *tympanic membrane*, where they are amplified and transmitted to the *malleus, incus,* and *stapes*. Vibration of the stapes against the *oval window* sets up pressure waves in the fluid in the *vestibular canal* within the *cochlea*, from which they are transmitted to the *tympanic canal* and then dissipated when they strike the *round window*. The pressure waves vibrate the *basilar membrane*, on which the *organ of Corti* is located within the *cochlear duct*. Tips of the hair cells arising from the organ of Corti are embedded in the *tectorial membrane*. When vibrated, they bend, triggering a depolarization, release of neurotransmitter, and initiation of action potentials in the sensory neurons of the *auditory nerve*, which leads to the *cerebral cortex*.

3. A motor neuron releases acetylcholine into the neuromuscular junction, initiating an action potential within the muscle fiber, which spreads into the cell through the transverse tubules. The action potential changes the permeability of the sarcoplasmic reticulum membrane, which releases calcium ions into the cytoplasm. The calcium binds with troponin, changes the shape of the tropomyosin–troponin complex, and exposes the myosin-binding sites of the actin molecules. Heads of myosin molecules in their high-energy configuration bind to these sites, forming cross-bridges. Myosin relaxes and its head bends, pulling the thin filament toward the center of the sarcomere. When myosin binds to an ATP, the cross-bridge breaks. Hydrolysis of the ATP returns the myosin to its high-energy configuration, and it binds further along the actin molecule. This sequence continues as long as there is ATP and until calcium is pumped

back into the sarcoplasmic reticulum, when the tropomyosin–troponin complex again blocks the actin sites.

Answers to Test Your Knowledge

Matching:

1. I	4. L	7. M	10. K
2. G	5. J	8. D	11. B
3. N	6. Q	9. A	12. P

Multiple Choice:

1. a	6. d	11. a	16. b
2. e	7. c	12. d	17. a
3. c	8. a	13. b	18. c
4. b	9. e	14. e	
5. d	10. b	15. a	

CHAPTER 46

AN INTRODUCTION TO ECOLOGY: DISTRIBUTION AND ADAPTATIONS OF ORGANISMS

■ INTERACTIVE QUESTIONS

46.1 a. Organism: physiological, behavioral, and morphological adaptations to the abiotic environment.
b. Population: growth and regulation of population size.
c. Community: interactions such as predator–prey and competition that regulate populations in the community.
d. Ecosystem: energy flow and chemical cycling.

46.2 a. South-facing slopes in the northern hemisphere are warmer and drier than north-facing slopes.
b. Air temperature drops with an increase in elevation, and high-altitude communities may be similar to communities in higher latitudes.
c. The windward side of a mountain range receives much more rainfall than the leeward side. The warm, moist air rising over the mountain releases moisture, and the drier, cooler air absorbs moisture as it descends the other side.

46.3 a. Regulators are able to tolerate more environmental variation and may have wider distributions, but their regulating mechanisms are energetically expensive.

b. A conformer spends little energy on homeostatic regulation, leaving more energy to devote to growth and reproduction.
c. Acclimation is a substantial though reversible physiological adaptation to environmental change.

46.4 a. desert **d.** temperate forest
b. grassland **e.** coniferous forest
c. tropical forest **f.** arctic and alpine tundra

46.5 a. + - -
b. - + +
c. + - -
d. - + -
e. + + +

46.6 a. 6
b. 7
c. 8
d. 1
e. 2
f. 3
g. 5
h. 4

Suggested Answers to Structure Your Knowledge

1. a. Ecology is the study of the relationships of organisms to their abiotic and biotic environments.

   b. Experimental manipulation of variables is done both in the laboratory and in the field, along with observational measurements of various physical factors and kinds and numbers of organisms. Mathematical models and computer simulations are useful for predicting the effects of variables, especially in situations where experimentation is impractical or impossible.

   c. Evolution, with the principles of natural selection and adaptation, is the key organizing theory applied to ecological study.

2. a. Biomes are characteristic communities, usually identified by the predominant vegetation, that range over broad geographical areas and result from abiotic and biotic interactions.

   b. Convergent evolution, common adaptations of different organisms to similar environments, accounts for similarities in life forms within geographically separated biomes.

Answers to Test Your Knowledge

Multiple Choice:

1. b	5. b	9. c	13. b
2. c	6. c	10. b	14. d
3. d	7. b	11. c	
4. a	8. e	12. e	

Matching:

1. C	4. E	7. B
2. H	5. A	8. F
3. D	6. G	

CHAPTER 47
POPULATION ECOLOGY

■ INTERACTIVE QUESTIONS

47.1 If 10 of the 30 voles captured in the second trapping were marked, she would estimate that $1/3$ of the vole population is marked. Since 25 voles were initially marked, the total population size is estimated at 75 ($25/0.333$). Another method is to use the formula: $N =$ (number marked × total second catch)/number marked recaptures.

47.2 a. Type I, humans and many large mammals
   b. Type II, some annual plants, *Hydra* and various invertebrates, some lizards and rodents
   c. Type III, many fishes and marine invertebrates such as oysters

47.3 a. positive (increasing)
   b. negative (decreasing)
   c. negative
   d. negative

47.4 a. exponential growth; $dN/dt = rN$
   b. logistic growth; $dN/dt = rN(K-N)/K$; K is 1500

47.5 a. nutrients, space for nests, accumulation of toxic wastes, predation, intrinsic limiting factors
   b. extremes in weather, natural disasters, fires

Suggested Answers to Structure Your Knowledge

1.

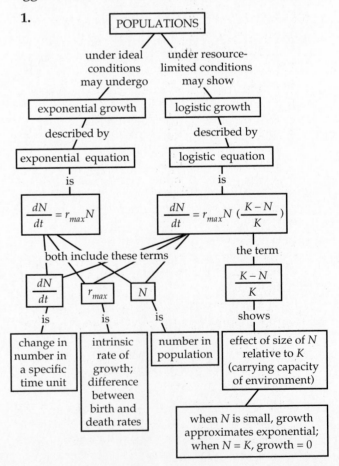

2. The age structure of a population affects population growth due to age-specific fecundity, or the differential reproductive capabilities of various age groups. The larger the proportion of the population in age groups with high birth rates, the greater the rate at which the population will grow. Population growth rate is likely to continue to be high if the age structure is skewed toward younger age groups that have not yet reached the age of greatest fecundity. The shorter the generation time (age at which reproduction first begins), the greater the growth rate. The sex structure of a population may affect growth rate, depending on the mating patterns

of the population. Survivorship influences growth rate in that the more individuals that live and reproduce, the more the population will grow. A population in which survival of offspring is usually low may show large fluctuations in population size depending on environmental conditions.

Answers to Test Your Knowledge

Multiple Choice:

1. a	4. e	7. c	10. e	13. a	16. c
2. c	5. a	8. d	11. e	14. b	17. a
3. b	6. d	9. a	12. b	15. a	18. d

CHAPTER 48
COMMUNITY ECOLOGY

■ INTERACTIVE QUESTIONS

48.1 Some key environmental factor, such as soil composition, may have changed abruptly.

48.2 a. Batesian mimicry
b. Müllerian mimicry

48.3 a. mutualism: mycorrhizae, flowering plants and pollinators, ants on acacia trees, protozoa in termites and ruminants
b. commensalism: cattle egrets and cattle that flush insects
c. predation and parasitism: animal predators, herbivores, endo- and ectoparasites
d. competition: resource partitioning by *Anolis* species, character displacement of finch beaks, interference competition of barnacle species

48.4 a. community A
b. community B

48.5 a. Species A
b. Species A was a keystone predator, keeping the population of C down so that B could compete. Alternatively, species A kept the population of B down; when A was removed, population B exploded, overutilized its resources, and died out.
c. The heterogeneous habitat allowed enough of the prey species to find refuge from the predator that all three species could coexist. In a simple habitat, the predator easily found the prey. As the predator population grew, it wiped out both species, and then starved.

48.6 The more disturbed community may be restricted to good colonizers typical of early successional stages, whereas in the less disturbed community, late-successional and highly competitive species would be dominant.

48.7 a. immigration—large island
b. immigration—small island
c. extinction—small island
d. extinction—large island
The small island equilibrium number is projected down from the intersection of lines b and c. The large island number is projected down from the intersection of a and d.

Suggested Answers to Structure Your Knowledge

1. a. species diversity
b. stability
c. trophic structure
d. competition
e. resource partitioning and character displacement
f. slightly different niches
g. predation
h. keystone predator
i. parasitism

2. Succession is the gradual transition in species composition in a community following a disturbance or creation of a new habitat. Early pioneers in the habitat may also inhibit other species until more competitive species are able to become established. Each stage alters the environment somewhat and may facilitate the transition to the next stage. Chance events, such as dispersion and disturbances, are important factors during succession. There is evidence that even communities that appear to be in final successional stages are in continual nonequilibrium, with species composition and diversity still changing.

CHAPTER 49
ECOSYSTEMS

■ INTERACTIVE QUESTIONS

49.1 a. primary producers (autotrophs): plants, phytoplankton
b. primary consumers: herbivores, zooplankton
c. secondary consumers: carnivores
d. tertiary consumers: carnivores
e. decomposers
A food chain shows the linear set of trophic levels in an ecosystem. A food web shows the complex feeding relationships in an ecosystem, in which higher level consumers may feed at several trophic levels and omnivores eat both producers and consumers.

49.2 Upwellings in these cold seas bring nitrogen and phosphorus to the surface. The lack of these nutrients limits productivity in tropical waters.

49.3 a. solar radiation; respiratory (metabolic) heat loss
b. $1/1000$; approximately 10 percent of the energy from one level gets passed to the next

49.4 a. nitrogen-fixing bacteria in root nodules
b. nitrogen-fixing soil bacteria
c. decomposers
d. ammonium (NH_4^+)
e. nitrifying bacteria
f. nitrates (NO_3^-)
g. denitrifying bacteria

Answers to Test Your Knowledge

Multiple Choice:

1. c	5. b	9. d	13. a
2. d	6. e	10. c	14. b
3. b	7. a	11. a	15. a
4. e	8. b	12. d	16. c

49.5 a. Tropical rain forests, because decomposition proceeds rapidly in the warm, wet climate.
b. Lakes and oceans; lack of oxygen slows decomposition and, unless there are upwellings, nutrients are not available.
c. Nutrients are not cycled and may leave the ecosystem through runoff.

49.6 Melting of ice caps and flooding of coastlines; change in precipitation patterns such that many currently important agricultural areas may become arid; changes in species composition or increased extinction rate as temperatures rise faster than communities can adjust.

49.7 a. habitat destruction
b. Biodiversity is a natural resource from which we obtain medicines, crops, fibers, and other products; many potentially valuable species will become extinct before they are known to scientists. Humans have evolved within the context of living communities, and it is unknown how well we will be able to adapt to changes in our ecosystem.

Suggested Answers to Structure Your Knowledge

1.

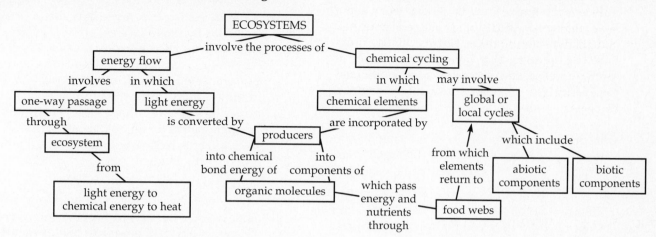

2. **a.** Deforestation can cause an increase in water runoff with an accompanying loss of soil and minerals. Clear-cutting of tropical forests results in loss of species diversity and a reduction in productivity of the area.

b. Dumping of wastes and runoff from agricultural lands has resulted in eutrophication of many lakes, killing fish and other organisms.

c. Toxic chemicals introduced into the environment have been incorporated into the food chain. As a result of biological magnification, these substances pose threats to top-level consumers.

d. An increase in atmospheric CO_2 from fossil fuel and wood combustion adds to the greenhouse effect. The resulting increase in temperature may have far-reaching effects on climate and sea level. Changes in CO_2 levels may also affect species composition of C_3 and C_4 plants

in natural communities and in agricultural crops.

e. Agriculture interrupts nutrient cycling and requires the addition of synthetic fertilizers, which upset the nitrogen cycle and can cause eutrophication.

f. Exotic species may outcompete native species and severely disrupt ecological balances.

g. Habitat destruction for industrial, agricultural, and urban uses reduces species diversity and increases extinction rates.

Answers to Test Your Knowledge

Multiple Choice:

1. b	4. c	7. b	10. c	13. e
2. a	5. d	8. c	11. d	14. b
3. c	6. e	9. a	12. b	15. a

CHAPTER 50
BEHAVIOR

■ INTERACTIVE QUESTIONS

50.1 a. The stimulus of an increase in day length may result in reproductive behavior.

b. Breeding is most successful during this time due to warm temperature and food supply.

50.2 a. An infant's smile and grasp are FAPs in response to visual or tactile stimuli.

b. A stronger response to an exaggerated sign stimulus.

50.3 a. imprinting

b. operant conditioning (associative)

c. habituation

d. play

e. observational learning

f. insight

g. classical conditioning (associative)

50.4 a. light or lack of light

b. day length

c. Experimental animals kept in constant light, dark, or day-length conditions continued to show rhythmic behaviors or physiological changes.

50.5 The sow bugs were showing negative phototaxis and a kinesis for relative humidity.

50.6 Both types of feeders would be expected to maximize energy intake over expenditure by somehow factoring in such things as the size of prey and relative abundance of different sizes; distance to the prey and the energy required to catch it; and risk of becoming prey while feeding. In addition, the generalist might compare the relative abundances of different types of prey, and foraging efficiency could be improved with the development of a search image for specific prey. The specialist often has anatomical and behavioral adaptations for feeding efficiently on one type of food item.

50.7 a. Members of the same species occupy the same niche and are thus competitors for resources such as food, territory, and mates. In agonistic encounters, one animal may establish its claim to these resources.
　　b. ritual behavior (threat displays, appeasement gestures), dominance hierarchies, and territories

50.8 a. female
　　b. Females usually have more parental investment in offspring due to the larger resource allocation in eggs and maternal care. Their reproductive fitness is improved by selecting mates that provide offspring with good genes or competent paternal care. A male has little resource commitment in sperm and may potentially improve his fitness by mating as often and with as many partners as possible.

50.9 a. plover fake injury display; female fireflies flashing male of another species and eating him; pregnant Asian monkey soliciting copulation with new dominant male
　　b. Mammals are mostly nocturnal; birds are diurnal.

50.10 a. a sibling
　　b. The individual shares more alleles in common with a sibling ($r = 0.5$); r for a grandparent is 0.25, and r for a cousin is 0.125.

Suggested Answers to Structure Your Knowledge

1. There is controversy over how much of animal behavior is innate and genetically programmed and how much is a product of experience and learning. Fixed-action patterns are clearly preprogrammed behavior, and many seemingly complex animal behaviors can be isolated into a series of FAPs. In other cases, genetics may set the parameters for an organism's behavior; however, experience can modify behavior, and learning is clearly evident.

2. According to the concept of Darwinian fitness, an animal's behavior should help to increase its chance of survival and reproduction of viable offspring. Survival behaviors that are genetically programmed would be most likely to be passed on, and species would evolve many innate behaviors for foraging, migrating, mating, and care of offspring. The evolution of cognitive ability would help individuals deal with novel situations. Social behaviors and communication may help reduce potentially harmful competition. Parental investment and certainty of paternity may result in differences in mating systems and parental care that influence an individual's reproductive success. Altruistic behavior has been explained on the basis of kin selection; the inclusive fitness of an animal increases if its altruistic behavior benefits related animals who may be carrying a high proportion of the same alleles.

Answers to Test Your Knowledge

Multiple Choice:

1. d	6. c	11. d
2. b	7. e	12. a
3. a	8. a	13. b
4. c	9. c	14. c
5. d	10. e	15. e